GROWTH AND FORM

ON
GROWTH AND FORM

BY

D'ARCY WENTWORTH THOMPSON

VOLUME II

Second edition, reprinted

Cambridge

at the University Press

1968

PUBLISHED BY
THE SYNDICS OF THE CAMBRIDGE UNIVERSITY PRESS

Bentley House, 200 Euston Road, London, N.W.1
American Branch: 32 East 57th Street, New York, N.Y. 10022

Standard Book Number: 521 06622 0
(set of two volumes)

First edition	1917
Second edition	1942
Reprinted	1952
	1959
	1963
	1968

591.134
T37o
67660
October 1969

First printed in Great Britain at the University Press, Cambridge
Reprinted in the Netherlands by De IJsel Press, Deventer

CHAPTER VII

THE FORMS OF TISSUES OR CELL-AGGREGATES

WE pass from the solitary cell to cells in contact with one another —to what we may call in the first instance "cell-aggregates," through which we shall be led ultimately to the study of complex tissues. In this part of our subject, as in the preceding chapters, we shall have to consider the effect of various forces; but, as in the case of the solitary cell, we shall probably find, and we may at least begin by assuming, that the agency of surface-tension is especially manifest and important. The effect of this surface-tension will manifest itself in surfaces *minimae areae*: where, as Plateau was always careful to point out, we must understand by this expression not an absolute but a relative minimum, an area, that is to say, which approximates to an absolute minimum as nearly as the circumstances and material exigencies of the case permit.

There are certain fundamental principles, or fundamental equations, besides those we have already considered, which we shall need in our enquiry; for instance, the case which we briefly touched on (on p. 426) of the angle of contact between the protoplasm and the axial filament in a Heliozoan, we shall now find to be but a particular case of a general and elementary theorem.

Let us re-state as follows, in terms of *Energy*, the general principle which underlies the theory of surface-tension or capillarity *.

When a fluid is in contact with another fluid, or with a solid or with a gas, a portion of the total energy of the system (that, namely, which we call *surface energy*) is proportional to the area of the surface of contact; it is also proportional to a coefficient which is specific for each particular pair of substances and is constant for these, save only in so far as it may be modified by changes of temperature or of electrical charge. Equilibrium, which is the condition of *minimum potential energy* in the system, will accordingly

---

* See Clerk Maxwell's famous article on "Capillarity" in the ninth edition of the *Encyclopedia Britannica*, revised by Lord Rayleigh in the tenth edition.

be obtained, *caeteris paribus*, by the utmost possible reduction of the surfaces in contact.

When we have three bodies in contact with one another the same is true, but the case becomes a little more complex. Suppose a drop of some fluid, A, to float on another fluid, B, while both are exposed to air, C. Here are three surfaces of contact, that of the drop with the fluid on which it floats, and those of air with the one and other of these two; and the whole surface-energy, E, of the system consists of three parts resident in these three surfaces,

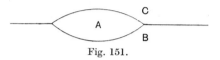

Fig. 151.

or of three specific energies, E_{AB}, E_{AC}, E_{BC}. The condition of equilibrium, or minimal potential energy, will be reached by contracting those surfaces whose specific energy happens to be large and extending those where it is small—contraction leading to the production of a "drop," and extension to a spreading "film." Floating on water, turpentine gathers into a drop, olive-oil spreads out in a film; and these, according to the several specific energies, are the ways by which the total energy of the system is diminished and equilibrium attained.

A drop will continue to exist provided its own two surface-energies exceed, per unit area, the specific energy of the water-air surface around: that is to say, provided (Fig. 151)

$$E_{AB} + E_{AC} > E_{BC}.$$

But if the one fluid happen to be oil and the other water, then the combined energy per unit-area of the oil-water and the oil-air surfaces together is less than that of the water-air surface:

$$E_{wa} > E_{oa} + E_{ow}.$$

Hence the oil-air and oil-water surfaces increase, the air-water surface contracts and disappears, the oil spreads over the water, and the "drop" gives place to a "film." In both cases the total surface-area is a minimum under the circumstances of the case, and always provided that no external force, such as gravity, complicates the situation.

The surface-energy of which we are speaking here is manifested in that contractile force, or tension, of which we have had so much to say*. In any part of the free water-surface, for instance, one surface-particle attracts another surface-particle, and the multitudinous attractions result in equilibrium. But a water-particle in the immediate neighbourhood of the drop may be pulled outwards,

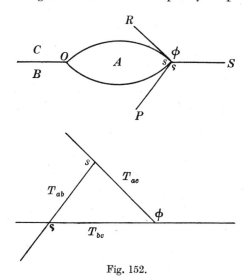

Fig. 152.

so to speak, by another water-particle, but find none on the other side to furnish the counter-pull; the pull required for equilibrium must therefore be provided by tensions existing in the *other two* surfaces of contact. In short, if we imagine a single particle placed at the very point of contact, it will be drawn upon by three different forces, whose directions lie in the three surface-planes and whose

---

* It can easily be proved (by equating the increase of energy stored in an increased surface with the work done in increasing that surface), that the tension measured per unit breadth, T_{ab}, is equal to the energy per unit area, E_{ab}. Surface-tensions are very diverse in magnitude, but all are positive; Clerk Maxwell conceived the existence of *negative* surface-tensions, but could not point to any certain instance. When blood-serum meets a solution of common salt, the two fluids hasten to mix, long streamers of the one running into the other; this remarkable phenomenon, first observed, by Almroth Wright (*Proc. R.S.* (B), xcii, 1921) and called by him "pseudopodial intertraction," was described by Schoneboom (*ibid.* (A), ci, 1922) as a case of negative surface-tension. But it is a diffusion-phenomenon rather than a capillary one.

magnitudes are proportional to the specific tensions characteristic of the three "interfacial" surfaces. Now for three forces acting at a point to be in equilibrium they must be capable of representation, in magnitude and direction, by the three sides of a triangle taken in order, in accordance with the theorem of the Triangle of Forces. So, if we know the form of our drop as it floats on the surface (Fig. 152), then by drawing tangents P, R, from O (the point of mutual contact), we determine the three angles of our triangle, and know therefore the relative magnitudes of the three surface-tensions proportional to its sides. Conversely, if we know the three tensions acting in the directions P, R, S (viz. T_{ab}, T_{ac}, T_{bc}) we know the three sides of the triangle, and know from its three angles the form of the section of the drop. All points round the edge of the drop being under similar conditions, the drop must be circular and its figure that of a solid of revolution*.

The principle of the triangle of forces is expanded, as follows, in an old seventeenth-century theorem, called Lamy's Theorem:

If three forces acting at a point be in equilibrium, each force is proportional to the sine of the angle contained between the directions of the other two. That is to say (in Fig. 152)

$$P : R : S = \sin \phi : \sin \rho : \sin s,$$

or
$$\frac{P}{\sin \phi} = \frac{R}{\sin \rho} = \frac{S}{\sin s}.$$

And from this, in turn, we derive the equivalent formulae by which each force is expressed in terms of the other two and of the angle between them: viz.

$$P^2 = R^2 + S^2 + 2RS \cos \phi, \text{ etc.}$$

From this and the foregoing, we learn the following important and useful deductions:

(1) The three forces can only be in equilibrium when each is less

* Bubbles have many beautiful properties besides the more obvious ones. · For instance, a floating bubble is always part of a sphere, but never more than a hemisphere; in fact it is always rather less, and a very small bubble is considerably less, than a hemisphere. Again, as we blow up a bubble, its thickness varies inversely as the square of its diameter; the bubble becomes a hundred and fifty times thinner as it grows from an inch in diameter to a foot. In an actual calculation we must always take account of the tensions *on both surfaces* of each film or membrane.

than the sum of the other two; otherwise the triangle is impossible. In the case of a drop of olive-oil on a clean water-surface, the relative magnitudes of the three tensions (at 15° C.) are nearly as follows:

Water-air surface		59
Oil-air	,,	25
Oil-water	,,	16

No triangle having sides of these relative magnitudes is possible, and no such drop can remain in existence*.

(2) The three surfaces may be all alike: as when two soap-bubbles are joined together on either side of a partition-film. The three tensions then are all co-equal, and the three angles are co-equal; that is to say, when three similar liquid surfaces, or films, meet together, they always do so at identical angles of 120°. Whether our two conjoined soap-bubbles be equal or unequal, this is still the invariable rule; because the specific tension of a particular surface is independent of form or magnitude.

(3) If all three surfaces be different, as when a fluid drop lies between water and air, the three surface-tensions will (in all likelihood) be different, and the two surfaces of the drop will differ in their amount of curvature.

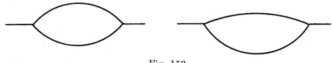

Fig. 153.

(4) If two only of the surfaces be alike, then two of the angles will be alike and the other will be unlike; and this last will be the difference between 360° and the sum of the other two. A particular case is when a film is stretched between solid and parallel walls, like a soap-film within a cylindrical tube. Here, so long as no external pressure is applied to either side, so long as both ends of the tube are open or closed, the angles on either side of the film will be equal, that is to say the film will set itself at right angles to the sides. Many years ago Sachs laid it down as a principle, which

---

* Nevertheless, if the water-surface be contaminated by ever so thin a film of oil, the oil-drop may be made to float upon it. See Rayleigh on Foam, *Collected Works*, III, p. 351.

has become celebrated in botany under the name of Sachs's Rule,
that one cell-wall always tends to set itself at right angles to another
cell-wall. But this rule only applies to the case we have just
illustrated; and such validity as it possesses is due to the fact that
among plant-tissues it commonly happens that one cell-wall has
become solid and rigid before another partition-wall impinges upon it.

(5) Another important principle arises, not out of our equations
but out of the general considerations which led to them. We saw in
the soap-bubble that at and near the point of contact between our
several surfaces, there is a continued balance of forces, carried so
to speak) across the interval; in other words, there is *physical
continuity* between one surface and another and it follows that the
surfaces merge one into another by a continuous curve.   Whatever

Fig. 154. Plateau's *bourrelet*,
in an algal filament. After
Berthold.   *a*   *b*
                                        Fig. 155.

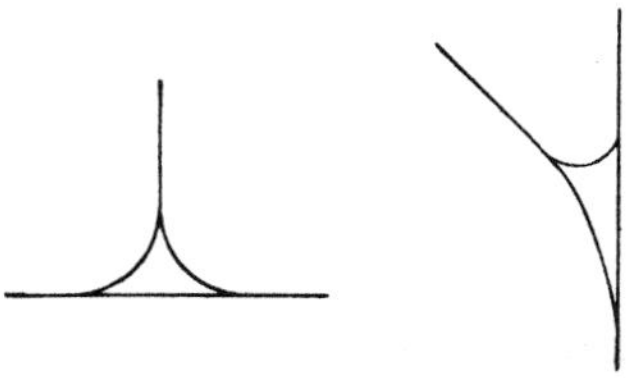

be the form of our surfaces and whatever the angle between them,
a small intervening curved surface is always there to bridge over
the line of contact; and this little fillet, or "bourrelet," as Plateau
called it, is big enough to be a common and conspicuous feature in
the microscopy of tissues (Fig. 154). A similar "bourrelet" is
clearly seen at the boundary between a floating bubble and the liquid
on which it floats: in which case it constitutes a "masse annulaire,"
whose mathematical properties and relation to the form of the
nearly hemispherical bubble have been investigated by van der
Mensbrugghe*. The superficial vacuoles in *Actinophrys* or *Actinosphaerium* present an identical phenomenon.

(6) It is a curious effect, or consequence, of the bourrelet that
a "horizontal" soap-film is never either horizontal or plane. For
the bourrelet at its edge is deformed by gravity, and the film is
correspondingly inclined upwards where it meets it (Fig. 155 *b*).

* Cf. Plateau, *op. cit.* p. 366.

(7) The bourrelet, a fluid mass connected with a fluid film, is no mere passive phenomenon but has its active influence or dynamical effect. This was pointed out by Willard Gibbs*, and Plateau's bourrelet is more often called, nowadays, "Gibbs's Ring." The ring is continuous in phase with the interior of the film, and fluid is sucked into it from the latter, which thins rapidly; and this, becoming a more potent factor of unrest than gravity itself, leads presently to the rupture of the film. Plateau's explanation of his bourrelet as a "surface of continuity" is thus but a part, and a small part of the story.

(8) In the succulent, or parenchymatous, tissue of a vegetable, the cells have their internal corners rounded off (Fig. 156) in a way which might suggest the bourrelet, but comes of another cause.

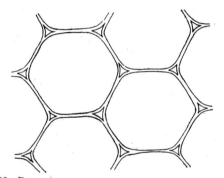

Fig. 156. Parenchyma of maize; shewing intercellular spaces.

Where the angles are rounded off the cell-walls tend to split apart from one another, and each cell seems tending to withdraw, as far as it can, into a sphere; and this happens, not when the tissue is young and the cell-walls tender and quasi-fluid, but later on, when cellulose is forming freely at the surface of the cell. The cell-walls no longer meet as fluid films, but are stiffening into pellicles; the cells, which began as an association of bubbles, are now so many balls, in solid contact or partial detachment; and flexibility and elasticity have taken the place of the capillary forces of an earlier and more liquid phase†.

* *Collected Works*, I, p. 309.

† J. H. Priestley, Cell-growth...in the flowering plant, *New Phytologist*, xxviii, pp. 54–81, 1929.

(9) Statically though not dynamically, that is to say as a line or surface of continuity in Plateau's sense, our bourrelet is analogous to the accumulation of sand seen where two nodal lines cross in a Chladni figure: "Vers les endroits où des lignes nodales se coupent, elles s'élargissent toujours, de sorte que la forme des parties vibrantes près de ces endroits n'est pas angulaire mais plus ou moins arrondie, souvent en forme d'hyperbole*." And in somewhat remoter analogy, we may look on the three *corpora Arantii* as so many *bourrelets*, helping to fill the angles where three semilunar valves meet at the base of the great arteries.

We may now illustrate some of the foregoing principles, constantly bearing in mind the principles set forth in our chapter on the Forms of Cells, and especially those relating to the pressure exercised by a curved film.

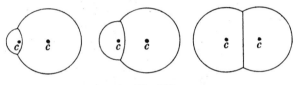

Fig. 157.

Let us look for a moment at the case presented by the partition-wall in a double soap-bubble. As we have just seen, the three films in contact (viz. the outer walls of the two bubbles and the partition-wall between) being all composed of the same substance and being all alike in contact with air, the three tensions must be equal, and the three films must, in all cases, meet at co-equal angles of 120°. But unless the two bubbles be of precisely equal size, and therefore of equal curvature, the tangents to the spheres will not meet the plane of their circle of contact at equal angles, and the partition-wall will of necessity be a curved, and indeed a spherical, surface; it is only plane when it divides two equal and symmetrical cells. It is obvious, from the symmetry of the figure, that the centres of the two bubbles and of the partition between are all on one and the same straight line.

The two bubbles exert a pressure inwards which is inversely

* E. F. F. Chladni, *Traité d'acoustique*, 1809, p. 127.

proportional to their radii: that is to say, $p : p' :: 1/r : 1/r'$; and the partition-wall must, for equilibrium, exert a pressure (P) which is equal to the difference between these two pressures, that is to say, $P = 1/R = 1/r' - 1/r = (r - r')/rr'$. It follows that the curvature of the partition must be just such as is capable of exerting this pressure, that is to say, $R = rr'/(r - r')$. The partition, then, is a portion of a spherical surface, whose radius is equal to the product, divided by the difference, of the radii of the two bubbles; if the two bubbles be equal, the radius of curvature of the partition is infinitely great, that is to say the partition is (as we have already seen) a plane surface.

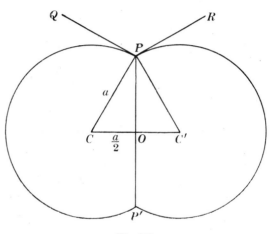

Fig. 158.

In the typical case of an evenly divided cell, such as a double and so-equal soap-bubble (Fig. 158), where partition-wall and outer walls are identical with one another and the same air is in contact with them all, we can easily determine the form of the system. For, at any point of the boundary of the partition, P, the tensions being equal, the angles QPP', RPP', QPR are all equal, and each is, therefore, an angle of 120°. But PQ, PR being tangents, the centres of the two spheres (or circular arcs in the figure) lie on lines perpendicular to them; therefore the radii CP, $C'P$ meet at an angle of 60°, and CPC' is an equilateral triangle. That is to say, the centre of each circle lies on the circumference of the other; the partition lies midway between the two centres; and the diameter of the partition-wall, PP', is $\dfrac{OP}{CP} = \sin 60° = \dfrac{\sqrt{3}}{2} = 0\cdot866$ times the diameter of each of the two cells. This gives us, then, the form of a combination of two co-equal spherical cells under uniform conditions.

By integrating between the known values of the meridian section and the plane partition, we should find each half of the double cell (or soap-bubble) to be equal to 27/32 of a complete sphere. Therefore the radius of curvature of each half of the divided bubble is greater than that of a sphere of equal volume in the ratio of:

$$\sqrt[3]{32} : \sqrt[3]{27} = 2.\sqrt[3]{4} : 3 = 1\cdot058 : 1 = 1 : 0\cdot945.$$

And the radius of the original sphere, before division, is to the radius of each half, or each product of cell-division, as

$$\sqrt[3]{54} : \sqrt[3]{32} = 3.\sqrt[3]{2} : 2.\sqrt[3]{4} = 1\cdot191 : 1 = 1 : 0\cdot84.$$

In the case of three co-equal and united bubbles (to which case we shall presently return), each is approximately five-sevenths of a whole sphere: and their radii, therefore, are to the radius of the whole sphere as

$$\sqrt[3]{7} : \sqrt[3]{5} = 1 : 0\cdot893 = 1 : (0\cdot945)^2.$$

When two co-equal bubbles coalesce, the internal pressure, due to the tension of the wall and varying inversely as its radius of curvature, is diminished in the ratio of $1 : 0\cdot945$, or say $5\frac{1}{2}$ per cent. And we begin to see, in the case of three bubbles, that the process proceeds in a geometrical progression, each new coalescence increasing the radius of curvature and diminishing the internal pressure, by a constant fraction of the whole. This and other simple corollaries may perchance, some day, be found useful to the biologist.

In the case of unequal bubbles, the curvature of their partition-wall is easily determined, and is shewn in Fig. 159. The three films meeting in P being (as before) identical films, the three tangents, PQ, PR, PS, meet at co-equal angles of 120°, and PS produced bisects the angle QPR. PQ, PR are tangents perpendicular to the radii CP, $C'P$; and $C''P$, the radius of the spherical partition PP', is found by drawing a perpendicular to PS in P. The centre C'' is, by the symmetry of the figure, in a straight line with C, C'.

Whether the partition be or be not a plane surface, it is obvious that its *line of junction* with the rest of the system lies in a plane, and is at right angles to the axis of symmetry. The actual curvature of the partition-wall is easily seen in optical section; but in surface view the line of junction is *projected* as a plane (Fig. 160), perpendicular to the axis, and this appearance has helped to lend support and authority to "Sachs's Rule."

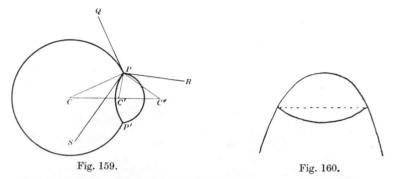

Fig. 159.   Fig. 160.

As soon as the tensions of the cell-walls become unequal, whether from changes in their own substance or in the substances with which they are in contact, then the form alters. If the tension along the partition P diminishes, the partition itself enlarges and the angle QPR increases: until, when the tension p is very small compared with q or r, the whole figure becomes a sphere, and the partition-wall, dividing it into two hemispheres, stands at right angles to the outer wall. This is the case when the outer wall of the cell is practically solid. On the other hand, if p begins to

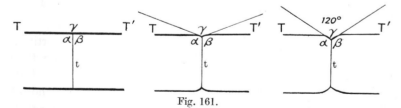

Fig. 161.

increase relatively to q and r, then the partition-wall contracts, and the two adjacent cells become larger and larger segments of a sphere, until at length the system becomes divided into two separate cells.

To put the matter still more simply, let the annexed diagrams (Fig. 161) represent a system of three films, one being a partition-wall running between the other two; and where the partition t meets the outer wall TT', let the several tensions, or the tractions exerted on a point at their meeting-place, be proportional to T, T' and t. Let α, β, γ be, as in the figure, the opposite angles. Then:

(1) If T be equal to T', and t be relatively insignificant, the angles α, β will be of 90°.

(2) If $T = T'$, but be a little greater than t, then t will exert an appreciable traction, and α, β will be more than 90°, say for instance, 100°.

(3) If $T = T' = t$, then α, β, γ will all equal 120°.

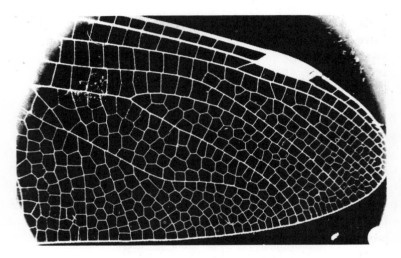

Fig. 162.   Part of a dragonfly's wing.

The outer walls of the two cells on either side of the partition will be straight, as well as continuous, in the first case, and more or less curved in the other two. We have a vivid illustration (if a somewhat crude one) of the first case in a section of honey: where the waxen walls, which meet one another at 120°, meet the wooden sides of the box at 90°.

The wing of a dragon-fly shews a seemingly complicated system of veins which the foregoing considerations help much to simplify. The wing is traversed by a few strong "veins," or ribs, more or less parallel to one another, between which finer veins make a meshwork of "cells," these lesser veins being all much of a muchness, and exerting tensions insignificant compared with those of the greater veins. Where (*a*) two ribs run so near together that only one row of cells lies between, these cells are quadrangular in form, their thin partitions meeting the ribs at right angles on either side. Where (*b*) two rows of cells are intercalated between a pair of ribs, one row fits into the other by angles of 120°, the result of co-equal tensions; but both meet the ribs at right angles, as in the former case. Where (*c*) the cell-rows are numerous, all their angles in common tend to be co-equal angles of 120°, and the cells resolve, consequently, into a hexagonal meshwork.

Many spherical cells, such as *Protococcus*, divide into two equal halves, separated by a plane partition. Among other lower Algae akin to *Protococcus*, such as the Nostocs and Oscillatoriae, in which the cells are embedded in a gelatinous matrix, we find a series of forms such as are represented in Fig. 163, which various conditions depend, according to what we have already learned, upon the relative magnitudes of the tensions at the surface of the cells and the boundary between them. In some cases (Fig. 163, B) the cells remain spherical, because they are merely embedded in the matrix, with no other physical continuity between them; even two soap-bubbles do not tend to unite, unless their surfaces be moist or we put a drop of soap-solution between them. In certain other cases, the system consists of a relatively thick-walled tube, subdivided by more delicate partitions, which latter then tend (as in D) to become plane septa, set at right angles to the walls. Or again, side-walls and septa may be all alike, or nearly so; and then the configuration (as in C, on Fig. 163) is that of a linear cluster of soap-bubbles*.

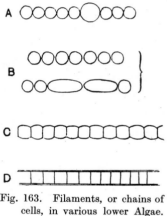

Fig. 163. Filaments, or chains of cells, in various lower Algae. (A) *Nostoc*; (B) *Anabaena*; (C) *Rivularia*; (D) *Oscillatoria*.

In the spores of liverworts, such as *Pellia*, the first partition

---

* Cf. Dewar, Studies on liquid films, *Proc. Roy. Inst.* 1918, p. 359.

(the equatorial partition in Fig. 165 *a*) divides the spore into two equal halves, and is therefore a plane surface normal to the surface of the cell. But the next partitions arise near to either end of the original spherical or elliptical cell, and each of these latter will likewise tend to set itself normally to the cell-wall—at least the

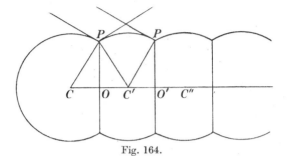

Fig. 164.

angles on either side of the partition will tend to be identical, and their magnitude will depend on the relative tensions of the cell-wall and the partition. The angles will be right angles if the cell-wall is solid or nearly so when the partition is formed; but they will be somewhat greater, if (in all probability) rigidity of the cell-wall has not been quite attained. In either case the partition itself will

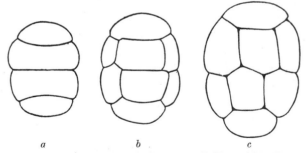

Fig. 165. Early development of a liverwort (*Pellia*). After Wildeman.

be part of a spherical surface, whose curvature will now correspond to the difference of pressures in the two chambers (or cells) which it serves to separate.

We have innumerable cases, near the tip of a growing filament for instance, where in like manner the partition-wall which cuts off the terminal, more or less conical, cell constitutes a spherical lens-shaped surface, set normally to the adjacent walls; and the centre of curvature is the meeting-point of two tangents to the cone. We find such a lenticular partition at the tips of the branches of many Florideae; in *Dictyota dichotoma*, as figured by Reinke, we have a succession of them. And by the way, where, in such cases as these, the tissues happen to be very transparent, we often have a puzzling confusion of lines (Fig. 166); one being the optical section

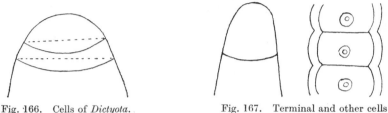

Fig. 166. Cells of *Dictyota*. After Reinke.

Fig. 167. Terminal and other cells of *Chara*.

of the curved partition-wall, the other being the straight linear projection of its outer edge to which we have already referred. In the conical terminal cell of *Chara*, we have the same lens-shaped curve; but a little lower down, where the sides of the shoot are approximately parallel, we have flat transverse partitions, and the form of the cells is, more or less, what we have been led to expect in the simple case of successive transverse partitions (Fig. 167)

In the young antheridia of *Chara* (Fig. 168), and in the geometrically similar case of the sporangium (or conidiophore) of *Mucor*, we easily recognise the hemispherical form of the septum which shuts off the large spherical cell from the cylindrical filament. Here, in the first phase of development, we should have to take into consideration the different pressures exerted by the single curvature of the cylinder and the double curvature of its spherical cap (p. 371); and we should find that the partition would have a somewhat low curvature, with a radius *less* than the diameter of the cylinder, which it would have exactly equalled but for the additional pressure inwards which it receives from the curvature of the large surrounding sphere. But as the latter continues to

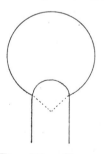

Fig. 168. Young antheridium of *Chara*.

grow its curvature decreases, and so likewise does the inward
pressure of its surface; and accordingly the little convex partition
bulges out more and more.

In the ordinary meristematic tissue of a plant, the new partition-
wall within a dividing cell will generally meet the old walls at right
angles to begin with, because its tension is usually small compared
to what theirs has become. But as the system grows and the old
wall strengthens, the tensions of all three walls become approximately the same; and they tend towards a new position of equilibrium, in which (as seen in optical section) they meet as before,
at co-equal angles of 120°*.

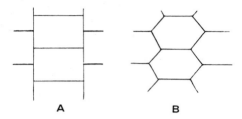

A   **B**

Fig. 169.   Cambium cells after division, altering from *A* to *B*.

The biological facts which the foregoing considerations go far to
explain and account for have been the subject of much argument
and discussion on the part of the botanists. Let me recapitulate,
in a very few words, the history of this long discussion.

Some seventy years ago, Hofmeister laid it down as a general,
but purely empirical, law that "The partition-wall stands always
perpendicular to what was previously the principal direction of
growth in the cell"—or, in most cases, perpendicular to the long
axis of the cell†. This contains an important truth; for it is as
much as to say that the cell tends to be divided by the smallest

    * J. H. Priestley, Studies...of cambium activity, *New Phytologist*, XXIX, p. 101,
1930. Cf. also J. J. Beijer, Vermehrung der radialen Reihen in Cambium, *Rec.
de trav. bot. Néerl.* XXIV, pp. 631–786, 1927.

    Hofmeister, *Pringsheim's Jahrb.* III, p. 272, 1863; *Hdb. d. physiol. Bot.* I,
p. 129, 1867; etc. Hofmeister adds the somewhat curious qualification: "Wohl-
bemerkt, nicht senkrecht zum grössten Durchmesser der Zelle, der mit der Richtung
des stärksten Wachstums nicht zusammenfallen braucht, und in sehr viel Fällen
in der That auch nicht mit ihr zusammenfällt."

partition capable of doing so. Ten years later, Sachs formulated his rule of "rectangular section," declaring that in all tissues, however complex, the cell-walls cut one another (at the time of their formation) at right angles*. Years before, Schwendener had found in the final results of cell-division a universal system of "orthogonal trajectories†"; and this idea Sachs further developed, introducing complicated systems of confocal ellipses and hyperbolae, and distinguishing between periclinal walls whose curves approximate to the peripheral contours, radial partitions which cut these at an angle of 90°, and finally anticlines, which stand at right angles to the other two.

Reinke (in 1880) was the first to throw doubt upon this explanation. He pointed out cases where the angle was not a right angle, but very definitely an acute one; and he saw in the commoner rectangular symmetry merely what he called a necessary, but *secondary*, result of growth‡.

Within the next few years a number of botanical writers were content to point out further exceptions to Sachs's rule§, and in some cases to show that the *curvatures* of the partition-walls, especially such cases of lenticular curvature as we have described, were by no means accounted for by either Hofmeister or Sachs; while within the same period, Sachs himself, and also Rauber, attempted to extend the main generalisation to animal tissues¶. The simple fact is that Sachs's rule is limited to those many cases where one cell-wall grows stiff or solid before another

* Sachs, Ueber die Anordnung d. Zellen in jüngsten Pflanzentheilen, *Verh. phys.-med. Gesellsch. Würzburg*, XI, pp. 219–242, 1877; Ueber Zellenanordnung u. Wachstum, *ibid.* XII, 1878; cf. *Arb. bot. Inst. Würzburg*, II, 1882; Ueber die durch Wachstum bedingte Verschiebung kleinster Theilchen in trajectorischen Curven, *Monatsb. k. Akad. Wiss. Berlin*, 1880; *Physiology of Plants*, chap. XXVII, Oxford, 1887.

† Schwendener, Bau u. Wachstum des Flechtenthallus, *Naturf. Gesellsch. Zürich*, 1860, pp. 272–296.

‡ Reinke, *Lehrbuch d. Botanik*, 1880, p. 519; Kienitz-Gerloff, *Botan. Ztg.* 1878, p. 58, had already shewn some exceptions to Sachs's rules, and ascribed them, vaguely, to "heredity." It was a time when *heredity* overruled everything, and when Sachs himself spoke of the difficulty of demonstrating the causes of any morphological phenomenon in any other way than "genetically": *Textbook*, 1882, p. 201.

§ E.g., Leitgeb, *Untersuchungen über die Lebermoose*, II, p. 4, Graz, 1881.

¶ Rauber, Neue Grundlegungen zur Kenntniss der Zelle, *Morphol. Jahrb.* VIII, pp. 279, 334, 1882.

impinges upon it; and, subject to this limitation, the rule is strictly true.

While these writers regarded the form and arrangement of the cell-walls as a biological phenomenon, with little if any direct relation to ordinary physical laws, or with but a vague reference to "mechanical conditions," the physical side of the case was soon urged by others, with more or less force and cogency. Indeed the general resemblance between a cellular tissue and a "froth" had been pointed out long before. Robert Hooke described the cells within the shaft of a feather as forming "a kind of solid or hardened froth, or a congeries of very small bubbles," and Grew described a parenchyma as made by "fermentation", "as we see Bread in Baking", and again as being "much the same thing, as to its construction which the froth of beer or eggs is." Later on, within the days of the cell-theory, Melsens made an "artificial tissue" by blowing into a solution of white of egg*.

In 1886, Berthold published his *Protoplasmamechanik*, in which he definitely adopted the principle of "minimal areas," and, following on the lines of Plateau, compared the forms of many cells and the arrangement of their partitions with those assumed under surface-tension by a system of "weightless films." But, as Klebs† pointed out, in reviewing the book, Berthold was so cautious as to stop short of attributing the biological phenomena to a mechanical cause. They remained for him, as they had done for Sachs, so many "phenomena of growth," or "properties of protoplasm."

In the same year, but while still unacquainted, apparently, with Berthold's work, Leo Errera published a short but very striking article‡ in which he definitely ascribed to the cell-wall (as Hofmeister had already done) the properties of a semi-liquid film, and drew from this as a logical consequence the deduction that it *must* assume the various configurations which the law of minimal areas imposes on the soap-bubble. So what we may call *Errera's Law* is formulated as follows: A cell-wall, at the moment of its

---

* *C.R.* xxxiii, p. 247, 1851; *Ann. de chimie et de phys.* (3), xxxiii, p. 170, 1851; *Bull. R. Acad. Belg.* xxiv, p. 531, 1857.

† Georg Klebs, *Biol. Centralbl.* vii, pp. 193–201, 1887.

‡ L. Errera, Sur une condition fondamentale d'équilibre des cellules vivantes, *C.R.* ciii, p. 822, 1886; *Bull. Soc. Belge de Microscopie*, xiii, Oct. 1886; *Recueil d'œuvres (Physiologie générale)*, 1910, pp. 201–205.

formation, tends to assume the form which would be assumed under the same conditions by a liquid film destitute of weight*.

Soon afterwards Chabry†, discussing the segmentation of the Ascidian egg, indicated many ways in which cells and cell-partitions repeat the surface-tension phenomena of the soap-bubble. He came to the conclusion that some, at least, of the embryological phenomena were purely physical, and the same line of investigation and thought was pursued and developed by Robert‡ in connection with the embryology of the Mollusca. Driesch also, in a series of papers, continued to draw attention to capillary phenomena in the segmenting cells of various embryos, and came to the conclusion, startling to the embryologists of the time, that the mode of segmentation was of little importance as regards the final result§.

Lastly de Wildeman¶, in a somewhat wider but also vaguer generalisation than Errera's, declared that "The form of the cellular framework of plants and also of animals depends, in its essential features, upon the forces of molecular physics."

Let us return to our problem of the arrangement of partition films. When we have three bubbles in contact, instead of two as in the case already considered, the phenomenon is strictly analogous to the former case. The three bubbles are separated by three partition surfaces, whose curvature will depend upon the relative size of the spheres, and which will be plane if the latter are all of equal size; but whether plane or curved, the three partitions will meet one another at angles of 120°, in an axial line. Various pretty geometrical corollaries accompany this arrangement. For instance, if Fig. 170 represent the three associated bubbles in a

---

* There was no lack of hearty antagonism to Berthold and Errera's views. Cf. (e.g.) Zimmermann, *Beitr. z. Morphologie und Physiologie der Pflanzenzelle*, Tübingen, 1891; Jost, *Vorlesungen über Pflanzenphysiologie*, 1904, p. 329, etc.; Giesenhagen, *Studien über Zelltheilungen im Pflanzenreiche*, 1905. Cf. also K. Habermehl, *Die mechanische Ursache für die regelmässige Anordnung der Teilungswände in Pflanzenzellen* (Inaug. Diss.), Kaiserslautern, 1909.

† L. Chabry, Embryologie des Ascidiens, *J. Anat. et Physiol.* xxiii, p. 266, 1887.

‡ H. Robert, Embryologie des Troques, *Arch. de Zool. expér. et gén.* (3), x, 1892.

§ "Dass der Furchungsmodus etwas für das Zukünftige unwesentliches ist," *Z. f. w. Z.* lv, 1893, p. 37. With this statement compare, or contrast, that of Conklin, quoted on p. 5; cf. also p. 287 (footnote).

¶ E. de Wildeman, Études sur l'attache des cloisons cellulaires, *Mém. Couronn. de l'Acad. R. de Belgique*, liii, 84 pp., 1893–94.

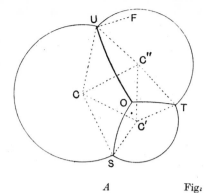
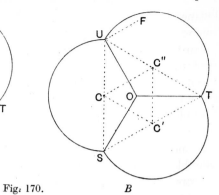

A   Fig. 170.   *B*

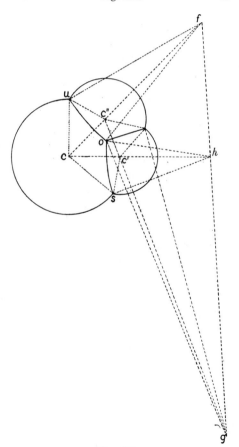

Fig. 171.

plane drawn through their centres, c, c', c'' (or what is the same thing, if it represent the base of three bubbles resting on a plane), then the lines uc, uc'', or sc, sc', etc., drawn to the centres from the points of intersection of the circular arcs, will always enclose an angle of 60°. Again (Fig. 171), if we make the angle $c''uf$ equal to 60°, and produce uf to meet cc'' in f, f will be the centre of the circular arc which constitutes the partition Ou; and further, the three points f, g, h, successively determined in this manner, will lie on one and the same straight line. In the case of three co-equal bubbles (as in Fig. 170, B), it is obvious that the lines joining their centres form an equilateral triangle: and consequently, that the centre of each circle (or sphere) lies on the circumference of the other two; it is also obvious that uf is now parallel to cc'', and accordingly that the centre of curvature of the partition is now infinitely distant, or (as we have already said) that the partition itself is plane.

The mathematician will find a more elegant way of dealing with our spherical bubbles and their associated interfaces by the method of spherical inversion. (i) Take three planes through a line, cutting one another at 60°, and invert from any point, and you have the case of two spherical bubbles fused, with their interface also spherical. (ii) Take the six planes projecting the edges of a regular tetrahedron from its centre, and you get by inversion the case of the three unequal bubbles and their three interfaces. (iii) Take these same planes with a bubble added centrally (thus adding a spherical tetrahedron), and inversion gives the general case of four fused bubbles and their six spherical partitions.

When we have four bubbles meeting in a plane (Fig. 172), they would seem capable of arrangement in two symmetrical ways: either (a) with four partition-walls intersecting at right angles, or (b) with five partitions meeting, three and three, at angles of 120°. The latter arrangement is strictly analogous to the arrangement of *three* bubbles in Fig. 170. Now, though both of these figures might seem, from their apparent symmetry, to be figures of equilibrium, yet in point of fact the latter turns out to be of stable and the former of unstable equilibrium. If we try to bring four bubbles into the form (a), that arrangement endures only for an instant; the partitions glide upon one another, an intermediate wall springs into existence, and the system assumes the form (b), with its two triple, instead of one quadruple, conjunction. In like manner, when four billiard-balls are packed close upon a table, two tend to come together and separate the other two.

Let us epitomise the Law of Minimal Areas and its chief clauses or corollaries in the particular case of an assemblage of fluid films, as was first done by Lamarle*. Firstly and in general: In every liquid system of thin films in stable equilibrium, the sum of the areas of the films is a minimum. From observation and experience, rather than by demonstration, it follows that (2) the area of *each* is a minimum under its own limiting conditions; and further that (3) the mean curvature of any film is constant throughout its whole area, null when the pressures are equal on either side and in other

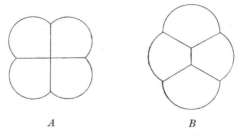

A   *B*

Fig. 172. A, an unstable arrangement of four cells or bubbles. B, the normal and stable configuration, showing the polar furrow.

cases proportional to their difference. Less obvious, very important, and likewise subject (but none too easily) to rigorous mathematical proof, are the next two propositions, both of which had been laid down empirically by Plateau: (4) the films meeting in any one edge are three in number; (5) the crests or edges meeting in any one corner are four in number, neither more nor less. Lastly, and following easily from these: (6) the three films meeting in a crest or edge do so at co-equal angles, and the same is true of the four edges meeting in a corner.

Wherever we have a true cellular complex, an arrangement of cells in actual physical contact by means of their intervening boundary walls, we find these general principles in force; we must only bear in mind that, for their easy and perfect recognition, we

* Ernest Lamarle, Sur la stabilité des systèmes liquides en lames minces, *Mém. de l'Acad. R. de Belgique*, XXXV, XXXVI, 1864–67.

must be able to view the object in a plane at right angles to the boundary walls. For instance, in any ordinary plane section of a vegetable parenchyma, we recognise the appearance of a "froth," precisely resembling that which we can construct by imprisoning a mass of soap-bubbles in a narrow vessel with flat sides of glass; in both cases we see the cell-walls everywhere meeting, by threes, at angles of 120°, irrespective of the size of the individual cells: whose relative size, on the other hand, determines the *curvature* of the partition-walls. On the surface of a honey-comb we have precisely the same conjunction, between cell and cell, of three boundary walls, meeting at 120°. In embryology, when we examine a segmenting egg, of four (or more) segments, we find in like manner, in the majority of cases if not in all, that the same principle is still exemplified. The four segments do not meet in a common centre, but each cell is in contact with two others; and the three, and only three, common boundary walls meet at the normal angle of 120°. A so-called *polar furrow**, the visible edge of a vertical partition-wall, joins (or separates) the two triple contacts, precisely as in Fig. 172, B, and so gives rise to a diamond-shaped figure, which was recognised more than a hundred years ago (in a newt or salamander) by Rusconi, and called by him a *tetracitula*.

That four cells, contiguous in a plane, tend to meet in a lozenge with three-way junctions and a "polar furrow" between the cells, is a geometrical theorem of wide bearing. The first four cells in a wasp's nest shew it neither better nor worse than do those of a segmenting ovum, or the ambulacral plates of a sea-urchin or the oosphere of *Oedogonium* giving birth to its four zoo-spores†. Going farther afield for an illustration, we find it in the molecules of a viscous liquid under shear: where a group of four

---

* It was so termed by Conklin in 1897, in his paper on Crepidula (*Journ. Morph.* XIII, 1897). It is the *Querfurche* of Rabl (*Morph. Jahrb.* v, 1879); the *Polarfurche* of O. Hertwig (*Jen. Zeitschr.* XIV, 1880); the *Brechungslinie* of Rauber (Neue Grundlage zur Kenntniss der Zelle, *Morph. Jahrb.* VIII, 1882); and the *cross-line* of T. H. Morgan (1897). It is carefully discussed by Robert, *op. cit.* p. 307 *seq.*

† Speaking of the complicated polygonal patterns in the test of the protozoon genus Peridinium, Barrows says: "In the experience of the writer no case has been found in which four sutures actually meet at one point. Cases which at first sight appeared as such, upon closer analysis in a favourable position have been resolved into two junction-points of three sutures each, etc." On skeletal variation in the genus Peridinium, *Univ. Calif. Publ.* 1918, p. 463.

molecules is supposed to slip from one lozenge-configuration to an opposite one, passing on the way through the simple cross or square —a configuration of "higher energy" and less stable equilibrium*.

The solid geometry of this four-celled figure is not without interest. If the two polar furrows (the one above and the other below) run criss-cross, the whole is a more or less flattened and distorted spherical tetrahedron. If they run parallel, then it is a four-sided lozenge with two curved quadrilateral faces, and two bilateral faces each bounded by two curved edges, like the "liths"

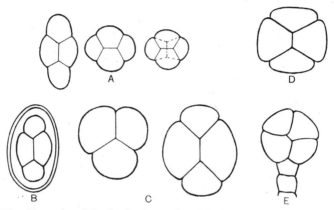

Fig. 173. Examples of the "polar furrow". A, Pollen-grains (tetrads) of *Neottia*. B, Egg of hookworm (*Ankylostoma*). C, First cells of a wasp's nest (*Polistes*). (From Packard, after Saussure.) D, Four-celled stage of Volvox: from Janet. E, Hair of Salvia, after Hanstein.

of an orange†. In either case the lozenge-configuration is under some restraint to keep its four cells in a plane; for a tetrahedral pile, or pyramid, of four spheres would be the simplest arrangement of all.

The polar furrow and the partition of which it forms an edge are, like all the edges and partitions in our associated cells, perfectly definite in dimensions and position; and to draw them to scale, in projection, is a simple matter. Taking the simplest case, when the radii of all four cells are equal to one another, let c, c', c'' and c''' be the centres of the four cells, Fig. 174. The centres of

* Cf. J. D. Bernal, *Proc. R.S.* (A), No. 914, p. 321, 1937.

† The geometer seldom takes account of such two-sided surfaces or facets; but in groups of cells or bubbles they are of common occurrence, and in the theory of polyhedra they fit in without difficulty with the rest (cf. *infra*, p. 737).

any two are related precisely as though two cells only were conjoined; the centres of three *contiguous* cells (as c, c' and c'') are related as though three only were concerned; and the centres of two opposite cells are situated symmetrically to one another. This is as much as to say that if there be two bubbles in contact the addition of a third does not disturb their symmetry; and if there be three in contact, the addition of a fourth leaves the first three likewise *in statu quo*. Thus the triangle $cc'c''$ is equilateral, as we already know. The partition *so* bisects the side cc'', and the angle $cc'c''$; and the point o is the centre of gravity of the triangle. Therefore $op = \frac{1}{2}oc''$ and $oo' = \frac{1}{3}c''c'''$.

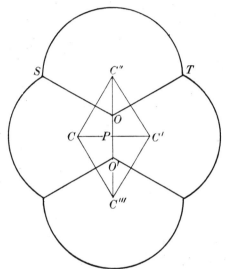

Fig. 174. The geometric symmetry of a system of four cells.

Again, in the triangle cpc'', where $cc'' = r$, $pc'' = \frac{\sqrt{3}}{2} r$, and oo' (the polar furrow) $= \frac{1}{\sqrt{3}} r$. Once again, in the triangle soc'', $sc = r$; and so (one of the partitions) $= \frac{2}{\sqrt{3}} r =$ twice oo'. The length of the polar furrow, then, as seen in vertical projection in a system of four co-equal cells, is (theoretically) just one-half that of the four intercellular partitions, and very nearly three-fifths that of a cell-radius.

It is worth while to remark that the universal phenomenon of a polar furrow gives *an appearance* of bilateral symmetry to every egg or embryo in its four-celled stage, no matter to what kind or class or organism it belongs.

In the four-celled stage of the frog's egg, Rauber (an exceptionally careful observer) shews us three alternative modes in which
the four cells may be found to be conjoined (Fig. 175).   In A we
have the commonest arrangement, which is that which we have
just studied and found to be the simplest theoretical one; that
namely where a straight polar furrow intervenes, and where the
partition-walls are conjoined at its extremities, three by three.
In B, we have again a polar furrow, which is now seen to be a
portion of the first "segmentation-furrow" by which the egg was
originally divided into two; the four-celled stage being reached by
the appearance of the two transverse furrows.   In this case, the
polar furrow is seen to be sinuously curved, and Rauber tells us that
its curvature gradually alters; as a matter of fact, it, or rather the

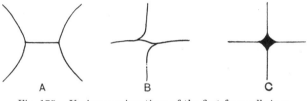

Fig. 175.   Various conjunctions of the first four cells in a
frog's egg.   After Rauber.

partition-wall corresponding to it, is gradually setting itself into a
position of equilibrium, that is to say of equiangular contact with
its neighbours, which position is already attained or nearly so in
A.   In C we have a very different condition, with which we shall
deal in a moment.

The polar furrow may be longer or shorter, and it may be so
minute as to be not easily discernible; but it is quite certain that
no simple and homogeneous system of fluid films such as we are
dealing with is in equilibrium without its presence.   In the accounts
given, however, by embryologists of the segmentation of the egg,
while the polar furrow is depicted in the great majority of cases,
there are others in which it has not been seen and some in which
its absence is definitely asserted*.   The cases where four cells lying

* Thus Wilson declared (*Journ. Morph.* VIII, 1895) that in *Amphioxus* the polar
furrow was occasionally absent, and Driesch took occasion to criticise and to
throw doubt upon the statement (*Arch. f. Entw. Mech.* I, p. 418, 1895).

in one plane meet *in a point*, such as were frequently figured by the older embryologists, are hard to verify and sometimes not easy to believe. Considering the physical stability of the other arrangement, the great preponderance of cases in which it is known to occur, the difficulty of recognising the polar furrow in cases where it is very small and unless it be specially looked for, and the natural tendency of the draughtsman to make an all but symmetrical structure appear wholly so, I was wont to attribute to error or imperfect observation all those cases where the junction-lines of four cells are represented (after the manner of Fig. 172, A) as a simple cross*. As a matter of fact, the simple cross is no very rare phenomenon, even in the frog's egg; but it is a transitory one, and unstable. Viscosity and friction may enable it to endure for a while, but the partitions inevitably shift into the stable, three-way, configuration. In such a case, the polar furrow manifests itself slowly and as it were laboriously; but in the more fluid soap-bubble it does so in the twinkling of an eye.

While a true four-rayed intersection, or simple cross, is theoretically impossible save as a transitory and unstable condition, there is another configuration which may closely simulate it, and which is common enough. There are plenty of faithful representations of segmenting eggs in which, instead of the triple junctions and polar furrow, the four cells (and also their more numerous successors) are represented as *rounded off*, and separated from one another by an empty space, or by a little drop of extraneous fluid, evidently not directly miscible with the fluid surface of the cells. Such is the case in the obviously accurate figure which Rauber gives (Fig. 175, C) of his third mode of conjunction in the four-celled stage of the frog's egg. Here Rauber is most careful to point out that the furrows do not simply "cross," or meet in a point, but are separated by a little space, which he calls the *Polgrübchen*, and asserts to be constantly present whensoever the polar furrow, or *Brechungslinie*, is not to be discerned. This little interposed space with its contained drop of fluid materially alters the case, and implies a new

---

* The same remark was made long ago by Driesch: "Das so oft schematisch gezeichnete Vierzellenstadium mit zwei sich in zwei Punkten scheidende Medianen kann man wohl getrost aus der Reihe des Existierenden streichen" (Entw. mech. Studien, *Z. f. w. Z.* LIII, p. 166, 1892). Cf. also his *Math. mechanische Bedeutung morphologischer Probleme der Biologie*, Jena, 59 pp., 1891.

condition of theoretical and actual equilibrium. For on the one hand, we see that now the four intercellular partitions do not meet *one another at all*; but really impinge upon four new and separate partitions, which constitute interfacial contacts not between cell and cell, but between the respective cells and the intercalated drop. And secondly, the angles at which these four little surfaces meet the four cell-partitions will be determined, in the usual way, by the balance between the respective tensions of these several surfaces. In an extreme case (as in some pollen-grains) it may be found that the cells under the observed circumstances are not truly in surface contact: that they are so many drops which touch but do not "wet" one another, and which are merely held together

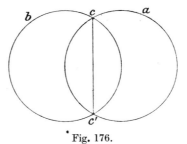

Fig. 176.

by the pressure of the surrounding envelope. But even supposing that they are in actual fluid contact, the case from the point of view of surface-tension presents no difficulty. In the case of the conjoined soap-bubbles, we were dealing with *similar* contacts and with *equal* surface-tensions throughout the system; but in the system of protoplasmic cells which constitute the segmenting egg we must make allowance for *inequality* of tensions, between the surfaces where cell meets cell and where on the other hand cell-surface is in contact with the surrounding medium—generally water or one of the fluids of the body. Remember that our general condition is that, in our entire system, the *sum of the surface energies* is a minimum; and, while this is attained by the *sum of the surfaces* being a minimum in the case where the energy is uniformly distributed, it is not necessarily so under non-uniform conditions. In the diagram (Fig. 176), if the energy per unit area be greater along the contact surface *cc'*, where cell meets cell, than along *ca* or cb, where cell-surface is in contact with the surrounding medium, these latter surfaces will tend to increase and the surface of cell-contact to diminish.   In short there will be the usual balance of forces between the tension along the surface cc', and the two opposing tensions along ca and cb.   If the former be greater than either of the other two, the outside angle will be less than $120°$; and if the tension along the surface cc' be as much or more than the sum of the other two, then the drops will merely touch one another, save for the possible effect of external pressure.   This is the explanation, in general terms, of the peculiar conditions obtaining in *Nostoc* and its allies (p. 477), and it also leads us to a consideration of the general properties and characters of a superficial or "epidermal" layer*.

While the inner cells of the honeycomb are symmetrically situated, sharing with their neighbours in equally distributed pressures or tensions, and therefore all tending closely to identity of form, the case is obviously different with the cells at the borders of the system.   So it is with our froth of soap-bubbles†.   The bubbles, or cells, in the interior of the mass are all alike in general character, and if they be equal in size are alike in every respect: as we see them in projection their sides are uniformly flattened, and tend to meet at equal angles of $120°$.   But the bubbles which constitute the outer layer retain their spherical surfaces (just as in the cells of a honeycomb), and these still tend to meet the partition-walls connected with them at constant angles of $120°$. This outer layer of bubbles, which forms the surface of our froth, constitutes after a fashion what we should call in botany an "epidermal" layer.   But in our froth of soap-bubbles we have, as

---

* A surface-layer always tends to have, *ipso facto*, a character of its own: a "skin" has such and such characteristics just because it is a skin.   The "Beilby layer" on a metallic surface is, in its own special way, a consequence of its own externality.

† A froth is a collocation of bubbles containing air; or in the language of colloid chemistry, an emulsion with air for its disperse phase.   The power of forming a froth is not the same as that of forming isolated bubbles; for some liquids, such as a solution of saponin, of gum arabic, of albumin itself, give a copious and lasting froth, but we find it hard to blow even a single tiny bubble with any of them. Something more than surface-tension seems necessary for the production and maintenance of a film: perhaps a certain amount of viscosity, to resist the tendency of surface-tension to tear the film asunder.

a rule, the same kind of contact (that is to say, contact with *air*)
both within and without the bubbles; while in our living cell, the
outer wall of the epidermal cell is exposed to air on the one side,
but is in contact with the protoplasm of the cell on the other: and
this involves a difference of tensions, so that the outer walls and
their adjacent partitions need no longer meet at precisely equal
angles of 120°. Moreover a chemical change, due perhaps to
oxidation or possibly also to adsorption, is very apt to affect the
external wall and lead to the formation of a "cuticle"; and this
process, as we have seen, is tantamount to a large increase of tension
in that outer wall, and will cause the adjacent partitions to impinge
upon it at angles more and more nearly approximating to 90°: the
bubble-like, or spherical, surfaces of the individual cells being more

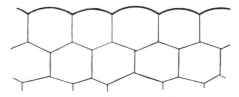

Fig. 177.   A froth, with its outer and inner cells or vesicles.

and more flattened in consequence. Lastly, the chemical changes
which affect the outer walls of the superficial cells may extend in
greater or less degree to their inner walls also: with the result that
these cells will tend to become more or less rectangular throughout,
and will cease to dovetail into the interstices of the next subjacent
layer. These then are the general characters which we recognise
in an epidermis; and we now perceive that its fundamental character
simply is that it lies outside, and that its physical characteristics
follow, as a matter of course, from the position which it occupies
and from the various consequences which that situation entails.

In the young shoot or growing point of a flowering plant botanists
(following Hanstein) find three cell-layers, and call them *dermatogen*,
periblem and *plerome*. The first is an epidermis, such as we have
just described. Its cells grow long as the shoot grows long; new
partitions cross the lengthening cell and tend to lie at right angles
to its hardening walls; and this epidermis, once formed, remains
a single superficial layer. The next few layers, the so-called peri-

blem, are compressed and flattened between the epidermis with its tense cuticle and the growing mass within; and under this restraint the cell-layers of the periblem also continue to divide in their own plane or planes. But the cells of the inner mass or plerome, lying in a more homogeneous field, tend to form "space-filling" poly-hedra, twelve- or perhaps fourteen-sided according to the freedom which they enjoy. In a well-known passage Sachs declares that the behaviour of the cells in the growing point is determined not by any specific characters or properties of their own, but by their position and the forces to which they are subject in the system of which they are a part*. This was a prescient utterance, and is abundantly confirmed†.

We have hitherto considered our cells, or our bubbles, as lying in a plane of symmetry, and have only considered their appearance as projected on that plane; but we must also begin to consider them as solids, whether they lie in a plane (like the four cells in Fig. 172), or are heaped on one another, like a froth of bubbles or a pile of cannon-balls. We have still much to do with the study of more complex partitioning in a plane, and we have the whole subject to enter on of the solid geometry of bodies in "close packing," or three-dimensional juxtaposition.

The same principles which account for the development of hexagonal symmetry hold true, as a matter of course, not only of *cells* (in the biological sense), but of any bodies of uniform size and originally circular outline, close-packed in a plane; and hence the hexagonal pattern is of very common occurrence, under widely varying circumstances. The curious reader may consult Sir Thomas Browne's quaint and beautiful account, in the *Garden of Cyrus*, of hexagonal, and also of quincuncial, symmetry in plants and animals, which "doth neatly declare how nature Geometrizeth, and observeth order in all things."

We come back to very elementary geometry. The first and simplest of all figures in plane geometry (with which for that reason Euclid begins his book) is the equilateral triangle; because three straight lines are the least number which enclose two-dimensional

---

* *Lectures on the Physiology of the Plant*, Oxford, 1887, p. 460, etc.
† Cf. J. H. Priestley in *Biol. Reviews*, III, pp. 1–20, 1928; U. Tetley in *Ann. Bot.* L, pp. 522–557, 1936; etc.

space, and three equal sides make the simplest of triangles. But it by no means follows that equilateral, or any other, triangles combine to form the simplest of polygonal associations or patterns. On the other hand, three straight lines meeting in a point are the least number by which we can subdivide or partition two-dimensional space; the simplest case of all is when the three partitions meet at co-equal angles, and a pattern of hexagons, so produced, is, geometrically speaking, the simplest of all ways in which a surface can be subdivided—the simplest of all two-dimensional "space-filling" patterns. So it comes to pass that we meet with a pattern of hexagons here and there and again and again, in all sorts of plane symmetrical configurations, from a soapy froth to the retinal pigment, from the cells of the honeycomb to the basaltic columns of Staffa and the Giant's Causeway.

We pass to solid geometry, and arrive by similar steps at an analogous result. Four plane sides are now the least number which enclose space, and (next to the sphere itself) the regular tetrahedron is the first and simplest of solids; but its simplicity is that of a solitary or isolated figure, and tetrahedra do not combine to fill space at all. But as the partitioning of an equilateral triangle was the first step towards the symmetrical partitioning of two-dimensional space, so we draw from the regular tetrahedron a first lesson in the partitioning of space of three dimensions; and as three lines meeting in a point were needed to partition two-dimensional space, so here, for three-dimensional space, we need four. The simplest case is, as before, when these meet at co-equal angles, but we do not see quite so easily what those four co-equal angles are.

For as the centre of symmetry of our equilateral triangle was defined by three lines bisecting its three angles and meeting one another in a point at co-equal angles of 120°, so in our regular tetrahedron four straight lines, running symmetrically inwards from the four corners, meet in a point at co-equal angles, and again define the centre of symmetry. If we make (as Plateau made) a wire tetrahedron, and dip it into soap-solution, we find that a film has attached itself to each of the six wires which constitute the little tetrahedral cage; that these six films meet, three by three, in four edges; and that these four edges meet at co-equal angles in a point, which is the *centroid*, or centre of symmetry, or centre of gravity, of the system.

This is the centre of symmetry not only for our tetrahedron, but for any close-packed tetrahedral aggregate of co-equal spheres; we meet with it over and over again, in a pile of cannon-balls, a froth of soap-suds, a parenchyma of cells, or the interior of the honeycomb. Moreover, in the actual demonstration by soap-films of this tetrahedral symmetry, we see realised all the main criteria laid down by Plateau and by Lamarle for a system *minimae areae*: three films and no more meet in an edge; four fluid edges and no more meet in a point, just as three wire edges and one fluid edge met in a point at each corner of the experimental figure. Lastly, the symmetry of the whole configuration is such that the three fluid films

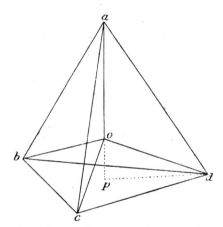

Fig. 178. A regular tetrahedron, with its centre of symmetry.

meeting in an edge, or the four fluid edges meeting in a point, all do so at co-equal angles.

In the plane configuration we saw without more ado that the angles of symmetry were the co-equal angles of 120°; but the four co-equal angles between the four edges which meet at the centre of our tetrahedron require a little more consideration. If in our figure of a regular tetrahedron (Fig. 178) o be the centroid, and we produce ao to p, the centre of the opposite side, bcd, it may be shewn that the line ap is so divided that $ao = 3op$ and $ao = bo = co = do$. For let four equal weights be put at the four corners of the tetrahedron, a, b, c, d. The resultant of the three at b, c, d is equivalent to $3W$ at p, the centre of symmetry of the equilateral triangle.

The resultant of all four is equal to the resultant of W at a, and $3W$ at p; it lies, therefore, on the straight line ap, and at the point o, such that $ao = 3op$. Therefore, in the triangle pod, as in the other three similar triangles in the figure, cos $pod = 1/3$, and cos $aod = -1/3$. Our tables tell us that the angle pod, whose trigonometrical value is the very simple one of cos $pod = 1/3$, has, in degrees and minutes to the nearest second, the seemingly less simple value of $70° 31' 43''$; and its supplement, the angle aod, has the corresponding value of $109° 28' 16''$.

This latter angle, then, of $109° 28' 16''$, or very nearly 109 degrees and a half, is the angle at which, in this and throughout *every other three-dimensional system* of liquid films, the edges of the partition-walls meet one another. It is the fundamental angle in simple homogeneous partitioning of three-dimensional space. It is an angle of statical equilibrium, an angle of close-packing, an angle of repose. In the simplest of carbon-compounds, the molecule of marsh-gas (CH_4), we may be sure that this angle governs the arrangement of the H-atoms; it determines the relation of the carbon-atoms one to another in a diamond—simplest of crystal-lattices; it defines the intersections of the bubbles in a froth, and of the cells in the honeycomb of the bee.

It is sometimes called the "tetrahedral angle"; it might be better called (for a reason we shall see presently) "Maraldi's angle." The whole story is less a physical than a mathematical one; for the phenomena do not depend on surface tension nor on any other physical force, but on such relations between surface and volume as are involved in the properties of space. If we take four little elastic balloons, half fill them with air, smear them with glycerine to lessen friction, place them in a bottle and exhaust the air therein, they will expand, adjust themselves together, and group themselves in a tetrahedral configuration, whose partition walls, edges and centre of symmetry are just those of our experiment of the soap-films.

This characteristic angle, though it leads in ordinary angular measurement to an endless decimal of a second, is nevertheless a very simple and perfectly definite magnitude. It is a strange property of Number that it fails to express certain simple and definite magnitudes, such as π, or $\sqrt{2}$, or $\sqrt{3}$, or this four-fold angle made by four lines meeting symmetrically in three-dimensional space. It is not these magnitudes that are peculiar, it is Number itself that is so! In all of these cases we have to import a new symbol; and in this case, when we draw it not from arithmetic but from trigonometry, and define our angle as cos − ⅓, nothing can or need be more precise or simpler. We may put the same thing a little differently, and say that Number itself fails us, now and then, to express what we want, although we have all the ten digits and their apparently endless permutations at our command. In such a deadlock, we have only to bring one new symbol, one new quantity, into use; and at once a wide new field is open to us.

Out of these two angles—the Maraldi angle of 109° etc., and the plane angle of 120°—we may construct a great variety of figures,

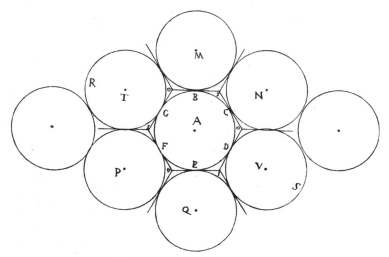

Fig. 179.   Diagram of hexagonal cells.   After Bonanni.

plane and solid, which become still more complex and varied when we consider associations of unequal as well as of co-equal cells, and thereby admit curved as well as plane intercellular partitions. Let us consider some examples of these, beginning with such as we need only consider in reference to a plane.

Let us imagine a system of equal cylinders, or equal spheres, in contact with one another in a plane, and represented in section by the equal and contiguous circles of Fig. 179. I borrow my figure from an old Italian naturalist, Bonanni (a contemporary of Borelli, of Ray and Willoughby, and of Martin Lister), who dealt with this matter in a book chiefly devoted to molluscan shells*.

* A. P. P. Bonanni, *Ricreatione dell' occhio e della mente, nell' Osservatione delle Chiocciole*, Roma, 1681.

It is obvious, as a simple geometrical fact, that each of these co-equal circles is in contact with six others around. Imagine the whole system under some uniform stress—of pressure caused by growth or expansion within the cells, or due to some uniformly applied constricting pressure from without. In these cases the six *points of contact* between the circles in the diagram will be extended into *lines,* representing *surfaces* of contact in the actual spheres or cylinders; and the equal circles of our diagram will be converted into regular and co-equal hexagons. The result is just the same so far as form is concerned—so long as we are concerned only with a morphological result and not with a physiological process—whatever be the force which brings the bodies together. For instance, the cells of a segmenting egg, lying within their vitelline membrane or within some common film or ectoplasm, are pressed together as they grow, and suffer deformation accordingly; their surface tends towards an *area minima,* but we need not even enquire, in the first instance, whether it be surface-tension, mechanical pressure, or what not other physical force, which is the cause of the phenomenon*.

The production by mutual interaction of polygons, which become regular hexagons when conditions are perfectly symmetrical, is beautifully illustrated by Bénard's *tourbillons cellulaires,* and also in some of Leduc's diffusion experiments. In these latter, a solution of gelatine is allowed to set on a plate of glass, and little drops of weak potassium ferrocyanide are then let fall at regular intervals upon the gelatine. Immediately each little drop becomes the centre of a system of diffusion currents, and the several systems conflict with and repel one another; so that presently each little area becomes the seat of a to-and-fro current system, outwards and back again, until the concentration of the field becomes equalised and the currents cease. When equilibrium is attained, and when the gelatin-layer is allowed to dry, we have an artificial tissue of

* The following is one of many curious corollaries to the principle of close-packing here touched upon. A circle surrounded by six similar circles, the whole bounded by a circle of three times the radius of the original one, forms a unit, so to speak, next in order after the circle itself. A round pea or grain of shot will pass through a hole of its own size; but peas or shot will not *run out* of a vessel through a hole less than *three times* their own diameter. There can be no freedom of motion among the close-packed grains when confronted by a smaller orifice. Cf. K. Takahasi, *Sci. Papers Inst. Chem., etc.,* Tokio, XXVI, p. 19. 1935.

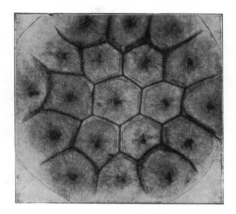

Fig. 180.　An "artificial tissue," formed by coloured drops of sodium chloride solution diffusing in a less dense solution of the same salt.　After Leduc.

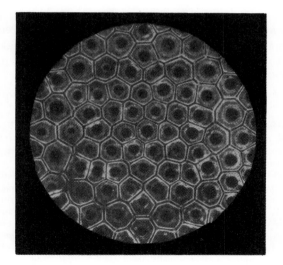

Fig. 181.　An artificial cellular tissue, formed by the diffusion in gelatine of drops of a solution of potassium ferrocyanide.　After Leduc.

hexagonal "cells," which simulate an organic parenchyma very closely; and by varying the experiment in ways which Leduc describes, we may imitate various forms of tissue, and produce cells with thick walls or with thin, cells in close contact or with wide intercellular spaces, cells with plane or with curved partitions, and so forth.

James Thomson (Kelvin's elder brother) had observed nearly sixty years ago a curious "tesselated structure" on a liquid surface, to wit, the soapy water of a wash-tub. The eddies and streaks of swirling water settled down into a cellular configuration, which continued for hours together to alter its details; small areoles disappeared, large ones grew larger, and subdivided into small ones again. With few and transitory exceptions three partitions and no more met at every node of the meshwork; and (as it seems to me) the subsequent changes were all due to such shifting of the lines as tended to make the three adjacent angles more and more nearly co-equal with one another: the obvious effect of this being to make the pattern more and more regularly hexagonal*.

In a not less homely experiment, hot water is poured into a shallow tin and a layer of milk run in below; on blowing gently to cool the water, holes, more or less close-packed and evenly inter-spaced, appear in the milk. They shew how cooling has taken place, so to speak, in spots, and the cooled water has descended in isolated columns†.

Bénard's "tourbillons cellulaires"‡, set up in a thin liquid layer,

---

* James Thomson, On a changing tesselated structure in certain liquids, *Proc. Glasgow Phil. Soc.* 1881–82; *Coll. Papers*, p. 136—a paper with which M. Bénard was not acquainted, but see Bénard's later note in *Ann. de Chim.* Dec. 1911.

† See Graham's paper, quoted below.

‡ H. Bénard, Les tourbillons cellulaires dans une nappe liquide, *Rev. génér. des Sciences*, XII, pp. 1261–1271, 1309–1328, 1900; *Ann. Chimie et Physique* (7), XXIII, pp. 62–144, 1901; *ibid.* 1911. Quincke had seen much the same long before: *Ann. d. Phys.* CXXXIX, p. 28, 1870. The "figures of de Heen" are an analogous electrical phenomenon; cf. P. de Heen, Les tourbillons et les projections de l'éther, *Bull. Acad. de Bruxelles* (3) XXXVII, p. 589, 1899; A. Lafay, *Ann. de Physique* (10), XIII, pp. 349–394, 1930. These various phenomena, all leading to a pattern of hexagons, have often been studied mathematically: cf. Rayleigh, *Phil. Mag.* XXXII, pp. 529–546, 1916, *Coll. Papers*, VI, p. 48; also Ann Pellew and R. V. Southwell, *Proc. R.S.* (A), CLXXVI, pp. 312–343, 1940. The hexagonal pattern is a particular case of stability, but not necessarily the simplest; it is only by experiment that we know it to be the permanent condition in an unlimited field.

are similar to but more elegant than James Thomson's tesselated patterns, and both of them are in their own way still more curious than M. Leduc's; for the latter depend on centres of diffusion artificially inserted into the system and determining the number and position of the "cells," while in the others the cells make themselves.   In Bénard's experiment a thin layer of liquid is warmed in a copper dish. The liquid is under peculiar conditions of instability, for the least fortuitous excess of heat here or there would suffice to start a current, and we should expect the whole system to be highly unstable and unsymmetrical. But if all be kept carefully uniform, small disturbances appear at random all over the system; a current ascends in the centre of each; and a "steady state," if not a stable equilibrium, is reached in time, when the descending currents, impinging on one another, mark out a "cellular system." If we set the fluid gently in motion to begin with, the first "cell-divisions" will be in the direction of the flow; long tubes appear, or "vessels," as the botanist would be apt to call them. As the flow slows down new cell-boundaries appear, at right angles to the first and at even distances from one another; parallel rows of cells arise, and this transitory stage of partial equilibrium or imperfect symmetry is such as to remind the botanist of his *cambium* tissues, which are, so to speak, a temporary phase of histological equilibrium. If the impressed motion be not longitudinal but rotary, the first lines of demarcation are spiral curves, followed by orthogonal inter-sections.

Whether we start with liquid in motion or at rest, symmetry and uniformity are ultimately attained. The cells draw towards uniformity, but four, five or seven-sided cells are still to be found among the prevailing hexagons. The larger cells grow less, the smaller enlarge or disappear; where four partition-walls happen to meet, they shift till only three converge; the sides adjust themselves to equal lengths, the angles also to equality. In the final stage the cells are hexagonal prisms of definite dimensions, which depend on temperature and on the nature and thickness of the liquid layer; molecular forces have not only given us a definite cellular pattern, but also a "fixed cell-size."

Solid particles in the fluid come to rest in symmetrical positions. If they be heavier they accumulate in little isolated heaps, each in the focus or axis of a cell; if they be lighter they drift to the boundaries, then towards the nodes, where they tend to form tri-radiate figures like so many "tri-radiate spicules". But if they be in very fine suspension a curious thing happens: for as they are carried round in the vortex, the lowermost layer of liquid, next to the solid floor, keeps free of particles; and this "dust-free coat*", rising in the axis of the cell and descending at its boundary-walls, surrounds an inner vortex to which the suspended particles are confined. The cell-contents have, so to speak, become differentiated into an "ectoplasm" and an "endoplasm"; and an analogy appears

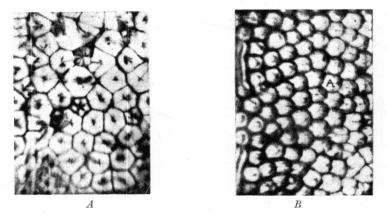

A   B

Fig. 182. Bénard patterns in smoke: A, at rest; B, under shear. After K. Chandra.

with the phenomenon of protoplasmic "rotation," where the outer layer of a cell tends to be free from granules. When bright glittering particles are used for the suspension (such as graphite or butterfly-scales) beautiful optical effects are obtained, deep shadows marking the outlines and the centres of the cells. Lastly, and this is by no means the least curious part of the phenomenon, the free surface of the liquid is not plane; but each little cell is found to be dimpled in the centre and raised at the edges, in a surface of very complex curvature†, and there is a curious pulsation in the flow, especially when waxes are used.

* Cf. Tyndall, *Proc. Roy. Inst.* vi, p. 3, 1870.
† The differences of level are of a very small order of magnitude, say $1\,\mu$ in a layer of spermaceti 1 mm. thick.

Ringing the changes on Bénard's experiment, we may use unstable layers of various liquids or gases, or even a thin layer of smoke between a hot plate and a cold (Fig. 182). The smoke will form waves and folds and rolling clouds; then, with increasing and more and more symmetrical instability, polygonal or hexagonal prisms; and all these configurations we may deform or "shear" by sliding one plate over the other. Familiar cloud-patterns, as of a dappled or mackerel sky, can be imitated in this way. When the hexagonal prisms have been developed it is found that a steady shear deforms them into a well-known curvilinear tesselated pattern (Fig. 183)

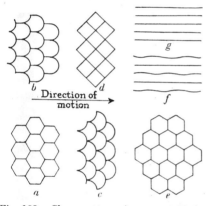

Fig. 183. Shear-patterns in an unstable layer of air or smoke. After Graham.

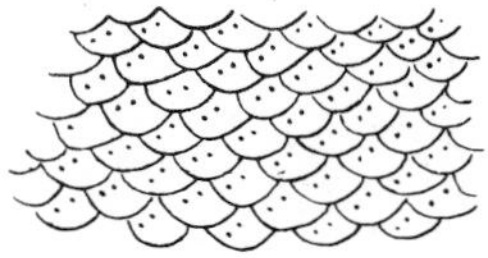

Fig. 184. Ambulacral plates of a sea-urchin (*Lepidesthes*), to illustrate the "shearing" of a pattern normally hexagonal. After Hawkins*.

and this may be sheared again into hexagons, oriented in an opposite direction to the first†. The very same pattern occurs now and then in organisms, as a deformation of what is normally a pattern of hexagons (Fig. 184).

* From H. L. Hawkins, *Phil. Trans.* (B), ccix, p. 383, 1920.

† Cf. A. Graham, Shear patterns in an unstable layer of air, *Phil. Trans.* (A), No. 714, 1933; Gilbert Walker and Phillips, *Q.J.R. Met. Soc.* lviii, p. 23, 1932; Mals, *Beitr. Phys. frei. Atmosph.* xvii, p. 45, 1930; H. Jeffreys, *Proc. R.S.* (A), cxviii, p. 195, 1928; Krishna Chandra, *ibid.* clxiv, pp. 231–242, 1938. After a steady state is reached it is found that in air or smoke the centre of each polygon is a funnel of descent, but it is an ascending column if the layer be liquid. Now the viscosity of a gas increases, and that of a liquid decreases, with rise of temperature, and the greater the viscosity the more stable is the layer. Accordingly, for a gas the upper, and for a liquid the lower layer is the more unstable, and it is there that in each case the flow begins.

We learn from these experiments of Bénard and others how similar distributions of force, and identical figures of equilibrium, may arise through different physical agencies. We see that patterns closely analogous to those of living cells and tissues may be due to very different causes; and we may be led to scrutinise anew, with an open mind, various histological configurations whose origin is doubtful or obscure. The chitinous shells of certain water-fleas (*Cladocera*) are beset with a roughly hexagonal pattern, and each little chitinous polygon is supposed to correspond to, and to be formed by, an underlying "hypodermis" cell*. But we presently discover that the existence of these hypodermis-cells is merely deduced from the polygons themselves and from a coincident

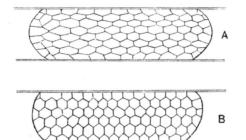

Fig. 185. Soap-froth under pressure. After Rhumbler.

distribution of pigment; it might not be amiss to look again into the development of the pattern, with an open mind as to the possibility of its being a purely physical phenomenon. Nor need we by any means assume that the calcareous prisms of a molluscan shell are necessarily derived from, or associated with, a like number of histological elements.

In a soap-froth imprisoned between two glass plates we have a symmetrical system of cells which appear in optical section (Fig. 185, B) as regular hexagons; but if we press the plates a little closer together the hexagons become deformed and flattened. The

---

* Cf. F. Claus, Zur Kenntniss...des feineren Baues der Daphniden, *Ztschr. f. wiss. Zool.* XXIII, XXVII, pp. 362–402, 1876: Ernest Warren, Relationship between size of cell and size of body in *Daphnia magna*, *Biometrika*, (3) II, pp. 255–259, 1902; Fritz Werner, Die Veränderung der Schalenform und der Zellenaufbau bei *Scapholeberis*, *Int. Revue der ges. Hydrobiologie*, pp. 1–20, 1923.

change is from a more to a less probable configuration—the entropy is diminished; and if we apply no further pressure the tension of the films adjusts itself again, and the system recovers its former symmetry*.

The epithelial lining of the blood-vessels shews a curious and beautiful pattern. The cells seem diamond-shaped, but looking closer we see that each is in contact (usually) with six others; they are not rhombs, or diamonds, but elongated hexagons, pulled out long by the growth of the vessel and the elastic traction of its walls. The sides of each cell are curiously waved, and a simple experiment explains this phenomenon. If we make a froth of white-of-egg upon a stretched sheet of rubber, the cells of the froth will tend to

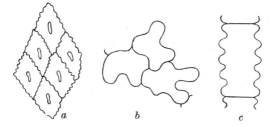

Fig. 186.   Sinuous outlines of epithelial cells.   *a*, endothelium of a blood-vessel; *b*, epidermis of *Impatiens*;   *c*, epidermal cells of a grass (*Festuca*).

assume their normal hexagonal pattern; but relax the elastic membrane, and the cell-walls are thrown into beautiful sinuous or wavy folds. The froth-cells cannot contract as the rubber does which carries them, nor can the epithelial cells contract as does the muscular coat of the blood-vessel; in both cases alike the cell-walls are obliged to fold or wrinkle up, from lack of power to shorten. The epithelial cells on the gills of a mussel† are wrinkled after the same fashion; but the more coarsely sinuous outlines of the epithelium in many plants is another story, and not so easily accounted for.

The hexagonal pattern is illustrated among organisms in countless cases, but those in which the pattern is perfectly regular, by

---

* That *everything* is passing all the while towards a "*more probable state*" is known as the "principle of Carnot," and is the most general of all physical laws or aphorisms.

† Cf. James Gray's *Experimental Cytology*, p. 252.

reason of perfect uniformity of force and perfect equality of the individual cells, are not so numerous. The hexagonal cells of the pigmented epithelium of the retina are a good example. Here we have a single layer of uniform cells, reposing on the one hand upon a basement membrane, supported behind by the solid wall of the sclerotic, and exposed on the other hand to the uniform fluid pressure of the vitreous humour. The conditions all point, and lead, to a symmetrical result: the cells, uniform in size, are flattened out to a uniform thickness by uniform pressure, and their reaction one upon another converts each flattened disc into a regular hexagon. An equally symmetrical case, one of the first-known examples of an "epithelium," is to be found on the inner wall of the amnion, where, as Theodor Schwann remarked, "die sechs-eckige Plättchen sind sehr schön und gross*."

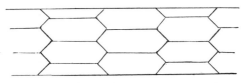

Fig. 187. Epidermis of *Girardia*. After Goebel.

In an ordinary columnar epithelium, such as that of the intestine, again the columnar cells are compressed into hexagonal prisms; but here the cells are less uniform in size, small cells are apt to be intercalated among the larger, and the perfect symmetry is lost accordingly. But obviously, wherever we have, in addition to the forces which tend to produce the regular hexagonal symmetry, some other component arising asymmetrically from growth or traction, then our regular hexagons will be distorted in various simple ways. Thus in the delicate epidermis of a leaf or young shoot we begin with hexagonal cells of exquisite regularity: on which, however, subsequent longitudinal growth may impose an equally simple and symmetrical deformation or polarity (Fig. 187).

In the growth of an ordinary dicotyledonous leaf, we see reflected in the form of its cells the tractions, irregular but on the whole longitudinal, which growth has superposed on the tensions of the

* *Untersuchungen*, p. 84; cf. Sydenham Society's translation, p. 75.

partition walls (Fig. 188). In the narrow elongated leaf of a mono-cotyledon, such as a hyacinth, the elongated, apparently quad-rangular cells of the epidermis appear as a necessary consequence of the simpler laws of growth which gave its simple form to the leaf as a whole. In all these cases alike, however, the rule still holds that only three partitions (in surface view or plane projection) meet in a point; and near their point of meeting the walls are manifestly curved for a little way, so as to permit the triple conjunction to take place at or near the co-equal angles of 120°, after the fashion described above.

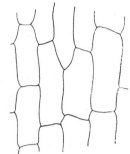

Briefly speaking, wherever we have a system of cylinders or spheres, associated together with sufficient mutual interaction to bring them into complete surface contact, there,

Fig. 188. Epidermal cells from leaf of *Elodea canadensis*. After Berthold.

in section or in surface view, we tend to get a pattern of hexagons.

In thickened cells or fibres of bast or wood, the "sclerenchyma" of vegetable histology, the hexagonal pattern is all but lost, and we see in cross-section the more or less *circular* transverse outlines of elongated and tapering cells. Looking closer we see that the primitive cell-walls preserve their angular contours, and shew much as usual an hexagonal pattern, with only such irregularities as follow from the unequal sizes of the associated cells. But when these primary walls are once laid down, the secondary deposits which follow them are under different conditions; and these obey the law of minimal areas in their own way, by filling up the angles of the primary cell and by continuing to grow inwards in concentric and more and more nearly circular rings.

While the formation of an hexagonal pattern on the basis of ready-formed and symmetrically arranged material units is a very common, and indeed the general way, it does not follow that there are not others by which such a pattern can be obtained. For instance, if we take a little triangular dish of mercury and set it vibrating (either by help of a tuning-fork, or by simply tapping on the sides) we shall have a series of little waves or ripples starting inwards from each of the three faces; and the intercrossing, or interference of these three sets of waves produces crests and hollows, and intermediate points of no disturbance, whose *loci* are seen as a beautiful pattern of minute hexagons. It is possible that the very minute and astonishingly regular pattern of hexagons which we see on the surface of many diatoms (Fig. 189) may be a phenomenon of this order*. The same may be the case also in Arcella, where an apparently hexagonal pattern is found not to consist of simple hexagons, but of "straight lines in three sets of parallels, the lines of each set making an angle of sixty degrees with those of the other two sets†." We must also bear in mind, in the case of the minuter forms, the large possibilities of optical illusion. For instance, in one of Abbe's "diffraction-plates," a pattern of dots, set at equal interspaces, is reproduced on a very minute scale by photography; but under certain conditions of microscopic illumination and focusing, these isolated dots appear as a pattern of hexagons.

A symmetrical arrangement of hexagons, such as we have just been studying, suggests various geometrical corollaries, of which the following may be a useful one. We sometimes desire to estimate the number of hexagonal areas or facets in some structure where these are numerous, such for instance as the cornea of an insect's eye, or in the minute pattern of hexagons on many diatoms. An approximate enumeration is easily made as follows.

For the area of a hexagon (if we call δ the short diameter, that namely which bisects two of the opposite sides) is $\delta^2 \times \sqrt{3}/2$, the area of a circle being $d^2 . \pi/4$. Then, if the diameter (d) of a circular area include n hexagons, the area of that circle equals $(n.\delta)^2 \times \pi/4$. And, dividing this number by the area of a single hexagon, we obtain for the number of areas in the circle, each equal to a hexagonal facet, the expression $n^2 \times \pi/4 \times 2/\sqrt{3} = 0.907n^2$, or $9/10 . n^2$, nearly.

This calculation deals, not only with the complete facets, but with the areas of the broken hexagons at the periphery of the circle. If we neglect these latter, and consider our whole field as consisting of successive rings of hexagons about a central one, we obtain a simpler rule. For obviously, around our central hexagon there stands a zone of six, and around these

---

* Cf. some of J. H. Vincent's photographs of ripples, in *Phil. Mag.* 1897–99; or those of F. R. Watson, in *Phys. Review*, 1897, 1901, 1916. The appearance will depend on the rate of the wave, and in turn on the surface-tension; with a low tension one would probably see only a moving "jabble." Cf. also Faraday, On the crispations of fluids resting upon a vibrating support, *Phil. Mag.* 1831, p. 299; and Rayleigh, *Sound*, II, p. 346, 1896. FitzGerald thought diatom-patterns might be due to electromagnetic vibrations (*Works*, p. 503, 1902); with which cf. W. D. Dye, Vibration-patterns of quartz plates, *Proc. R.S.* (A), CXXXVIII, p. 1, 1932. Dye's Fig. 17, which he calls "one of the most beautiful types of minor vibration met with in discs", is closely akin to the diatom *Orthoneis splendida*. In both cases two nodal systems, conjugate to one another, are based on two foci near the ends of an elliptical plate; but bands in the experimental plate are further broken up into rows of dots in the diatom. See also Max Schultze, Die Struktur der Diatomeenschale verglichen mit gewissen aus Fluorkiesel künstlich darstellbaren Kieselhäuten, *Verh. naturh. Ver. Bonn*, XX, pp. 1–42, 1863; *Trans. Microsc. Soc.* (N.S.), XI, pp. 120–136, 1863; H. J. Slack, *Monthly Microsc. Journ.* 1870, p. 183.

† J. A. Cushman and W. P. Henderson, *Amer. Nat.* XL, pp. 797–802, 1906.

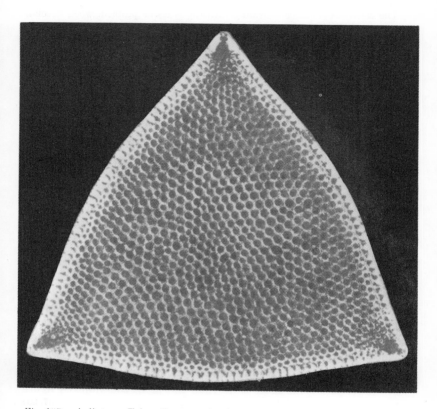

Fig. 189. A diatom, *Triceratium* sp., shewing pattern of hexagons; ×300. From O. Prochnow, *Formenkunst der Natur*.

a zone of twelve, and around these a zone of eighteen, and so on. And the total number, excluding the central hexagon, is accordingly:

For one zone	6	$= 3 \times 1 \times 2 = 6 \times 1$	
,, two zones	18	2×3	3
,, three zones	36	3×4	6
,, four zones	60	4×5	10
,, five zones	90	5×6	15

and so forth. If N be the number of zones, and if we add one to the above numbers for the odd central hexagon, then the rule is that the total number $H = 3N(N+1)+1$. Thus, if in a preparation of a fly's cornea I can count twenty-five facets in a line from a central one, the total number in the entire field is $(3 \times 25 \times 26) + 1 = 1951$ *.

The electrical engineer is dealing with the selfsame problem when he finds he can pack $6 + 18 + 36 + \ldots$ wires around a central wire, to form a multiple cable of 1, 2 and 3 concentric strands. He counts them by the same formula, in the simpler form of $6t + 1$: where t is a "triangular number," 1, 3, 6, 10, etc., corresponding to the number of strands. Thus $1951 = 6 \times 325 + 1$; 325 being the triangular number of 25, $1 + 2 + 3 + \ldots + 25$.

We have many varied examples of this principle among corals, wherever the polypes are in close juxtaposition, with neither empty space nor accumulations of matrix between their adjacent walls. *Favosites gothlandica*, for instance, furnishes us with an excellent example. In the great genus Lithostrotion we have some species which are "massive" and others which are "fasciculate." In other words, in some the long cylindrical corallites are closely packed together, and in others they are separate and loosely bundled (Fig. 190); in the former the corallites are squeezed into hexagonal prisms, while in the latter they retain their cylindrical form. Where the polypes are comparatively few, and so have room to spread, the mutual pressure ceases to work or only tends to push them asunder, letting them remain circular in outline (e.g. *Thecosmilia*). Where they vary gradually in size, as for instance in *Cyathophyllum hexagonum*, they are more or less hexagonal but are not regular hexagons; and where there is greater and more irregular variation in size, the cells will be *on the average* hexagonal, but some will have fewer and some more sides than six, as in the annexed figure of *Arachnophyllum* (Fig. 192). Where larger and smaller cells, corresponding to two different kinds of zooids, are mixed together, we may get various results. If the larger cells are numerous enough

* This estimate neglects not merely the broken hexagons, but all those whose centres lie between the circle and a hexagon inscribed in it. The discrepancy is considerable, but a correction is easily made. It will be found that the numbers arrived at by the two methods are approximately as $6 : 5$. For more detailed calculations see a paper by H.M. (? H. Munro) in *Q.J.M.S.* VI, p. 83, 1858. The methods of enumeration used by older writers, especially by Leeuwenhoek, are sometimes curious and interesting; cf. Hooke, *Micrographia*, 1665, p. 176; Leeuwenhoek, *Arcana naturae*, 1695, p. 477; *Phil. Trans.* 1698, p. 169; *Epist. physiolog.* 1719, p. 342; Swammerdam, *Biblia Naturae*, 1737, p. 490. Leeuwenhoek found, or estimated, 3181 facets on the cornea of a scarab, and 8000 on that of a fly; M. Puget, about the same time, found 17,325 in that of a butterfly. See also Karl Leinemann, *Die Zahl der Facetten in den...Coleopteren*, Hildesheim, 1904.

to be more or less in contact with one another (e.g. various Monti-
culiporae) they will be irregular hexagons, while the smaller cells
between them will be crushed into all manner of irregular angular

Fig. 190.　*Lithostrotion Martini.*
After Nicholson.

Fig. 191.　*Cyathophyllum hexagonum.*
From Nicholson, after Zittel.

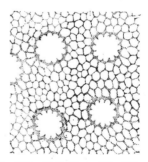

Fig. 192.　*Arachnophyllum pentagonum.*
After Nicholson.

Fig. 193.　*Heliolites.*　After Woods.

forms. If on the other hand the large cells are comparatively few
and are large and strong-walled compared with their smaller neighbours, then the latter alone will be squeezed into hexagons while
the larger ones will tend to retain their circular outline undisturbed
(e.g. *Heliopora*, *Heliolites*, etc. (Fig. 193)).

When, as happens in certain corals, the peripheral walls or "thecae" of the individual polypes remain undeveloped but the radiating septa are formed and calcified, then we obtain new and beautiful mathematical configurations (Fig. 194). For the radiating septa are no longer confined to the circular or hexagonal bounds of a polypite, but tend to meet and become confluent with their neighbours on every side; and, tending to assume positions of equilibrium, or of minimum, under the restraints to which they are subject, they fall into congruent curves, which correspond in a striking manner to lines running in a common field of force between a number of secondary centres. Similar patterns may be produced in various ways by the play of osmotic or magnetic forces;

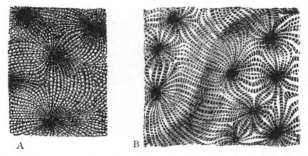

A   B

Fig. 194.   Surface-views of corals with undeveloped thecae and confluent septa. A, *Thamnastraea*; B, *Comoseris*.   From Nicholson, after Zittel.

and a very curious case is to be found in those complicated forms of nuclear division known as triasters, polyasters, etc., whose relation to a field of force Hartog in part explained*. It is obvious that in our corals these curving septa are all orthogonal to the non-existent hexagonal boundaries; and, as the phenomenon is due to the imperfect development, or non-existence, of a thecal wall, it is not surprising that we find identical configurations among various corals, or families of corals, not otherwise related to one another. We find the same or very similar patterns displayed, for instance, in Synhelia (*Oculinidae*), in Phillipsastraea (*Rugosa*), in Thamnastraea (*Fungida*), and in many more.

* Cf. M. Hartog, The dual force of the dividing cell, *Science Progress* (N.S.), I, Oct. 1907, and other papers.   Also Baltzer, *Mehrpolige Mitosen bei Seeeigeleiern*, Diss., 1908.

A beautiful hexagonal pattern is seen in the male and female cones of *Zamia*, where the scales which bear the pollen-sacs or the ovules are crowded together, and are so formed and circumstanced that they cannot protrude and overlap.   They become compressed accordingly into regular hexagons, smaller and more regular in the male cone than in the female, in which latter the cone as a whole has tended to grow more in breadth than in length, and the hexagons are somewhat broader than they are long.   In a cob of maize the hexagonal form of the grains, such as should result from close-packing and mutual compression, is exhibited faintly if at all; for growth and elongation of the spike itself has relieved, or helped to relieve, the mutual pressure of the grains.

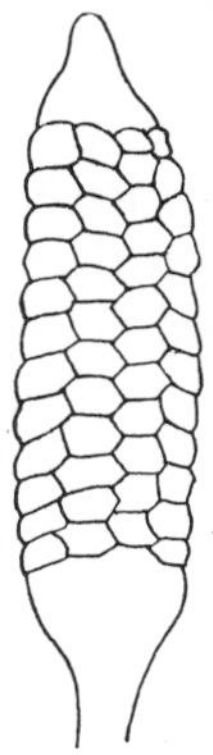

Fig. 195.
Female cone
of *Zamia*.

The pine-cone shews a simple, but unusual mode of close-packing.   The spiral arrangement causes each scale to lie, to right and to left, on two principal spirals; it has close neighbours on four sides, and mutual compression leads to a square or rhomboidal, instead of an hexagonal, configuration*.   On the other hand, the scales of the larch-cone overlap: therefore they are not subject to compression, but grow more freely into leaf-like curves.

The story of the hexagon leads us far afield, and in many directions, but it begins with something simpler even than the hexagon.   We have seen that in a soapy froth three films, and three only, meet in an edge, a phenomenon capable of explanation by the law of *areae minimae*.   But the conjunction, three by three, of almost any assemblage of partitions, of cracks in drying mud, of varnish on an old picture, of the various cellular systems we have described, is a general tendency, to be explained more simply still.   It would be a complex pattern indeed, and highly improbable, were all the cracks (for instance) to meet one another six by six; four by four would be less so, but still too much; and three by three is nature's way, simply because it is the simplest and the least.   When the partitions meet three by three, the angles by which they do so may vary indefinitely, but their *average* will be 120°; and if all be *on the*

---

* In some small, few-scaled cones the packing remains incomplete, and the scales are four-, five-, or six-sided, as the case may be.

average angles of 120°, the polygonal areas must, on the average, be *hexagonal*. This, then, is the simple *geometrical* explanation, apart from any physical one, of the widespread appearance of the pattern of hexagons.

If the law of minimal areas holds good in a "cellular" structure, as in a froth of soap-bubbles or in a vegetable parenchyma, then not merely on the average, but actually at every node, three partition-walls (in plane projection) meet together. Under perfect symmetry they do so at co-equal angles of 120°, and the assemblage consists

Fig. 196.   Cracks in drying mud; a thread encircles and marks out a "polar furrow"; cf. p. 487. From R. H. Wodehouse.

(in plane projection) of co-equal hexagons; but the angles may vary, the cells be unequal, and the hexagons interspersed with other polygonal figures. Nevertheless, so long as three partition-walls and only three meet together, the cells are, *ipso facto*, on the average hexagonal*.

We may count the cells if we please. A section of Cycas-petiole gave the following numbers:

Number of sides	3	4	5	6	7	8	9
,, ,, instances	0	8	97	207	96	9	0

Mean number of sides: 6·00.

The fine emulsion of an Agfa plate shews a beautiful polygonal pattern which obeys the law of the triple node; and a patch of a

* A more elaborate proof is given by W. C. Goldstein, On the average number of sides of polygons of a net. *Ann. Math.* (2), xxxii, pp. 149–153, 1931.

thousand cells in such a plate has been found, like the Cycas-petiole, to average out at six sides each, precisely*.

The cracking of a fine varnish may illustrate (as in Fig. 197) the same phenomenon. A little water in the varnish tends to accumulate between the cells, interferes with their close-packing, and complicates the arrangement of their partitions.

The horny plates which form the carapace of a tortoise (different as the case may seem) still obey the two guiding principles (1) that the polygonal boundaries meet in three-way nodes, and (2) that the three angles tend towards equality, always provided that no alien influences interfere. These principles are of the widest application; the carapace of a Eurypterid, the dermal armour of an Old Red

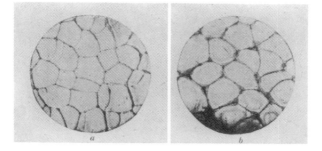

Fig. 197. Cellular patterns in varnish. *a*, dissolved in dry acetone; *b*, containing a little water.

Sandstone fish like Hugh Miller's *Asterolepis*, exhibit them at a glance†. The carapace of our tortoise is formed of a bony framework of ribs and vertebrae, overlaid by superficial plates of horn or tortoiseshell; it is these latter with which we are about to deal. They are arranged in three rows down the back, and a marginal row of smaller plates surrounds the others; there are (normally) twenty-four plates in the marginal series, and five large ones in the median longitudinal row. With these few facts, and our general principles

* F. T. Lewis, Polygons in a film...and the pattern of simple epithelium, *Anat. Record*, L, pp. 235–265, 1931.

† The all but universal law of the triple corner, or triradiate suture, is now and then enough to give a deceptive likeness to very different things. When Cope marred his brilliant classification of the Ostracoderm fishes by seeing in the carapace of Bothriolepis a likeness to the dorsal plates of the tunicate Chelyosoma, it was this and this alone which led him astray.

in hand, how far can we go towards depicting the carapace? The
five plates of the median row *must* alternate with their lateral
neighbours, and four plates in each lateral row will, accordingly, be
the simplest case or most probable number. If at each three-way
junction between median and lateral plates the angles tend to
equality, it follows that the median plates become converted into
more or less regular, or at least symmetrical, hexagons. As to the
twenty-four marginal plates, let us put one in front* and one
behind, leaving eleven for each side; and let us see to it carefully

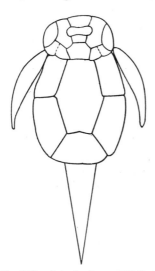

Fig. 198. *Asterolepis*: an Old Red
Sandstone fish. After Traquair.

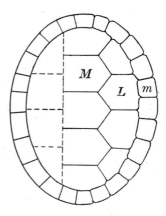

Fig. 199. Horny carapace of a
tortoise; diagrammatic.

that the sutures between these do not coincide but *alternate*
with those of the lateral row. Here we begin to meet with conditions of restraint analogous to those of the surface layer of a
froth, for the long marginal cells must remain marginal, and their
sides must continue to run more or less parallel, or more or less
perpendicular, to the edge of the shell; only in the immediate

---

* The Old World tortoises have twenty-five marginal plates, those of the
New World lack the anterior median, or "nuchal" plate. This difference is a
biological accident, it has neither mathematical interest nor functional significance;
it exemplifies the aphorism that whatsoever is possible Nature will sooner or
later do.

neighbourhood of each corner will the sides tend to curve in, so forming a notch whose curved sides have tangents approximately 120° apart, again just as in our projection of the surface of a froth (p. 494). Already these considerations lead us to a fair sketch of, or first approximation to, the carapace of a tortoise; but we may go on a little further. The horny plates and the bony carapace below must grow at such a rate as to keep pace, more or less exactly, with one another; but it does not follow that they will keep time precisely. If the horny plates grow ever so little faster than the bones below, they will fail to fit, will overcrowd one another, and will be forced to bulge or wrinkle. Both of these things they often, and even characteristically, do; the wrinkles appear in orderly, parallel folds, pointing to alternate periods, or spurts, of faster and slower growth; and the characteristic patterns which ensue are the visible expression of these differential growth-rates.

In all this we assume that the plates are lying in one and the same plane or even surface, abutting against one another as they grow, and so crowding and squeezing one another into the form of straight-edged polygons. The result will be very different if they overlap, after the manner of slates on a roof: the difference is what we have seen to exist between the cones of *Pinus* and of *Larix*. The overlapping edges will be free to grow into natural, rounded curves; each plate, uncrowded and unconstrained, will stay smooth and unwrinkled; the number and order of the plates will be the same as before—but the shell will be no longer that of a tortoise, but of the turtle from which "tortoise-shell" is obtained.

A snow-crystal is a very beautiful example of hexagonal symmetry. It belongs to another order of things to those we have been speaking of: for in substance it is a solid, and in form it is a crystal, and its own intrinsic molecular forces build it up in its own way. But (as we have mentioned once before) it is an exquisite illus‧tration of Nature's way of producing infinite variety from the permutations and combinations of a single type. The snowflake is a crystal formed by sublimation, that is to say by precipitation from a vapour without passing through a liquid phase. It begins as a tiny hexagon, the making of which tends to use up the vapour near by; the angles of the hexagon jut out, so to speak, into regions of greater, or less depleted, vapour-pressure, and at these corners further crystallisation will next set in. The hexagon will tend to grow out into a six-rayed star; and later and more slowly the material for further crystallisation will make its way between the rays, and begin to build side-growths on them *.

The basaltic column

Hexagonal patterns are by no means confined to the organic world. The basalt of Staffa and the Giant's Causeway shews a wonderful array of prismatic columns of irregular size and form,

Fig. 200. Basalt at Giant's Causeway. By Mr R. Welch, Belfast.

but mostly hexagonal; so also does the frozen soil of Spitzbergen; starch sets on cooling into analogous prisms, but in a ruder fashion as on a smaller scale; and all these are due to simple forces in a simple field, namely to tension, or shrinkage, in a horizontal mass or layer. Imagine a sheet or "sill" of intrusive basalt, thrust in as a molten mass between older rocks. It is gradually chilled by the cold air above or by the rocks on either side, and its inner mass, cooling slower than the outer layer, contracts slowly. Nothing hinders its vertical contraction, rather is this helped by its own weight and by the load above; but no further lateral contraction can take place without splitting the mass, once the basalt sets hard. Con-

* Cf. Gerald Seligmann, *Nature*, 26 June, 1937, p. 1090.

traction, however, does take place, irresistibly, and it may be that long cracks appear; the strain being so far relieved, the next cracks will tend to take place at right angles to the first. But more commonly rupture is delayed until considerable strain-energy has been stored up; once started, it proceeds explosively from a number of centres, and shatters the whole mass into prismatic fragments. However quickly and explosively the cracks succeed one another each relieves an existing tension, and the next crack will give relief in a different direction to the first. When one crack meets another it will seldom cross it, for the strain which led to the former fracture does not extend into the new field. In short the cracks will be found to meet one another three by three, and therefore at angles *on the average* of 120°, and the columns will be *on the average* hexagonal. For the making of a prismatic structure all that is required is more or less uniform tensile strain in the two dimensions of a horizontal plane; uniform tension in three dimensions would have given rise to a cellular structure, of which the hexagonal "causeway" is the two-dimensional analogue.

The columnar structure is accompanied by sundry secondary phenomena. The vertical columns tend to break across on further contraction, and exhibit rounded or basin-shaped ends, fitting together in a shallow ball-and-socket; this beautiful configuration has only lately been explained*. When cooling has caused the mass to split into vertical columns, air or it may be water enters the rifts and further cools or quenches the now solid but still glowing basalt. Each column tends to be chilled all round while still hot within; but the hot unshrunken mass within checks or hinders the contraction of the cooler outer layers. Thus unequal cooling causes vertical as well as horizontal tensions; and just as these last are relieved by the existing cracks, so new rifts appear crosswise to the column, and relieve the vertical tensions.

If the cooling come downward from above, then, at any given level, the column will always be cooler above it than below; the

* F. W. Preston, On ball-and-socket jointing in basalt prisms, *Proc. R.S.* (B), CVI, pp. 87–92, 1930; A study of the rupture of glass, *Trans. Soc. of Glass Technology*, 1926, p. 263. On the general subject of prismatic structure in igneous rocks, see also Robert Mallet, *Phil. Mag.* (4), L, pp. 122–135, 201–226, 1875; James Thomson, *Trans. Geol. Soc. Glasgow*, March, 1877; *Coll. Works*, p. 422; R. B. Sosman, *Journ. Geology*, XXIV, p. 215, 1916.

part above will shrink the more; and this will set up a horizontal shear in the interior of the column, in addition to the existing vertical tension. The principal stress, compounded of shear and tension, will be neither vertical nor horizontal, but inclined obliquely between the two. Now in a brittle substance, such as glass or basalt, an advancing fracture tends to advance at right angles to the principal tension. Where the surface of the column meets the cool air the tension is parallel to the face, and the fissure enters at right angles to the face, that is to say, horizontally; it is for this reason that the ball-and-socket joint is found to have *a square lip*. Once inside the boundary, however, the advancing rift finds itself in a region where the principal stress is inclined,

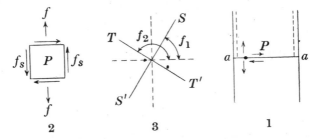

Fig. 201.   (1) Diagram of the vertical and shearing stresses in a shrinking column of basalt.   (2) The same in the neighbourhood of the point P.   (3) SS' resultant stress, and TT' direction of rupture.   After F. W. Preston.

slightly, to the vertical; the crack consequently bends down, the downward tilt increases for a short distance and then approaches the horizontal again, and the opposing surfaces of "ball and socket" are thus defined. The crack often fails to complete its journey, and leaves a core of rock unbroken in the middle of the bowl.

The curved sides of the basin, its square lip, its flattened centre, often incomplete, are thus all explained; and whether it be convex or concave, dome or basin, merely depends on whether the cooling or quenching came from above or from below.

The basin, or bowl, will always be a shallow one. For if at a point P, within the column, the vertical tension be f, and the horizontal shear-stress be f_s, then the direction of the principal planes will be $2\phi - \tan^{-1}(-2f_s/f)$; so that, since f and f_s are both positive, ϕ_1 lies ween $45°$ and $90°$, while ϕ_2 lies between $135°$

and 180°. Hence the rift, dipping down at the angle TT', will never dip at a greater angle than 45°, and generally much less. Now a spherical bowl whose lip is at 45° to the horizontal has a depth of $\dfrac{\sqrt{2}-1}{2}$ times its lip diameter, or approximately one-fifth. And one-fifth of the diameter of the column is, approximately, the depth of the deepest bowl.

The hexagonal pattern and the three-way corners on which it is based are characteristic (as we have already explained) of a condition of symmetry or uniformity under which a partition is as likely to arise in any one direction as in any other, and no series of partitions has precedence over the rest. We have seen,

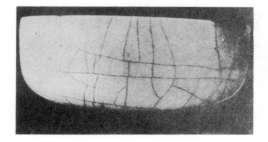

Fig. 202. Crackles on a porcelain bowl. From H. Hukusima.

in the dragon-fly's wing and in cambium-tissue, how different is the result when primary partitions are first established and con-solidated, to be followed by a secondary and a weaker set. The "crackles" on a porcelain bowl look somewhat like a cellular epithelium; but the porcelain has been under strain in more ways than one. The plastic clay was first shaped upon a wheel, and potential stress-energy so acquired is stored up even in the finished ware; again, as the ware cools and shrinks after it is drawn from the kiln the glaze is apt to cool quicker and shrink more than the paste below, and tension-energy is stored up till the glaze ruptures and the cracks appear. Various rates of cooling and of contraction, the nature of paste and glaze, the shape of the ware, and even the way in which the potter worked the clay, may all influence the pattern of the crackle. The primary crack will be perpendicular to the main tension; secondaries will tend to be more or less orthogonal to the primary cracks; two secondaries on opposite sides of a primary will tend to be near together, though not opposite; and spiral cracks are often to be seen, the remote effect of plasticity under the potter's wheel. The net result is that certain primary cracks appear, related (more or less clearly) to the circular or spiral shaping of the clay upon the wheel; and each primary crack so relieves the tension in one direction that the secondaries tend to follow in a direction at right angles to the first. While co-equal angles of 120° were likely to occur in certain symmetrical cases, leading to a simultaneous pattern of hexagons, in other cases suc-

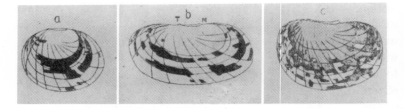

Fig. 203. Colour-patterns of kidney-beans with diagrammatic contour-lines added; *a*, *b*, Japanese "quail-beans"; *c*, scarlet-runner. After M. Hirata.

cessive partitions or cracks tend to be at right angles to one another; and Sachs's Law becomes truer of the porcelain than of the plant*.

In this latter case, and doubtless in many more, we are dealing not with a random pattern, but with one *based* on systematic and predetermined lines. The apparently confused or random pattern of a kidney-bean comes under the same class of configurations, inasmuch as it also is based on an underlying polarity, whose centre of symmetry is in the stalk or "hilus." For simplicity's sake, imagine the bean round like a pea, and its surface mapped out, orthogonally, by two sets of boundary-lines, radial and concentric. Then suppose an asymmetry of growth to be introduced so that the round pea grows into the ellipsoid of a bean; and suppose that the whole system of boundary-lines is subject to the same conformal transformation—which elliptic functions might help

* See H. Hukusima, Cracks upon the glazed surface of ceramic wares, *Sci. Papers, Inst. of Chem. and Phys. Research*, Tokyo, XXVII, pp. 235–243, 1935.

us to define. The colour-pattern of the bean will then be found following the direction of the boundary-lines, and occupying areas or patches corresponding to parts of the orthogonal system. The lines are equipotential lines, or akin thereto. If we varnish an elastic bag, dry it and expand it, the varnish will tend to crack along the same orthogonal boundaries*.

The bee's cell

The most famous of all hexagonal conformations, and one of the most beautiful, is the bee's cell. As in the basalt or the coral, we have to deal with an assemblage of co-equal cylinders, of circular section, compressed into regular hexagonal prisms; but in this case we have two layers of such cylinders or prisms, one facing one way and one the other, and a new problem arises in connection with their inner ends. We may suppose the original cylinders to have spherical ends†, which is their normal and symmetrical way of terminating; then, for closest packing, it is obvious that the end of any one cylinder in the one layer will touch, and fit in between, the ends of three cylinders in the other. It is just as when we pile round-shot in a heap; we begin with three, a fourth fits into its nest between the three others, and the four form a "tetrad," or regular tetragonal arrangement.

Just as it was obvious, then, that by mutual pressure from the *sides* of six adjacent cells any one cell would be squeezed into a hexagonal prism, so is it also obvious that, by mutual pressure against the *ends* of three opposite neighbours, the end of each and every cell will be compressed into a trihedral pyramid. The three sides of this pyramid are set, *in plane projection*, at co-equal angles of 120° to one another; but the three apical angles (as in the analogous case already described of a system of soap-bubbles) are,

---

* M. Hirata, Coloured patches in kidney-beans, *Sci. Papers, Inst. Chem. Research, Tokyo*, XXVI, pp. 122–135, 1936.

† In the combs of certain tropical bees the hexagona structure is imperfect and the cells are not far removed from cylinders. They are set in tiers, not contiguous but separated by little pillars of wax, and the base of each cell is a portion of a sphere. They differ from the ordinary honeycomb in the same sort of way as the fasciculate from the massive corals, of which we spoke on p. 512. Cf. Leonard Martin, Sur les Mélipones de Brésil, *La Nature*, 1930, pp. 97–100.

by the geometry of the case*, co-equal angles of 109° and so many minutes and seconds.

If we experiment, not with cylinders but with spheres, if for instance we pile bread-pills together and then submit the whole to a uniform pressure, as we shall presently find that Buffon did : each ball (like the seeds in a pomegranate, as Kepler said) will be in contact with *twelve* others—six in its own plane, three below and three above, and under compression it will develop twelve plane surfaces. It will repeat, above and below, the conditions to which the bee's cell is subject at one end only; and, since the sphere is symmetrically situated towards its neighbours on all sides, it follows that the twelve plane sides to which its surface has been reduced will be all similar, equal and similarly situated. Moreover, since we have produced this result by squeezing our original spheres close together, it is evident that the bodies so formed completely fill space. The regular solid which fulfils all these conditions is the *rhombic dodecahedron*. The bee's cell is this figure incompletely formed; it represents, so to speak, one-half of that figure, with its apex and the six adjacent corners proper to the rhombic dodecahedron, but six sides continued, as a hexagonal prism, to an open or unfinished end†.

The bee's comb is vertical and the cells nearly horizontal, but sloping slightly downwards from mouth to floor; in each prismatic cell two sides stand vertically, and two corners lie above and below. Thus for every honeycomb or "section" of honey, there is one and only one "right way up"; and the work of the hive is so far controlled by gravity. Wasps build the other way, with the cells upright and the combs horizontal; in a hornet's nest, or in that of *Polistes*, the cells stand upright like the wasp's, but their mouths look downwards in the hornet s nest and upwards in the wasp's.

What Jeremy Taylor called "the discipline of bees and the rare fabric of honeycombs" must have attracted the attention and excited the admiration of mathematicians from time immemorial. "Ma maison est construite," says the bee in the *Arabian Nights*, "selon

---

* The dihedral angle of 120° is, physically speaking, the essential thing; the Maraldi angle, of 109°, etc., is a geometrical consequence. Cf. G. Césaro, Sur la forme de l'alvéole de l'abeille, *Bull. Acad. R. Belgique (Sci.)*, 1920, p. 100.

† See especially Haüy, the crystallographer; Sur le rapport des figures qui existe entre l'alvéole des abeilles et le grenat dodécaèdre, *Journ. d'hist. naturelle*, II, p. 47, 1792.

les lois d'une sévère architecture; et Euclidos lui-même s'instruirait en admirant la géométrie de ses alvéoles*." Ausonius speaks of the *geometrica forma favorum*, and Pliny tells of men who gave a lifetime to its study.

Pappus the Alexandrine has left us an account of its hexagonal plan, and drew from it the conclusion that the bees were endowed with "a certain geometrical forethought"†. "There being, then, three figures which of themselves can fill up the space round a point, viz. the triangle, the square and the hexagon, the bees have wisely selected for their structure that which contains most angles, suspecting indeed that it could hold more honey than either of the

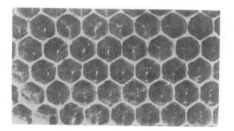

Fig. 204. Portion of a honeycomb. After Willem.

other two‡." Erasmus Bartholin was apparently the first to suggest that the hypothesis of "economy" was not warranted, and that the hexagonal cell was no more than the necessary result of equal pressures, each bee striving to make its own little circle as large as possible.

The investigation of the ends of the cell was a more difficult matter than that of its sides, and came later. In general terms the arrangement was doubtless often studied and described: as for

* Ed. Mardrus, xv, p. 173.

† φυσικὴν γεωμετρικὴν πρόνοιαν. Pappus, Bk. v; cf. Heath, *Hist. of Gk. Math.* ii, p. 589. St Basil discusses τὴν γεωμετρίαν τῆς σοφωτάτης μελίσσης: *Hexaem.* viii, p. 172 (Migne); Virgil speaks of the *pars divinae mentis* of the bee, and Kepler found the bees *animâ praeditas et geomet. riae suo modo capaces.*

‡ This was according to the "theorem of Zenodorus." The use by Pappus of "economy" as a guiding principle is remarkable. For it means that, like Hero with his mirrors, he had a pretty clear adumbration of that principle of *minima*, which culminated in the *principle of least action*, which guided eighteenth-century physics, was generalised (after Fermat) by Lagrange, inspired Hamilton and Maxwell, and reappears in the latest developments of wave-mechanics.

instance, in the *Garden of Cyrus*: "And the Combes themselves
so regularly contrived that their mutual intersections make three
Lozenges at the bottom of every Cell; which severally regarded
make three Rows of neat Rhomboidall Figures, connected at the
angles, and so continue three several chains throughout the whole
comb." Or as Réaumur put it, a little later on: "trois cellules
accolées laissent un vuide pyramidal, précisément semblable à celui
de la base d'une autre cellule tournée en sens contraire."

Kepler had deduced from the space-filling symmetry of the honeycomb that its angles must be those of the rhombic dodecahedron;
and Swammerdam also recognised the same geometrical figure in
the base of the cell*. But Kepler's discovery passed unnoticed,
and Maraldi the astronomer, Cassini's nephew, has the credit of
ascertaining for the first time the shape of the rhombs and of the
solid angle which they bound, while watching the bees in "les ruches
vitrées dans le jardin de M. Cassini attenant l'Observatoire de
Paris†." The angles of the rhomb, he tells us, are 110° and 70°:
"Chaque base d'alvéole est formée de trois rombes presque toujours
égaux et semblables, qui, *suivant les mesures que nous avons prises*,
ont les deux angles obtus chacun de 110 degrès, et par conséquent
les deux aigus chacun de 70 degrès." Further on (p. 312), he
observes that on the magnitude of the angles of the three rhombs
at the base of the cell depends that of the basal angles of the six
trapezia which form its sides; and it occurs to him to ask what
must these angles be, if those of the floor and those of the sides be
equal one to another. The solution of this problem is that "les
angles aigus des rombes étant de 70 degrès 32 minutes, et les obtus
de 109 degrès 28 minutes, ceux des trapèzes qui leur sont contigus
doivent être aussi de la même grandeur." And lastly: "Il résulte
de cette grandeur d'angle non seulement une plus grande facilité et
simplicité dans la construction, à cause que par cette manière les
abeilles n'employent que deux sortes d'angles, mais il en résulte
encore une plus belle simétrie dans la disposition et dans la figure

---

* Kepleri *Opera omnia*, ed. Fritsch, v, pp. 115, 122, 178, vii, p. 719, 1864;
Swammerdam, *Tractatus de apibus* (observations made in 1673).

† Obs. sur les abeilles, *Mém. Acad. R. Sciences* (1712), 1731, pp. 297–331.
Sir C. Wren had used "transparent bee-hives" long before; see his Letter concerning
that pleasant and profitable invention, etc., in S. Hartlib's *Reformed Common-
Wealth of Bees*, 1655.

de l'Alvéole." In short, Maraldi takes the two principles of simplicity and mathematical beauty as his sure and sufficient guides.

The next step was that which had been foreshadowed long before by Pappus. Though Euler had not yet published his famous dissertation on curves *maximi minimive proprietate gaudentes*, the idea of *maxima* and *minima* was in the air as a guiding postulate, an heuristic method, to be used as Maraldi had used his principle of simplicity. So it occurred to Réaumur, as apparently it had not done to Maraldi, that a minimal configuration, and consequent economy of material in the waxen walls of the cell, might be at the root of the matter: and that, just as the close-packed hexagons gave the minimal extent of boundary in a plane, so the figure determined by Maraldi, namely the rhombic dodecahedron, might be that which employs the minimum of surface for a given content: or which, in other words, should hold the most honey for the least wax. "Convaincu que les abeilles employent le fond pyramidal qui mérite d'être préféré, j'ai soupçonné que la raison, ou une des raisons, qui les avoit décidées était l'épargne de la cire; qu'entre les cellules de même capacité et à fond pyramidal, celle qui pouvait être faite avec moins de matière ou de cire étoit celle dont chaque rhombe avoit deux angles chacun d'environ 110 degrés, et deux chacun d'environ 70°." He set the problem to Samuel Koenig, a young Swiss mathematician: Given an hexagonal cell terminated by three similar and equal rhombs, what is the configuration which requires the least quantity of material for its construction? Koenig confirmed Réaumur's conjecture, and gave 109° 26′ and 70° 34′ as the angles which should fulfil the condition; and Réaumur then sent him the Mémoires de l'Académie for 1712, where Koenig was "agreeably surprised" to find: "que les rombes que sa solution avait déterminé, avait à deux minutes près* les angles que M. Maraldi avait trouvés *par des mesures actuelles* à chaque rhombe des cellules d'abeilles.... Un tel accord entre la solution et les mesures actuelles a assurément de quoi surprendre." Koenig asserted that the bees had solved a problem beyond the reach of

---

* The discrepancy was due to a mistake of Koenig's, doubtless misled by his tables, in the determination of $\sqrt{2}$; but Koenig's own paper, sent to Réaumur, remained unpublished and his method of working is unknown. An abridged notice appears in the *Mém. de l'Acad.* 1739, pp. 30–35.

the old geometry and requiring the methods of Newton and Leibniz. Whereupon Fontenelle, as Secrétaire Perpétuel, summed up the case in a famous judgment, in which he denied intelligence to the bees but nevertheless found them blindly using the highest mathematics by divine guidance and command*.

When Colin Maclaurin studied the honeycomb in Edinburgh, a few years after Maraldi in Paris, he proceeded to solve the problem without using "any higher Geometry than was known to the Antients," and he began his account by saying: "These bases are formed from Three equal Rhombus's, the obtuse angles of which are found to be the doubles of an Angle that often offers itself to mathematicians in Questions relating to Maxima and Minima.†" It was an angle of 109° 28′ 16″, with its supplement of 70° 31′ 44″. And this angle of the bee's cell determined by Maraldi, Koenig and Maclaurin in their several ways, this angle which has for its cosine 1/3 and is double of the angle which has for its tangent $\sqrt{2}$, is on the one hand an angle of the rhombic dodecahedron, and on the other is that very angle of simple tetrahedral symmetry which the soap-films within the tetrahedral cage spontaneously assume, and whose frequent appearance and wide importance we have already touched upon‡.

That "the true theoretical angles were 109° 28′ and 70° 32′, *precisely corresponding with the actual measurement of the bee's cell*," and that the bees had been "proved to be right and the mathematicians wrong," was long believed by many. Lord Brougham

* La grande merveille est que la détermination de ces angles passe de beaucoup les forces de la Géométrie commune, et n'appartient qu'aux nouvelles Méthodes fondées sur la Théorie de l'Infini. Mais à la fin les Abeilles en sçauraient trop, et l'excès de leur gloire en est la ruine. Il faut remonter jusqu'à une Intelligence infinie, qui les fait agir aveuglément sous ses ordres, sans leur accorder de ces lumières capables de s'accroître et de se fortifier par elles-mêmes, qui font l'honneur de notre Raison." *Histoire de l'Académie Royale*, 1739, p. 35.

† Colin Maclaurin, On the bases of the cells wherein the bees deposit their honey, *Phil. Trans.* XLII, pp. 561–571, 1743; also in the Abridgement, VIII, pp. 709–713, 1809; it was characteristic of Maclaurin to use geometrical methods for wellnigh everything, even in his book on Fluxions, or in his famous essay on the equilibrium of spinning planets. Cf. also Lhuiller, Mémoire sur le minimum du cire des alvéoles des Abeilles, et en particulier sur un *minimum minimorum* relatif à cette matière, *Nouv. Mém. de l'Acad. de Berlin*, 1781 (1783), pp. 277–300. Cf. Castillon, *ibid.* (commenting on Lhuiller); also Ettore Carruccio, Notizie storiche sulla geometria delle api, *Periodico di Mathematiche*, (4) XVI, pp. 35–54, 1936.

‡ *Supra*, p. 497. The faces of a regular octahedron meet at the same angle.

helped notably to spread these and other errors, and his writings
on the bee's cell contain, according to Glaisher, "as striking
examples of bad reasoning as are often to be met with in writings
relating to mathematical subjects." The fact is that, were the
angles and facets of the honeycomb as sharp and smooth, and as
constant and uniform, as those of a quartz-crystal, it would still be
a delicate matter to measure the angles within a minute or two of
arc, and a technique unknown in Maraldi's day would be required
to do it. The minute-hand of a clock (if it move continuously)
moves through one degree of arc in ten seconds of time, and through
an angle of two minutes in one-third of a second;—and this last is
the angle which Maraldi is supposed to have measured. It was
eighty years after Maraldi had told Réaumur what the angle was
that Boscovich pointed out for the first time that to ascertain the
angle to the nearest minute by direct admeasurement of the waxen
cell was utterly impossible. Yet Réaumur had certainly believed,
and apparently had persuaded Koenig, that Maraldi's determinations, first and last, were the result of measurement; and Fontenelle,
the historian of the Academy, epitomising Koenig's paper, speaks
of "les mesures actuelles de M. Maraldi," of the bees being in
error to the trifling extent of 2′, and of the *grande merveille* of their
so nearly solving a problem belonging to the higher geometry.
Boscovich, in a long-forgotten note, rediscovered by Glaisher, puts
the case in a nutshell: "Mirum sane si Maraldus ex observatione
argulum aestimasset intra minuta, quod in tam exigua mole fieri
utique non poterat. At is (ut satis patet ex ipsa ejus determina-
tione) affirmat se invenisse angulos circiter 110° et 70°, nec minuta
eruit ex observatione sed ex equalitate angulorum pertinentium ad
rhombos et ad trapezia; ad quam habendam Geometria ipsa docuit
requiri illa minuta*." Indeed he goes on to say the wonder is that
the angles could be measured even within a few *degrees*, variable
and irregular as they are seen to be, and as even Réaumur† knew

---

* In his note *De apium cellulis*, appended to the philosophical poem of Benedict
Stay, II, pp. 498–504, Romae, 1792.

† *Op. cit.* v, p. 382. Several authors recognised that the cells are far from
identical, and do no more than approximate to an average or ideal angle: e.g.
Swammerdam in the *Biblia Naturae*, II, p. 379; G. S. Klügel, Grösstes u. Kleinstes,
in *Mathem. Wörterb.* 1803; Castillon, *op. cit.*; and especially Jeffries Wyman,
Notes on the cells of the bee, *Proc. Amer. Acad. Sci. and Arts*, VII, 1868.

well they were. The old misunderstanding was at last explained and corrected by Leslie Ellis; and better still by Glaisher, in a little-known but very beautiful paper*. For these two mathematicians shewed that, though Maraldi's account of his "measurements" led to misunderstanding, yet he had really done well and scientifically when he eked out a rough observation by finer theory, and deemed himself entitled thereby to discuss the cell and its angles in the same precise terms that he would use as a mathematician in speaking of its geometrical prototype†.

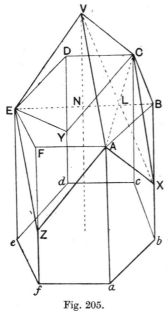

Fig. 205.

Many diverse proofs‡ have been given of the minimal character of the bee's cell, some few, like Maclaurin's, purely geometrical, others arrived at by help of the calculus. The following seems as simple as any:

ABCDEF, abcdef, is a right prism upon a regular hexagonal base. The corners B, D, F are cut off by planes through the lines AC, CE, EA, meeting in a point V on the axis VN of the prism, and intersecting Bb, Dd, Ff, in X, Y, Z. The volume of the figure thus formed is the same as that of the original prism with its hexagonal ends: for, if the axis cut the hexagon $ABCDEF$ in N, the volumes $ACVN$, $ACBX$ are equal.

It is required to find the inclination to the axis of the faces forming the

---

* Leslie Ellis, On the form of bees' cells, in *Mathematical and other Writings*, 1863, p. 353; J. W. L. Glaisher, do., *Phil. Mag.* (4), xlvi, pp. 103–122, 1873.

† The learned and original Kieser, in his *Mémoire sur l'organisation des plantes*, 1812, p. iv, gives advice to the same effect: "Il est indispensable de se former, avant de dessiner, une idée de l'objet dans sa plus grande perfection, et de dessiner selon cette idée, et non pas l'objet plus ou moins imparfait, plus ou moins altéré par le scalpel. Voilà la méthode qu'ont suivi Haller, Albinus et tous les autres grands anatomistes....Mais il faut employer pour cela la plus grande précaution, la circonspection la plus tranquille pour l'observation, etc."

‡ Cf. Koenig, Lhuiller and Boscovich, *opp. cit.*; H. Hennessy, *Proc. R.S.* xxxix, p. 253, 1885; xli, pp. 442, 443, 1886; xlii, pp. 176, 177, 1887.

trihedral angle at V, such that the surface of the whole figure may be a minimum.

Let the angle NVX, which is the inclination of the plane of the rhombus to the axis of the prism, $= \theta$; the side of the hexagon, as AB, $= s$; and the height, as Aa, $= h$.

Then $AC = 2s \cos 30°$, $= s \sqrt{3}$. And, from inspection of the triangle LXB,

$$VX = \frac{s}{\sin \theta}.$$

Therefore the area of the rhombus

$$VAXC = \frac{s^2 \sqrt{3}}{2 \sin \theta}.$$

And the area of $\quad AabX = \frac{s}{2} (2h - \tfrac{1}{2}VX \cos \theta),$

$$= \frac{s}{2} (2h - \tfrac{1}{2}s \cot \theta).$$

Therefore the total area of the figure

$$= \text{the hexagonal base } abcdef + 3s (2h - \tfrac{1}{2}s \cot \theta) + 3 \frac{s^2 \sqrt{3}}{2 \sin \theta}.$$

Therefore $\qquad \dfrac{d \,(\text{area})}{d\theta} = \dfrac{3s^2}{2} \left(\dfrac{1}{\sin^2 \theta} - \dfrac{\sqrt{3} \cos \theta}{\sin^2 \theta} \right).$

But this expression vanishes, or $\dfrac{d \,(\text{area})}{d\theta} = 0$, when $\cos \theta = \dfrac{1}{\sqrt{3}}$,

that is to say, when $\qquad \theta = 54° \, 44' \, 8'' = \tfrac{1}{2} (109° \, 28' \, 16'')$.

Such then are the conditions under which the total area of the figure has its minimal value.

The following is, in substance, Maclaurin's elementary but somewhat lengthy proof of the minimal properties of the bee's cell, using "no higher Geometry than was known to the Antients."

Let $ABCD$, $abcd$, represent one-half of a right prism on a regular hexagonal base; and let $AabE$, $EbcC$ be the trapezial portions of two adjacent sides, to which one of the three rhombs, $AECe$, is fitted.

Let O be the centre of the hexagon, of which AB, BC are adjacent sides; join AC and OB, intersecting in P. Then, because $AOC = ABC$, and $BE = Oe$, the solid $AECB = AeCO$; whence it appears that the solid *content* of the whole cell will be the same, wherever the point E be taken in Bb, and will in fact be identical with the content of the hexagonal prism. We have then to enquire where E is to be taken in Bb, in order that the combined *surfaces* of the rhomb and of the two trapezia may be a minimum.

Because Ee is perpendicular to AC in P, the area of the rhombus $= PE.AC$; and the area of the two trapezia $= (Aa + Eb) \times BC$. The total area in question, then, is $PE.AC + 2BC.Bb - BC.BE$. But $BC.Bb$ is constant; so the question remains, When is $PE.AC - BC.BE$ a minimum?

Let a point L be so taken in Bb that $BL : PL :: BC : AC$. From the centre P, in the plane PBE, describe the circular arc ER, meeting PL in R; and on PL let fall the perpendiculars ES, BT.

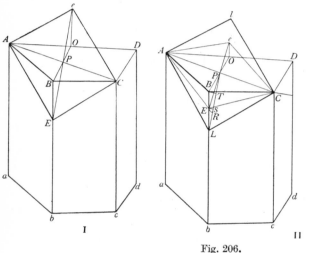
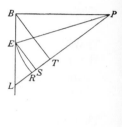

Fig. 206.

The triangles LES, LBT, LPB are all similar. Therefore

$$LS : LE :: LT : LB :: LB : LP,$$

and (by hypothesis) $:: BC : AC$.

Hence $(LT - LS) : (LB - LE) :: BC : AC$,

i.e. $ST : BE :: BC : AC$.

Therefore $ST.AC = BE.BC$

and consequently, $PE.AC - BE.BC = PE.AC - ST.AC$

$$= AC(PE - ST).$$

But $PE = PR$, therefore $AC(PR - ST) = AC(PT + RS)$.

But AC and PT do not vary, while RS varies with the position of E. Accordingly, $AC\,(PT + RS)$, or $PE\,.\,AC - BE\,.\,BC$, is a minimum when RS vanishes: that is to say, when E coincides with L.

"Therefore $ALCl$ is the *Rhombus* of the most advantageous Form in respect of Frugality, when BL is to PL as BC to AC."

Again, since $OB = BC$, and $OP = PB$, $BC^2 = 4PB^2$, and $PC^2 = 3PB^2$, and $AC = 2PC = 2\,\sqrt{3}\,.\,PB$.

Therefore $BC : AC :: 2PB : 2\,\sqrt{3}\,.\,PB :: 1 : \sqrt{3},$

and, by hypothesis, $BL : PL :: BC : AC$

$$= 1 : \sqrt{3}$$

or $PL : PB :: \sqrt{3} : \sqrt{2},$

and $PB : PC :: 1 : \sqrt{3}.$

Therefore $PL : PC :: 1 : \sqrt{2}.$

"That is, the angle CLP is that whose Tangent is to the *Radius* as $\sqrt{2}$ is to 1, or as 1·4142135 to 1·0000000; and therefore is of 54° 44′ 08″, and consequently the Angle of the *Rhombus* of the Best Form is that of 109° 28′ 16″."

When we have thus ascertained that the characteristic angles of the rhombs are 109° 28′ 16″ and its supplement 70° 31′ 44″, the cosine of which latter angle is 1/3, the construction of a model is of the easiest.

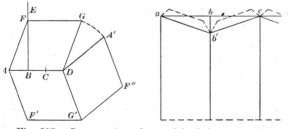

Fig. 207.   Construction of a model of the bee's cell

On AD make $AB = BC = CD$. Let $AF = AD$ meet the perpendicular BE in F. Then the angle BAF (whose cosine = 1/3, or whose tangent = $2\sqrt{2}$) = 70° 31′ 44″. Complete the rhomb $ADGF$, and repeat three times as indicated. Make a developed hexagonal prism with sides ab, bc, = BF. Cut away angles $bb'a$, $bb'c$, etc., = BAF. Fold, and attach together.

A soap-bubble, or soap-film, assumes a minimal configuration instantaneously*, however small the saving of surface-area may be. But after learning that the bee's cell has undoubted minimal properties, we should like to know what saving is actually obtained by substituting a rhomboidal pyramid for a plane base in the hexagonal prism. It turns out, after all, to be a small matter! The calculation was first made by Maclaurin and by Lhuiller, in both cases briefly but correctly. Lhuiller stated that the whole amount used in the bee's cell was to that required for a flat-topped prismatic cell of equal volume as $25 + \sqrt{6}$ (or 27·45) to 28, the saving being thus a little more than 2 per cent. of the whole quantity of wax required †. Glaisher recalculated the values, taking the cell part by part. Assuming, with Lhuiller, that the radius of the inscribed circle of the hexagon is to the depth of the prismatic cell, when the latter has the same capacity as the real cell, as $1\frac{1}{5}$ to 5, then, taking the side of the hexagon as unity, we have for the same depth (viz. the longest side of the trapezium in the real cell) the value $\dfrac{25\sqrt{3}}{12}$; and then (to three places of decimals):

Area of the three rhombs, $\frac{9}{4}\sqrt{2}$ $\qquad = 3\cdot182$ $\qquad$ (i).

,, ,, ,, six triangles, $\frac{3}{4}\sqrt{2}$ $\qquad = 1\cdot061$ $\qquad$ (ii).

,, ,, ,, six sides of the equivalent prismatic cell, $\frac{25}{2}\sqrt{3}$ $\qquad = 21\cdot651$ $\qquad$ (iii).

,, ,, ,, hexagonal base, $\frac{3}{2}\sqrt{3}$ $\qquad = 2\cdot598$ $\qquad$ (iv).

The whole surface of the real cell, accordingly,

$$= \text{(i)} + \text{(iii)} - \text{(ii)} = 23\cdot772;$$

---

* For the most part *instantaneously*; but sometimes, when there are two positions of nearly equal potential energy, the film "creeps" from the less to the more advantageous of the two.

† We must take into account the depth of the cell, or assume a value for it, if we are to estimate the percentage saving of wax on the whole construction. But (as Dr G. T. Bennett says) the whole saving is on the roof, and the height of the house does not matter; the question rather is, what is saved on the rhomboidal sloping roof compared with a flat one? If the short axis of the rhombs be 2 units (the edge of the cube), then 3 rhombs have area $6\sqrt{2}$, the wall-saving is $2\sqrt{2}$, while the flat hexagonal top is $4\sqrt{3}$. So the actual saving is the difference between $4\sqrt{2}$ and $4\sqrt{3}$—which looks much less negligible! But it is only on a small portion of the work.

and that of the flat-bottomed hexagonal prism

$$= (iii) + (iv) = 24 \cdot 249;$$

and

$$\frac{24 \cdot 249}{23 \cdot 772} = \frac{102}{100}, \text{ or a little more};$$

so that the saving in the former case amounts to about 2 per cent., as Lhuiller had found it to be.

Glaisher sums up the matter as follows: "As the result of a tolerably careful examination of the whole question, I may be permitted to say that I agree with Lhuiller in believing that the economy of wax has played a very subordinate part in the determination of the form of the cell; in fact I should not be surprised if it were acknowledged hereafter that the form of the cell had been determined by other considerations, into which the saving of wax did not enter (that is to say did not enter sensibly; of course I do not mean that the amount of wax required was a matter of absolute indifference to the bees). The fact of all the dihedral angles being 120° is, it is not unlikely, the cause that determined the form of the cell." This last fact, that in such a cell every plane cuts every other plane at an angle of 120°, was known both to Klügel and to Boscovich; it is no mere corollary, but the root of the matter. It is, as Glaisher indicates, the fundamental physical principle of construction from which the apical angles of 109° follow as a geometrical corollary. And it is curious indeed to see how the obtuse angle of the rhomb, and its cosine $-\frac{1}{3}$, drew attention all the while; but the dihedral angle of 120° of the rhombohedron, and the inclination of its three short diagonals at 90° to one another, got rare and scanty notice.

Darwin had listened too closely to Brougham and the rest when he spoke of the bee's architecture as "the most wonderful of known instincts"; and when he declared that "beyond this stage of perfection in architecture natural selection could not lead; for the comb of the hive-bee, as far as we can see, is *absolutely perfect* in economising labour and wax."

The minimal properties of the cell and all the geometrical reasoning in the case postulate cell-walls of uniform tenuity and edges which are mathematically straight. But the walls, and still more their edges, are always thickened; the edges are never accurately straight, nor the cells strictly horizontal. The base is always thicker than the side-walls; its solid angles are by no means sharp, but filled up with curving surfaces of wax, after the fashion, but more coarsely, of Plateau's bourrelet. Hence the Maraldi angle is seldom or never attained; the mean value (according to Vogt) is no more than 106·7° for the workers, and 107·3° for the drones. The hexagonal angles of the prism are fairly constant; about 4° is the limit of departure, and about 1·8° the mean error, on either side.

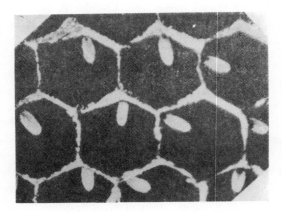

Fig. 208.   Brood-comb, with eggs.

The bee makes no economies; and whatever economies lie in the theoretical construction, the bee's handiwork is not fine nor accurate enough to take advantage of them*.

The cells vary little in size, so little that Thévenot, a friend of Swammerdam's, suggested using their dimensions as a modulus or standard of length; but after all, the constancy is not so great as has been supposed. Swammerdam gives measurements which work out at 5·15 mm. for the mean diameter of the worker-cells, and 7 mm. for those of the drones; Jeffries Wyman found mean values for the worker-cells from 5·1 to 5·2 mm.; Vogt, after many careful measurements, found a mean of 5·37 mm. for the worker-

* All this Heinrich Vogt has abundantly shewn, in part by making casts of the interior of the cells, as Castellan had done a hundred years before. See his admirable paper on the Geometrie und Oekonomie der Bienenzelle, in *Festschrift d. Universität Breslau*, 1911, pp. 27–274.

cells, with an insignificant difference in the various diameters, and a mean of 6·9 for the drone-cells, with their horizontal diameter somewhat in excess, and averaging 7·1 mm. A curious attempt has been made of late years by Italian bee-keepers to let the bees work on a larger foundation, and so induce them to build larger cells; and some, but by no means all, assert that the young bees reared in the larger cells are themselves of larger stature*.

That the beautiful regularity of the bee's architecture is due to some automatic play of the physical forces, and that it were fantastic to assume (with Pappus and Réaumur) that the bee intentionally seeks for a method of economising wax, is certain; but the precise manner of this automatic action is not so clear. When the hive-bee builds a solitary cell, or a small cluster of cells, as it does for those eggs which are to develop into queens, it makes but a rude construction. The queen-cells are lumps of coarse wax hollowed out and roughly bitten into shape, bearing the marks of the bee's jaws like the marks of a blunt adze on a rough-hewn log.

Omitting the simplest of all cases, when (among some humble-bees) the old cocoons are used to hold honey, the cells built by the "solitary" wasps and bees are of various kinds. They may be formed by partitioning off little chambers in a hollow stem; they may be rounded or oval capsules, often very neatly constructed out of mud or vegetable fibre or little stones, agglutinated together with a salivary glue; but they shew, except for their rounded or tubular form, no mathematical symmetry. The social wasps and many bees build, usually out of vegetable matter chewed into a paste with saliva, very beautiful nests of "combs"; and the close-set papery cells which constitute these combs are just as regularly hexagonal as are the waxen cells of the hive-bee. But in these cases (or nearly all of them) the cells are in a single row; their sides are regularly hexagonal, but their ends, for want of opponent forces, remain simply spherical.

In *Melipona domestica* (of which Darwin epitomises Pierre Huber's description) "the large waxen honey-cells are nearly spherical, nearly equal in size, and are aggregated into an irregular mass."

---

* Cf. (*int. al.*) H. Gontarsi, Sammelleistungen von Bienen aus vergrösserten Brutzellen, *Arch. f. Bienenkunde*, XVI, p. 7, 1935; A. Ghetti, Celli ed api piu grandi, *IV Congresso nazion. della S.A.I.* 1935.

But the spherical form is only seen on the outside of the mass; for inwardly each cell is flattened into "two, three or more flat surfaces, according as the cell adjoins two, three or more other cells. When one cell rests on three other cells, which from the spheres

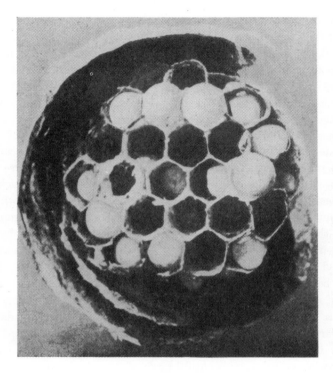

Fig. 209.   An early stage of a wasp's nest.   Observe the spherical caps, and the irregular shape of the peripheral cells.   After R. Bott.

being nearly of the same size is very frequently and necessarily the case, the three flat surfaces are united into a pyramid; and this pyramid, as Huber has remarked, is manifestly a gross imitation of the three-sided pyramidal base of the cell of the hive-bee*."

* *Origin of Species*, ch. VIII (6th ed., p. 221).   The cells of various bees, humble-bees and social wasps have been described and mathematically investigated by K. Müllenhoff, *Pflüger's Archiv*, XXXII, p. 589, 1883; but his many interesting results are too complex to epitomise.   For figures of various nests and combs see (e.g.) von Büttel-Reepen, *Biol. Centralbl.* XXXIII, pp. 4, 89, 129, 183, 1903.

We had better be content to say that it depends on the same elementary geometry.

The question is, To what particular force are we to ascribe the plane surfaces and definite angles which define the sides of the cell in all these cases, and the ends of the cell in cases where one row meets and opposes another? We have seen that Bartholin suggested, and it is still commonly believed, that this result is due to mere physical pressure, each bee enlarging as much as it can the cell which it is a-building, and nudging its wall outwards till it fills every intervening gap, and presses hard against the similar efforts of its neighbour in the cell next door*.

That the bee, if left to itself, "works in segments of circles," or in other words builds a rounded and roughly spherical cell, is an old contention† which some recent experiments of M. Victor Willem amplify and confirm‡. M. Willem describes vividly how each cell begins as a little hemispherical basin or "cuvette," how the workers proceed at first with little apparent order and method, laying on the wax roughly like the mud when a swallow builds; how presently they concentrate their toil, each burying its head in its own cuvette, and slowly scraping, smoothing and ramming home; how those on the other side gradually adjust themselves

* Darwin had a somewhat similar idea, though he allowed more play to the bee's instinct or conscious intention. Thus, when he noticed certain half-completed cell-walls to be concave on one side and convex on the other, but to become perfectly flat when restored for a short time to the hive, he says: "It was absolutely impossible, from the extreme thinness of the little plate, that they could have effected this by gnawing away the convex side; and I suspect that the bees in such cases stand on opposite sides and push and bend the ductile and warm wax (which as I have tried is easily done) into its proper intermediate plane, and thus flatten it." Huber thought the difference in form between the inner and the outer cells a clear proof of intelligence; it is really a direct proof of the contrary. And while cells differ when their situations and circumstances differ, yet over great stretches of comb extreme uniformity, unbroken by any sign of individual differences, is the strikingly mechanical characteristic of the cells.

† It is so stated in the *Penny Cyclopedia*, 1835, Art. "Bees"; and is expounded by Mr G. H. Waterhouse (*Trans. Entom. Soc., London*, II, p. 115, 1864) in an article of which Darwin made good use. Waterhouse shewed that when the bees were given a plate of wax, the separate excavations they made therein remained hemispherical, or were built up into cylindrical tubes; but cells in juxtaposition with one another had their party-walls flattened, and their forms more or less prismatic.

‡ Victor Willem, L'architecture des abeilles, *Bull. Acad. Roy. de Belgique* (5), XIV, pp. 672–705, 1928.

to their opposite neighbours; and how the rounded ends of the cells fashion themselves into the rhomboidal pyramids, "à la suite de l'amincissement progressif des cloisons communes, et des pressions antagonistes exercées sur les deux faces de ces cloisons."

Among other curious and instructive observations, M. Willem has watched the bees at work on the waxen "foundations" now commonly used, on which a rhomboidal pattern is impressed with a view to starting the work and saving the labour of the bees. The bees (he says) disdain these half-laid foundations of their cells; they hollow out the wax, erase the rhombs, and turn the pyramidal hollows into hemispherical "cuvettes" in their usual way; and the vertical walls which they raise, more or less on the lines laid down for them, are not hexagonal but cylindrical to begin with. "La forme plane, en facettes, tant de prismes que des fonds, n'est obtenue que plus tard, progressivement, comme résultat de retouches, d'enlèvements et de pressions exercées sur les cloisons qui s'amincissent, par des groupes d'ouvrières opérant face à face, de manière antagoniste."

But when all is said and done, it is doubtful whether such *retouches, enlèvements* and *pressions antagonistes*, such mechanical forces intermittently exercised, could produce the nearly smooth surfaces, the all but constant angles and the close approach to a minimal configuration which characterise the cell, whether it be constructed by the bee of wax or by the wasp of papery pulp. We have the properties of the material to consider; and it seems much more likely to me that we have to do with a true tension effect: in other words, that the walls assume their configuration when in a semi-fluid state, while the watery pulp is still liquid or the wax warm under the high temperature of the crowded hive. In the first few cells of a wasp's comb, long before crowding and mutual pressure come into play, we recognise the identical configurations which we have seen exhibited by a group of three or four soap-bubbles, the first three or four cells of a segmenting egg. The direct efforts of the wasp or bee may be supposed to be limited, at this stage, to the making of little hemispherical cups, as thin as the nature of the material permits, and packing these little round cups as close as possible together. It is then conceivable, and indeed probable, that the symmetrical tensions of the semi-fluid films should suffice (however retarded by viscosity) to bring the whole system into equilibrium, that is to say into the configuration which the comb actually assumes.

The remarkable passage in which Buffon discusses the bee's cell and the hexagonal configuration in general is of such historical importance, and tallies so closely with the whole trend of our enquiry, that before we leave the subject I will quote it in full*: "Dirai-je encore un mot: ces cellules des abeilles, tant vantées, tant admirées, me fournissent une preuve de plus contre l'enthousiasme et l'admiration; cette figure, toute géométrique et toute régulière qu'elle nous paraît, et qu'elle est en effet dans la spéculation, n'est ici qu'un résultat mécanique et assez imparfait qui se trouve souvent dans la nature, et que l'on remarque même dans les productions les plus brutes; les cristaux et plusieurs autres pierres, quelques sels, etc., prennent constamment cette figure dans leur formation. Qu'on observe les petites écailles de la peau d'une roussette, on verra qu'elles sont hexagones, parce que chaque écaille croissant en même temps se fait obstacle et tend à occuper le plus d'espace qu'il est possible dans un espace donné: on voit ces mêmes hexagones dans le second estomac des animaux ruminans, on les trouve dans les graines, dans leurs capsules, dans certaines fleurs, etc. Qu'on remplisse un vaisseau de pois, ou plûtot de quelque autre graine cylindrique, et qu'on le ferme exactement après y avoir versé autant d'eau que les intervalles qui restent entre ces graines peuvent en recevoir; qu'on fasse bouillir cette eau, tous ces cylindres deviendront de colonnes à six pans. On y voit clairement la raison, qui est purement mécanique; chaque graine, dont la figure est cylindrique, tend par son renflement à occuper le plus d'espace possible dans un espace donné, elles deviennent donc toutes nécessairement hexagones par la compression réciproque. Chaque abeille cherche à occuper de même le plus d'espace possible dans un espace donné, il est donc nécessaire aussi, puisque le corps des abeilles est cylindrique, que leurs cellules sont hexagones—par la même raison

---

* Buffon, *Histoire naturelle*, iv, p. 99, Paris, 1753. Bonnet criticised Buffon's explanation, on the ground that his description was incomplete; for Buffon took no account of the Maraldi pyramids. Not a few others discovered impiety in his hypotheses, and some dismissed them with the remark that "philosophical absurdities are the most difficult to refute"; cf. W. Smellie, *Philosophy of Natural History*, Edinburgh, 1790, p. 424.

des obstacles réciproques. On donne plus d'esprit aux mouches dont les ouvrages sont les plus réguliers; les abeilles sont, dit-on, plus ingénieuses que les guêpes, que les frélons, etc., qui savent aussi l'architecture, mais dont les constructions sont plus grossières et plus irrégulières que celles des abeilles: on ne veut pas voir, ou l'on ne se doute pas, que cette régularité, plus ou moins grande, dépend uniquement du nombre et de la figure, et nullement de l'intelligence de ces petites bêtes; plus elles sont nombreuses, plus il y a des forces qui agissent également et s'opposent de même, plus il y a par conséquent de contrainte mécanique, de régularité forcée, et de perfection apparente dans leurs productions*."

Of parenchymatous cells

Just as Bonanni and other early writers sought, as we have seen, to explain hexagonal symmetry on mechanical principles, so other early naturalists, relying more or less on the analogy of the bee's cell, endeavoured to explain the cells of vegetable parenchyma; and to refer them to the rhombic dodecahedron or garnet-form, which solid figure, in close-packed association, was believed in their time, and long afterwards, to enclose space with a minimal extent of surface.

* Among countless papers on the bee's cell, see John Barclay and others in *Ann. of Philosophy*, IX, X, 1817; Henry Lord Brougham, in *Dissertations...connected with Natural Theology*, app. to Paley's Works, I, pp. 218–368, 1839; *C.R. Acad. Sci. Paris*, XLVI, pp. 1024–1029, 1858; *Tracts, Mathematical and Physical*, 1860, pp. 103–121, etc.; E. Carruccio, Note storiche sulla geometria delle api, *Periodico di Matem.* (4), XVI, 20 pp., 1936; G. Césaro, Sur la forme de l'alvéole des abeilles, *Bull. Acad. Roy. Belg.* (Sci.), Avril 10, 1929; Sam. Haughton, On the form of the cells made by various wasps and by the honey-bee, *Proc. Nat. Hist. Soc. Dublin*, III, pp. 128–140, 1863; *Ann. Mag. Nat. Hist.* (3), XI, pp. 415-429, 1863; A. R. Wallace, Remarks on the foregoing paper, *ibid.* XII, p. 33; J. O. Hennum, *Arch. f. Math. u. Vidensk.*, Christiania, IX, p. 301, 1884; F. Huber, *Nouv. obs. sur les abeilles*, II, p. 475, 1814; F. W. Hultmann, *Tidsskr. f. Math.*, Uppsala, I, p. 197, 1868; John Hunter, Observations on bees, *Phil. Trans.* 1792, pp. 128–195; Jacob, *Nouv. Ann. de Math.* II, p. 160, 1843; G. S. Klügel, Mathem. Betrachtungen üb. d. kunstreichen Bau d. Bienenzellen, *Hannoversches Mag.* 1772, pp. 353–368; Léon Lalanne, Note sur l'architecture des abeilles, *Ann. Sc. Nat. Zool.* (2), XIII, pp. 358–374, 1840; B. Powell, *Proc. Ashmol. Soc.* I, p. 10, 1844; K. H. Schellbach, *Mathem. Lehrstunde: Lehre v. Grössten u. Kleinsten*, 1860, pp. 35–37; Sam. Sharpe, *Phil. Mag.* IV, pp.,19–21, 1828; J. E. Siegwart, Die Mathematik im Dienste d. Bienenzucht, *Schw. Bienenzeitung*, III, 1880; O. Terquem, *Nouv. Ann. de Math.* XV, p. 176, 1856; C. M. Willick, On the angle of dock-gates and the bee's cell, *Phil. Mag.* (4) XVIII, p. 427, 1859; *C.R.* LI, p. 633, 1860; Chauncy Wright, *Proc. Amer. Acad. Arts and Sci.* IV, p. 432, 1860.

We have mentioned both Hooke and Grew*, and we have just heard Buffon engaged in such speculations; but the matter was more elaborately treated near the beginning of last century by Dieterich George Kieser†, an ingenious friend and colleague of the celebrated Lorenz Oken. Kieser clearly understood that the cell has not a shape of its own, but merely one impressed on it by physical forces and defined by mathematical laws. In his *Mémoire sur l'organisation des plantes*, he gives an admirable historical account of the work of Malpighi, Hooke, Grew, John Hill and other early microscopists; and then he says "La forme des cellules est variée dans

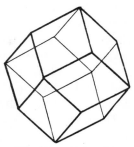

Fig. 210.   A rhombic dodecahedron.

les plantes différentes, mais il y a des formes principales, fondées sur les lois des mathématiques, que la nature suit toujours dans ses formations....La forme la plus commune est celle que prennent nécessairement des globules rondes ou allongées, pressées ensemble, celle des corps hexagonaux à parois quadrilatérales, ou d'une colonne très courte hexagone, coupée horizontalement d'en haut et d'en bas." Here we have, briefly described and sufficiently accounted for, the configuration of what we call a "pavement epithelium," or other simple association of cells in a single layer.

But another passage (from the same author's *Phytotomie*) is worth quoting at length, where he deals with cells in the mass, that is to say with the three-dimensional problem. "Die nach mathematische Gesetzen bestimmte als nothwendige Grundform der Zelle der vollkommenen Zellengewebe ist das langgezogene Rhombododekaheder. ...Mathematisch liegt das Beweis dass diese Figur die Grundform der vollkommenen Zellengewebe sei darin, dass unter allen mathematischen Körpern welche durch Zusammensetzung einen soliden Körper ohne Zwischenräume bilden, das Rhombododekaheder die einzige ist welche mit der wenigsten Masse des Umkreises den grössten Raum einschliesst. Sollte also aus dem Globus—dem

* R. Hooke, *Micrographia*, 1665, pp. 115–116; Nehemiah Grew, *Anatomy of Plants*, 1682, pp. 64, 76, 120.

† D. G. Kieser, *Mémoire*, etc., Haarlem, 1814, p. 89; *Phytotomie, oder Grundzüge der Anatomie der Pflanzen*, Jena, 1815, p. 4.

ursprünglichsten Schleimbläschen der Pflanzenzelle—ein eckiger Körper gebildet werden, so musste dieser das Rhombododekaheder sein, weil dieser im Hinsicht des Minimums der Masse zu dem Maximum des eingeschlossenen Raumes dem Globus am nächsten liegt. Als die Urform der Pflanzenzelle ist nicht Globus sondern Ellipsoide, daher muss das Dodekaheder, welche die Grundform der eckigen Pflanzenzelle ist, auch aus dem Ellipsoide entstanden sein. Das Rhombododekaheder wird also vom unten nach oben gestreckt, und die Grundform der eckigen Pflanzenzelle ist das in perpendiculärer Richtung längsgestreckte Rhombododekaheder."

These views and speculations of Kieser's, now all but forgotten, were by no means neglected in their day. Oken accepted them, and taught them*; Schleiden remarks that "the form of cells frequently passes into that of the rhombic dodecahedron, so beautifully determined, *à priori*, by Kieser†"; and De Candolle thought it necessary to warn his readers that cells are not as geometrically regular as published figures might lead one to believe‡.

The same principles apply to various orders of magnitude, and close-packing may be seen even in the inner contents of a cell. In vitally stained "goblet-cells," the mucin gathers into clumps or droplets, of which each appears in optical section to be surrounded by six more. When fixed they draw together, appear in optical section to be hexagonal, and we may take it that they have become, to a first approximation, rhombic dodecahedra§.

These then, and such as these, were the not unimportant speculations on the forms of cells by men who early grasped the fact that form had a physical cause and a mathematical significance. But their conception of the phenomenon was of necessity limited to the play of the mechanical forces; for Plateau's *Statique des Liquides* had not yet shewn what the capillary forces can do, nor opened a way thereby for Berthold and for Errera.

A very beautiful hexagonal symmetry as seen in section, or dodecahedral as viewed in the solid, is presented by the pith of certain rushes (e.g. *Juncus effusus*), and somewhat less diagram-

* Oken, *Physiophilosophy* (Ray Society), 1847, p. 209.

† *Müller's Archiv*, 1838, p. 146.

‡ *Organogénie végétale*, I, p. 13, 1827.

§ E. S. Duthie, in *Proc. R.S.* (B), CXIII, pp. 459–463, 1933.

matically by the pith of the banana. The cells are stellate, and
the tissue has the appearance in section of a network of six-rayed
stars (Fig. 211), linked together by the tips of the rays, and separated
by symmetrical, air-filled intercellular spaces, which give its snow-
like whiteness to the pith. In thick sections, the solid twelve-rayed
"star-dodecahedra" may be very beautifully seen under the
binocular microscope. They are not difficult to understand.
Imagine, as before, a system of equal spheres in close contact, each
one touching its twelve neighbours, six of them in the equatorial

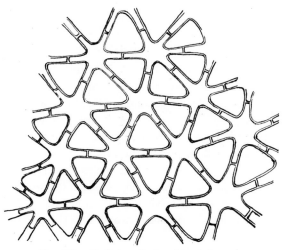

Fig. 211.   Stellate cells in pith of *Juncus*.

plane; and let the cells be not only in contact, but become attached
at the points of contact. Then, instead of each cell expanding so
as to encroach on and fill up the intercellular spaces, let each tend
to shrink or shrivel up by the withdrawal of fluid from its interior.
The result will be to enlarge the intercellular spaces; the attachments
of each cell to its neighbours will remain fixed, but the walls between
these points of attachment will be withdrawn in a symmetrical
fashion towards the centre. As the final result we have the star-
dodecahedron, which appears in plane section as a six-rayed figure.
It is necessary not only that the pith-cells should be attached to
one another, but also that the outermost should be attached to a
boundary wall, to preserve the symmetry of the system. What

actually occurs in the rush is tantamount to this, but not absolutely identical. It is not so much the pith-cells which tend to shrink within a boundary of constant size, but rather the boundary wall which continues to expand after the pith-cells which it encloses have ceased to grow or to multiply. The points of attachment on the surface of each little pith-cell are drawn asunder, but the content of the cell does not correspondingly increase; and the remaining portions of the surface shrink inwards, accordingly, and gradually constitute the complicated figure which Kepler called a star-dodecahedron, which is still a symmetrical figure, and is still a surface of minimal area under the new and altered conditions.

The tetrakaidekahedron

A few years after the publication of Plateau's book, Lord Kelvin shewed, in a short but very beautiful paper*, that we must not hastily assume from such arguments as the foregoing that a close-packed assemblage of rhombic dodecahedra will be the true and general solution of the problem of dividing space with a minimum partitional area, or will be present in a liquid "foam," in which the general problem is completely and automatically solved. The general mathematical solution of the problem (as we have already indicated) is, that every interface or partition-wall must have constant mean curvature throughout; that where these partitions meet in an edge, they must intersect at angles such that equal forces, in planes perpendicular to the line of intersection, shall balance; that no more than three such interfaces may meet in a line or edge, whence it follows (for symmetry) that the angle of intersection of all surfaces or facets must be 120°; and that neither more nor less than four edges meet in a point or corner. An assemblage of rhombic dodecahedra goes far to meet the case. It fills space; its surfaces or interfaces are planes, and therefore surfaces of constant curvature throughout; and they meet together at angles of 120°. Nevertheless, the proof that the rhombic dodecahedron (which we find exemplified in the bee's cell) is a figure of minimal area is not a comprehensive proof; it is limited to certain conditions, and

---

* Sir W. Thomson, On the division of space with minimum partitional area, *Phil. Mag.* (5), XXIV, pp. 503–514, Dec. 1887; cf. *Baltimore Lectures*, 1904, p. 615; *Molecular tactics of a crystal* (Robert Boyle Lecture), 1894, pp. 21–25.

practically amounts to no more than this, that of the ordinary space-filling solids with all sides plane and similar, this one has the least surface for its solid content.

The rhombic dodecahedron has six tetrahedral angles and eight trihedral angles. At each of the latter three, and at each of the former six, dodecahedra meet in a point in close packing; and four edges meet in a point in the one case and eight in the other. This is enough to shew that the conditions for minimal area are not rigorously met. In one of Plateau's most beautiful experiments*, a wire cube is dipped in soap-solution. When lifted out, a film is seen to pass inwards from each of the twelve edges of the cube, and these twelve films meet, three by three, in eight edges, running inwards from the eight corners of the cube; but the twelve films and their eight edges do not meet in a point, but are grouped around a small central quadrilateral film (Fig. 212). Two of the eight edges run to each corner of the little square, and, with the two sides of the square itself, make up the four edges meeting in a point which the theory of area minima requires. We may sub-

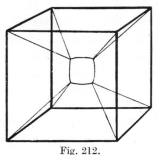

Fig. 212.

stitute (by a second dip) a little cube for the little square; now an edge from each corner of the outer cube runs to the corresponding corner of the inner one, and with the three adjacent edges of the little cube itself the number four is still maintained. Twelve films, and eight edges meeting in a point, were essentially unstable; but the introduction of the little square or cube meets most of the conditions of stability which Plateau was the first to lay down. One more condition has to be met, namely the equality of angles at which the four edges meet in each conjunction. These co-equal "Maraldi angles" at each corner of the square can only be constructed by help of a slight curvature of the sides, and the little square is seen to have its sides curved into circular arcs accordingly; moreover its size and shape, as that of all the other films in the system, are perfectly definite. It is all one, according to the

* Also discovered independently by Sir David Brewster, *Trans. R.S.E.* XXIV, p. 505, 1867; XXV, p. 115, 1869.

symmetry of the figure, to which side of the skeleton cube the square lies parallel; wherever it may be, if we blow gently on it, then (as M. Van Rees discovered) it alters its place and sets itself parallel now to one and now to another of the paired faces of the cube.

The skeleton cube, like the tetrahedron which we have already studied, is only one of many interesting cases; for we may vary the shape of our wire cages and obtain other and not less beautiful configurations. An hexagonal prism, if its sides be square or nearly so, gives us six vertical triangular films, whose apices meet the corners of a horizontal hexagon*; also six pairs of truncated triangles, which link the top and bottom edges of the cage to the sides of the median hexagon. But if the height of the hexagonal prism be increased, the six vertical films become curvilinear triangles, with sides concave towards the apex; and the twelve remaining films, which spring from the top and bottom of the hexagon, are curved surfaces, looking like a sort of hexagonal hourglass†.

There is a deal of elegant geometry in these various configurations. Lamarle shewed that if, in a figure represented by our wire cage, we suppress (in imagination) one face and all the other faces adjoining it, then the faces which remain are those which appear in the centre of the figure after the cage has been withdrawn from the soap-solution. Thus, in a cube, we suppress one face and the four adjacent to it; only one remains, and it reappears as the central square in the middle of the new configuration; in the tetrahedron, when we have suppressed one face and the three adjacent to it, there is nothing left—save a median point, corresponding to the opposite corner. In a regular dodecahedron, if we suppress one pentagonal face and its five neighbours, the other half of the whole figure remains; and the dodecahedral cage, after immersion in the soap, shews a central and symmetrical group of six pentagons‡.

Moreover, while the cage is carrying its configuration of films, we may blow a bubble within it, and so insert a new polyhedron

---

* The angles of a hexagon are too big, as those of a square were too small, to form the Maraldi angles of symmetry; hence the sides of the hexagon are found to be concave, as those of the square bulged out convexly.

† Cf. Dewar, *op. cit.* 1918.

‡ That is to say, if nF_m be a polyhedron (of n m-faced sides), the corresponding wire cage will exhibit $(n - m + 1)$ F_m as central fenestrae.

within the old, and set it in place of the former fenestra. The inner polyhedral bubble so produced may be of any dimensions, but it resembles the outer polyhedral cage precisely, except in the curvature of its sides; it has all its faces spherical, and all of equal radius of curvature; its edges are either arcs of circles or straight lines. Later on, we shall see that there is no small biological interest attaching to these configurations.

Lord Kelvin made the remarkable discovery that the square fenestra with the four quadrilateral films impinging on its sides, in Plateau's experiment, represented the one-sixth part of a symmetrical figure; that this figure when complete was bounded by six squares and eight hexagons; that by means of an assemblage of these

Fig. 213.   A set of 14-hedra, to shew close-packing.   From F. T. Lewis.

fourteen-sided figures, or "tetrakaidekahedra," space is filled and homogeneously partitioned—into equal, similar and similarly situated cells—with an economy of surface in relation to volume even greater than in an assemblage of rhombic dodecahedra*.

The tetrakaidekahedron, in its most generalised case, is bounded by three pairs of equal and opposite quadrilateral faces, and four pairs of equal and opposite hexagonal faces, neither the quadrilaterals nor the hexagons being necessarily plane. In its simplest case, with all its facets plane and equilateral, it is Kelvin's "ortho-tetrakaidekahedron"; and also (though Kelvin was unaware of the fact) one of the thirteen semi-regular and isogonal polyhedra, or "Archimedean bodies." In a particular case, the quadrilaterals are plane surfaces with curved edges, but the hexagons are slightly

* Kelvin, *Boyle Lecture* and *Baltimore Lectures*. In the first of these Kelvin described the plane-faced tetrakaidekahedron; in the second he shewed how that figure must have its faces warped and edges curved to fulfil all the conditions of minimal area.

curved "anticlastic" surfaces; and these latter have at every point equal and opposite curvatures, and are surfaces of minimal curvature for a boundary of six curved edges. This figure has the remarkable property that, like the plane rhombic dodecahedron, it so partitions space that three faces meeting in an edge do so everywhere at co-equal angles of 120°; and, unlike the rhombic dodecahedron, four edges meet in each point or corner at co-equal angles of 109° 28′*.

We may take it as certain that, in a homogeneous system of fluid films like the interior of a froth of soap-bubbles, where the films are perfectly free to glide or turn over one another and are of approximately co-equal size, the mass is actually divided into cells of this remarkable conformation: and the possibility of such a configuration being present even in the cells of an ordinary vegetable parenchyma was suggested in the first edition of this book. It is all a question of *restraint*, of degrees of mobility or fluidity. If we squeeze a mass of clay pellets together, like Buffon's peas, they come out, or all the inner ones do, in neat garnet-shape, or rhombic dodecahedra. But a young student once shewed me (in Yale) that if you wet these clay pellets thoroughly, so that they slide easily on one another and so acquire a sort of pseudo-fluidity in the mass, they no longer come out as regular dodecahedra, but with square and hexagonal facets recognisable as those of ill-formed or half-formed tetrakaidekahedra.

Dr F. T. Lewis has made a long and careful study of various vegetable parenchymas, by simple maceration, wax-plate recon-

---

* Von Fedorow had already described (in Russian), unaware that Archimedes had done so, the same figure under the name of cubo-octahedron, or hepta-parallelohedron, limited however to the case where all the faces are plane and regular. This cubo-octahedron, together with the cube, the hexagonal prism, the rhombic dodecahedron and the "elongated dodecahedron," constitute the five plane-faced, parallel-sided figures by which space is capable of being completely filled and uniformly partitioned; the series so forming the foundation of Von Fedorow's theory of crystalline structure—though the space-fillers are not all, and cannot all be, crystalline forms. All of these figures, save the hexagonal prism, are related to and derivable from the cube; so we end by recognising two principal types, cubic and hexagonal. We have learned to recognise the dodecahedron, and we may find in still closer packing the cubo-octahedron, in a parenchyma; the elongated dodecahedron is, essentially, the figure of the bee's cell; the cube we have, in essence, in cambium-tissue; the hexagonal prism, dwarf or tall, simple or recognisably deformed, we see in every epithelium.

struction and otherwise, and has succeeded in shewing that the tetrakaidekahedral form is closely approached, or even attained, in certain simple and homogeneous tissues. After reconstructing a large model of the cells of elder-pith, he finds that the fourteen-sided figure clearly manifests itself as the characteristic or typical form to which the cells approximate, in spite of repeated cell-divisions and consequent inequalities of size. Counting in a hundred cells the number of contacts which each made with its neighbours, that is to say the total number both of actual and potential facets, Lewis found that 74 per cent. of the cells were either 12, 13, 14, 15 or 16-sided, 56 per cent. either 13, 14 or 15-sided, and that the average

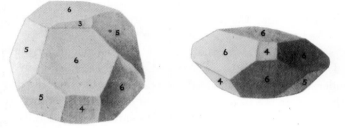

Fig. 214.   Reconstructed models of cells of elder-pith, shewing a certain approximation to 14-hedral form.   From F. T. Lewis.

number of facets or contacts was, in this instance, just 13·96. These figures indicate the general symmetry of the cells, their departure from the dodecahedral, and their tendency towards the tetra-kaidekahedral, form*.

But after all, the geometry of the 14-hedron, displayed to per-fection by our soap-films in the twinkling of an eye, is only roughly developed in an organic structure, even one so delicate as elder-pith; the conditions are no longer simple, for friction, viscosity and

---

* F. T. Lewis, The typical shape of polyhedral cells in vegetable parenchyma, and the restoration of that shape following cell-division, *Proc. Amer. Acad. of Arts and Sci.* LVIII, pp. 537–552, 1923, and other papers. See also (*int. al.*) J. W. Marvin, The aggregation of orthis-tetrakaidekahedra, *Science*, LXXXIII, p. 188, 1936; E. B. Metzger, An analysis of the orthotetrakaidekahedron, *Bull. Torrey Bot. Club*, LIV, pp. 341–348, 1927. Professor van Iterson of Delft tells me that *Asparagus Sprengeri* (a common greenhouse plant) is a good subject for shewing the 14-hedral cells.

solidification have vastly complicated the case. We get a curious and an unexpected variant of the same phenomenon in the microscopic foam-like structure assumed as molten metal cools. If these foam-cells were again 14-hedra, their facets would all be either squares or hexagons; but pentagonal facets are commoner than either, and the cells often approach closely to the form of a regular *pentagonal* dodecahedron! The edges of this figure meet at angles of 108°, not far from the characteristic Maraldi angle of 109° 28 ; and the faces meet at an angle not far removed from 120°. A slight curvature of the sides is enough to turn our pentagonal dodecahedron into a possible figure of equilibrium for a foam-cell. We cannot close-pack pentagonal dodecahedra, whether equal or unequal, so as to fill space; but still the figure may be, and seems to be, common, interspersed among the polyhedra of various shapes and sizes which are packed together in a metallic foam*.

A somewhat similar result, and a curious one, was found by Mr J. W. Marvin, who compressed leaden small-shot in a steel cylinder, as Buffon compressed his peas; but this time the pressure on the plunger ran from 1000 to 35,000 lb. or nearly twenty tons to the square inch. When the shot was introduced carefully, so as to lie in ordinary close packing, the result was an assemblage of regular rhombic dodecahedra, as might be expected and as Buffon had found. But the result was very different when the shot was poured at random into the cylinder, for the average number of facets on each grain now varied with the pressure, from about 8·5 at 1000 lb. to 12·9 at 10,000 lb., and to no less than 14·16 facets or contacts after all interstices were eliminated, which took the full pressure of 35,000 lb. to do. An average of just over fourteen facets might seem to indicate a tendency to the production of tetrakaidekahedra, just as in the froth of soap-bubbles; but this is not so. The squeezed grains are irregular in shape, and pentagonal facets are much the commonest, just as we found them to be in the microscopic structure of a once-molten metal. At first sight it might seem that, though the experiment has something to teach us about random packing in a limited space, it has no biological significance; but it is curious to find that the pith-cells of

* Cf. Cecil H. Desch, The solidification of metals from the liquid state, *Journ. Inst. of Metals*, XXII, p. 247, 1919.

Eupatorium have a similar average configuration, with the same predominance of pentagonal facets*.

We learn, in short, from Lewis and from Marvin that the mechanical result of mutual pressure, even in an assemblage of co-equal spheres, is more varied and more complex than we had supposed. The two simple and homogeneous configurations—the rhombo-dodecahedral and tetrakaidekahedral assemblages—are easily and commonly produced, the one by the compression of solid spheres in ordinary close-packing, the other when a liquid system of spheres or bubbles is free to slide and glide into a packing which is closer still. Between these two configurations there is no other symmetrical or homogeneous arrangement possible; but random packing and degrees of compression leave their random effects, among which are traces here and there of regular shape and symmetry.

As a froth has its histological lessons for us, throwing light on the structure of a parenchyma, so may we draw an illustration or two from the analogous characteristics of an emulsion. Both alike are "states of aggregation"; both are "two-phase systems," one phase being dispersed and the other the medium of dispersion. Both phases are liquid in the emulsion, in the froth the dispersed phase is a gas; our living tissue is, so far, more likely to be an emulsion than a froth. The concept widens. A colony of bacteria, the blood corpuscles in their plasma, the filaments of an alga, the heterogeneous texture of any ordinary tissue, may all be brought under the general concept of "phase systems," and share the common character that one phase exposes a large "interface" to the other. If we take milk as a simple emulsion, we see its liquid oil-globules dispersed in a watery medium and rounded by surface tension into spheres. The watery medium, as is usual in such emulsions, contains dissolved substances which tend to lower the interfacial tension; for were that tension high the globules would tend to be larger and their aggregate surface less. Suppose the "phase-ratio" to alter, the globules becoming more numerous and the disperse medium less and less, the globules will be close-packed

* J. W. Marvin, The shape of compressed lead-shot, etc.; *Amer. Journ. of Botany*, XXVI, pp. 280–288, 1939; Cell-shape studies in the pith of Eupatorium, *ibid.* pp. 487–504.

at last.   Then each (provided they be of equal size) will be in touch
with twelve neighbours; and if the spheres were solid—were the
system not an emulsion but a "suspension"—the matter would end
here.   But our liquid globules are capable of deformation, and the
points of contact are flattened in still closer packing into planes.
They become polyhedral, and tend to take the form of rhombic
dodecahedra, or it may be even of 14-hedra, and the dispersion-
medium is reduced to mere films or pellicles between.   At the stage
of mere twelve-point contact, the spherules constitute about 74 per
cent., and the disperse medium 26 per cent., of the whole.   But in
the final stage the phase-ratio has so altered that the disperse-
medium is but a small fraction of the whole, the thin film to which
it has been reduced has the appearance of a cell-membrane separating
the cells, and the microscopic structure of the whole corresponds to
the cellular configuration of a parenchymatous tissue*.

Of certain groupings of cells

It follows from all that we have said that the problems connected
with the conformation of cells, and with the manner in which a
given space is partitioned by them, soon become complex; and
while this is so even when all our cells are equal and symmetrically
placed, it becomes vastly more so when cells varying even slightly
in size, in hardness, rigidity or other qualities, are packed together.
The mathematics of the case very soon become too hard for us,
but in its essence the phenomenon remains the same.   We have
little reason to doubt, and no just cause to disbelieve, that the
whole configuration, for instance of an egg in the advanced stages
of segmentation, is accurately determined by simple physical laws:
just as much as in the early stages of two or four cells, during which
early stages we are able to recognise and demonstrate the forces
and their effects.   But when mathematical investigation has become
too difficult, physical experiment can often reproduce the phenomena which Nature exhibits, and which we are striving to comprehend.   In an admirable research, M. Robert not only shewed
some years ago that the early segmentation of the egg of *Trochus*
(a marine univalve mollusc) proceeded in accordance with the laws

* Cf. E. Hatschek, Homogeneous partitionings, etc., *Phil. Mag.* xxxiii, p. 83,
1917.

of surface-tension, but he also succeeded in imitating by means of soap-bubbles one stage after another of the developing egg.

M. Robert carried his experiments as far as the stage of sixteen cells, or bubbles. It is not easy to carry the artificial system quite so far, but in the earlier stages the experiment is easy; we have merely to blow our bubbles in a little dish, adding one to another, and adjusting their sizes to produce a symmetrical system. One of the simplest and prettiest parts of his investigation concerned the "polar furrow" of which we have spoken on p. 489. On blowing four little contiguous bubbles he found (as we may all find with the greatest ease) that they form a symmetrical system, two in contact with one another by a laminar film, and two which are elevated a little above the others and are separated by the length of the aforesaid lamina. The bubbles are thus in contact three by three, their partition-walls making with one another equal angles of 120°. The upper and lower edges of the intermediate lamina (the lower one visible through the transparent system) constitute the two polar furrows of the embryologist (Fig. 215, 1–3). The lamina itself is plane when the system is symmetrical, but it responds by a corresponding curvature to the least inequality of the bubbles on either side. In the experiment, the upper polar furrow is usually a little shorter than the lower, but parallel to it; that is to say, the lamina is of trapezoidal form: this lack of perfect symmetry being due (in the experimental case) to the lower portion of the bubbles being somewhat drawn asunder by the tension of their attachments to the sides of the dish (Fig. 215, 4). A similar phenomenon is usually found in *Trochus*, according to Robert, and many other observers have likewise found the upper furrow to be shorter than the one below. In the various species of the genus *Crepidula*, Conklin asserts that the two furrows are equal in *C. convexa*, that the upper one is the shorter in *C. fornicata*, and that the upper one all but disappears in *C. plana*; but we may well be permitted to doubt, without the evidence of very special investigations, whether these slight physical differences are actually characteristic of, and constant in, particular *species*. Returning to the experimental case, Robert found that by withdrawing a little air from, and so diminishing the bulk of the two terminal bubbles (i.e. those at the ends of the intermediate lamina), the upper polar furrow was caused to elongate, till it became equal in length to the lower; and by continuing the process it became the longer in its turn. These two conditions have again been described by investigators as characteristic of this embryo or that; for instance in *Unio*, Lillie has described the two furrows as gradually altering their respective lengths*; and Wilson (as Lillie

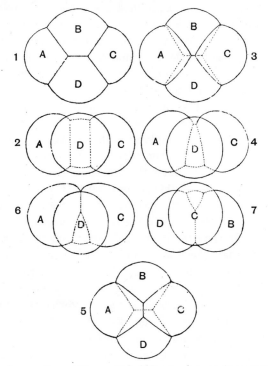

Fig. 215. Aggregations of four soap-bubbles, to shew various arrangements of the intermediate partition and polar furrows. After Robert.

remarks) had already pointed out that "the reduction of the apical cross-furrow, as compared with that at the vegetative pole in molluscs and annelids, 'stands in obvious relation to the different size of the cells produced at the two poles†'"

When the two lateral bubbles are gradually reduced in size, or the two terminal ones enlarged, the upper furrow becomes shorter

* F. R. Lillie, Embryology of the Unionidae, *Journ. Morph.* x, p. 12, 1895.
† E. B. Wilson, The cell-lineage of Nereis, *Journ. Morph.* vi, p. 452, 1892.

and shorter; and at the moment when it is about to vanish, a new furrow makes its instantaneous appearance in a direction perpendicular to the old one; but the inferior furrow, constrained by its attachment to the base, remains unchanged, and it looks as though our two polar furrows, which were formerly parallel, were now at right angles to one another. But in fact, the geometry of the whole system is entirely altered. Before, two furrows left each end of one polar furrow for *the same end* of the other polar furrow, and the two cells at either end were shaped like "liths" of an orange. Under the new arrangement, two furrows leave each end of one for *the two ends* of another. The figure is now divided by six similar furrows into four similar curvilinear triangles*; it has become (approximately) a spherical tetrahedron, and the four cells into which it is divided are four similar and symmetrical figures, also tetrahedral, all meeting in a point at the centroid of the figure. Such a four-celled embryo, described as having two polar furrows arranged in a cross, has often been seen and figured by the embryologists. Robert himself found this condition in *Trochus*, as an occasional or exceptional occurrence: it has been described as normal in *Asterina* by Ludwig, in *Branchipus* by Spangenberg, and in *Podocoryne* and *Hydractinia* by Bunting.

So, by slight and delicate modifications, we pass through many, and perhaps through *all*, of the possible arrangements of external furrows and internal partitions which divide the four cells from one another in a four-celled egg or embryo; and many, or most, or possibly *all* of these arrangements have been more or less frequently observed in the four-celled stages of various embryos. And all these configurations, which the embryologists have witnessed and described, belong to that large class of phenomena whose distribution among embryos, or among organisms in general, bears no relation to the boundaries of zoological classification; through molluscs, worms, coelenterates, vertebrates and what not, we meet with now one and now another, in a medley which defies classification. They are not "vital phenomena," or "functions" of the organism, or special characteristics of this organism or that, but purely physical

---

* That the sphere can be symmetrically divided into four equilateral triangles, after the manner of these embryos (or of many pollen-grains), is an elementary fact of great importance in geometry and trigonometry.

phenomena. The kindred but more complicated phenomena analogous
to the polar furrow, which arise when a larger number of cells
than four are associated together, we shall deal with in the next
chapter.

Having shewn that the capillary phenomena are patent and
unmistakable during the earlier stages of embryonic development,
but soon become more obscure and less capable 'of experimental
reproduction in the later stages when the cells have increased in
number, various writers including Robert himself have been inclined
to argue that the physical phenomena die away, and are over-
powered and cancelled by agencies of a different order.   Here we
pass into a region where observation and experiment are not at
hand tó guide us, and where a man's trend of thought, and way of
judging the whole evidence in the case, must shape his philosophy.
We must always remember that even in a froth of soap-bubbles
we can apply an exact analysis only to the simplest cases and
conditions; we cannot describe, but can only imagine, the forces
which in such a froth control the respective sizes, positions and
curvatures of the innumerable bubbles· and films of which it consists; but our knowledge is enough to leave us assured that what
we have learned by investigation of the simplest cases includes the
principles which determine the most complex.   In the case of the
growing embryo we know from the beginning that surface-tension
is only one of the physical forces at work; and that other forces,
including those displayed within the interior of each living cell, play
their part in the determination of the system.   But we have no
evidence whatsoever that at this point, or that point, or at any, the
dominion of the physical forces over the material system gives place
to a new condition where agencies at present unknown to the
physicist impose themselves on the living matter, and become
responsible for the conformation of its material fabric.

Before we leave for the present the subject of the segmenting
egg, we may take brief note of two associated problems: viz.
(1) the formation and enlargement of the segmentation cavity, or
central interspace around which the cells tend to group themselves
in a single layer, and (2) the formation of the gastrula, that is to
say (in a typical case) the conversion by "invagination," of the

one-layered ball into a two-layered cup. Neither problem is free from difficulty, and all we can do meanwhile is to state them in general terms, introducing some more or less plausible assumptions.

The former problem is comparatively easy, as regards the tendency of a segmentation cavity to *enlarge*, when once it has been established. We may then assume that subdivision of the cells is due to the appearance of a new-formed septum within each cell, that this septum has a tendency to shrink under surface-tension, and that these changes will be accompanied on the whole by a diminution of surface-energy in the system. This being so, it may be shewn that the volume of the divided cells must be less than it was prior to division, or in other words that part of their contents must exude during the process of segmentation*. Accordingly, the case where the segmentation cavity enlarges and the embryo developes into a hollow blastosphere may, under the circumstances, be simply described as the case where that outflow or exudation from the cells of the blastoderm is directed on the whole inwards.

The physical forces involved in the invagination of the cell-layer to form the gastrula have been repeatedly discussed †, but the several explanations are conflicting, and are far from clear. There is, however, a certain homely phenomenon which goes some way, perhaps a long way, to explain this remarkable configuration. An ordinary gelatine lozenge, or jujube, has (like the developing gastrula) a more or less spherical form, depressed or dimpled at one side; this is a very noteworthy conformation, and it arises, automatically, by the shrinkage of a sphere. Were the initial sphere of gelatine perfectly homogeneous, and so situated as to shrink with absolute uniformity, it would merely shrink into a smaller sphere; it does nothing of the kind. There is always some part or other which shrinks *a little more* than the rest‡; and the dimple so formed goes on increasing, until at last a very perfect cup-shaped figure is formed. I imagine that the gastrula is formed in much the

---

* Professor Peddie has given me this interesting result, but the mathematical reasoning is too lengthy to be set forth here.

† Cf. Bütschli, *Arch. f. Entw. Mech.* v, p. 592, 1897; Rhumbler, *ibid.* xiv, p. 401, 1902; Assheton, *ibid.* xxxi, p. 46, 1910.

‡ Just as there may be some small part which shrinks a little *less*. But this we should not distinguish from the common case where one small part *grows a little more*, and so "produces a *bud*," as in the yeast-cell on p. 363.

same way, save only that the initial dimple, instead of being fortuitous, has its constant place, determined by the physico-chemical heterogeneity of the embryo. We may even go one step further, and see (or imagine we see) in the formation of the gastrula a physico-chemical or physiological turning-point, the segmentation cavity being due (as we have seen) to an inward flow, and a reversal of the current leading to that shrinkage which produces the gastrula.

Fig. 216. Effect of shrinkage on a globule of gelatine.
After E. Hatschek.

A note on shrinkage

We have dealt much with *growth*, but the fact is that negative growth, or shrinkage, is also an important matter; and just as we find a whole series of phenomena to be based on the extension or expansion of bubbles, vesicles, etc., so there is another series, physically alike and mathematically identical, which depend on the shrinkage of a solid or semi-fluid mass. After all, growth and its converse go hand in hand, and a special case of shrinkage is that surface-tension to which all the Plateau configurations are due. One clear case, the gastrula, we have touched on, and we have discussed another which led to the stellate dodecahedra of the Rush.

As a cube of gelatine, or of paraffin, dries, and shrinks, it alters its shape in a remarkable way*. Its corners become more salient, its sides become concave; its cross-section has the form of a four-

* Emil Hatschek, *Kolloid Ztschr.* xxxv, pp. 67–76, 1924; *Nature*, 1st Nov. 1924.

rayed star with rounded angles. The block has dried unequally;
its corners and its edges were naturally the first to dry*, and the
twelve dried and hardened edges began to play the part of the wire
frame in Plateau's soap-bubble experiment. The shrinking cube is
tending towards the identical configuration shewn in Fig. 212; it is
a minimal configuration, partially realised in a coarse material, but
realisable to perfection in a film.

A shrinking cylinder (as Plateau knew) shews various
phenomena, depending on its proportions. A low, squat cylinder
begins to show a pulley-like groove—a catenoid—around its periphery,
precisely like the soap-film between its two wire rings in Fig. 108;
and as the groove deepens, the plane surfaces of the cylinder also begin

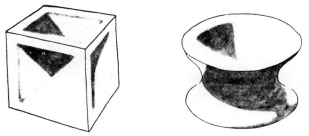

Fig. 217.   Shrinkage of cube and cylinder.

to dimple in. They become spherical as they grow more concave, and
the deepening groove of the pulley passes from a catenoid to a nodoid
curve—so at least theory tells us †, for, beautiful as the experimental
configurations are, they hardly lend themselves to precise measurements of curvature. But this shrunken cylinder is now wonderfully
like the "amphicoelous" vertebra of a cartilaginous fish, the simplest
and most "primitive" vertebra of all. A series of cracks, or splits,
around the circular groove in the vertebra seem to be a final result
of irregular shrinkage, not shewn in the more homogeneous gelatine.

A long cylinder, or thread, of gelatine tends to become fluted,
with three or more ribs or folds, and it is in this way that threads

* Just as, conversely, the prominent parts of a crystal tend to grow more
rapidly than the rest in a super-saturated solution, and to dissolve more rapidly
in one below saturation; cf. O. Lehmann, Ueber das Wachstum der Krystalle,
Ztschr. f. Krystallogr. I, p. 453.

† See p. 369.

of viscose, or artificial silk, tend likewise to have a ridged or fluted structure, and gain in lustre thereby. The subject is new, and hardly ripe for full discussion; but it holds out promise (as it seems to me) of many biological lessons and illustrations.

We glanced in passing at such "shrinkage-patterns" as are found, for instance, on the little shells of *Lagena*, or on those other hanging drops which constitute Emil Hatschek's artificial medusae; it is no small subject. A stretched elastic membrane, circular or spherical, remains spherical or circular when we let its tension relax; but if, to begin with, we coat the rubber with a pliant but non-elastic material such as wax, the waxen layer, failing to con-

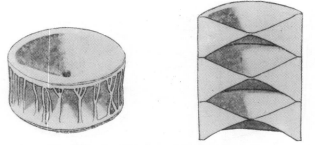

Fig. 218.   Amphicoelous vertebrae of a shark.

tract, is thrown into more or less characteristic folds. In a dried pea the seed has shrunken through loss of moisture, and the loose outer coat wrinkles up*. The pretty pattern of a poppy-seed arises in the same way; but so do the wrinkles on an old man's withered skin. When our experimental elastic with its non-contractile coat is suffered to contract, the first sign of the coat's inability to keep pace is the appearance of little domes, or hummocks, or blisters; and soon from each of these there run out folds, which tend to fork, and the angles between the three branches tend to equalise. They tend, in simple and symmetrical cases, to form a pattern of hexagons, with occasional pentagons or quadrilaterals between; but where the surface is larger and the coat more flexible the folds form an irregular network, still with the various anticlines mostly meeting in three-

* The difference between a smooth and a wrinkled pea, familiar to Mendelians, merely depends, somehow, on amount and rate of shrinkage.

way nodes*. Nature will ring the changes on the resultant patterns, according as the surface be plane or curved, spherical or cylindrical, coarse or fine, fragile or tough. But on these general lines very many structures, both regular and irregular, spines, bristles, ridges, tubercles and wrinkled patterns, bid fair to find their physical or mechanical interpretation; and it is in the more or less hardened parts of plant or animal that we find them one and all displayed. On the egg of a butterfly, on the grooved and dotted elytron of a beetle, on the notched forehead of a scarab, in the saw-like teeth on a grasshopper's leg, in the little lines of dotted tubercles on the shell of a *Rissoa*, more crudely in the lozenged bark of elm or pine, we see a very few of this innumerable class of "shrinkage-patterns."

* The fact that such triplets of divergent ridges or crests are not a feature in the topography of mountain-ranges is a strong argument against the view that general shrinkage accounts for the pattern of the earth's crust. Cf. A. J. Bull, The pattern of a contracting earth, *Geolog. Mag.* LXIX, pp. 73–75, 1932; A. E. B. de Chancourtois, *C.R.* LIX, p. 348, 1903.

CHAPTER VIII

THE FORMS OF TISSUES OR CELL-AGGREGATES (*continued*)

THE problems which we have been considering, and especially that of the bee's cell, belong to a class of "isoperimetrical" problems, which deal with figures whose surface is a minimum for a definite content or volume. Such problems soon become difficult*, but we may find many easy examples which lead us towards the explanation of biological phenomena; and the particular subject which we shall find most easy of approach is that of the division, in definite proportions, of some definite portion of space, by a partition-wall of minimal area. The theoretical principles so arrived at we shall then attempt to apply, after the manner of Berthold and Errera, to the biological phenomena of cell-division.

This investigation may be approached in two ways: by considering the partitioning off from some given space or area of one-half (or some other fraction) of its content; or again, by dealing with the partitions necessary for the breaking up of a given space into a definite number of compartments.

If we begin with the simple case of a cubical cell, it is obvious that, to divide it into two halves, the smallest partition-wall is one which runs parallel to, and midway between, two of its opposite sides. If we call a the length of one of the edges of the cube, then a^2 is the area, alike of one of its sides and of the partition which we have interposed parallel thereto. But if we now consider the bisected cube, and wish to divide the one-half of it again, it is obvious that another partition parallel to the first, so far from being the smallest possible, is twice the size of a cross-partition perpendicular to it; for the area of this new partition is $a \times a/2$. And again, for a third bisection, our next partition must be perpendicular to the other two, and is obviously a little square, with an area of $(\frac{1}{2}a)^2 = \frac{1}{4}a^2$.

---

* Minkowski and others have shewn how hard'it is, for instance, to prove the seemingly obvious proposition that the sphere, of all figures, has the greatest volume for a given surface; cf. (e.g.) T. Bonneson, *Les problèmes des isopérimètres et des isépiphanes,* Paris, 1929. For a historical account of this class of problems, see G. Enestrom, in *Bibl. Math.* 1888.

From this we may draw the simple rule that, for a rectangular body or parallelepiped to be bisected by means of a partition of minimal area, (1) the partition must cut across the longest axis of the figure; and (2) in successive bisections, each partition must run at right angles to its immediate predecessor.

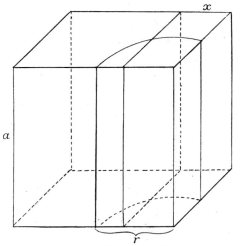

Fig. 219.   After Berthold.

We have already spoken of "Sachs's Rules," which are an empirical statement of the method of cell-division in plant-tissues; and we may now set them forth as follows:

(1)   The cell tends to divide into two co-equal parts.

(2)   Each new plane of division tends to intersect the preceding plane of division at right angles.

The first of these rules is a statement of physiological fact, not without its exceptions, but so generally true that it will justify us in limiting our enquiry for the most part to cases of equal subdivision.   That it is by no means universally true for cells generally is shewn, for instance, by such well-known cases as the unequal segmentation of the frog's egg.   It is true, when the dividing cell is homogeneous and under the influence of symmetrical forces; but it ceases to be true when the field is no longer dynamically symmetrical, as when the parts differ in surface tension or internal pressure, or, speaking generally, in their chemico-physical properties and conditions. This latter condition, of asymmetry of field, is frequent in segmenting eggs*, and it then covers or includes the principle upon which Balfour laid stress as leading to "unequal" or to "partial" segmentation of the egg—viz. the unequal or asymmetrical distribution of protoplasm and of food-yolk.

The second rule, which also has its exceptions, is true in a large number of cases, and owes its validity, as we may judge from the illustration of the repeatedly bisected cube, to the guiding principle of minimal areas. It is in short subordinate to a much more important and fundamental rule, due not to Sachs but to Errera; that (3) the incipient partition-wall of a dividing cell tends to be such that *its area is the least possible by which the given space-content can be enclosed.*

Let us return to the case of our cube, and suppose that, instead of bisecting it, we desire to shut off some small portion only of its volume. It is found in the course of experiments upon soap-films, that if we try to bring a partition-film too near to one side of a cubical (or rectangular) space it becomes unstable, and is then easily shifted to a new position in which it constitutes a curved cylindrical wall cutting off one corner of the cube. It still meets the sides of the cube at right angles (for reasons which we have already considered); and, as we may see from the symmetry of the case, it constitutes one-quarter of a cylinder. Our plane transverse partition had always the same area, wherever it was placed, viz. a^2; and it is obvious that a cylindrical wall, if it cut off a small corner, may be much less than this. We want, accordingly, to determine what volume might be partitioned off with equal economy of wall-space in one way as the other, that is to say, what area of cylindrical

---

* M. Robert (*loc. cit.* p. 305) has compiled a long list of cases among the molluscs and the worms, where the initial segmentation of the egg proceeds by equal or unequal division. The two cases are about equally numerous. But like most other writers of his time, he would ascribe this equality or inequality rather to a provision for the future than to a direct effect of immediate physical causation: "Il semble assez probable, comme on l'a dit souvent, que la plus grande taille d'un blastomère est liée à l'importance et au développement précoce des parties du corps qui doivent en naître: il y aurait là une sorte de reflet des stades postérieures du développement sur les premières phénomènes, ce que M. Ray Lankester appelle *precocious segregation.* Il faut avouer pourtant qu'on est parfois assez embarrassé pour assigner une cause à pareilles différences."

wall would be neither more nor less than the area a^2. The calculation is easy:

The *surface-area* of a cylinder of length a is $2\pi r \cdot a$, and that of our quarter-cylinder is, therefore, $a \cdot \pi r/2$; and this being, by hypothesis, $= a^2$, we have $a = \pi r/2$, or $r = 2a/\pi$.

The *volume* of a cylinder of length a is $a\pi r^2$, and that of our quarter-cylinder is $a \cdot \pi r^2/4$, which (by substituting the value of r) is equal to a^3/π.

Now precisely this same volume is, obviously, shut off by a transverse partition of area a^2 if the third side of the rectangular space be equal to a/π; and this fraction,

if we take $a = 1$, is equal to $0\cdot318\ldots$, or rather less than one-third. And, as we have just seen, the radius, or side, of the corresponding quarter-cylinder will be twice that fraction, or equal to $0\cdot636$ times the side of the cubical cell.

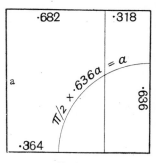

Fig. 220.

If then, in the process of division of a cubical cell, it so divide that the two portions be not equal in volume but that one portion be anything less than about three-tenths of the whole or three-sevenths of the other portion, there will be a tendency for the cell to divide, not by means of a plane transverse partition, but by means of a curved, cylindrical wall cutting off one corner of the original cell; and the part so cut off will be one-quarter of a cylinder.

By a similar calculation we can shew that a *spherical* wall, cutting off one solid angle of the cube and constituting an octant of a sphere, would likewise be of less area than a plane partition as soon as the volume to be enclosed was not greater than about one-quarter of the original cell*. But while both the cylindrical wall and the

* The principle is well illustrated in an experiment of Sir David Brewster's (*Trans. R.S.E.* xxv, p. 111, 1869). A soap-film is drawn over the rim of a wine-glass, and then covered by a watch-glass. The film is inclined or shaken till it becomes attached to the glass covering, and it then immediately changes place, leaving its transverse position to take up that of a spherical segment extending from one side of the wine-glass to its cover, and so enclosing the same volume of air as formerly but with a great economy of surface, precisely as in the case of our spherical partition cutting off one corner of a cube.

spherical wall would be of less area than the plane transverse partition after that limit (of one-quarter volume) was passed, the cylindrical would still be the better of the two up to a further limit. It is only when the volume to be partitioned off is no greater than

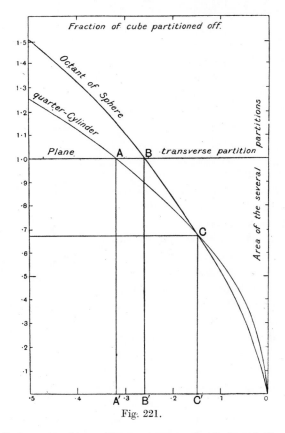

Fig. 221.

about 0·15, or somewhere about one-seventh of the whole, that the spherical cell-wall in a corner of the cubical cell, that is to say the octant of a sphere, is definitely of less area than the quarter-cylinder. In the accompanying diagram (Fig. 221) the relative areas of the three partitions are shewn for all fractions, less than one-half, of the divided cell.

In this figure, we see that the plane transverse partition, whatever fraction of the cube it cut off, is always of the same dimensions, that is to say is always equal to a^2, or $=1$. If one-half of the cube have to be cut off, this plane transverse partition is much the best, for we see by the diagram that a cylindrical partition cutting off an equal volume would have an area about 25 per cent. and a spherical partition would have an area about 50 per cent. greater. The point A in the diagram corresponds to the point where the cylindrical partition would begin to have an advantage over the plane, that is to say (as we have seen) when the fraction to be cut off is about one-third, or 0·318 of the whole. In like manner, at B the spherical octant begins to have an advantage over the plane; and it is not till we reach the point C that the spherical octant becomes of less area than the quarter-cylinder.

The case we have dealt with is of little practical importance to the biologist, because the cases in which a cubical, or rectangular, cell divides unequally and unsymmetrically are apparently few; but we can find, as Berthold pointed out, a few examples, as in the hairs within the reproductive "conceptacles" of certain Fuci (*Sphacelaria*, etc., Fig. 222), or in the "paraphyses" of mosses (Fig. 226). But it is of great theoretical importance: as serving to introduce us to a large class of cases in which, under the guiding principle of minimal areas, the shape and relative dimensions of the original cavity lead to cell-division in very definite and sometimes

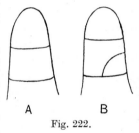

A   B
Fig. 222.

unexpected ways. It is not easy, nor indeed possible, to give a general account of these cases, for the limiting conditions are somewhat complex and the mathematical treatment soon becomes hard. But it is easy to comprehend a few simple cases, which carry us a good long way; and which will go far to persuade the student that, in other cases which we cannot fully master, the same guiding principle is at the root of the matter.

The bisection of a solid (or its subdivision in other definite proportions) soon leads us into a geometry which, if not necessarily difficult, is apt to be unfamiliar; but in such problems we can go some way, and often far enough for our purpose, if we merely consider the plane geometry of a side or section of our figure. For instance, in the case of the cube which we have just been considering, and in the case of the plane and cylindrical partitions by which it has been divided, it is obvious, since these two partitions extend symmetrically from top to bottom of our cube, that we need only have considered

the manner in which they subdivide the *base* of the cube; in short
the problem of the solid, up to a certain point, is contained in our
plane diagram of Fig. 221. And when our particular solid is a
solid of revolution, then it is equally obvious that a study of its
plane of symmetry (that is to say any plane passing through its
axis of rotation) gives us the solution of the whole problem. The
right cone is a case in point, for here the investigation of its modes
of symmetrical subdivision is completely met by an examination
of the isosceles triangle which constitutes its plane of symmetry.

The bisection of an isosceles triangle by a line which shall be the
shortest possible is an easy problem; for it is obvious that, if the
triangle be low, a vertical partition will be shortest; if it be high,
a horizontal one; if it be equilateral, the partition may run parallel
to any side; and if it be right-angled, the partition may bisect the
right angle or run parallel to either side equally well.

Let ABC be an isosceles triangle of which A is the apex; it may
be shewn that, for its shortest line of bisection, we are limited to
three cases: viz. to a vertical line AD, bisecting the angle at A and
the side BC; to a transverse line parallel to the base BC; or to an
oblique line parallel to AB or to AC. The lengths of these partition
lines follow at once from the magnitudes of the angles of our triangle.
We know, to begin with, since the areas of similar figures vary as
the squares of their linear dimensions, that, in order to bisect the
area, a line parallel to one side of our triangle must always have
a length equal to $1/\sqrt{2}$ of that side. If then, we take our base,
BC, in all cases of a length $= 2$, the transverse partition, EF, drawn
parallel to it will always have a length equal to $2/\sqrt{2}$, or $= \sqrt{2}$.
The vertical partition, AD, since $BD = 1$, will always equal $\tan \beta$;
and the oblique partition, GH, being equal to $AB/\sqrt{2}, = 1/\sqrt{2} \cos \beta$.
If then we call our vertical, transverse and oblique partitions V,
T, and O, we have $V = \tan \beta$; $T = \sqrt{2}$; and $O = 1/\sqrt{2} \cos \beta$, or

$$V : T : O = \tan \beta/\sqrt{2} : 1 : 1/2 \cos \beta.$$

And, working out these equations for various values of β, we soon
see that the vertical partition (V) is the least of the three until
$\beta = 45°$, at which limit V and O are each equal to $1/\sqrt{2} = 0\cdot707$;
that O then becomes the least of the three, and remains so until

$\beta \coloneqq 60°$, when $\cos \beta = 0.5$, and $O = T$; after which T (whose value always $= 1$) is the shortest of the three partitions. And, as we have

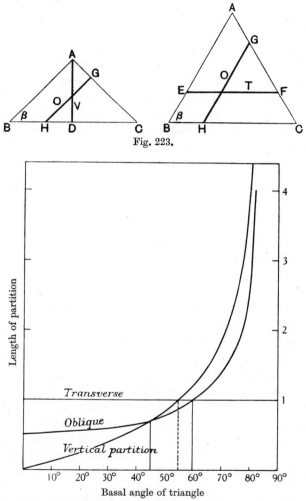

Fig. 223.

Fig. 224.   Comparative length of the partitions, transverse, oblique or vertical, bisecting an isosceles triangle.

seen, these results are at once applicable, not only to the case of the plane triangle, but also to that of the conical or pyramidal cell.

In like manner, if we have a spheroidal body less than a hemisphere, such for instance as a low, watchglass-shaped cell (Fig.
225, A), it is obvious that the smallest partition by which we can
divide it into two halves is (as in our flattened disc) a median
vertical one; and likewise, the hemisphere itself can be bisected
by no smaller partition meeting the walls at right angles than that
median one which divides it into two similar quadrants of a sphere.
But if we produce our hemisphere into a more elevated conical
body, or into a cylinder with spherical cap, there comes a point
where a transverse horizontal partition will bisect the figure with
less area of partition-wall than a median vertical one (C). And
furthermore, there will be an intermediate region, a region where
height and base have their relative dimensions nearly equal (as

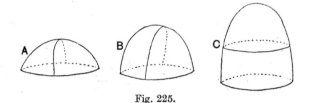

Fig. 225.

in B), where an oblique partition will be better than either the
vertical or the transverse; though here the analogy of our triangle
does not suffice to give us the precise limiting values.

We need not examine these limitations in detail, but we must
look at the curvatures which accompany the several conditions. We
have seen that a film tends to set itself at equal angles to the surface
which it meets, and therefore, when that surface is a solid, to meet
it (or its tangent) at right angles. Our *vertical* partition is, therefore, a plane surface, everywhere normal to the original cell-walls.
But in the taller, conical cell with transverse partition, the latter
still meets the opposite sides of the cell at right angles, and it
follows that it must itself be curved; moreover, since the tension,
and therefore the curvature, of the partition is everywhere uniform,
it follows that its curved surface must be a portion of a sphere,
concave towards the apex of the original cell. In the intermediate
case, where we have an oblique partition meeting both the base
and the curved sides of the mother-cell, the contact must still be

everywhere at right angles: provided we continue to suppose that
the walls of the mother-cell (like those of our diagrammatic cube)
have become practically rigid before the partition appears, and are
therefore not affected and deformed by the tension of the latter.
In such a case, and especially when the cell is elliptical in cross-
section or still more complicated in form, the partition may have
to assume a complex curvature in order to remain a surface of
minimal area.

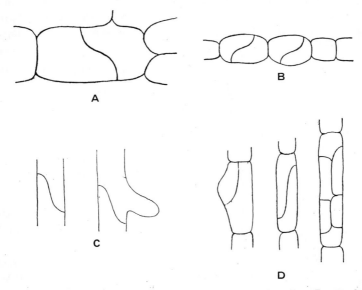

Fig. 226. S-shaped partitions: A, *Taonia atomaria* (after Reinke); B, paraphyses
of *Fucus*; C, rhizoids of moss; D, paraphyses of *Polytrichum*.

While in very many cases the partitions (like the walls of the
original cell) will be either plane or spherical, a more complex
curvature will sometimes be assumed.   It will be apt to occur when
the mother-cell is irregular in shape, and one particular case of
such asymmetry will be that in which (as in Fig. 227) the cell has
begun to branch before division takes place.   And again, whenever
we have a marked internal asymmetry of the cell, leading to irregular
and anomalous modes of division, in which the cell is not necessarily
divided into two equal halves and in which the partition-wall may

assume an oblique position, then equally anomalous curvatures will tend to make their appearance*.

Suppose an oblong cell to divide by means of an oblique partition (as may happen through various causes or conditions of asymmetry), such a partition will still have a tendency to set itself at right angles to the rigid walls of the mother-cell: and it follows that our oblique partition, throughout its whole extent, will assume the form of a complex, saddle-shaped or anticlastic surface.

Many such partitions of complex or double curvature exist, but they are not always easy of recognition, nor do they often appear in a *terminal* cell. We may see them in the roots (or rhizoids) of

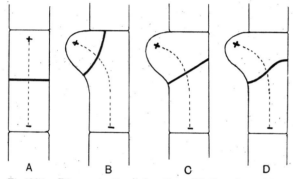

Fig. 227.   Diagrammatic explanation of **S**-shaped partition.

mosses, especially at the point of development of a new rootlet (Fig. 226, C); and again among mosses, in the "paraphyses" of the male plants (e.g. in *Polytrichum*), we find more or less similar partitions (D). They are frequent also among Fuci, as in the hairs or paraphyses of *Fucus* itself (B). In *Taonia atomaria*, as figured in Reinke's memoir on the Dictyotaceae of the Gulf of Naples†, we see, in like manner, oblique partitions, which on more careful examination are seen to be curves of double curvature (Fig. 226, A).

The physical cause and origin of these **S**-shaped partitions is somewhat obscure, but we may attempt a tentative explanation. When we assert a tendency for the cell to divide transversely to its long axis, we are not only stating empirically that the partition

---

* Cf. Wildeman, *Attache des cloisons,* etc., pls. 1, 2.
† *Nova Acta K. Leop. Akad.* xi, 1, pl. iv.

tends to appear in a small, rather than a large cross-section of the cell: but we are also ascribing to the cell a longitudinal *polarity* (Fig. 227, A), and implicitly asserting that it tends to divide (just as the segmenting egg does), by a partition transverse to its polar axis. Such a polarity may conceivably be due to a chemical asymmetry, or anisotropy, such as we have learned of (from Macallum's experiments) in our chapter on Adsorption. Now if the chemical concentration, on which this anisotropy or polarity (by hypothesis) depends, be unsymmetrical, one of its poles being as it were deflected to one side where a little branch or bud is being (or about to be) given off—all in precise accordance with the adsorption phenomena described on p. 460—then our "polar axis" would necessarily be a curved axis, and the partition, being constrained (again *ex hypothesi*) to arise transversely to the polar axis, would lie obliquely to the *apparent* axis of the cell (as in B or C). And if the oblique partition be so situated that it has to meet the *opposite* walls (as in C), then, in order to do so symmetrically (i.e. either perpendicularly, as when the cell-wall is already solidified, or at least at equal angles on either side), it is evident that the partition, in its course from one side of the cell to the other, must necessarily assume a more or less **S**-shaped curvature (D).

The complex curvature of the partition-walls in such cases as these may be illustrated by the following experiment. Set two plates of glass (as in Fig. 228) in a wire frame, so that they may lie parallel or at any angle to one another; and dip the whole thing in soap-solution, so that a sheet of film is formed between the two plates and is framed by the two wires which carry them. The film is, of course, a surface of minimal area; its

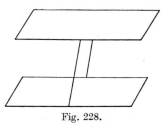

Fig. 228.

mean curvature is constant everywhere, and (since the film is an open surface with identical pressure on both sides) the mean curvature is everywhere *nil*. A related condition is that the film must meet its solid framework, glass or wire, everywhere at right angles or "orthogonally"; and this last constraint leads to curvatures of extreme complexity, which continually vary as we rotate one plate on the plane of the other.

As a matter of fact, while we have abundant simple illustrations of the principles which we have now begun to study, apparent exceptions to this simplicity, due to an asymmetry of the cell itself or of the system of which the single cell is a part, are by no means rare. We know that in cambium-cells division often takes place parallel to the long axis of the cell, though a partition of much less area would suffice if it were set cross-ways: and it is only when a considerable disproportion has been set up between the length and breadth of the cell that the balance is in part redressed by the appearance of a transverse partition. It was owing to such exceptions that Berthold was led to qualify and even to depreciate the importance of the law of minimal areas as a factor in cell-division, after he himself had done so much to demonstrate and elucidate it*. He was deeply and rightly impressed by the fact that other forces besides surface tension, both external and internal to the cell, play their part in determining its partitions, and that the answer to our problem is not to be given in a word. How fundamentally important it is, however, in spite of all conflicting tendencies and apparent exceptions, we shall see better and better as we proceed.

But let us leave the exceptions and consider the simpler and more general phenomena. And let us leave the case of the cubical, quadrangular or cylindrical cell, and examine that of a spherical cell and of its successive divisions, or the still simpler case of a circular, discoidal cell.

When we attempt to investigate mathematically the place and form of a partition of minimal area, it is plain that we shall be dealing with comparatively simple cases wherever even one dimension of the cell is much less than the other two. Where two dimensions are small compared with the third, as in a thin cylindrical filament like that of *Spirogyra*, we have the problem at its simplest; for it is obvious, then, that the partition must lie transversely to the long axis of the thread. But even where one dimension only is relatively small, as for instance in a flattened plate, our problem

* Cf. *Protoplasmamechanik*, p. 229: "Insofern liegen also die Verhältnisse hier wesentlich anders als bei der Zertheilung hohler Körperformen durch flüssige Lamellen. Wenn die Membran bei der Zelltheilung die von dem Prinzip der kleinsten Flächen geforderte Lage und Krümmung annimmt, so werden wir den Grund dafür in andrer Weise abzuleiten haben."

is so far simplified that we see at once that the partition cannot be parallel to the extended plane, but most cut the cell, somehow, at right angles to that plane. In short, the problem of dividing a much flattened solid becomes identical with that of dividing a simple *surface* of the same form.

There are a number of small algae growing in the form of small flattened discs, and consisting (for a time at any rate) of but a single layer of cells, which, as Berthold shewed, exemplify this comparatively simple problem; and we shall find presently that it is admirably illustrated in the cell-divisions which occur in the egg of a frog or a sea-urchin, when it is flattened out under artificial pressure. These same little algae which serve to exemplify the partitioning of a disc also illustrate, now and then, a curious feature of its contour. Such a small green alga as *Castagna* (Fig. 229) shews, and many Desmids shew just as well, a sinuous border running out into rounded crenations or lobes. This is a surface-tension phe-nomenon. A little milk poured over an apple-pie gives a homely illustration of the same sinuous outlines; a drop on a greasy plate spreads in the same uneven

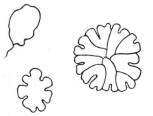

Fig. 229. *Castagna polycarpa.* Swarm-spore and young plants. After Berthold.

way, and does so indeed unless the utmost care be taken to ensure absolute cleanliness and surface equilibrium*.

Fig. 230† represents younger and older discs of the little alga *Erythrotrichia discigera*; and it will be seen that in all stages save the first we have an arrangement of cell-partitions which looks somewhat complex, but into which we must attempt to throw some light and order. Starting with the original single, and flattened, cell, we have no difficulty with the first two cell-divisions; for we know that no bisecting partitions can possibly be shorter than the two diameters, which divide the cell into halves and into quarters. We have only to remember that, for the sum total of partitions to

---

* Cf. Quincke's "Ausbreitungserscheinungen," in *Poggendorff's Annalen*, cxxxix, p. 37, 1870; also Tomlinson's papers in *Phil. Mag.* viii–xxxix; and Van der Mensbrugghe, *Mém. Cour. de l'Acad. R. Belgique*, xxxiv, 1870; xxxvii, 1873.

† From Berthold's *Monograph of the Naples Bangiaceae*, 1882.

be a minimum, three only must meet in a point; and therefore, the four quadrantal walls must shift a little, producing the usual little median partition, or cross-furrow, instead of one common central point of junction. This intermediate partition, however,

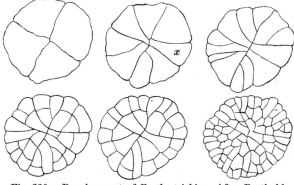

Fig. 230.   Development of *Erythrotrichia*.   After Berthold.

will be small, and to all intents and purposes we may deal with the case as though we had now to do with four equal cells, each one of them a perfect quadrant; so our problem is, to find the shortest line which shall divide the quadrant of a circle into two halves of

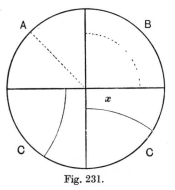

Fig. 231.

equal area. A radial partition (Fig. 231, A), starting from the apex of the quadrant, is at once excluded, for the reason just referred to; our choice must lie between two modes of division such as are illustrated in Fig. 231, where the partition is either (as in B) concentric with the outer border of the cell, or else (as in C) cuts that outer border; in other words, our partition may (B) cut *both* radial walls, or (C) may cut *one* radial wall and the periphery. These are the two methods of division which Sachs called, respectively, (B) *periclinal*, and (C) *anticlinal**. We may either treat the walls of the dividing quadrant as already solidified, or at least as having a tension compared with which that of the incipient partition film is inconsiderable; in either case the new partition must meet the old wall, on either side, at right angles, and (its own tension and curvature being everywhere uniform) must take the form of a circular arc.

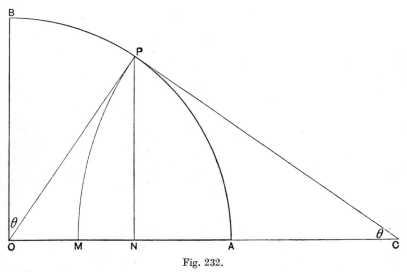

Fig. 232.

We find that a flattened cell which is approximately a quadrant of a circle invariably divides after the manner of Fig. 231, C, that is to say, by an approximately circular, *anticlinal* wall, and this we now recognise in the eight-celled stage of *Erythrotrichia* (Fig. 230); let us then consider that Nature has solved our problem, and let us work out the actual geometric conditions.

Let the quadrant *OAB* (in Fig. 232) be divided into two parts of equal area, by the circular arc *MP*. It is required to determine

---

* There is, I think, some ambiguity or disagreement among botanists as to the use of this latter term: the sense in which I am using it, viz. for any partition which meets the outer or peripheral wall at right angles (the strictly *radial* partition being for the present excluded), is, however, clear.

(1) the position of P upon the arc of the quadrant, that is to say the angle BOP; (2) the position of the point M on the side OA; and (3) the length of the arc MP in terms of a radius of the quadrant.

(1) Draw OP; also PC a tangent, meeting OA in C; and PN, perpendicular to OA. Let us call a a radius; and θ the angle at C, which is equal to OPN, or POB. Then

$$CP = a \cot \theta; \quad PN = a \cos \theta; \quad NC = CP \cos \theta = a \cdot \cos^2 \theta / \sin \theta.$$

The area of the portion PMN

$$= \tfrac{1}{2} CP^2 \, \theta - \tfrac{1}{2} PN \cdot NC$$
$$= \tfrac{1}{2} a^2 \, \theta \cot^2 \theta - \tfrac{1}{2} a \cos \theta \cdot a \cos^2 \theta / \sin \theta$$
$$= \tfrac{1}{2} a^2 \, (\theta \cot^2 \theta - \cos^3 \theta / \sin \theta).$$

And the area of the portion PNA

$$= \tfrac{1}{2} a^2 \, (\pi/2 - \theta) - \tfrac{1}{2} ON \cdot NP$$
$$= \tfrac{1}{2} a^2 \, (\pi/2 - \theta) - \tfrac{1}{2} a \sin \theta \cdot a \cos \theta$$
$$= \tfrac{1}{2} a^2 \, (\pi/2 - \theta - \sin \theta \cdot \cos \theta).$$

Therefore the area of the whole portion PMA

$$= a^2/2 \, (\pi/2 - \theta + \theta \cot^2 \theta - \cos^3 \theta / \sin \theta - \sin \theta \cdot \cos \theta)$$
$$= a^2/2 \, (\pi/2 - \theta + \theta \cot^2 \theta - \cot \theta),$$

and also, by hypothesis, $= \tfrac{1}{2} \cdot$ area of the quadrant, $= \pi a^2/8$.
Hence θ is defined by the equation

$$a^2/2 \, (\pi/2 - \theta + \theta \cot^2 \theta - \cot \theta) = \pi a^2/8,$$

or
$$\pi/4 - \theta + \theta \cot^2 \theta - \cot \theta = 0.$$

We may solve this equation by constructing a table (of which the following is a small portion) for various values of θ.

θ	$\pi/4$	$-\theta$	$-\cot \theta$	$+\theta \cot^2 \theta$	$=x$
34° 34′	0·7854	$-$ 0·6033	$-$ 1·4514	$+$ 1·2709 $=$	0·0016
35′	0·7854	0·6036	1·4505	1·2700	0·0013
36′	0·7854	0·6039	1·4496	1·2690	0·0009
37′	0·7854	0·6042	1·4487	1·2680	0·0005
38′	0·7854	0·6045	1·4478	1·2671	0·0002
39′	0·7854	0·6048	1·4469	1·2661	$-$0·0002
40′	0·7854	0·6051	1·4460	1·2652	$-$0·0005

We see accordingly that the equation is solved (as accurately as need be) when θ is an angle somewhat over 34° 38′, or say

34° 38½′. That is to say, a quadrant of a circle is bisected by a circular arc cutting the side and the periphery of the quadrant at right angles, when the arc is such as to include (90° − 34° 38′), i.e. 55° 22′ of the quadrantal arc. This determination of ours is practically identical with that which Berthold arrived at by a rough and ready method, without the use of mathematics. He simply tried various ways of dividing a quadrant of paper by means of a circular arc, and went on doing so till he got the weights of his two pieces of paper approximately equal. The angle, as he thus determined it, was 34·6°, or say 34° 36′.

(2)   The position of M on the side of the quadrant OA is given by the equation $OM = a \operatorname{cosec} \theta - a \cot \theta$; the value of which expression, for the angle which we have just discovered, is 0·3028. That is to say, the radius (or side) of the quadrant will be divided by the new partition into two parts, in the proportions, nearly, of three to seven.

(3)   The length of the arc MP is equal to $a\theta \cot \theta$; and the value of this for the given angle is 0·8751. This is as much as to say that the curved partition-wall which we are considering is shorter than a radial partition in the proportion of $8\frac{3}{4}$ to 10, or seven-eighths, almost exactly.

But we must also compare the length of this curved anticlinal partition-wall (MP) with that of the concentric, or periclinal, one (RS, Fig. 233) by which the quadrant might also be bisected. The length of this partition is obviously equal to the arc of the quadrant (i.e. the peripheral wall of the cell) divided by $\sqrt{2}$; or, in terms of the radius, $= \pi/2 \sqrt{2} = 1\cdot111$. So that, not only is the anticlinal partition (such as we actually find in nature) notably the best, but the periclinal one, when it comes to dividing

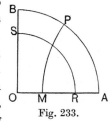

Fig. 233.

an entire quadrant, is very considerably larger even than a radial partition.

The two cells into which our original quadrant is now divided are equal in volume, but of very different shapes; the one is a triangle (MAP) with circular arcs for two of its sides, and the other is a four-sided figure ($MOBP$), which we may call approximately oblong. How will they continue to divide? We cannot say as

yet how the triangular portion ought to divide; but it is obvious
that the least possible partition-wall which shall bisect the other
must run across the long axis of the oblong, that is to say periclinally.
This is precisely what tends actually to take place.   In the following
diagrams (Fig. 234) of a frog's egg dividing under pressure, that
is to say when reduced to the form of a flattened plate, we see,
firstly (A), the division into four quadrants (by the partitions 1, 2);
secondly (B), the division of each quadrant by means of an anticlinal circular arc (3, 3), cutting the peripheral wall of the quadrant
approximately in the proportions of three to seven;   and thirdly

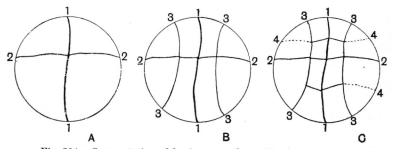

Fig. 234.   Segmentation of frog's egg, under artificial compression.
After Roux.

(C), we see that of the eight cells (four triangular and four oblong)
into which the whole egg is now divided, the four which we have
called oblong now proceed to divide by partitions transverse to
their long axes, or roughly parallel to the periphery of the egg.

The question how the other, or triangular, portion of the divided
quadrant will next divide leads us to a well-defined problem which
is only a slight extension, making allowance for the circular arcs,
of that elementary problem of the triangle we have already considered. We know now that an entire quadrant (in order that its
bisecting wall shall have the least possible area) must divide by
means of an anticlinal partition, but how about any smaller sectors
of circles? It is obvious in the case of a small prismatic sector,
such as that shewn in Fig. 235, that a *periclinal* partition is the
least by which we can bisect the cell; we want, accordingly, to
know the limits below which the periclinal partition is always the

best, and above which the anticlinal arc has the advantage, as in the case of the whole quadrant.

This may be easily determined; for the preceding investigation is a perfectly general one, and the results hold good for sectors of any other arc, as well as for the quadrant, or arc of 90°. That is to say, the length of the partition-wall MP is always determined by the angle θ, according to our equation $MP = a\theta \cot \theta$; and the angle θ has a definite relation to α, the angle of arc.

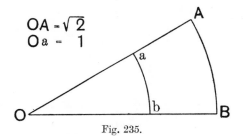

$OA = \sqrt{2}$
$Oa = 1$

Fig. 235.

Moreover, in the case of the periclinal boundary, RS (Fig. 233) (or ab, Fig. 235), we know that, if it bisects the cell,

$$RS = a \cdot \alpha/\sqrt{2}.$$

Accordingly, the arc RS will be just equal to the arc MP when

$$\theta \cot \theta = \alpha/\sqrt{2}.$$

When $\theta \cot \theta > \alpha/\sqrt{2}$, or $MP < RS$,

then division will take place as in RS, or periclinally.

When $\theta \cot \theta < \alpha/\sqrt{2}$, or $MP > RS$,

then division will take place as in MP, or anticlinally.

In the accompanying diagram (Fig. 236), I have plotted the various magnitudes with which we are concerned, in order to exhibit the several limiting values. Here we see, in the first place, the curve marked α, which shews on the (left-hand) vertical scale the various possible magnitudes of that angle (viz. the angle of arc of the whole sector which we wish to divide), and on the horizontal scale the corresponding values of θ, or the angle which determines the point on the periphery where it is cut by the partition-wall,

MP. Two limiting cases are to be noticed here: (1) at 90° (point *A* in diagram), because we are at present only dealing with arcs no greater than a quadrant; and (2), the point (*B*) where the angle θ

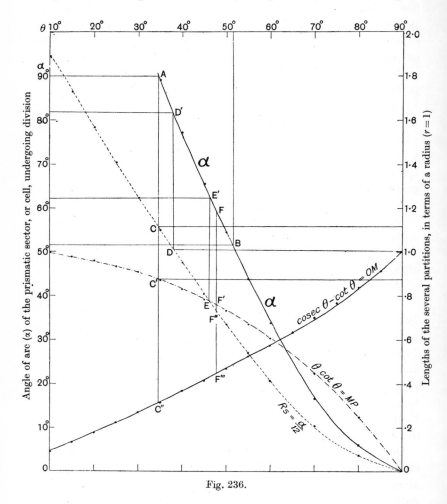

Fig. 236.

comes to equal the angle α, for after that point the construction becomes impossible, since an anticlinal bisecting partition-wall would be partly outside the cell. The only partition which, after that point, can possibly exist is a periclinal one; and this point, as our diagram shews us, occurs when the two angles (α and θ) are both rather under 52°.

Next I have plotted on the same diagram, and in relation to the same scale of angles, the corresponding lengths of the two partitions, viz. RS and MP, their lengths being expressed (on the right-hand side of the diagram) in terms of the radius of the circle (a), that is to say the side wall, OA, of our cell.

The limiting values here are (1), C, C', where the angle of arc is 90°, and where, as we have already seen, the two partition-walls have the relative magnitudes of $MP : RS = 0.875 : 1.111$: (2) the point D, where RS equals unity, that is to say where the periclinal partition has the same length as a radial one; this occurs when α is rather under 82° (cf. the points D, D'): (3) the point E, where RS and MP intersect, that is to say the point at which the two partitions, periclinal and anticlinal, are of the same magnitude; this is the case, according to our diagram, when the angle of arc is just over $62\frac{1}{2}$°. We see from this that what we have called an anticlinal partition, as MP, is only likely to occur in a triangular or prismatic cell whose angle of arc lies between 90° and $62\frac{1}{2}$°; in all narrower or more tapering cells the periclinal partition will be of less area, and will therefore be more and more likely to occur.

The case (F) where the angle α is just 60° is of some interest. Here, owing to the curvature of the peripheral border, and the consequent fact that the peripheral angles are somewhat greater than the apical angle α, the periclinal partition has a very slight and almost imperceptible advantage over the anticlinal, the relative proportions being about as $MP : RS :: 0.73 : 0.72$. But if the triangle be a plane equiangular triangle, bounded by circular arcs, then we see that there is no longer any distinction at all between our two partitions; MP and RS are now identical.

On the same diagram, I have inserted the curve for values of $\mathrm{cosec}\ \theta - \cot \theta = OM$, that is to say the distances from the centre, along the side of the cell, of the starting-point (M) of the anticlinal partition. The point C'' represents its position in the case of a quadrant, and shews it to be (as we have already said) about 3/10 of the length of the radius from the centre. If on the other hand our cell be an equilateral triangle, then we have to read off the point on this curve corresponding to $\alpha = 60°$; and we find it at the point F''' (vertically under F), which tells us that the partition now starts 45/100, or nearly halfway, along the radial wall.

The foregoing considerations carry us a long way in our investigation of the simpler forms of cell-division. Strictly speaking they are limited to the case of flattened cells, in which we can treat the problem as though we were partitioning a plane surface. But it is obvious that, though they do not teach us the whole conformation of the partition which divides a more complicated solid into two halves, yet, even in such a case they so far enlighten us as to tell us the appearance presented in one plane of the actual solid. And, as this is all that we see in a microscopic section, it follows that the results we have arrived at will help us greatly in the interpretation of microscopic appearances, even in comparatively complex cases of cell-division.

Let us now return to our quadrant cell ($OAPB$), which we have found to be divided into a triangular and a quadrilateral portion, as in Figs. 233 or 237; and let us now suppose the whole system to grow, in a uniform fashion, as a prelude to further subdivision. The whole quadrant, growing uniformly (or with equal radial increments), will still remain a quadrant, and it is obvious, therefore, that for every new increment of size, more will be added to the margin of its triangular portion than to the narrower margin of the quadrilateral; and the increments will be in proportion to

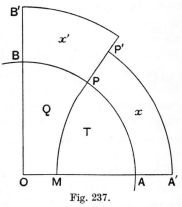

Fig. 237.

the angles of arc, viz. 55° 22′ : 34° 38′, or as 0·96 : 0·60, i.e. as 8 : 5. Accordingly, if we may assume (and the assumption is a very plausible one), that, just as the quadrant itself divided into two halves after it got to a certain size, so each of its two halves will reach the same size before again dividing, it is obvious that the triangular portion will be doubled in size, and therefore ready to divide, a considerable time before the quadrilateral part. To work out the problem in detail would lead us into troublesome mathematics; but if we simply assume that the increments are proportional to the increasing radii of the circle, we have the following equations:

Call the triangular cell T, and the quadrilateral Q (Fig. 237); let the radius, OA, of the original quadrantal cell $= a = 1$; and let the increment which is required to add on a portion equal to T (such as $PP'A'A$) be called x, and let that required, similarly, for the doubling of Q be called x'.

Then we see that the area of the original quadrant

$$= T + Q = \tfrac{1}{4}\pi a^2 = 0.7854a^2,$$

while the area of T $= Q = 0.3927a^2.$

The area of the enlarged sector, $P'OA'$,

$$= (a + x)^2 \times (55° \; 22') \div 2 = 0.4831 \, (a + x)^2,$$

and the area OPA

$$= a^2 \times (55° \; 22') \div 2 = 0.4831a^2.$$

Therefore the area of the added portion, T',

$$= 0.4831 \, \{(a + x)^2 - a^2\}.$$

And this, by hypothesis,

$$= T = 0.3927a^2.$$

We get, accordingly, since $a = 1$,

$$x^2 + 2x = 0.3927/0.4831 = 0.810,$$

and, solving,

$$x + 1 = \sqrt{1.81} = 1.345, \text{ or } x = 0.345.$$

Working out x' in the same way, we arrive at the approximate value, $x' + 1 = 1.517$.

This is as much as to say that, supposing each cell tends to divide into two halves when (and not before) its original size is doubled, then, in our flattened disc, the triangular cell T will tend to divide when the radius of the disc has increased by about a third (from 1 to 1·345), but the quadrilateral cell, Q, will not tend to divide until the linear dimensions of the disc have increased by fully a half (from 1 to 1·517).

The case here illustrated is of no small importance. For it shews us that a uniform and symmetrical growth of the organism (symmetrical, that is to say, under the limitations of a plane surface, or plane section) by no means involves a uniform or symmetrical growth of the individual cells, but may under certain conditions actually lead to inequality among these; and this phenomenon (or to be quite candid, this hypothesis, which is due to Berthold) is independent of any change or variation in surface tensions, and is essentially different from that unequal segmentation (studied by Balfour) to which we have referred on p. 568.

After this fashion we might go on to consider the manner, and the order of succession, in which subsequent cell-divisions should tend to take place, as governed by the principle of minimal areas.

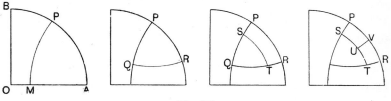

Fig. 238.

The calculations would grow more difficult, and the results got by simple methods would grow less and less exact; at the same time some of the results would be of great interest, and well worth our while to obtain. For instance, the precise manner in which our triangular cell, T, would next divide would be interesting to know, and a general solution of this problem is certainly troublesome to calculate. But in this particular case we see that the width of the triangular cell near P (Fig. 238) is so obviously less than that near either of the other two angles, that a circular arc cutting off that angle is bound to be the shortest possible bisecting line; and that, in short, our triangular cell will tend to subdivide, just like the original quadrant, into a triangular and a quadrilateral portion.

But the case will be different next time, because in this new triangle, PRQ, the least width is near the innermost angle, that at Q; and the bisecting circular arc will therefore be opposite to Q, or (approximately) parallel to PR. The importance of this fact is at once evident; for it means to say that there comes a time when, whether by the division of triangles or of quadrilaterals, we find only quadrilateral cells adjoining the periphery of our circular disc. In the subsequent division of these quadrilaterals, the partitions will arise transversely to their long axes, that is to say, *radially* (as U, V); and we shall consequently have a superficial or peripheral layer of quadrilateral cells, with sides approximately parallel, that is to say what we are accustomed to call *an epidermis*. And this epidermis or superficial layer will be in clear contrast with the more irregularly shaped cells, the products of triangles and quadrilaterals, which make up the deeper, underlying layers of tissue.

In following out these theoretic principles, and others like to them, in the actual division of living cells, we must bear in mind certain conditions and qualifications. In the first place, the law of minimal area and the other rules which we have arrived at are not absolute but relative: they are links, and very important links, in a chain of physical causation; they are always at work, but their effects may be overridden and concealed by the operation of other forces. Secondly, we must remember that, in most cases, the cell-system which we have in view is constantly increasing in magnitude by active growth; and by this means the form and also the proportions of the cells are continually altering, of which phenomenon we have already had an example. Thirdly, we must carefully remember that, until our cell-walls become absolutely solid and rigid, they are always apt to be modified in form owing to the tension of the adjacent walls; and again, that so long as our partition films are fluid or semifluid, their points and lines of contact with one another may shift, like the shifting outlines of a system of soap-bubbles. This is the physical cause of the movements frequently seen among segmenting cells, like those to which Rauber called attention in the segmenting ovum of the frog, and like those more striking movements or accommodations which give rise to a so-called "spiral" type of segmentation.

Bearing in mind these considerations, let us see what our flattened disc is likely to look like, after a few successive cell-divisions. In Fig. 239 *a*, we have a diagrammatic representation of our disc, after it has divided into four quadrants, and each quadrant into a triangular and a quadrilateral portion; but as yet, this figure has scarcely anything like the normal look of an aggregate of living cells. But let us go a little further, still limiting ourselves to the consideration of the eight-celled stage. Wherever one of our radiating partitions meets the peripheral wall, there will (as we know) be a mutual tension between the three convergent films, which will tend to set their edges at equal angles to one another, angles that is to say of 120°. In consequence of this, the outer wall of each individual cell will (in this surface view of our disc) be an arc of a circle of which we can determine the centre by the method used on p. 485; and, furthermore, the narrower cells, that is to say the quadrilaterals, will have this outer border somewhat more

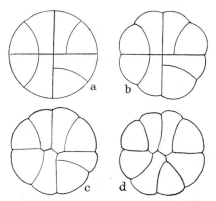

Fig. 239.   Diagram of flattened or discoid cell dividing into octants: to shew
gradual tendency towards a position of equilibrium.

curved than their broader neighbours. We arrive, then, at the condition shewn in Fig. 239 b. Within the cell, also, wherever wall meets wall, the angle of contact must tend, in every case, to be an angle of 120°; in no case may more than three films (as seen in section) meet in a point (c); and this condition, of the partitions meeting three by three and at co-equal angles, will involve the curvature of some, if not all, of the partitions (d) which to begin with we treated as plane. To solve this problem in a general way is no easy matter; but it is a problem which Nature solves in every case where, as in the case we are considering, eight bubbles or eight cells meet together in a plane or curved surface. An approximate solution has been given in d; and it will at once be recognised that this figure has vastly more resemblance to an aggregate of living cells than had the diagram of *a*, with which we began.

Just as we have constructed in this case a series of purely diagrammatic or schematic figures, so will it be possible as a rule to diagrammatise, with but little alteration, the complicated appearances presented by any ordinary aggregate of cells. The accompanying little figure (Fig. 240), of a germinating spore of a Liverwort (*Riccia*), after a drawing of D. H. Campbell's, scarcely needs further explanation: for it is well-nigh a typical diagram of the method of space-partitioning which we are now considering. The same is equally true of any one of Hanstein's figures of the hairs on a leaf-bud*, or Berthold's of the small discoid algae. Let us look again at our figures of *Erythrotrichia* or *Chaetopeltis* from Berthold's *Monograph*, and redraw some of the earlier stages.

Fig. 240.

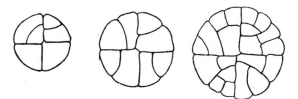

Fig. 241.   Embryo-stages of *Chaetopeltis orbicularis*.   After Berthold.

In the following diagrams (Fig. 242) the new partitions, or those just about to form, are in each case outlined; and in the next succeeding stage they are shewn after settling down into position, and after exercising their respective tractions on the walls previously laid down.   It is clear, I think, that these four diagrammatic figures represent all that is shewn in the first five stages drawn by Berthold from the plant itself; but the correspondence cannot in this case be precisely accurate, for the reason that Berthold's figures are taken from different individuals, and so are not strictly and consecutively continuous.   The last of the six drawings in Fig. 230 is already too complicated for diagrammatisation, that is to say it is too complicated for us to decipher with certainty the order of appearance of the numerous partitions which it contains.   But in Fig. 243 I shew one more diagrammatic figure, of a disc which has

* *Bot. Zeitung*, xxvi, p. 11, xi, xii, 1868.

divided, according to the theoretical plan, into about sixty-four
cells; and making due allowance for the changes which mutual
tensions and tractions bring about, increasing in complexity with
each succeeding stage, we can see, even at this advanced and

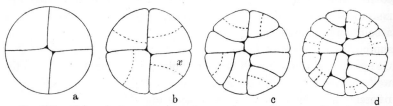

Fig. 242.   Theoretical arrangement of successive partitions in a discoid cell;
for comparison with Figs. 230 and 241.

complicated stage, a very considerable resemblance between the
actual picture and the diagram which we have here constructed
in obedience to a few simple rules.

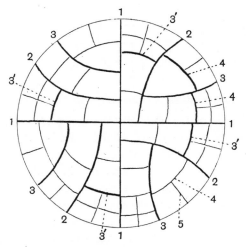

Fig. 243.   Theoretical division of a discoid cell into sixty-four chambers: no
allowance being made for the mutual tractions of the cell-walls.

In like manner, in the annexed figures representing sections
through a young embryo of a moss, we have little difficulty in
discerning the successive stages which must have intervened between
the two stages shewn: so as to lead from the just divided or dividing
quadrants (a), to the stage (b) in which a well-marked epidermal

layer surrounds an at first sight irregular agglomeration of "funda-mental tissue".

In the last paragraph but one, I have spoken of the difficulty of so arranging the meeting-places of a number of cells that at each junction only three cell-walls shall meet in a point, and all three shall meet at equal angles of 120°. As a matter of fact, the problem is soluble in a number of ways; that is to say, when we have a number of cells enclosed in a common boundary, say eight as in the case considered, there are various ways in which their walls may meet internally, three by three, at equal angles; and these differences will entail differences also in the curvature of the walls, and consequently in the shape of the cells. The question is some-

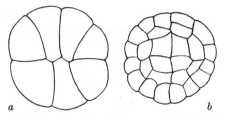

a   *b*

Fig. 244. Sections of embryo of a moss. After Kienitz-Gerloff.

what complex; it has been dealt with by Plateau, and treated mathematically by M. Van Rees*.

If within our boundary we have only three cells all meeting internally, they must meet in a point; furthermore, they tend to do so to equal angles of 120°, and there is an end of the matter. If we have four cells, then, as we have already seen, the conditions are satisfied by interposing a little intermediate wall, the two extremities of which constitute the meeting-points of three cells each, and the upper edge of which marks the "polar furrow." In the case of five cells, we require *two* little intermediate walls, and two polar furrows; and we soon arrive at the rule that, for n cells, we require $n - 3$ little longitudinal partitions (and corre-sponding polar furrows), connecting the triple junctions of the cells†; and these little walls, like all the rest within the system, tend to

* *Cit.* Plateau, *Statique des Liquides*, I, p. 358.

† There is an obvious analogy between this rule for the number of internal partitions within a polygonal system, and Lamarle's rule for the number of "free films" within a polyhedron. *Vide supra*, p. 550.

incline to one another at angles of 120°. Where we have only one
such wall (in the case of four cells), or only two (in the case
of five cells), there is no room for ambiguity. But where we have
three little connecting-walls, in the case of six cells, we can
arrange them in three different ways, as Plateau* found his six

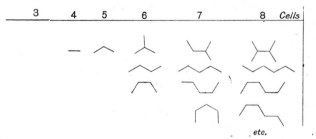

Fig. 245. Various possible arrangements of intermediate positions,
in groups of 3, 4, 5, 6, 7 or 8 cells.

soap-films to do (Fig. 246). In the system of seven cells, the four
partitions can be arranged in four ways; and the five partitions
required in the case of eight cells can be arranged in no less than

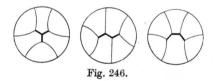

Fig. 246.

twelve different ways†. It does not follow that these various
arrangements are all equally good; some are known to be more
stable than others, and some are hard to realise in actual experiment.

Examples of these various arrangements meet us at every turn,
in all sorts of partitioning, whether there be actual walls or mere

* Plateau experimented with a wire frame or "cage", in the form of a low
hexagonal prism. When this was plunged in soap-solution and withdrawn upright,
a vertical film occupied its six quadrangular sides and nothing more. But when
it was drawn out sideways, six films starting from the six vertical edges met somehow in the middle, and divided the hexagon into six cells. Moreover the partition-
films automatically solved the problem of meeting one another three-by-three, at
co-equal angles of 120°; and did so in more ways than one, which could be controlled
more or less, according to the manner and direction of lifting the cage.

† Plateau, on Van Rees's authority, says thirteen; but this is wrong—unless
he meant to include the case where one cell is wholly surrounded by the seven
others.

rifts and cracks in a broken surface. The phenomenon is in the first instance mathematical, in the second physical; and the limited number of possible arrangements appear and reappear in the most diverse fields, and are capable of representation by the same diagrams. We have seen in Fig. 196 how the cracks in drying mud exhibit to perfection the polar furrow joining two three-way nodes, which is the characteristic feature of the four-celled stage of a segmenting egg.

The possible arrangements of the intermediate partitions becomes a question of permutations. Let us call the flexure between two consecutive furrows a or b, according to its direction, right or left; and let a triple conjunction be called c. Then the three possible arrangements in a system of six cells are aa, ab, c; the four in a system of seven cells are aaa, aab, aba, ac; and the twelve possible arrangements in a system of eight cells are as follows*:

a	aac	f	$aabb$
b	abc	g	$aaba$
c	acb	h	$aaab$
d	aca	i	$aaaa$
e	bcb	j	$abab$
		k	$abba$
	l	cc	

We may classify, and may denote or symbolise, these several arrangements in various ways. In the following table we see: A, the twelve arrangements of the five intermediate partitions which are necessary to enable all the boundary walls of a plane assemblage of eight cells (none being "insular") to meet in three-way junctions; B, the literal permutations which symbolise the same; C, the number of sides (other than the external boundary) which in each case each cell possesses, i.e. the number of contacts each makes with its neighbours. The total number of contacts (as we shall see presently) is 26, and the mean number 3·25; if we take the departures from the mean, and sum them irrespective of sign, the sum is shewn under D.

---

* I believe that Kirkman, in a paper of more than 80 years ago, said that the number of 8-sided convex, Eulerian polyhedra, with trihedral corners, was thirteen.

	A	B	C	D
a		aac	222 33 44 6	8·5
e		aca	222 33 44 6	8·5
c		acb	222 33 4 55	8·5
d		bcb	222 33 4 55	8·5
b		abc	222 3 444 5	8·0
j		abab	22 33 4444	6·0
f		aabb	22 3333 55	7·0
g		aaba	22 333 44 5	6·5
k		abba	22 333 44 5	6·5
h		aaab	22 3333 4 6	7·0
i		aaaa	22 33333 7	7·5
l		cc	2222 44 55	10·0

Nine cells may be arranged in twenty-seven ways. In higher series the numbers increase very rapidly, but the cells will tend to overlap, and so introduce a new complication*.

We may draw help from the theory of polyhedra (in an elementary way) if we treat our group of eight cells (none of them "insular") as part of a polyhedron, to be completed by one eight-sided cell,

* Max Brückner states the number of possible arrangements of thirteen cells, with trihedral junctions, as nearly 50,000; of sixteen, nearly 30 millions; and of eighteen, "bereits über einige Billionen" (*Proc. Math. Congress*, Bologna, 1930, vol. IV, p. 11). It is plain that the study of "cell-lineage," or the mapping out in detail of the cell-arrangements after repeated cell-divisions, is only possible under severe limitations.

serving (so to speak) as a base under all the rest. Then, calling F_3, F_4, etc., the number of three-sided and four-sided facets, we re-classify our twelve configurations as follows:

Polyhedral arrangements of eight cells (none of them "insular"), considered as part of a nine-faced polyhedron, whose ninth face is octagonal.

	F_3	F_4	F_5	F_6	F_7	F_8
l	4	—	2	2	—	1
ae	3	2	2	—	1	1
cd	3	2	1	2	—	1
b	3	1	3	1	—	1
i	2	5	—	—	—	2
h	2	4	1	—	1	1
f	2	4	—	2	—	1
gk	2	3	2	1	—	1
j	2	2	4	—	—	1

It is of interest, and of more than mere mathematical interest, to know, not only that these possible arrangements are few, but that they are strictly defined as to the number and form of the respective faces. For we know that we are limited to three-way corners or nodes; and, that being so, the following simple rule holds for the facets—a rule which we shall use later on in still more curious circumstances, and which may be easily verified in any line of the foregoing table:

$$3F_3 + 2F_4 + F_5 \pm 0.F_6 - F_7 - 2F_8 - \text{etc.} = 12.$$

We may produce and illustrate all these configurations by blowing bubbles in a dish and here (Fig. 247) is the complete series, up to seven cells. They correspond precisely to the diagrams shewn on p. 596, and their resemblance to embryological diagrams is only cloaked a little by the circular outline, the artificial boundary of the system. Of the twelve eight-celled arrangements, four seem unstable; these include the one case (i) where one cell of the eight is in contact with all seven others, and the three cases (a, e, h) where one is in contact with six others. The reason of this insta-bility is, I imagine, that the internal angles cannot be angles of 120°, as equilibrium demands, unless the sides be curved, and convex inwards; but this implies a combined pressure from without on the large cell in the middle. While it adjusts its walls, then, to the required angles, the large cell tends to close up, to lose hold of the boundary of the system, and to become an island-cell entirely surrounded by the rest.

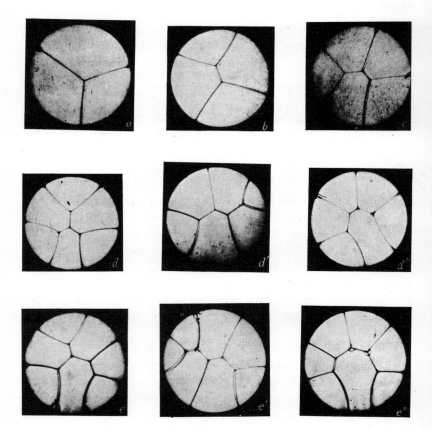

Fig. 247. Group of soap-bubbles, blown in Petri dishes. *a*, *b*, *c*, the normal partitioning of groups of three, four or five cells or bubbles. *d*, the three ways of partitioning a group of six cells or bubbles. *e*, three of the four ways of partitioning a group of seven cells.

Among the published figures of embryonic stages and other cell aggregates, we only discern the little intermediate partitions in cases where the investigator has drawn carefully just what lay before him, without any preconceived notions as to radial or other symmetry; but even in other cases we can often recognise, without much difficulty, what the actual arrangement was whereby the cell-walls met together in equilibrium. I suspect that a leaning towards Sachs's Rule, that one cell-wall tends to set itself at right angles to another cell-wall (a rule whose strict limitations and narrow range of application we have already considered) is responsible for many inaccurate or incomplete representations of the mutual arrangement of associated cells.

In the accompanying series of figures (Figs. 248–255) I have set forth a few aggregates of eight cells, mostly from drawings of

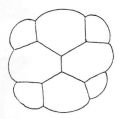

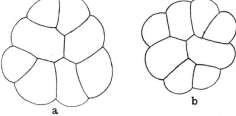

Fig. 248. Segmenting egg of *Trochus*. After Robert.

Fig. 249. Two views of segmenting egg of *Cynthia partita*. After Conklin.

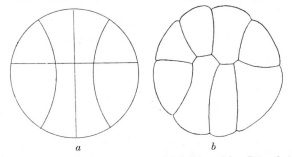

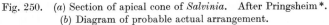

Fig. 250. (*a*) Section of apical cone of *Salvinia*. After Pringsheim *.
(*b*) Diagram of probable actual arrangement.

segmenting eggs. In some cases they shew clearly the manner in which the cell-walls meet one another, always by three-way junctions, at angles of about 120°, more or less, and always with the help of five intermediate boundary walls within the eight-celled system; in other cases I have added a slightly altered drawing, so as to shew, with as little change as possible, the arrangement of boundaries

* This, like many similar figures, is manifestly drawn under the influence of Sachs's theoretical views, or assumptions, regarding orthogonal trajectories, coaxial circles, confocal ellipses, etc.

which may have existed, and given rise to the appearance which the observer drew. These drawings may be compared with the diagrams on p. 598, in which the twelve possible arrangements of five inter-mediate partitions for a system of eight cells have been set forth.

It will be seen that Robert-Tornow's figure of the segmenting egg of *Trochus* (Fig. 248) clearly shews the cells grouped after the fashion of *l*; while Conklin's figure of the ascidian egg (*Cynthia*) shews equally clearly the arrangement *e*. A sea-urchin egg segmenting

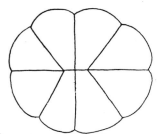

Fig. 251.   Egg of *Pyrosoma*.
After Korotneff.

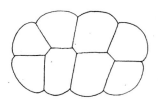

Fig. 252.   Egg of *Echinus*, segmenting
under pressure. After Driesch.

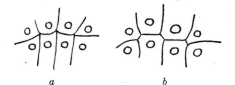

a   b

Fig. 253.   (a) Part of segmenting egg of Cephalopod (after Watase);
(b) probable actual arrangement.

under pressure, as figured by Driesch, scarcely wants any modifica-tion of the drawing to appear in one case as type *f*, in another as *g*. Turning to a botanical illustration, we have a figure of Pringsheim's shewing an eight-celled stage in the apex of the young cone of *Salvinia*: it is ill drawn, but may be referable, as in my diagram, to type *f*; after it is figured a very different object, a segmenting egg of the ascidian *Pyrosoma*, after Korotneff, also, but still more doubtfully, referred to *f*. In the cuttlefish egg there is again some uncertainty, but it is probably referable to *g*. Lastly, I have copied from Roux a curious figure of the frog's egg, viewed from the animal pole; it is obviously inaccurate, but may perhaps belong to type *e*. Of type *i*, in which the five partitions form

four re-entrant angles, that is to say a figure representing the five
sides of a hexagon, and one cell is in touch with seven others,
I have found no examples among published figures of segmenting
eggs.   It is obvious enough, without more ado, that these phenomena

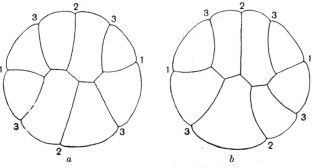

Fig. 254.   (a) Egg of *Echinus*; (b) do. of *Nereis*, under pressure.
After Driesch.

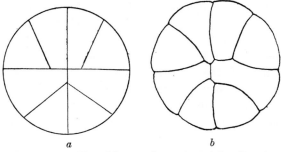

Fig. 255.   (a) Egg of frog, under pressure (after Roux);
(b) probable actual arrangement.

are in the strictest and completest way common to both plants and
animals, in which respect they tally with, and further extend, the
fundamental conclusions laid down by Schwann wellnigh a hundred
years ago, in his *Mikroskopische Untersuchungen über die Ueberein-
stimmung in der Struktur und dem Wachsthum der Thiere und
Pflanzen**.

But now that we have seen how a certain limited number of
types of eight-celled segmentation (or of arrangements of eight
cell-partitions) appear and reappear here and there throughout the
whole world of organisms, there still remains the very important

* Berlin, 1839; Sydenham Society, 1847.

question, whether *in each particular organism* the conditions are such as to lead to one particular arrangement being predominant, characteristic, or even invariable. In short, is a particular arrangement of cell-partitions to be looked upon (as the published figures of the embryologist are apt to suggest) as a *specific character*, or at least a constant or normal character, of the particular organism? The answer to this question is a direct negative, but it is only in the work of the most careful and accurate observers that we find it revealed. Rauber (whom we have more than once had occasion to quote) was one of those embryologists who recorded just what he saw, without prejudice or preconception; as Boerhaave said of Swammerdam, *quod vidit id asseruit*. Now Rauber has put on record a considerable number of variations in the arrangement of the first eight cells, which form a discoid surface about the dorsal (or "animal") pole of the frog's egg. In a certain number of cases these figures are identical with one another in type, identical (that is to say) save for slight differences in magnitude, relative proportions, or orientation. But I have selected (Fig. 256) six diagrammatic figures, which are all *essentially different*, and these diagrams seem to me to bear intrinsic evidence of their accuracy: the curvatures of the partition-walls and the angles at which they meet agree closely with the requirements of theory, and when they depart from theoretical symmetry they do so only to the slight extent which we might expect in a material system*. Of these six illustrations, two are exceptional. In Fig. 256, 5, we observe that one of the eight cells is *insular*, and surrounded by the other seven. This is a perfectly natural condition, and represents, like the rest, a phase of partial or conditional equilibrium; but it is not included in the series we are now considering, which is restricted to the case of eight cells extending outwards to a common boundary. The condition shewn in Fig. 256, 6, is

* Such preconceptions as Rauber entertained were all in a direction likely to lead him away from such phenomena as he has faithfully depicted. Rauber had no idea whatsoever of the principles by which we are guided in this discussion, nor does he introduce at all the analogy of surface-tension, or any other purely physical concept; but he was deeply under the influence of Sachs's rule of rectangular intersection, and he was accordingly disposed to look upon the configuration represented above in Fig. 256, 6, as the most typical or primitive. His articles on *Thier und Pflanze*, in *Biol. Cbt.* IV, 1881, tell us much about this and other biological theories of his time.

again peculiar, and is probably rare, but it is included under the cases considered on p. 491, in which the cells are not in complete fluid contact but are separated by little droplets of extraneous matter; it needs no further comment.   But the other four cases are beautiful diagrams of space-partitioning, similar to those we have just been considering, but so exquisitely clear that they need no modification, no "touching-up," to exhibit their mathematical regularity.   It will easily be recognised that in Fig. 256, 1 and 2,

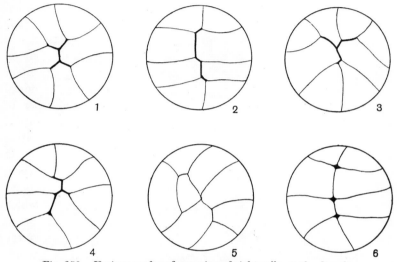

Fig. 256.   Various modes of grouping of eight cells, at the dorsal or epiblastic pole of the frog's egg.   After Rauber.

we have the arrangements corresponding to l and g, and in 3 and 4 to c in our table on p. 598.   One thing stands out as very certain indeed: that the elementary diagram of the frog's segmenting egg given in textbooks of embryology—in which the cells are depicted as uniformly symmetrical and more or less quadrangular bodies—is entirely inaccurate and grossly misleading*.

* Cf. Rauber, Neue Grundlegungen z. K. der Zelle, *Morphol. Jahrb.* VIII, p. 273, 1883: "Ich betone noch, dass unter meinen Figuren diejenige gar nicht enthalten ist, welche zum Typus der Batrachierfurchung gehörig am meisten bekannt ist....Es haben so ausgezeichnete Beobachter sie als vorhanden beschrieben, dass es mir nicht einfallen kann, sie überhaupt nicht anzuerkennen."   See also O. Hertwig, Ueber den Werth d. erste Furchungszelle für die Organbildung des Embryo, *Arch. f. Anat.* XLIII, 1893;   here O. Hertwig maintains that there is no such thing as "cellular homology."

We begin to realise the remarkable fact, which may even appear a startling one to the biologist, that all possible groupings or arrangements whatsoever of eight cells in a single layer or surface (none being submerged or wholly enveloped by the rest) are referable to one or another of twelve types or patterns; and that all the thousands and thousands of drawings which diligent observers have made of such eight-celled embryos or blastoderms, or other eight-celled structures, animal or vegetable, anatomical, histological or

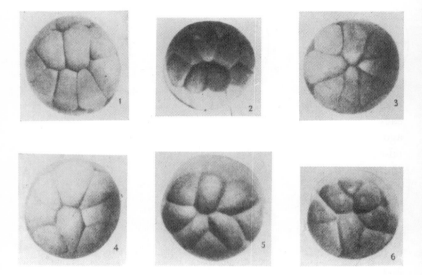

Fig. 257.   Photographs of frogs' eggs, shewing various arrangements, or partitionings, of the first eight cells.

embryological, are one and all of them representations of some one or other of these twelve types—or rather for the most part of less than the whole twelve; for a certain small number are essentially unstable, and have at best but a transitory and evanescent existence. But that even the unstable cases should now and then be seen is not to be wondered at: when viscidity and friction, and in general the imperfect fluidity of the system, retard the adjustment of the cells and delay the advent of equilibrium.

As soon as we realise that the number of cell-patterns, for instance in a segmenting egg, is strictly limited, we want to know how many patterns actually occur and in what proportions they do so in a random sample of identical eggs. Some years ago Mr Martin Adamson photographed more than a thousand frogs' eggs in my laboratory, all at the stage shewing an eight-celled group of epiblastic cells: with the remarkable result that every one of the twelve possible arrangements was found to occur, but some were common and some rare, and the following were their comparative frequencies:

Type	Frequency	Type	Frequency
c	19·0 %	d	6·7 %
j	17·0	h	6·6
b	12·8	f	5·1
g	10·3	k	3·0
a	7·8	l	2·4
e	6·9	i	1·8

In six separate batches of eggs (combined in the above list) one or other of the first two types (c or j) was always the commonest; and the first four taken together made up from 50 to 80 per cent. of each separate sample. On the other hand, when Roux, many years ago, shewed how various cell-configurations might be simulated by oil-drops*—as we have done by means of soap-bubbles—he found that the type i was essentially unstable, the large drop with its seven contacts easily slipping into the centre of the system, and there taking up a stable position of equilibrium. That the latter is the more stable, and therefore the more probable, configuration, seems obvious enough; and indeed type i seems so obviously unstable that we are not surprised to find it at the bottom of Martin Adamson's list of frequencies. The order in which the rest occur is by no means so easy of explanation.

There is a point worth considering in regard to the number of contacts between cell and cell. In a system of eight cells, all reaching the boundary and all with three-way junctions, there are, besides the eight peripheral boundary-walls, thirteen internal partitions, or $2(n-2)+1$; the number of interfacial *contacts* is double that number, or twenty-six; and the mean number of contacts for each cell is 26/8, or 3·25. But, looking at the diagrams in Fig. 259 (which represent three out of our twelve possible arrangements of

---

* Roux's experiments were performed with drops of paraffin suspended in dilute alcohol, to which a little calcium acetate was added to form a soapy pellicle over the drops and prevent them from reuniting with one another.

eight cells), we see that, in type j, two cells are each in contact with
two others, two with three others, and four each with four other
cells; in type l, four cells are each in contact with two, two with

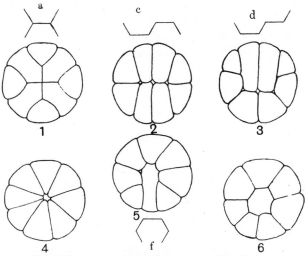

Fig. 258.   Aggregations of oil-drops.   After Roux.
Nos. 5, 6 represent successive changes in a single system.

four and two with five; and in type i, two are in contact with two,
four with three and one with no less than seven.   And if we sum
up, irrespective of sign, the differences from the mean in these three
cases, the sum amounts in j to 6, in i to 7·5, and in l to no less than

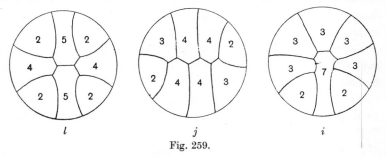

Fig. 259.

10.   We might expect to find in such arrangements, that the commonest and most stable types were those in which the cell-contacts
were most evenly distributed, and the fact that j is (according to
Martin Adamson's results) one of the commonest, and l one of the

rarest of all looks like supporting the conjecture. Moreover, in all the commonest types we have a more or less equable division; but on the other hand, the number of contacts in type f is just the same as in b, but the latter occurred thrice as often, and k, which is as equable as any, was one of the least frequent of all. Coincidences are weighed down by discrepancies, and we are left pretty much in the dark as to why some types are much commoner than others.

The rules and principles which we have arrived at from the point of view of surface tension have a much wider bearing than is at once suggested by the problems to which we have applied them; for in this study of a segmenting egg we are on the verge of a subject adumbrated by Leibniz, studied more deeply by Euler, and greatly developed of recent years. It is the *Geometria Situs* of Gauss, the *Analysis Situs* of Riemann, the Theory of Partitions of Cayley, of Spatial Complexes or Topology of Johann Benedict Listing*. It begins with regions, boundaries and neighbourhoods, but leads to abstruse developments in modern mathematics. Leibniz had pointed out† that there was room for an analysis of mere position, apart from magnitude: "je croy qu'il nous faut encor une autre analyse, qui nous exprime directement *situm*, comme l'Algèbre exprime *magnitudinem*." There were many things to which the new *Geometria Situs* could be applied. Leibniz used it to explain the game of solitaire, Euler to explain the knight's move on the chess-board, or the routes over the bridges of a town. Vandermonde created a *géometrie de tissage*‡, which Leibniz himself had foreseen, to describe the intricate complexity of interwoven threads in a satin or a brocade§. Listing, in a famous paper‖, admired by Maxwell, Cayley and Tait, gave a new name to this new "algorithm," and shewed its application to the curvature of a twining stem or tendril,

---

* Cf. Clerk Maxwell, On reciprocal figures, *Trans. R.S.E.* xxvi, p. 9, 1870.

† In a letter to Huygens, Sept. 8, 1679; see *Hugenii Exercitationes math. et philos.*, etc., ed. Uylenbroeck, p. 9, 1833.

‡ Remarques sur les problèmes de situation, *Mém. Acad. Sci. Paris* (1771), 1774, p. 566.

§ A problem developed by many eminent mathematicians, and which Edouard Lucas shewed to be intimately related to the construction of Magic Squares: *Récréations mathém.* i, p. xxii, 1891.

‖ Vorstudien über Topologie, *Göttinger Studien*, i, pp. 811–75, 1847; Der Census räumlicher Complexe, *ibid.* x, pp. 97 *seq.*, 1861.

the aestivation of a flower, the spiral of a snail-shell, the scales on a fir-cone, and many other common things. The theory of "spatial complexes," as illustrated especially by knots, is a large part of the subject.

Topological analysis seems somewhat superfluous here; but it may come into use some day to describe and classify such complicated, and diagnostic, patterns as are seen in the wings of a butterfly or a fly. Let us look for a moment at how the topologist might begin to study one of our groups of cells; he would probably

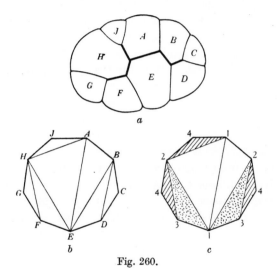

Fig. 260.

call it an island divided into n counties, all maritime (i.e. none encircled by the rest), and having inland none but three-way junctions*. Here (in Fig. 260 a) is an island with nine counties; and here (b) is a 9-gon, whose corners represent the same counties, and the lines connecting these (whether sides or chords) represent the contacts between. The polygon is now divided by six chords into seven triangles. Three of these are peripheral, BCD, FGH, HJA; mark their vertices, C, G, J, each with the symbol 4, and obliterate these three triangles (as in c). The remaining polygon has two peripheral triangles BDE, EFH; obliterate these, after marking

* This, like many another thing, comes from my good friend Dr G. T. Bennett.

their vertices with the symbol 3.   There remains the quadrilateral
ABEH, containing two peripheral triangles *ABE*, *EHA*; mark *B*
and *H* each with the symbol 2.   The residual points *A*, *E* are to be
marked 1 and 1.   The polygon *ABCDEFGHJ* may now be read
off: 1 2 4 3 1 3 4 2 4; and this formula, resulting from the triangula-
tion, defines completely the system of chords and the topology of the
"island."   This is one of the twenty-seven cases of a nine-celled
arrangement; and here are our twelve arrangements of eight cells,
recatalogued under the new method:

a	1 1 2 3 1 3 2 3		*g*	1 2 3 4 1 2 4 3
b	1 1 3 2 1 3 2 3		*h*	1 2 3 4 1 3 4 2
c	1 2 3 1 2 3 1 3		*i*	1 2 3 4 1 4 3 2
d	1 2 3 1 3 1 3 2		*j*	1 2 4 3 1 2 4 3
e	1 2 3 1 3 2 1 3		*k*	1 2 4 3 1 3 4 2
f	1 2 3 4 1 2 3 4		*l*	1 3 2 3 1 3 2 3

The crucial point for the biologist to comprehend is, that in a
closed surface divided into a number of faces, the arrangement of
all the faces, lines and points in the system is capable of analysis,
and that, when the number of faces or areas is small, the number
of possible arrangements is small also.   This is the simple reason
why we meet in such a case as we have been discussing (viz. the
arrangement of a group or system of eight cells) with the same few
types recurring again and again in all sorts of organisms, plants as
well as animals, and with no relation to the lines of biological
classification: and why, further, we find similar configurations
occurring to mark the symmetry, not of cells merely, but of the
parts and organs of entire animals.   The phenomena are not
"functions," or specific characters, of this or that tissue or organism,
but involve general principles, even "properties of space," which
lie within the province of the mathematician.

The theory of space-partitioning, to which the segmentation of
the egg gives us an easy practical introduction, is illustrated in
innumerable ways, some simple, some extremely complicated, in
other fields of natural history; and some serve the better to illustrate
the mathematical, and others the physical groundwork of the
phenomenon.

Very beautiful instances are to be found in insects' wings. In the dragonfly's wing (which we have already spoken of on p. 476) we see at first sight a vague assemblage of reticulate cells; but their arrangement is both orderly and simple. The long narrow wing is stiffened by longitudinal "veins," which in front lie near and parallel, for reasons well known to the student of aerodynamics*, but become remote and divergent over the rest of the wing; finer veinlets, running between the veins, break up the surface into cells or areolae. Where two large veins run parallel, and so near together that there is only room for one row of cells between, the walls of these meet the large veins at right angles, for the reason that the

Fig. 261.   Wing of "demoiselle" dragonfly (*Agrion*).

tension in these latter is much greater than their own; and this happens nearly all over the delicate wings of the little dragonflies called "demoiselles." But in the big dragonflies (*Aeschna*), and in general wherever there is space enough between two strong veins to hold a double row of cells, the walls of these intercalate with one another at co-equal angles of 120°, while still impinging at right angles on the strong longitudinal partitions. Wherever, as in the hinder parts of the wing, the great veins are few, the cells numerous, and their walls equally delicate, then the reticulum of cells becomes an hexagonal network of all but perfect regularity. In a cicada and in many others there is less contrast between great veins and small; the cells are few, the veins meet neither orthogonally nor at co-equal angles, and the shape of the cells suggests a common deformation under strain. In this last case, and generally in flies, bees and butterflies, the few cells form a complex space-arrangement, simplified

---

* Sir George Cayley was the first to shew that in a sail—or wing—set at an acute angle to the wind, the centre of pressure lay near the front edge, which had, therefore, to be supported or stiffened (*Nicholson's Journal*, xxv, 1810).

only by the condition that the walls impinge on one another three by three; and this being so, the assemblage includes a number of small intermediate partitions analogous to the "polar furrows" of embryology. We have seen how complex such configurations become as the cells increase in number; and another source of complexity comes in when the veins are of varying thickness and unequal tension, and hence meet one another at varying angles.

The entomologist is much concerned with the number and arrangement of these veins. In Fig. 263 we shew three forewings of a certain stonefly, which serve first to shew how constantly the veins meet in three-way junctions; and then we notice how the

Fig. 262.　Forewing of cicada.

three wings are not exactly alike, for all their close resemblance, because in two of them there is one cell less than in the third, *a* being confluent with *b* in one of these cases and with *c* in the other*. In other words, the veinlet *ab* has gone amissing in the one, and *ac* in the other, and in each case the remaining veinlet has sprung into a position of equilibrium. This is one of more than two hundred variations which have been recorded in the wing-veins of this one insect; it might seem superfluous to look for more.

The lower algae shew us many beautiful patterns or collocations of cells, sometimes very complicated, as in *Volvox* or *Hydrodictyon*. A simpler case is that of *Gonium*. Of its sixteen cells four commonly form a square, being so thrust apart out of closer packing by accumulated intercellular substance; the other twelve are grouped around these four, and obey the rule that three cells and no more meet at each node or point of contact (Fig. 264). The twelve cells

---

* From Arthur Willey, Graded mutations in wings of a stonefly (*Allocapnia pygmaea* Burm.), *Nature*, July 17, 1937.

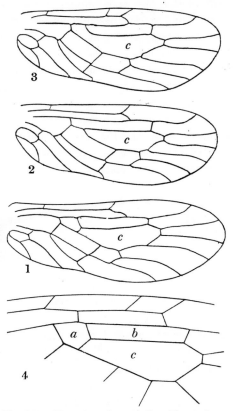

Fig. 263.   Forewing of a stonefly, *Allocapnia* sp.
1–3 from A. Willey.

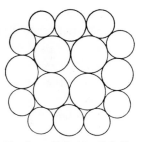

Fig. 264.   The four-sided, 16-celled disc of *Gonium*,
a minute algae.   After Harper, diagrammatic.

are thus in three series, four in touch with six neighbours each, four with four, and four with three; and this arrangement leads to curved contour-lines which may be seen in some of Cohn's figures. Sometimes the interstitial substance is not enough to keep the four cells apart; then they come together in the usual rhomb or lozenge, and the rest group themselves around in the simplest and most symmetrical way. That the whole arrangement is as compact as possible under the conditions, that it is in accord with the principle of minimal areas, and is "such as would result from surface-tension and adhesion between viscous colloidal globules" is now well known to botanists *.

The tiny plates which form the microscopic shell of a Peridinian illustrate over again by their various collocations the principles which we have been studying in the partition-walls of the segmenting egg; and if the one case has shewn us pitfalls in the way of the embryologist, the other shews how the systematist, in his endless task of describing the forms and patterns of things, may sometimes base distinctions on what seem trivial differences from the physical or mathematical point of view.

On the upper half (or *epitheca*) of the globular test of a Peridinium we have fourteen little plates, or fourteen "cells," to use the word in a mathematical rather than a histological sense, whose boundary-walls always meet in three-way nodes. We may reproduce the identical arrangement of the Peridinial plates by blowing bubbles in a saucer; but to deal with so many bubbles at once needs more patience than do the other similar experiments which we have described. That the cells are fourteen in number is, from the physical point of view, the merest accident, but from the zoologist's it is a criterion of the genus; when there are more cells or fewer, the organism is called by another name. The number of possible arrangements of fourteen polygonal cells, linked by three-way nodes, is very large; but the "characters of the genus" exclude many of the variants. Many of them occur—it is quite possible that all occur—in Nature; but they are not called *Peridinium*. The following arrangement defines the genus. There is a central or apical cell, around which are grouped six others; of these six, one extends to the boundary of the figure, that is, to the equator of the globular

---

* R. A. Harper, The colony in *Gonium*, *Trans. Amer. Microsc. Soc.* XXXI, pp. 65–84, 1912; cf. F. Cohn, in *Nova Acta Acad. C.L.C.*, XXIV, p. 101, 1854.

shell, and seven other "equatorial cells" complete the boundary. In other words, the apical hexagon is surrounded by two concentric rows, originally of six cells each, whereof one cell of the inner row has (as it were) burst through a cell of the outer row, so reaching the boundary itself, and so dividing into two the equatorial cell which it encroached on and bisected.

In any such collocation as this, the number of sides and the number of nodes or corners are strictly determined; there are here fourteen cells, all conjoined by three-way nodes, and it follows that there are just 39 separate walls or edges, and just 26 nodes or corners. Many of these last are already defined for us; for six of them are the corners of the central hexagon, eight lie on the equator, and six more are at the inner ends of the radial partitions which separate the equatorial cells from one another. Six remain to be determined, those, namely, where the partitions running outwards from the apical cell meet the walls of the equatorial cells. The diagram (Fig. 265) shews us two sets of radiating partitions, six running inwards from the equator (a, b, c, d, e, f) and six running outwards from the central hexagon ($A...F$), those of the one set being nearly opposite to those of the other; but near as they may be they never meet, for to do so would be to make a four-way node, which

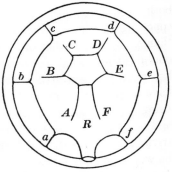

Fig. 265.   Dorsal view of a *Peridinium*: diagrammatic.

theory forbids and which observation tells us does not occur. In every case, one partition must be slewed a little to one side or other of its opposite neighbour; and the whole range of possible variations depends on whether the shift be to the one side or to the other. We have six pairs of partitions, and in each of the six there is this possible alternative of right or left; there are therefore 2^6 or 64 possible variations in all. Whether all six may vary, I do not know; there is no obvious reason why they should not. But alternative variation does occur in the two anterior and two posterior pairs of partitions; and these four give us 2^4 or 16 possible arrangements.

When the partitions A, F meet the equatorial cells on the *hither* side of the partitions a, f, then, obviously, the large cell R is an irregular hexagon, and is in contact with two equatorial cells only; such an arrangement is said to define the genus *Orthoperidinium*. When A, F happen to fall on the *farther* sides of a, f, the large cell has eight sides, and is in contact with four equatorial cells; we have the new genus *Paraperidinium*. When A falls within a, but F falls beyond f, we have the genus *Metaperidinium*. There remains one alternative case, the converse of the last, to which the systematists have not given a name.

The same physical phenomenon occurs at the opposite pole of the disc, where the partitions C, D may fall within or without, or one within and one without the positions of c, d; where, in other words, the intercalated cell CD is in contact with one, with three, or with two equatorial cells. Jörgensen, seeing these three types occurring both among the Orthoperidinia and the Metaperidinia, draws the conclusion that this character is more primitive, or more ancestral, than that by which Ortho- and Metaperidinia are separated from one another, a phylogenetic deduction concerning which topology has nothing to say*. Within the restricted genus *Peridinium* we have at present two sub-genera and seven sub-groups of these, this being the number of the 64 possible arrangements so far recognised and named. These may have a certain constancy or stability; and trivial as their differences may seem to the physicist, they may still be worth the naturalist's while to study and record.

Another case, geometrically akin but biologically very different, is to be found in the little diatoms of the genus *Asterolampra*, and their immediate congeners†. In *Asterolampra* we have a little disc, in which we see (as it were) radiating spokes of one material alternating with intervals occupied on the flattened wheel-like disc by another (Fig. 266). The spokes vary in number, but the general appearance is in a high degree suggestive of the Chladni figures produced by the vibration of a circular plate. The spokes broaden out towards the centre, and interlock by visible junctions, which

* E. Jörgensen, Ueber Planktonproben, *Svenska Hydrogr. Biol. Komm. Skrifter*, IV, 1913.

† See K. R. Greville, Monograph of the genus *Asterolampra*, *Q.J.M.S.* VIII, (Trans.), pp. 102–124, 1860; cf. *ibid.* (n.s.), II, pp. 41–55, 1862.

generally obey the rule of triple intersection, and accordingly exemplify the partition-figures with which we have been dealing. But whereas we have found the particular arrangement in which one cell is in contact with all the rest to be unstable, according to Roux's oil-drop experiments, and to be conspicuous by its absence from our diagrams of segmenting eggs, here in *Asterolampra*, on the other hand, it occurs frequently, and is indeed the commonest arrangement (Fig. 266, B). In all probability, we are entitled to consider this marked difference natural enough. For we may suppose that in *Asterolampra* (unlike the case of the segmenting egg) the tendency is to perfect radial symmetry, all the spokes emanating from a point in the centre: such a condition would be

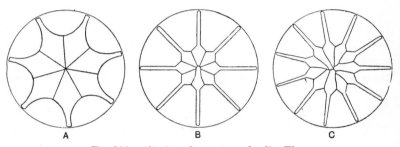

Fig. 266.   (A) *Asterolampra marylandica* Ehr.;
(B, C) *A. variabilis* Grev.   After Greville.

eminently unstable, and would break down under the least asymmetry. A very simple, perhaps the simplest case, would be that one single spoke should differ slightly from the rest, and should so tend to be drawn in amid the others, these latter remaining similar and symmetrical among themselves. Such a configuration would be vastly less unstable than the original one in which all the boundaries meet in a point; and the fact that further progress is not made towards other configurations of still greater stability may be sufficiently accounted for by viscosity, rapid solidification, or other conditions of restraint. A perfectly stable condition would of course be obtained if, as in the case of Roux's oil-drop (Fig. 257, 6), one of the cellular spaces passed into the centre of the system, the other partitions radiating outwards from its circular wall to the periphery of the whole system. Precisely such a condition occurs among our diatoms; but when it does so, it is looked upon as the mark and characterisation of the *allied genus Arachnoidiscus*.

A simple case, introductory to others of a more complex kind, is that of the radial canals of the Medusae. Here, in certain cases (e.g. *Eleutheria*), the usual arrangement of eight radial canals is not seldom modified, as for example, when two or more of them arise

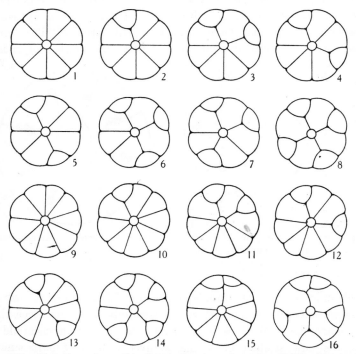

Fig. 267. Variations observed in the canal-system of a medusoid (*Eleutheria*); after Hans Lengerich. 1–8, the eight possible arrangements of eight radial canals; 9–16, some observed instances of nine radial canals.

not separately but by bifurcation*. We then have just eight possible arrangements, as shewn in Fig. 267, 1–8, and of these eight no less than six have been actually observed. The other two are just as likely to occur, and we may take it that they also will in due time be recorded. It is yet another simple illustration of the

* Hans Lengerich, Verzweigungsarten der Radialkanäle bei *Eleutheria*, *Zool. Jahrbuch*, 1922, p. 325.

aphorism that whatsoever is possible, that Nature does, all in her own time and way; what things Nature does not do are the things which are mathematically impossible or are barred by physical conditions. It is possible for our little Medusa to develop nine radial canals instead of eight, and so at times she does; again they may be simple or branched, and trifurcations as well as bifurcations may appear. So we may extend our list of possible permutations and combinations, and find as before that a fair proportion of these possible arrangements have been observed already. There are many other Medusae (e.g. *Willsia*), where the number of radial

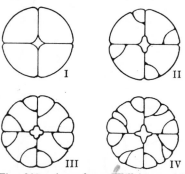

Fig. 268. A medusa (*Willsia ornata*) shewing, diagrammatically, the order of development of the numerous radial canals. After Mayor.

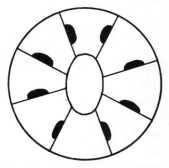

Fig. 269. Section of Alcyonarian polyp.

canals much exceeds the simple symmetry of four or eight; and in these we may sometimes see, very beautifully, how the successive canals arrange themselves according to the same principles which we have now studied in so many diverse cases of partitioning (Fig. 268)*.

In a diagrammatic section of an Alcyonarian polyp (Fig. 269), we have eight chambers set, symmetrically, about a ninth, which constitutes the "stomach." In this arrangement there is no difficulty, for it is obvious that, throughout the system, three boundaries meet (in plane section) in a point. In many corals we have as

* Such branching canals are characteristic of the Dendrostaurinae, a subfamily of the Oceanidae, a family of Anthomedusae; and very much the same occur in a certain subfamily of Leptomedusae. See Mayor's *Medusae of the World*, I, p. 190.

simple or even simpler conditions, for the radiating calcified partitions either converge upon a central chamber, or fail to meet it and end freely. But in a few cases, the partitions or "septa" converge to meet *one another*, there being no central chamber on which they may impinge; and here the manner in which contact is effected becomes complicated, and involves problems identical with those which we are now studying.

In the great majority of corals we have as simple or even simpler conditions than those of *Alcyonium*; for as a rule the calcified partitions or septa of the coral either converge upon a central chamber (or central "columella"), or else fail to meet it and end freely. In the latter case the problem of space-partitioning does not arise; in the former, however numerous the septa be, their separate contacts with the wall of the central chamber comply with our fundamental rule according to which three lines and no more meet in a point, and from this simple and symmetrical arrangement there is little tendency to variation. But in a few cases, the septal partitions converge to meet *one another*, there being no central chamber on which they may impinge; and here the manner in which contact is effected becomes complicated, and involves problems of space-partitioning identical with those which we are now studying. In the genus *Heterophyllia* and in a few allied forms we have such conditions, and students of the Coelenterata have found them very puzzling. McCoy*, their first discoverer, pronounced these corals to be "totally unlike" any other group, recent or fossil; and Professor Martin Duncan, writing a memoir on *Heterophyllia* and its allies†, described them as "paradoxical in their anatomy."

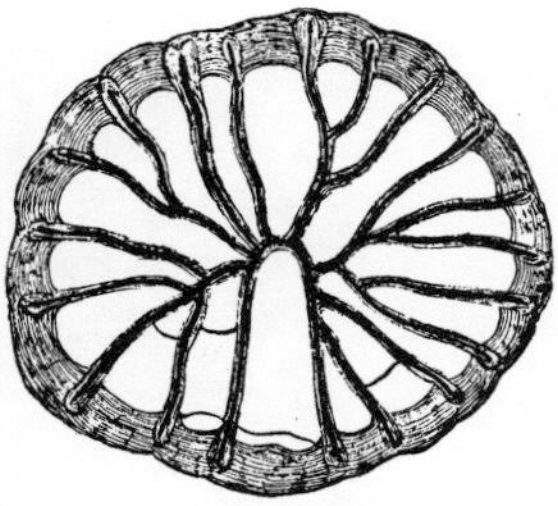

Fig. 270.   *Heterophyllia angulata*.   After Nicholson.

The simplest or youngest Heterophylliae known have six septa (as in Fig. 271, *A*); in the case figured, four of these septa are conjoined two and two, thus forming the usual triple junctions together with their intermediate partition-walls; and in the case of the other two we may fairly assume that their proper and original

* *Ann. Mag. N.H.* (2), III, p. 126, 1849.
† *Phil. Trans.* CLVII, pp. 643–656, 1867.

arrangement was that of our type 6 *b* (Fig. 245), though the central intermediate partition has been crowded out by partial coalescence. When with increasing age the septa become more numerous, their arrangement becomes exceedingly variable; for the simple reason that, from the mathematical point of view, the number of possible arrangements, of 10, 12 or more cellular partitions in triple contact, tends to increase with great rapidity, and there is little to choose among many of them in regard to symmetry and equilibrium. But while, mathematically speaking, each particular case among the

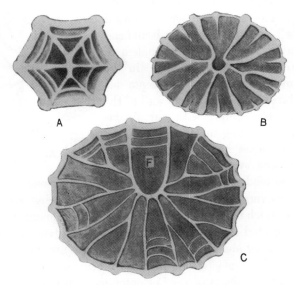

Fig. 271.   *Heterophyllia* sp.   After Martin Duncan.

multitude of possible cases is an orderly and definite arrangement, from the purely biological point of view on the other hand no law or order is recognisable; and so McCoy described the genus as being characterised by the possession of septa "destitute of any order of arrangement, but irregularly branching and coalescing in their passage from the solid external walls towards some indefinite point near the centre where the few main lamellae irregularly anastomose."

In the two examples figured (Fig. 271 B, C), both comparatively simple ones, it will be seen that, of the main chambers, one is in each case an unsymmetrical one; that is to say, there is one chamber which is in contact with a greater number of its neighbours than any other, and which at an earlier stage must have had contact with them all; this was the case of our type i, in the eight-celled system (p. 598). Such an asymmetrical chamber (which may occur in a system of any number of cells greater than six) constitutes what is known to students of the Coelenterata as a "fossula"; and we may recognise it not only here, but also in *Zaphrentis* and its allies, and in a good many other corals besides. Moreover, certain corals are described as having more than one fossula: this appearance being naturally produced under certain of the other asymmetrical variations of normal space-partitioning. Where a single fossula occurs, we are usually told that it is a symptom of "bilaterality"; and this is in turn interpreted as an indication of a higher grade of organisation than is implied in the purely "radial symmetry" of the commoner types of coral. The mathematical aspect of the case gives no warrant for this interpretation.

Let us carefully notice (lest we run the risk of confusing two distinct problems) that the space-partitioning of *Heterophyllia* by no means agrees with the details of that which we have studied in (for instance) the case of the developing disc of *Erythrotrichia*: the difference simply being that *Heterophyllia* illustrates the general case of cell-partitioning as Plateau and Van Rees studied it, while in *Erythrotrichia*, and in our other embryological and histological instances, we have found ourselves justified in making the additional assumption that each new partition divided a cell into *co-equal parts*. No such law holds in *Heterophyllia*, whose case is essentially different from the others: inasmuch as the chambers whose partition we are discussing in the coral are mere empty spaces (empty save for the mere access of sea-water); while in our histological and embryological instances, we were speaking of the division of a cellular unit of living protoplasm. Accordingly, among other differences, the "transverse" or "periclinal" partitions, which were bound to appear at regular intervals and in definite positions, when co-equal bisection was a feature of the case, are comparatively few and irregular in the earlier stages of *Heterophyllia*, though they begin to appear in numbers after the main, more or less radial, partitions have become numerous, and when accordingly thes radiating partitions come to bound narrow and almost parallel-sided interspaces; then it is that the transverse or periclinal partitions begin to come in, and form what the student of the Coelenterata calls the "dissepiments" of the coral. We need go no further into the configuration and anatomy of the corals; but it seems to me beyond a doubt that the whole question of the complicated arrangement of septa and dissepiments throughout the group (including the curious vesicular or bubble-like tissue of the Cyathophyllidae and the general structural plan of the *Tetracoralla*, such as *Streptoplasma* and its allies) is well worth investigation from the physical and mathematical point of view, after the fashion which is here slightly adumbrated.

The method of dividing a circular, or spherical, system into eight parts, equal as to their areas but unequal in their peripheral boundaries, is probably of wide biological application; that is to say, without necessarily supposing it to be rigorously followed, the typical configuration which it yields seems to recur again and

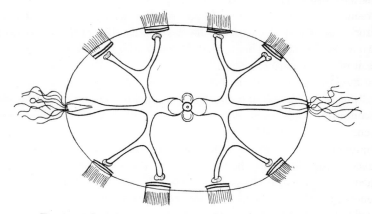

Fig. 272. Diagrammatic section of a Ctenophore (*Eucharis*).

again, with more or less approximation to precision, and under widely different circumstances. I am inclined to think, for instance, that the unequal division of the surface of a Ctenophore by its meridian-like ciliated bands is a case in point (Fig. 272). Here, if we imagine each quadrant to be twice bisected by a curved anticline, we shall get what is apparently a close approximation to the actual position of the ciliated bands. The case however is complicated by the fact that the sectional plan of the organism is never quite circular, but always more or less elliptical. One point, at least, is clearly seen in the symmetry of the Ctenophores; and that is that the radiating canals which pass outwards to correspond in position with the ciliated bands have no common centre, but diverge from one another by repeated bifurcations, in a manner comparable to the conjunctions of our cell-walls.

In the early development of the shell (or "test") of a sea-urchin*, each interambulacral area consists of a lozenge of four plates, in the familiar configuration assumed by four cells or bubbles, the polar furrow lying in the direction of a "radius" of the shell. A fifth plate, or "cell," presently fits itself in between the third and fourth, that is to say between the terminal plate and one of the lateral ones, and in doing so thrusts the former to one side; a sixth intercalates itself between the fourth and fifth, and so on alternately. An ambulacrum consists of two columns of calcareous plates, which fit into one another in the usual way by

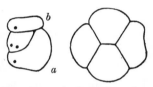

Fig. 273. A "triad" and a "lozenge": stages in the development of the ambulacral and interambulacral plates of a sea-urchin. After I. Gordon.

sutures set at angles of 120°. Each plate consists of three platelets; a "primary" plate (*a*) is succeeded by a smaller and narrower secondary plate (*b*); the squarish primary has one corner cut off by a curved partition, to form a "demiplate", and the whole is called by students of this group an "echinoid triad". Though we do not know precisely how the partitions arise, nor can we prove by measurement their obedience to the laws of maxima and minima, yet their general analogy to the principles we have explained is sufficiently obvious.

I am even inclined to think that the same principle helps us to understand the arrangement of the skeletal rods of a larval Echinoderm, and the complex conformation of the larva which is brought about by the presence of these long, slender skeletal

* Isabella Gordon, The development of the calcareous test of *Echinus miliaris*, *Phil. Trans.* (B), No. 214, p. 282, 1926; etc.

radii*. In Fig. 274 I have divided a circle into its four quadrants,
and have bisected each quadrant by a circular arc (*BC*), passing from
radius to periphery, as in the foregoing cases of cell-division (*e.g.*
p. 590); and I have again bisected, in a similar way, the triangular

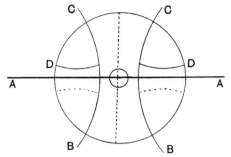

Fig. 274.   Diagrammatic arrangement of partitions, represented by skeletal
rods, in a larval Echinoderm (*Ophiura*).

halves of each quadrant (*D, D*).   I have also inserted a small circle in
the middle of the figure, concentric with the large one.   If now we
imagine the partition-lines in the figure to be replaced by solid
rods, we shall have at once the frame-work of an Ophiurid (Pluteus)

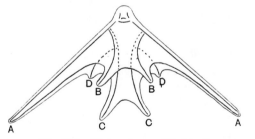

Fig. 275.   Pluteus-larva of Ophiurid.

larva.   Let us imagine all these arms to be bent symmetrically
downwards, so that the plane of the paper is transformed into
a conical surface with curved sides; let a membrane be spread,
umbrella-like, between the outstretched skeletal rods, and let

---

* J. Loeb has shewn (*Amer. Journ. Physiol.* VII, p. 441, 1900) that the sea-urchin's
egg can be reared for a time in a balanced solution of sodium. potassium and
calcium chloride, developing no spicules and so forming no pluteus larva; on
adding sodium carbonate the spicules are laid down and the pluteus larva takes
shape accordingly.

its margin loop from rod to rod in curves which are possibly
catenaries but are more probably portions of an "elastic curve,"
and the outward resemblance to a Pluteus larva is now complete.
By various slight modifications, by altering the relative lengths of
the rods, by modifying their curvature or by replacing the curved
rod by a tangent to itself, we can ring the changes which lead us
from one known type of Pluteus to another. The case of the
Bipinnaria larvae of Echinids is certainly analogous, but it becomes
very much more complicated; we have to do with a more complex
partitioning of space, and I confess that I am not yet able to
represent the more complicated forms in so simple a way.

There are a few notable exceptions (besides the various unequally
segmenting eggs) to the general rule that in cell-division the mother-
cell tends to divide into equal halves; and one of these exceptional
cases is to be found in connection with the development of
"stomata" in the leaves of plants*. The epidermal cells by which
the leaf is covered may be of various shapes; sometimes, as in a
hyacinth, they are oblong, but more often they have an irregular
shape in which we can recognise, more or less clearly, a distorted
or imperfect hexagon. In the case of the oblong cells, a transverse
partition will be the least possible, whether the cell be equally or
unequally divided, unless (as we have already seen) the space to
be cut off be a very small one, not more than about three-tenths
the area of a square based on the *short* side of the original rectangular
cell. As the portion usually cut off is not nearly so small as this,
we get the form of partition shewn in Fig. 276, and the cell so cut
off is next bisected by a partition at right angles to the first; this
latter partition splits, and the two last-formed cells constitute the
so-called "guard-cells" of the stoma. In other cases, as in Fig. 277,
there will come a point where the minimal partition necessary to
cut off the required fraction of the cell-content is no longer a

* We know more about the physical activities of the stomata than about the
mechanics of their development. It is known that the rate of gaseous diffusion
through apertures *of their order of magnitude* is inversely proportional to the
diameters of the apertures; and this law, by which the sufficient entry of carbonic
acid through the stomata is fully accounted for, is (like Pfeffer's work on natural
semi-permeable membranes) one of the notable cases where physiology has enlarged
the boundaries of physical science. Cf. Horace T. Brown, Some recent work on
diffusion, *Proc. Roy. Instit.* March, 1901.

transverse one, but is a portion of a cylindrical wall (2) cutting off
one corner of the mother-cell. The cell so cut off is now a certain
segment of a circle, with an arc of approximately 120°; and its
next division will be by means of a curved wall cutting it into a

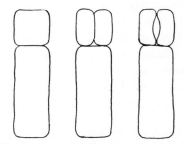

Fig. 276.   Diagrammatic development of stomata in hyacinth.

triangular and a quadrangular portion (3). The triangular portion
will continue to divide in a similar way (4, 5), and at length (for
a reason which is not yet clear) the partition wall between the
new-formed cells splits, and again we have the phenomenon of a

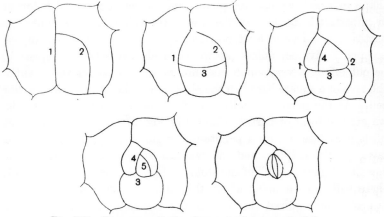

Fig. 277.   Diagrammatic development of stomata in *Sedum*.
(Cf. fig. in Sachs's *Botany*, 1882, p. 103.)

"stoma" with its attendant guard-cells. In Fig. 277 are shewn the
successive stages of division, and the changing curvatures of the
various walls which ensue as each subsequent partition appears,
and introduces a new tension into the system. Among the oblong

cells of the epidermis in the hyacinth the stomata will be found arranged in regular rows, while they will be irregularly distributed over the surface of the leaf in such a case as we have depicted in *Sedum*.

As I have said, the mechanical cause of the split which constitutes the orifice of the stoma is not quite clear. It may be directly due to the subepidermal air-space which the stoma communicates with, for an air-surface on both sides of the delicate epidermis might well cause such an alteration of tensions that the two halves of the dividing cell would tend to part company. In Professor Macallum's experiments, which we have briefly discussed in our short chapter on Adsorption, it was found that large quantities of potassium gathered together along the outer walls of the guard-cells of the stoma, thereby indicating a low surface-tension along these outer walls. The tendency of the guard-cells to bulge outwards is so far explained, and it is possible that, under the existing conditions of restraint, we may have here a force tending, or helping, to split the two cells asunder. It is clear enough, however, that the last stage in the development of a stoma is, from the physical point of view, not yet properly understood*. It is noteworthy, and Nägeli took note of it wellnigh a hundred years ago, that the stomatal mother-cells remain small while the others grow, and also that they only divide once for all, while their neighbours divide and divide again, to produce the lateral or accessory guard-cells.

In all our foregoing examples of the development of a "tissue" we have seen that the process consists in the *successive* division of cells, each act of division being accompanied by the formation of a boundary-surface, which, whether it become at once a solid or semi-solid partition or whether it remain semi-fluid, exercises in all cases an effect on the position and the form of the boundary which comes into being with the next act of division. In contrast to this general process stands the phenomenon known as "free cell-formation," in which, out of a common mass of protoplasm, a number of separate cells are *simultaneously*, or all but simul-taneously, differentiated; and the case is all the more interesting

---

* Botanische Beiträge, *Linnaea*, XVI, p. 238, 1842. Cf. Garreau, Mém. sur les stomates, *Ann. Sc. Nat., Bot.* (4), I, p. 213, 1854.

when the daughter-cells remain, for a time at least, within the
envelope of the mother-cell.   It sometimes happens, to begin with,
that a number of mother-cells are formed simultaneously, and
that the content of each divides, by successive divisions, into four
"daughter-cells."   These daughter-cells tend to group themselves,
just as would four soap-bubbles, into a "tetrad," the four cells
forming a spherical tetrahedron.   For the system of four bodies
is in perfect symmetry.   The four cells are closely packed within
the cell-wall of the mother-cell; their outer walls divide the
sphere into four equiangular triangles; their inner walls meet
three-by-three in an edge, and the four edges converge in the
geometrical centre of the system; and these partition walls and
their respective edges meet one another everywhere at co-equal
angles.   This is the typical mode of development of pollen-grains,
common among monocotyledons and all but universal among
dicotyledonous plants.   By a loosening of the surrounding tissue
and an expansion of the cavity, or anther-cell, in which they lie,
the pollen-grains afterwards fall apart, and their individual form
will depend upon whether or no their walls have solidified before
this liberation takes place.   For if not, then the separate grains will
be free to assume a spherical form as a consequence of their own
individual and unrestricted growth; but if they become set or rigid
prior to the separation of the tetrad, then they will conserve more or
less completely the plane interfaces and sharp angles of the elements
of the tetrahedron.   The latter is apparently the case in the pollen-
grains of *Epilobium* (Fig. 278, 1) and in many others.   In the passion-
flower (2) we have an intermediate condition: in which we can still see
an indication of the facets where the grains abutted on one another
in the tetrad, but the plane faces have been swollen by growth into
spheroidal or spherical surfaces.   In heaths and in azaleas the four
cells of the tetrad remain attached together, and form a compound
tetrahedral pollen-grain.   Six furrows correspond to the six edges
of the tetrahedron, and each is continued across a pair of cells; they
are formed (I take it) along lines of weakness at the edges of the
tetrahedron, and they make three furrows upon each one of the
four coherent grains, just as we see them on a large number of
ordinary separate and non-coherent pollen-grains.   On the other
hand, there may easily be cases where the tetrads of daughter-cells

fail to assume, even temporarily, the tetrahedral form: cases, in a general way, where the four cells escape from the confinement of their envelope, and fall into a looser, less close-packed arrangement*. The figures given by Goebel of the development of the pollen of *Neottia* (3, *a–e*: all the figures referring to grains taken from a single anther) illustrate this to perfection, and it will be seen that,

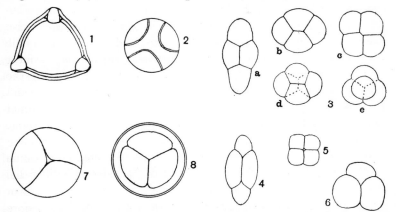

Fig. 278.   Various pollen-grains and spores (after Berthold, Campbell, Goebe and others). (1) *Epilobium*; (2) *Passiflora*; (3) *Neottia*; (4) *Periploca graeca*; (5) *Apocynum*; (6) *Erica*; (7) spore of *Osmunda*; (8) tetraspore of *Callithamnion*.

Fig. 279.   Pollen of bulrush (*Typha*).   After Wodehouse.

when the four cells lie in a plane, they conform exactly to our typical diagram of the first four cells in a segmenting ovum; physically, as well as biologically, the tetrads *a–d* and the tetrad *e* are "allelomorphs" of one another.   Again in the bulrush (Fig. 279),

* Cf. C. Nägeli, *Zur Entwicklungsgeschichte des Pollens bei den Phanerogamen*, 36 pp., Zürich, 1842; Hugo Fischer, *Vergleichende Morphologie der Pollenkörner*, Berlin, 1890; see also, for many and varied illustrations, R. P. Wodehouse's beautiful book on *Pollen*, 574 pp., New York, 1935, and earlier papers.

the four cells remain attached to one another, and lie upon a level with a "polar furrow" well displayed. Occasionally, though the four cells lie in a plane, the diagram seems to fail us, for the cells appear to meet in a simple cross (as in 5); but here we soon perceive that the cells are not in complete interfacial contact, but are kept apart by a little intervening drop of fluid or bubble of air. The spores of ferns (7) for the most part develop in much the same way as pollen-grains; they also very often retain traces of the shape which they assumed as members of a tetrahedral figure, and the same is equally true of liverworts. Among the "tetraspores" (8) of the Florideae, or red seaweeds, we have a condition which is in every respect analogous. The same thing happens in certain simple algae allied to *Protococcus*: where four daughter-cells, confined within a mother-cell, form a spherical tetrahedron, much like a spore of *Osmunda* on a smaller scale*.

Here again it is obvious that, apart from differences in actual magnitude, and apart from superficial or "accidental" differences (referable to other physical phenomena) in the way of colour, texture and minute sculpture or pattern, a very small number of diagrammatic figures will sufficiently represent the outward forms of all the tetraspores, four-celled pollen-grains, and other four-celled aggregates which are known or are even capable of existence. And it is equally obvious that the resemblance of these things, to this extent, is a matter of physical and mathematical symmetry, and carries no proof of near relationship or common ancestry.

We have been dealing hitherto (save for some slight exceptions) with the partitioning of cells on the assumption that the system either remains unaltered in size or else that growth has proceeded uniformly in all directions. But we extend the scope of our enquiry greatly when we begin to deal with *unequal growth*, with cells so growing and dividing as to produce a greater extension along some one axis than another. And here we come close in touch with that great and still (as I think) insufficiently appreciated generalisation of Sachs, that the manner in which the cells divide is *the result*, and not the cause, of the form of the dividing structure: that the form of the mass is caused by its growth as a whole, and

* Cf. A. Pascher, *Arch. f. Protozoenk.* LXXVI, p. 409; LXXVII, p. 195, 1932.

is not a resultant of the growth of the cells individually considered *.
Such asymmetry of growth may be easily imagined, and may
conceivably arise from a variety of causes. In any individual cell,
for instance, it may arise from molecular asymmetry of the structure
of the cell-wall, giving it greater rigidity in one direction than
another, while all the while the hydrostatic pressure within the
cell remains constant and uniform. In an aggregate of cells, it
may very well arise from a greater chemical, or osmotic, activity
in one than another, leading to a localised increase in the fluid
pressure, and to a corresponding bulge over a certain area of the
external surface. It might conceivably occur as a direct result of
preceding cell-divisions, when these are such as to produce many
peripheral or concentric walls in one part and few or none in another,
with the obvious result of strengthening the boundary wall here
and weakening it there; that is to say, in our dividing quadrant,
if its quadrangular portion subdivide by periclines, and the
triangular portion by oblique anticlines (as we have seen to be
the natural tendency), then we might expect that external growth
would be more manifest over the latter than over the former areas.
As a direct and immediate consequence of this we might expect
a tendency for special outgrowths, or "buds," to arise from the
triangular rather than from the quadrangular cells; and this turns
out to be not merely a tendency towards which theoretical considerations point, but a widespread and important factor in the
morphology of the cryptogams. But meanwhile, without enquiring
further into this complicated question, let us simply take it that,
if we start from such a simple case as a round cell which has divided
into two halves or four quarters (as the case may be), we shall at
once get bilateral symmetry about a main axis, and other secondary
results arising therefrom, as soon as one of the halves, or one of
the quarters, begins to shew a rate of growth in advance of the
others; for the more rapidly growing cell, or the peripheral wall
common to two or more such rapidly growing cells, will bulge out,
and may finally extend into a cylinder with rounded end. This
latter very simple case is illustrated in the development of a

---

* Sachs, *Pflanzenphysiologie* (*Vorlesung* XXIV), 1882; cf. Rauber, Neue Grundlegungen zur Kenntniss der Zelle, *Morphol. Jahrb.* VIII, p. 303 *seq.*, 1883;
E. B. Wilson, Cell-lineage of *Nereis*, *Journ. Morph.* VI, p. 448, 1892; etc.

pollen-tube, where the rapidly growing cell develops into the elongated cylindrical tube, and the slow-growing or quiescent part remains behind as the so-called "vegetative" cell or cells.

Just as we have found it easier to study the segmentation of a circular disc than that of a spherical cell, so let us begin in the same way, by enquiring into the divisions which will ensue if the disc tend to grow, or elongate, in some one particular direction instead of in radial symmetry. The figures which we shall then obtain will not only apply to the disc, but will also represent, in all essential features, a projection or longitudinal section of a solid body, spherical to begin with, preserving its symmetry as a solid of revolution, and subject to the same general laws as we study in the disc *.

(1) Suppose, in the first place, that the axis of growth lies symmetrically in one of the original quadrantal cells of a segmenting disc; and let this growing cell elongate with comparative rapidity before it subdivides. When it does divide, it will necessarily do so by a transverse partition, concave towards the apex of the cell: and, as further elongation takes place, the cylindrical structure which will be developed thereby will tend to be again and again subdivided by similar transverse partitions (Fig. 280). If at any time, through this process of concurrent elongation and subdivision, the apical cell become equivalent to, or less than, a hemisphere, it will next divide by means of a longitudinal, or vertical partition; and similar longitudinal partitions will arise in the other segments of the cylinder, as soon as it comes about that their length (in the direction of the axis) is less than their breadth.

But when we think of this structure in the solid, we at once perceive that each of these flattened segments, into which our cylinder divided to begin with, is equivalent to a flattened circular disc; and its further division will accordingly tend to proceed like

* In the following account I follow closely on the lines laid down by Berthold; *Protoplasmamechanik*, cap. vii. Many botanical phenomena identical and similar to those here dealt with are elaborately discussed by Sachs in his *Physiology of Plants* (chap. xxvii, pp. 431–459, Oxford, 1887), and in his earlier papers, Ueber die Anordnung der Zellen in jüngsten Pflanzentheilen, and Ueber Zellenanordnung und Wachsthum (*Arb. d. botan. Inst. Würzburg*, 1877/78). But Sachs's treatment differs entirely from that which I adopt and advocate here: his explanations being based on his "law" of rectangular succession, and involving complicated systems of confocal conics, with their orthogonally intersecting ellipses and hyperbolas.

any other flattened disc, namely into four quadrants, and afterwards
by anticlines and periclines in the usual way. A section across the
cylinder, then, will tend to shew us precisely the same arrangements
as we have already so fully studied in connection with the typical
division of a circular cell into quadrants, and of these quadrants
into triangular and quadrangular portions, and so on.

But there are other possibilities to be considered, in regard to
the mode of division of the elongating quasi-cylindrical portion, as
it gradually develops out of the growing and bulging quadrantal
cell; for the manner in which this latter cell divides will simply
depend upon the form it has assumed before each successive act

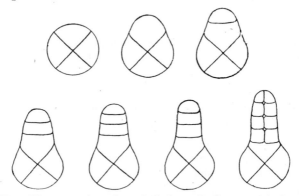

Fig. 280. Diagrammatic, or hypothetical, result of asymmetrical growth.

of division takes place, that is to say upon the ratio between its
rate of growth and the frequency of its successive divisions. For,
as we have already seen, if the growing cell attain a markedly
oblong or cylindrical form before division ensues, then the partition
will arise transversely to the long axis; if it be but a little more
than a hemisphere, it will divide by an oblique partition; and if
it be less than a hemisphere (as it may come to be after successive
transverse divisions) it will divide by a vertical partition, that is
to say by one coinciding with its axis of growth. An immense
number of permutations and combinations may arise in this way,
and we must confine our illustrations to a small number of cases.
The important thing is not so much to trace out the various
conformations which may arise, but to grasp the fundamental
principle: which is, that the forces which dominate the *form* of

each cell regulate the manner of its subdivision, that is to say the form of the new cells into which it subdivides; or in other words, the form of the growing organism regulates the form and number of the cells which eventually constitute it. The complex cell-network is not the cause but the result of the general configuration, which latter has its essential cause in whatsoever physical and chemical processes have led to a varying velocity of growth in one direction as compared with another.

In the annexed figure of an embryo of *Sphagnum* we see a mode of development almost precisely corresponding to the hypothetical case which we have just described—the case, that is to say, where one of the four original quadrants of the mother-cell is the chief agent in future growth and development. We see at the base of our first figure (*a*), the three stationary, or undivided quadrants, one of which has further slowly divided in the stage *b*. The active quadrant has grown quickly into a cylindrical structure, which inevitably divides, in the next place, into a series of transverse partitions; and accordingly, this mode of development carries with it the presence of a single "apical cell." whose lower wall is a spherical surface with its convexity downwards. Each cell of the subdivided cylinder now appears as a more

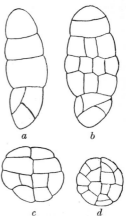

Fig. 281. Development of *Sphagnum*. After Campbell.

or less flattened disc, whose mode of further subdivision we may prognosticate according to our former investigation, to which subject we shall presently return.

(2) In the next place, still keeping to the case where only one of the original quadrant-cells continues to grow and develop, let us suppose that this growing cell falls to be divided when by growth it has become just a little greater than a hemisphere; it will then divide, as in Fig. 282, 2, by an oblique partition, in the usual way, whose precise position and inclination to the base will depend entirely on the configuration of the cell itself, save only, of course, that we may have also to take into account the possibility of the division being into two unequal halves. By our hypothesis, the

growth of the whole system is mainly in a vertical direction, which
is as much as to say that the more actively growing protoplasm,
or at least the strongest osmotic force, will be found near the apex;
where indeed there is obviously more external surface for osmotic
action. It will therefore be that one of the two cells which contains,
or constitutes, the apex which will grow more rapidly than the
other, and which therefore will be the first to divide; and indeed
in any case, it will usually be this one of the two which will tend
to divide first, inasmuch as the triangular and not the quadrangular
half is bound to constitute the apex*. It is obvious that (unless

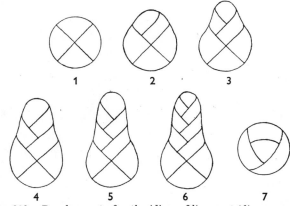

Fig. 282.　Development of antheridium of liverwort (diagrammatic).

the act of division be so long postponed that the cell has become
quasi-cylindrical) it will divide by another oblique partition, starting
from, and running at right angles to, the first. And so division
will proceed by oblique alternate partitions, each one tending to
be, at first, perpendicular to that on which it is based and also to
the peripheral wall; but all these points of contact soon tending,
by reason of the equal tensions of the three films or surfaces which
meet there, to form angles of 120°. There will always be a single
apical cell, of a·triangular form. The developing antheridium of a
liverwort (*Riccia*) is ,a typical example of such a case. In Fig. 283
which represents a "gemma" of a moss, we see just the same thing;
with this addition, that here the lower of the two original cells has
grown even more quickly than the other, constituting a long cylin-

* Cf. p. 590.

drical stalk, and dividing in accordance with its shape by means of transverse septa. In all such cases the cells may continue to sub-divide, and the manner in which they do so must depend upon their own proportions; and in all cases there will sooner or later be a tendency to the formation of periclinal walls, cutting off an epidermal layer of cells, as Fig. 284 illustrates very well.

The method of division by means of oblique partitions is a common one in the case of "growing points"; for it evidently includes all cases in which the act of cell-division does not lag far behind that elongation which is determined by the specific rate of

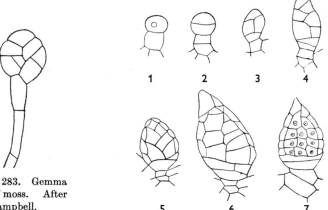

Fig. 283. Gemma
of moss. After
Campbell.

Fig. 284. Development of antheridium
of *Riccia*. After Campbell.

growth. And it is also obvious that, under a common type, there must here be included a variety of cases which will, at first sight, present a very different appearance one from another. For instance, in Fig. 285 which represents a growing shoot of *Selaginella*, and somewhat less diagrammatically in the young embryo of *Junger-mannia* (Fig. 286), we have the appearance of an almost straight vertical partition running up in the axis of the system, and the primary cell-walls are set almost at right angles to it—almost transversely, that is to say, to the outer walls and to the long axis of the structure. We soon recognise, however, that the difference is merely a difference of degree. The more remote the partitions are, that is to say the greater the velocity of growth relatively to that of division, the less abrupt will be the alternate kinks or curvatures of the portions which lie along the axis, and the more will these portions appear to constitute a single unbroken wall.

(3) But an appearance nearly, if not quite, indistinguishable from this may be got in another way, namely, when the original growing cell is so nearly hemispherical that it is actually divided

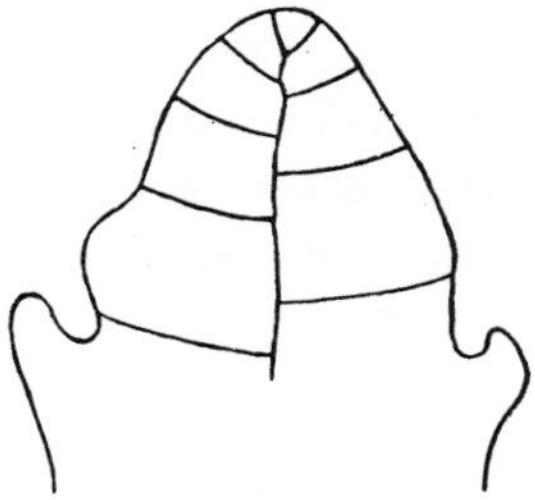

Fig. 285.   Section of growing shoot
of *Selaginella*, diagrammatic.

Fig. 286.   Embryo of *Jungermannia*.
After Kienitz-Gerloff.

by a vertical partition into two quadrants, and when from this vertical partition, as it elongates, lateral partition-walls arise on either side.   Then, by the tensions exercised by these, the vertical partition

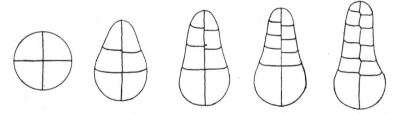

Fig. 287.

will be bent into little portions set at 120° one to another, and the whole will come to look just like that which, in the former case, was made up of parts of many successive oblique partitions (Fig. 287).

Let us now, in one or two cases, follow out a little further the stages of cell-division whose beginnings we have studied in the last paragraphs.   In the antheridium of *Riccia*, after successive oblique partitions have produced the longitudinal series of cells shewn in Fig. 284, 4, it is plain that the next partitions will arise periclinally, that is to say parallel to the outer wall, which coincides with the short axis of the oblong cells.   The effect is to produce an epidermal layer, whose cells subdivide further by partitions perpendicular to the surface, that is to say crossing the flattened cells by their shortest diameter. The inner mass consists of cells which are still more or less oblong, or which become so in process of growth; and these again divide, parallel to their short axes, into squarish cells, which as usual, by the mutual tension of their walls, become hexagonal as seen in a plane section. There is a clear distinction, then, in form as well as in position, between the outer covering-cells and those which lie within this envelope; the latter are reduced to a condition which fulfils the mechanical function of a protective coat, while the former undergo less modification, and become the actively living, reproductive elements.

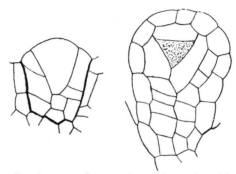

Fig. 288.   Development of sporangium of *Osmunda*.   After Bower.

In Fig. 288 is shewn the development of the sporangium of a fern (*Osmunda*). We may trace here the common phenomenon of a series of oblique partitions, built alternately on one another, and cutting off a conspicuous triangular apical cell. Over the whole system an epidermal layer is formed, in the manner we have described; and in this case it covers the apical cell also, owing to the fact that it was of such dimensions that, at one stage of growth, a periclinal partition wall, cutting off its outer end, was indicated as of less area than an anticlinal one. This periclinal wall cuts down the apical cell to the proportions, very nearly, of an equilateral triangle, but the solid form of the cell is obviously that of a tetrahedron with curved faces; and accordingly, the least possible partitions by which further subdivision can be effected will run successively parallel to its four sides (or its three sides when we confine ourselves to the appearances as seen in section). The effect is to cut off on each side of the apical cell a characteristically flattened cell, oblong as seen in section, still leaving a triangular (or strictly speaking, a tetrahedral) one in the centre. The oblong cells, which constitute no specific structure and perform no specific physiological function*, but which merely represent certain directions in space towards which the whole system of partitioning has gradually led, are called by botanists the "tapetum." The active growing tetrahedral cell which lies between them, and from which in a sense every other cell in the system has been either directly or indirectly segmented off, still manifests its vigour and activity, and becomes, by internal subdivision, the mother-cell of the spores.

In all these cases, for simplicity's sake, we have merely considered the appearances presented in a single longitudinal plane of optical section. But it is not difficult to interpret from these appearances what would be seen in another plane, for instance in a transverse section. In our first example, for instance, that of the developing embryo of *Sphagnum* (Fig. 281 *c*, *d*), we see that, at appropriate levels, the cells of the original cylindrical row have divided into transverse rows of four, and then of eight cells. We may be sure that the four cells represent, approximately, quadrants of a cylindrical disc, the four cells, as usual, not meeting in a point, but intercepted by a small intermediate partition. Again, where we have a plate of eight cells, we may well imagine that the eight octants are arranged in what we have found to be the way naturally resulting from the division of four quadrants, that is to say into alternately triangular and quadrangular portions; and this is found by means of sections to be the case. The figure is precisely comparable to our previous diagrams of the arrangement of eight cells in a dividing disc, save only that, in two cases, the cells have already undergone a further subdivision.

It follows that we are apt to meet with this characteristic figure, in one or other of its possible and strictly limited variations, in the cross-sections of many growing structures, just as we have already

---

* This is not to say that Nature makes *no use* of the tapetal cells. In the end they break down and contribute to the growth of the spore-mother-cell: very much as the "superfluous" eggs in a fly's ovary contribute yolk-material to the developing ovum.

seen it appear in cases where the entire system consists of eight
cells only.   For example, we have it in a section of a young embryo
of a moss (*Phascum*), and again, in a section of an embryo of a
fern (*Adiantum*).   In Fig. 290, shewing a section through a growing
frond of a sea-weed (*Girardia*), we have a case where the partitions
forming the eight octants have conformed to the usual type;   but
instead of the usual division by periclines of the four quadrangular
spaces, these latter are dividing by means of oblique septa, apparently
owing to the fact that the cell is not dividing into two equal, but
into two unequal portions.   In this last figure we have a peculiar
look of stiffness or formality, such that it appears at first to bear
little resemblance to the rest.   The explanation is of the simplest.
The mode of partitioning differs little (except to some slight extent

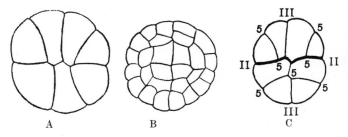

Fig. 289.   (A, B) Sections of younger and older embryos of *Phascum*;
(C) do. of *Adiantum*.   After Kienitz-Gerloff.

in the way already mentioned) from the normal type;   but in this
case the partition walls are so thick and become so soon comparatively solid and rigid, that the secondary curvatures due to their
successive mutual tractions are here imperceptible.

A curious and beautiful case, apparently aberrant but which
would doubtless be found conforming strictly to physical laws if
only we clearly understood the actual conditions, is indicated in
the development of the antheridium of a fern, as described by
Strasbürger.   Here the antheridium develops from a single cell,
which (Fig. 291) has grown to something more than a hemisphere;
and the first partition, instead of stretching transversely across the
cell, as we should expect it to do if the cell were actually spherical,
has as it were sagged down to come in contact with the base, and
so to develop into an annular partition, running round the lower

margin of the cell. The phenomenon is precisely identical to that
bisection of a quadrant by means of a circular arc, of which we
spoke on p. 581, and the annular film is very easy to reproduce
by means of a soap-bubble in the bottom of a cylindrical dish or
beaker. The next partition is a periclinal one, concentric with the
outer surface of the young antheridium; and this in turn is followed
by a concave partition which cuts off the apex of the original cell:
but which becomes connected with the second, or periclinal partition
in precisely the same annular fashion as the first partition did with
the base of the little antheridium. The result is that, at this stage,
we have four cell-cavities in the little antheridium: (1) a central

Fig. 290. Section through frond of
Girardia sphaceldria. After Goebel.

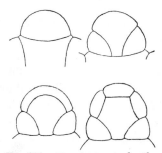

Fig. 291. Development of antheridium of *Pteris*. After Strasbürger.

cavity; (2) an annular space around the lower margin; (3) a narrow
annular or cylindrical space around the sides of the antheridium;
and (4) a small terminal or apical cell. It is evident that the
tendency, in the next place, will be to subdivide the flattened
external cells by means of anticlinal partitions, and so to convert
the whole structure into a single layer of epidermal cells, surrounding
a central cell within which, in course of time, the antherozoids are
developed.

The foregoing account deals only with a few elementary phenomena, and may seem to fall far short of an attempt to deal in general
with "the forms of tissues." But it is the principle involved, and
not its ultimate and very complex results, that we can alone attempt
to grapple with. The stock-in-trade of mathematical physics, in
all the subjects with which that science deals, is for the most part

made up of simple, or simplified, cases of phenomena which in their actual and concrete manifestations are usually too complex for mathematical analysis; hence, even in physics, the full mechanical explanation of a phenomenon is seldom if ever more than the "cadre idéal" towards which our never-finished picture extends. When we attempt to apply the same methods of mathematical physics to our biological and histological phenomena, we need not wonder if we be limited to illustrations of a simple kind, which cover but a small part of the phenomena with which histology has to do.   But yet it is only relatively that these phenomena to which we have found the method applicable are to be deemed simple and few.   They go already far beyond the simplest phenomena of all, such as we see in the dividing *Protococcus*, and in the first stages, two-celled or four-celled, of the segmenting egg. They carry us into stages where the cells are already numerous, and where the whole conformation has become by no means easy to depict or visualise, without the help and guidance which the phenomena of surface-tension, the laws of equilibrium and the principle of minimal areas are at hand to supply.   And so far as we have gone, and so far as we can discern, we see no sign of the guiding principles failing us, or of the simple laws ceasing to hold good.

CHAPTER IX

ON CONCRETIONS, SPICULES, AND SPICULAR SKELETONS

THE deposition of inorganic material in the living body, usually in the form of calcium salts or of silica, is a common phenomenon. It begins by the appearance of small isolated particles, crystalline or non-crystalline, whose form has little relation or none to the structure of the organism; it culminates in the complex skeletons of the vertebrate animals, in the massive skeletons of the corals, or in the polished, sculptured and mathematically regular molluscan shells. Even among very simple organisms, such as diatoms, radiolarians, foraminifera or sponges, the skeleton displays extraordinary variety and beauty, whether by reason of the intrinsic form of its elementary constituents or the geometric symmetry with which these are interconnected and arranged.

With regard to the form of these various structures (and this is all that immediately concerns us here), we have to do with two distinct problems, which merge with one another though they are theoretically distinct. For the form of the spicule or other skeletal element may depend solely on its chemical nature, as for instance, to take a simple but not the only case, when it is purely crystalline; or the inorganic material may be laid down in conformity with the shapes assumed by cells, tissues or organs, and so be, as it were, moulded to the living organism; and there may well be intermediate stages in which both phenomena are simultaneously at work, the molecular forces playing their part in conjunction with the other forces inherent in the system.

So far as the problem is a purely chemical one we must deal with it very briefly indeed: all the more because special investigations regarding it have as yet been few, and even the main facts of the case are very imperfectly known. This at least is clear, that the phenomena with which we are about to deal go deep into the subject of colloid chemistry, and especially that part of the science which deals with colloids in connection with surface phenomena. It is to the special student of the chemistry and physics of the colloids that we must look for the elucidation of our problem*.

In the first and simplest part of our subject, the essential problem is the problem of crystallisation in presence of colloids. In the cells of plants true crystals are found in comparative abundance, and consist, in the majority of cases, of calcium oxalate. In the stem and root of the rhubarb for instance, in the leaf-stalk of *Begonia* and in countless other cases, sometimes within the cell, sometimes in the substance of the cell-wall, we find large and well-formed crystals of this salt; their varieties of form, which are extremely numerous, are simply the crystalline forms proper to the salt itself, and belong to the two systems, cubic and monoclinic, in one or other of which, according to the amount of water of crystallisation, this salt is known to crystallise. When calcium oxalate crystallises according to the latter system (as it does when its molecule is combined with two molecules of water), the microscopic crystals have the form of fine needles, or "raphides"; these are very common in plants, and may be artificially produced when the salt is crystallised out in presence of glucose or of dextrin†.

Calcium carbonate, on the other hand, when it occurs in plant-cells, as it does abundantly (for instance in the "cystoliths" of the Urticaceae and Acanthaceae, and in great quantities in *Melobesia* and the other calcareous or "stony" algae), appears in the form of fine rounded granules, whose inherent crystalline structure is only revealed (like that of a molluscan shell) under polarised light. Among animals, a skeleton of carbonate of lime occurs under a multitude of forms, of which we need only mention a few of the most conspicuous. The spicules of the calcareous sponges are triradiate, occasionally quadriradiate, bodies, with pointed rays, not crystalline in outward form but with a definitely crystalline internal

---

* There is much information regarding the chemical composition and mineralogical structure of shells and other organic products in H. C. Sorby's Presidential Address to the Geological Society (*Proc. Geol. Soc.* 1879, pp. 56–93); but Sorby failed to recognise that association with "organic" matter, or with colloid matter whether living or dead, introduced a new series of purely physical phenomena.

† Julien Vesque, Sur la production artificielle de cristaux d'oxalate de chaux semblables à ceux qui se forment dans les plantes, *Ann. Sc. Nat. (Bot.)* (5) XIX, pp. 300–313, 1874.

structure; we shall return again to these, and find for them what would seem to be a satisfactory explanation of their form. Among the Alcyonarian zoophytes we have a great variety of spicules*, which are sometimes straight and slender rods, sometimes flattened and more or less striated plates, and still more often disorderly aggregations of micro-crystals, in the form of rounded or branched concretions with rough or knobby surfaces† (Figs. 292, 298). A third type, presented by several very different things, such as a

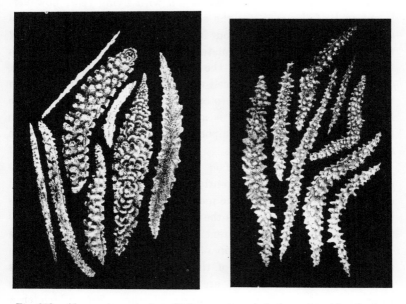

Fig. 292. Alcyonarian spicules: *Siphonogorgia* and *Anthogorgia*. After Studer.

pearl or the ear-bone of a bony fish, consists of a more or less rounded body, sometimes spherical, sometimes flattened, in which the calcareous matter is laid down in concentric zones, denser and clearer layers alternating with one another. In the development of the molluscan shell and in the calcification of a bird's egg or a crab's shell, small spheroidal bodies with similar concentric striation make their appearance; but instead of remaining separate they become

* Cf. Kölliker, *Icones Histologicae*, 1864, p. 119, etc.
† In rare cases, these shew a single optic axis and behave as individual crystals: W. J. Schmidt, *Arch. f. Entw. Mech.* li. pp. 509–551, 1922.

crowded together, and in doing so are apt to form a pattern of hexagons. In some cases the carbonate of lime, on being dissolved away by acid, leaves behind it a certain small amount of organic residue; in many cases other salts, such as phosphates of lime, ammonia or magnesia, are present in small quantities; and in most cases, if not all, the developing spicule or concretion is somehow so associated with living cells that we are apt to take it for granted that it owes its form to the constructive or plastic agency of these.

The appearance of direct association with living cells, however, is apt to be fallacious; for the actual *precipitation* takes place, as a rule, not in actively living, but in dead or at least inactive tissue*; that is to say in the "formed material" or matrix which accumulates round the living cells, or in the interspaces between these latter, or, as often happens, in the cell-wall or cell-membrane rather than within the substance of the protoplasm itself. We need not go the length of asserting that this is a rule without exception; but, so far as it goes, it is of great importance and to its consideration we shall presently return†.

Cognate with this is the fact that, at least in some cases, the organism can go on, in apparently unimpaired health, when stinted or even wholly deprived of the material of which it is wont to make its spicules or its shell. Thus the eggs of sea-urchins reared in lime-free water develop, in apparent health and comfort, into larvae which lack the usual skeleton of calcareous rods: and in which, accordingly, the long arms of the Pluteus larva, which the rods should support and extend, are entirely absent‡. Again, when foraminifera are kept for generations in water from which they gradually exhaust the lime, their shells grow hyaline and transparent, and dwindle to a mere chitinous pellicle; on the other hand,

---

* In an interesting paper by Robert Irvine and Sims Woodhead on the Secretion of carbonate of lime by animals (*Proc. R.S.E.* xv, pp. 308–316; xvi, pp. 324–351, 1889–90) it is asserted (p. 351) that "lime salts, of whatever form, are deposited *only* in vitally inactive tissue."

† The tube of *Teredo* shews no trace of organic matter, but consists of irregular prismatic crystals: the whole structure "being identical with that of small veins of calcite, such as are seen in thin sections of rocks" (Sorby, *Proc. Geol. Soc.* 1879, p. 58). This, then, would seem to be a somewhat exceptional case of a shell laid down completely outside of the animal's external layer of organic substance.

‡ Cf. Pouchet and Chabry, *C.R. Soc. Biol. Paris* (9), I, pp. 17–20, 1889; *C.R. Acad. Sci.* cviii, pp. 196–198, 1889.

in the presence of excess of lime their shells become much altered, are strengthened with various ridges or "ornaments," and come to resemble other varieties and even "species*."

The crucial experiment, then, is to attempt the formation of similar spicules or concretions apart from the living organism. But however feasible the attempt may be in theory, we must be prepared to encounter many difficulties; and to realise that, though the reactions involved may be well within the range of physical chemistry, yet the actual conditions of the case may be so complex, subtle and delicate that only now and then, and only in the simplest of cases, has it been found possible to imitate the natural objects successfully. Such an attempt is part of that wide field of enquiry through which Stéphane Leduc and other workers have sought to produce, by synthetic means, forms similar to those of living things; but it is a circumscribed and well-defined part of that wider investigation†.

When we find ourselves investigating the forms assumed by chemical compounds under the peculiar circumstances of association with a living body, and when we find these forms to be characteristic or recognisable, and somehow different from those which the same substance is wont to assume under other circumstances, an analogy, captivating though perhaps remote, presents itself to our minds between this subject of ours and certain synthetic problems of the organic chemist. There is doubtless an essential difference, as well as a difference of scale, between the visible form of a spicule or con-

---

* Cf. Heron-Allen, *Phil. Trans.* (B), ccvi, p. 262, 1915.

† Leduc's artificial growths were mostly obtained by introducing salts of the heavy metals or alkaline earths into solutions which form with them a "precipitation-membrane"—as when we introduce copper sulphate into a ferrocyanide solution. See his *Mechanism of Life*, 1911, ch. x, for copious references to other works on the "artificial production of organic forms." Closely related to Leduc's experiments are those of Denis Monnier and Carl Vogt, Sur la fabrication artificielle des formes des éléments organiques, *Journ. de l'Anat.* xviii, pp. 117–123, 1882; cf. Moritz Traube, Zur Geschichte der mechanischen Theorie des Wachstums der organischen Zelle, *Botan. Ztg.* xxxvi, 1878. Cf. also A. L. Herrera, Sur les phénomènes de vie apparente observés dans les émulsions de carbonate de chaux dans la silice gélatineuse, *Mem. Soc. Alzate, Mexico*, xxvi, 1908; Los Protobios, *Boll. de la Dir. de Estud. Biolog., Mexico*, i, pp. 607–631, and other papers. Also (*int. al.*) R. S. Lillie and E. N. Johnston, Precipitation-structures simulating organic growth, *Biol. Bull.* xxxiii, p. 135, 1917; xxxvi, pp. 225–272, 1919; *Scientific Monthly*, Feb. 1922, p. 125; H. W. Morse, C. H. Warren and J. D. H. Donnay, Artificial spherulites, etc., *Amer. Jl. of Sci.* (5) xxiii, pp. 421–439, 1932.

cretion and the hypothetical form of an individual molecule. But molecular form is a very important concept; and the chemist has not only succeeded, since the days of Wöhler, in synthetising many substances which are characteristically associated with living matter, but his task has included the attempt to account for the molecular *forms* of certain "asymmetric" substances—glucose, malic acid and many more—as they occur in Nature. These are bodies which, when artificially synthetised, have no optical activity, but which, as we actually find them in organisms, turn (when *in solution*) the plane of polarised light in one direction rather than the other; thus dextroglucose and laevomalic acid are common products of plant metabolism, but dextromalic acid and laevoglucose do not occur in Nature at all. The optical activity of these bodies depends, as Pasteur shewed eighty years ago*, upon the form, right-handed or left-handed, of their molecules, which molecular asymmetry further gives rise to a corresponding right- or left-handedness (or enantiomorphism) in the crystalline aggregates. It is a distinct problem in organic or physiological chemistry, and by no means without its interest for the morphologist, to discover how it is that Nature, for each particular substance, habitually builds up, or at least selects, its molecules in a one-sided fashion, right-handed or left-handed as the case may be. It will serve us no better to assert that this phenomenon has its origin in "fortuity" than to repeat the Abbé Galiani's saying, *"les dés de la nature sont pipés."*

The problem is not so closely related to our immediate subject that we need discuss it at length; but it has its relation, such as it is, to the general question of *form* in relation to vital phenomena, and it has its historic interest as a theme of long-continued discussion. According to Pasteur, there lay in the molecular asymmetry of the natural bodies and their symmetry when artificially produced, one of the most deep-seated differences between vital and non-vital phenomena: he went further, and declared that "this was perhaps the *only* well-marked line of demarcation that can at present [1860] be drawn between the chemistry of dead and of living matter." Nearly forty years afterwards the same theme was pursued and

* Lectures on the molecular asymmetry of natural organic compounds, *Chemical Soc. of Paris*, 1860; also in Ostwald's *Klassiker d. exact. Wiss.* No. 28, and in *Alembic Club Reprints*, No. 14, Edinburgh, 1897; cf. G. M. Richardson, *Foundations of Stereochemistry*, New York, 1901.

elaborated by Japp in a celebrated lecture*, and the distinction still has its weight, I believe, in the minds of many chemists. "We arrive at the conclusion," said Professor Japp, "that the production of single asymmetric compounds, or their isolation from the mixture of their enantiomorphs, is, as Pasteur firmly held, the prerogative of life. Only the living organism, or the living intelligence with its conception of asymmetry, can produce this result. Only asymmetry can beget asymmetry." In these last words (which, so far as the chemist and the biologist are concerned, we may acknowledge to be true†) lies the crux of the difficulty.

Observe that it is only the first beginnings of chemical asymmetry that we need discover; for when asymmetry is once manifested, it is not disputed that it will continue "to beget asymmetry." A plausible suggestion is at hand, which if it were confirmed and extended would supply or at least sufficiently illustrate the kind of explanation that is required. We know that when ordinary non-polarised light acts upon a chemical substance, the amount of chemical action is proportionate to the amount of light absorbed. We know in the second place‡ that light circularly polarised is absorbed in certain cases in different amounts by the right-handed or left-handed varieties of an asymmetric substance. And thirdly, we know that a portion of the light which comes to us from the sun is already plane-polarised light, which becomes in part circularly polarised, by reflection (according to Jamin) at the surface of the sea, and then rotated in a particular direction under the influence of terrestrial magnetism. We only require to be assured that the relation between absorption of light and chemical activity will continue to hold good in the case of circularly polarised light; that is to say that the formation of some new substance or other, under the influence of light so polarised, will proceed asymmetrically in consonance with the asymmetry of the light itself; or conversely,

* F. R. Japp, Stereochemistry and vitalism, *Brit. Ass. Rep.* (Bristol), 1898, p. 813; cf. also a voluminous discussion in *Nature*, 1898–99.

† They represent the general theorem of which particular cases are found, for instance, in the asymmetry of the ferments (or *enzymes*) which act upon asymmetrical bodies, the one fitting the other, according to Emil Fischer's well-known phrase, as lock and key. Cf. his Bedeutung der Stereochemie für die Physiologie, *Z. f. physiol. Chemie*, v, p. 60, 1899, and various papers in the *Ber. d. d. chem. Ges.* from 1894.

‡ Cf. Cotton, *Ann. de Chim. et de Phys.* (7), viii, pp. 347–432 (cf. p. 373), 1896.

that the asymmetrically polarised light will tend to more rapid decomposition of those molecules by which it is chiefly absorbed. This latter proof is said to be furnished by Byk*, who asserts that certain tartrates become unsymmetrical under the continued influence of the asymmetric rays. Here then we seem to have an example, of a particular kind and in a particular instance, an example limited but yet crucial if confirmed, of an asymmetric force, non-vital in its origin, which might conceivably be the starting-point of that asymmetry which is characteristic of so many organic products.

The mysteries of organic chemistry are great, and the differences between its processes or reactions as they are carried out in the organism and in the laboratory are many†; the actions, catalytic and other, which go on in the living cell are of extraordinary complexity. But the contention that they are different in kind from ordinary chemical operations, or that in the production of single asymmetric compounds there is actually, as Pasteur maintained, a "prerogative of life," would seem to be no longer tenable. Our historic interest in the whole question is increased by the fact, or the great probability, that "the tenacity with which Pasteur fought against the doctrine of spontaneous generation was not unconnected with his belief that chemical compounds of one-sided symmetry could not arise save under the influence of life‡." But the question whether spontaneous generation be a fact or not does not depend upon theoretical considerations; our negative response is based, and is soundly based, on repeated failures to demonstrate its occurrence. Many a great law of physical science, not excepting gravitation itself, has no higher claim on our acceptance.

Let us return from this digression to the general subject of the forms assumed by certain chemical bodies when deposited or precipitated within the organism, and to the question of how far these forms may be artificially imitated or theoretically explained.

* A. Byk, Zur Frage der Spaltbarkeit von Racemverbindungen durch zirkular-polarisiertes Licht, ein Beitrag zur primären Entstehung optisch-activer Substanzen, Zeitsch. f. physikal. Chemie, XLIX, pp. 641–687, 1904. It must be admitted that positive evidence on these lines is still awanting.

† Cf. (int. al.) Emil Fischer, Untersuchungen über Aminosäuren, Proteine, etc. Berlin, 1906.

‡ Japp, loc. cit. p. 828.

Mr George Rainey, of St Thomas's Hospital (of whom we have spoken before), and Professor P. Harting, of Utrecht, were the first to deal with this specific problem. Rainey published, between 1857 and 1861, a series of valuable and thoughtful papers to shew that shell and bone and certain other organic structures were formed "by a process of molecular coalescence, demonstrable in certain artificially formed products*." Harting, after thirty years of experimental work, published in 1872 a paper, which has become classical, entitled *Recherches de morphologie synthétique, sur la production artificielle de quelques formations calcaires organiques* †; his aim was to pave the way for a "morphologie synthétique," as Wöhler had laid the foundations of a "chimie synthétique" by his classical discovery forty years before.

Rainey and Harting used similar methods—and these were such as other workers have continued to employ—partly with the direct object of explaining the genesis of organic forms and partly as an integral part of what is now known as Colloid Chemistry. The gist of the method was to bring some soluble salt of lime, such as the chloride or nitrate, into solution within a colloid medium, such as gum, gelatine or albumin; and then to precipitate it out in the form of some insoluble compound, such as the carbonate or oxalate. Harting found that, when he added a little sodium or potassium carbonate to a concentrated solution of calcium chloride in albumin, he got at first a gelatinous mass, or "colloid precipitate": which slowly transformed by the appearance of tiny microscopic particles,

---

* George Rainey, On the elementary formation of the skeletons of animals, and other hard structures formed in connection with living tissue, *Brit. and For. Med. Ch. Rev.* xx, pp. 451–476, 1857; published separately with additions, 8vo, London, 1858. For other papers by Rainey on kindred subjects see *Q.J.M.S.* vi (*Tr. Microsc. Soc.*), pp. 41–50, 1858; vii, pp. 212–225, 1859; viii, pp. 1–10, 1860; i (n.s.), pp. 23–32, 1861. Cf. also W. Miller Ord, *On the influence exercised by colloids upon crystalline form*, pp. x, 179, 1874; cf. also *Q.J.M.S.* xii, pp. 219–239, 1872; also the early but still interesting observations of Mr Charles Hatchett, Chemical experiments on zoophytes; with some observations on the component parts of membrane, *Phil. Trans.* 1800, pp. 327–402. For early references to sclerites formed in cells, see (e.g.) L. Selenka, *Z.f.w.Z.* xxxiii, p. 45, 1879 and R. Semon, *Mitth. Zool. St. Neapel*, vii, p. 288, 1886 (both in holothurians); Blochmann, *Die Epithelfrage bei Cestoden u. Trematoden*, Hamburg, 1896; also Leger's Observations on crystals of calcium oxalate in the cysts of *Lithocystis Schneideri*, *A.M.N.H.* (6), xviii, p. 479, 1895.

† Cf. *Q.J.M.S.* xii, pp. 118–123, 1872.

shewing, as they grew larger, the typical Brownian movement. So far, much the same phenomena were witnessed whether the solution were albuminous or not, and similar appearances indeed had been witnessed and recorded by Gustav Rose, so far back as 1837*; but in the later stages the presence of albuminoid matter made a great difference. Now, after a few days, the calcium carbonate was seen to be deposited in the form of large rounded concretions, each with a more or less distinct central nucleus and with a surrounding structure at once radiate and concentric; the presence of concentric zones or lamellae, alternately dark and clear, was especially characteristic. These round "calcospherites" shewed a tendency to aggregate in layers, and then to assume polyhedral, often regularly hexagonal, outlines. In this latter condition they closely resemble the early stages of calcification in a molluscan (Fig. 296), or still more in a crustacean shell†; while in their isolated condition they

---

* Cf. Quincke, Ueber unsichtbare Flüssigkeitsschichten, etc., *Ann. der Physik* (4), VII, pp. 631–682, 701–744, 1902.

† See for instance other excellent illustrations in Carpenter's article "Shell," in Todd's *Cyclopœdia*, IV, pp. 556–571, 1847–49. According to Carpenter, the shells of the mollusca (and also of the crustacea) are "essentially composed of *cells*, consolidated by a deposit of carbonate of lime in their interior." That is to say, Carpenter supposed that the spherulites or calcospherites of Harting were, to begin with, just so many living protoplasmic cells. Soon afterwards, however, Huxley pointed out that the mode of formation, while at first sight "irresistibly suggesting a cellular structure...is in reality nothing of the kind," but "is simply the result of the concretionary manner in which the calcareous matter is deposited"; *ibid.* art. "Tegumentary organs," V, p. 487, 1859. Quekett (*Lectures on Histology*, II, p. 393, 1854, and *Q.J.M.S.* XI, pp. 95–104, 1863) supported Carpenter; but Williamson (Histological features in the shells of the Crustacea, *Q.J.M.S.* VIII, pp. 35–47, 1860) amply confirmed Huxley's view, which in the end Carpenter himself adopted (*The Microscope*, 1862, p. 604). A like controversy arose later in regard to corals. Mrs Gordon (M. M. Ogilvie) asserted that the coral was built up "of successive layers of calcified cells, which hang together at first by their cell-walls, and ultimately, as crystalline changes continue, form the individual laminae of the skeletal structures" (*Phil. Trans.* CLXXXVII, p. 102, 1896): whereas von Koch had figured the coral as formed out of a mass of "Kalkconcremente" or "crystalline spheroids," laid down outside the ectoderm, and precisely similar both in their early rounded and later polygonal stages (though von Koch was not aware of the fact) to the calcospherites of Harting (Entw. d. Kalkskelettes von Astroides, *Mitth. Zool. St. Neapel*, III, pp. 284–290, pl. XX, 1882). Lastly, W. H. Bryan finds all ordinary corals (*Hexacoralla*) to be mineral aggregates formed by "spherulitic crystallisation," due in turn to the presence of a colloid matrix secreted by certain areas of ectoderm; see *Proc. R.S. Queensland*, LII, pp. 41–53, 1940; *Univ. of Queensland Papers, Geology*, II, 4 and 5, 1941. Cf. J. E. Duerden, On *Siderastraea*, *Carnegie Inst. Washington*, 1904, p. 34.

Fig. 293.   Calcospherites, or concretions of calcium carbonate,
deposited in white of egg.
After Harting.

Fig. 294.   A single
calcospherite, with
central "nucleus,"
and striated, irides-
cent border. After
Harting.

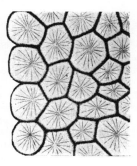

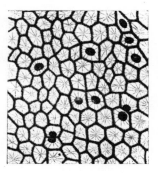

Fig. 295.   Later stages in the same experiment.

A   B

Fig. 296.   A, Section of shell; B, Section of hinge-tooth of *Mya*.
After Carpenter.

closely resemble the little calcareous bodies in the tissues of a trematode or a cestode worm, or in the oesophageal glands of an earthworm*.

When the albumin was somewhat scanty, or when it was mixed with gelatine, and especially when a little phosphate of lime was added to the mixture, the spheroidal globules tended to become rough, by an outgrowth of spinous or digitiform projections; and

Fig. 297.   Large irregular calcareous concretions, or spicules, deposited in a piece of dead cartilage, in presence of calcium phosphate.   After Harting.

in some cases, but not without the presence of the phosphate†, the result was an irregularly shaped knobby spicule, precisely similar to those which are characteristic of the *Alcyonaria*‡.

The rough spicules of the *Alcyonaria* are extraordinarily variable in shape and size, as, looking at them from the chemist's or the physicist's point of view, we should expect them to be.   Partly upon the form of these spicules, and partly on the general form or mode of branching of the entire colony of polyps, a vast number of separate "species" have been based by systematic

* Cf. Claparède, *Z.f.w.Z.* xix, p. 604, 1869.   On the structure of the molluscan shell, see O. B. Boggild, *K. Vidensk. Selsk. Skr.*, Kjöbenh., (9) ii, 1930.   On nacre, or mother-of-pearl, see Brewster, *Treatise on Optics*, 1853, p. 137; Schmidt, *Die Bausteine der Tierkörper in polarisirtem Licht*, Bonn, 1924.   Also S. Ruma Swamy, *Proc. Ind. Acad. Sci.* (A), i, p. 871, 1935; P. S. Srinivasam, *ibid.* v, pp. 464–483, 1937; and, on the specific qualities of the nacre in the several divisions of the Mollusca, Sir C. V. Raman, *ibid.* pp. 559, etc., 1935.

† On the deposition of phosphates in organisms, cf. Pauli u. Samec, *Biochem. Ztschr.* xvii, p. 235, 1909; *Wiener mediz. Wochenschr.* 1910, pp. 2287–2292.

‡ Spicules much like those of the *Alcyonaria* occur also in a few sponges; cf. (e.g.), Vaughan Jennings, *Journ. Linn. Soc.* xxiii, p. 531, pl. 13, fig. 8, 1891.

zoologists.   But it is now admitted that even in specimens of a single species, from one and the same locality, the spicules may vary immensely in shape and size: and Professor S. J. Hickson declared that after many years of laborious work in striving to determine species of these animal colonies, he felt "quite convinced that we have been engaged in a more or less fruitless task*."

The formation of a tooth is a phenomenon of the same order.   That is to say, "calcification in both dentine and enamel is in great part a physical phenomenon; the actual deposit in both tissues occurs in the form of calcospherites, and the process in mammalian tissue is identical in every point with

Fig. 298.   Additional illustrations of alcyonarian spicules: *Eunicea*.   After Studer.

the same process occurring in lower organisms†."   The ossification of bone, we may be sure, is in the same sense and to the same extent a physical phenomenon.

The typical structure of a calcospherite is no other than that of a pearl, nor does it differ essentially from that of the otolith of a mollusc or of a bony fish.   (The otoliths of the elasmobranch fishes, like those of reptiles and birds, are not developed after this fashion, but are true crystals of calc-spar.)

The effect of surface-tension is manifest throughout these pheno mena.   It is by surface-tension that ultra-microscopic particles are brought together in the first floccular precipitate or coagulum; by

* *Mem. Manchester Lit. and Phil. Soc.* LX, p. 11, 1916.

† J. H. Mummery, On calcification in enamel and dentine, *Phil. Trans.* (B), CCV, pp. 95–111, 1914.

the same agency the coarser particles are in turn agglutinated into
visible lumps; and the form of the calcospherites, whether it be
that of the solitary spheres or that assumed in various stages of
aggregation (e.g. Fig. 300)*, is likewise due to the same agency.

From the point of view of colloid chemistry the whole phenomenon is important and significant; and not the least significant
part is this tendency of the solidified deposits to assume the form
of "spherulites" and other rounded contours.   In the phraseology

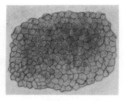

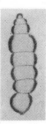

Fig. 299.   A "crust" of close-packed
calcareous concretions, precipitated
at the surface of an albuminous
solution.   After Harting.

Fig. 300.   Aggregated calcospherites.   After Härting.

of that science, we are dealing with a *two-phase* system, which
finally consists of solid particles in suspension in a liquid—a *disperse
phase* in a *dispersion medium*.   In accordance with a rule first
recognised by Ostwald, when a substance begins to separate out
from a solution, so making its appearance as a *new phase*, it always
makes its appearance first as a liquid†.   Here is a case in point.
The minute quantities of material, on their way from a state of
solution to a state of "suspension," pass through a liquid to a solid
form; their temporary sojourn in the former leaves its impress in
the rounded contours which surface-tension brought about while the
little aggregate was still labile or fluid: while coincidently with this
surface-tension effect, crystallisation tends to take place throughout
the little liquid mass, or in such portions of it as have not yet consolidated and crystallised.

* The artificial concretion represented in Fig. 300 is identical in appearance
with the concretions found in the kidney of *Nautilus*, as figured by Willey (*Zoological
Results*, p. lxxvi, Fig. 2, 1902).

† This rule, undreamed of by Errera, supports and justifies his cardinal
assumption (of which we have had so much to say in discussing the forms of cells
and tissues) that the *incipient* cell-wall behaves as, and indeed actually is, a liquid
film (cf. p. 482).

Where we have simple aggregates of two or three calcospherites the resulting figure is that of so many contiguous soap-bubbles. In other cases composite forms result which are not so easily explained, but which, if we could only account for them, would be of very great interest to the biologist. For instance, when smaller calcospheres seem, as it were, to invade the substance of a larger one, we get curious conformations which somewhat resemble the outlines of certain diatoms (Fig. 301). Another curious formation, which Harting calls a "conostat," is of frequent occurrence, and in it we

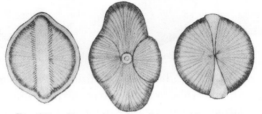

Fig. 301. Composite calcospheres. After Harting.

see at least a suggestion of analogy with the configuration which, in a protoplasmic structure, we have spoken of as a "collar-cell." The conostats, which are formed in the surface layer of the solution, consist of a portion of a spheroidal calcospherite, whose upper part is continued into a thin spheroidal collar of somewhat larger radius than the solid sphere; but the precise manner in which the collar is formed, possibly around a bubble of gas, possibly about a vortex-like diffusion-current, is not obvious.

Among these various phenomena, the concentric striation of the calcospherite has acquired a special interest and importance*. It is part of a phenomenon now widely known under the name of "Liesegang's Rings†."

* Cf. Harting, op. cit. pp. 22, 50: "J'avais cru d'abord que ces couches concentriques étaient produites par l'alternance de la chaleur ou de la lumière, pendant le jour et la nuit. Mais l'expérience, expressément instituée pour examiner cette question, y a répondu négativement."

† R. E. Liesegang, Ueber die Schichtungen bei Diffusionen, Leipzig, 1907, and earlier papers. A periodic precipitate is said to have been first noticed (on filter-paper) by Runge, in 1885; cf. Quincke, Ueber unsichtbare Flüssigkeitsschichten, Ann. d. Physik (4), VII, pp. 643–7, 1902. On a very minute periodicity in the so-called Hookham's crystals, formed by crystallising copper sulphate and salicin in strong syrup, see Rayleigh, Collected Papers, VI, p. 661: "There is much here," says Rayleigh, "to excite admiration and perplexity."

If we dissolve, for instance, a little potassium bichromate in gelatine, pour it on to a glass plate, and after it is set pour upon it a drop of silver nitrate solution, there appears in the course of a few hours the phenomenon of Liesegang's rings. At first the silver

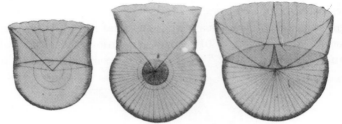

Fig. 302. Conostats. After Harting.

forms a central patch of abundant reddish-brown chromate precipitate; but around this, as the silver nitrate diffuses slowly through the gelatine, the precipitate no longer comes down continuously, but forms a series of concentric rings or zones, beautifully

Fig. 303. Liesegang's rings. After Leduc.

regular, which alternate with clear interspaces of jelly and stand farther and farther apart in a definite ratio as they recede from the centre*. For a discussion of the *raison d'être* of this phenomenon, the student will consult the textbooks of physical and colloid chemistry. But, speaking generally, we may say that the appearance

* It is now known that periodic precipitation may be exhibited even in aqueous solutions, and that what the gel does is to enlarge the intervals, and to enhance the phenomenon, by affecting the rate or relative rates of diffusion. Cf. H. W. Morse, *Journ. Phys. Chem.* 1931.

of Liesegang's rings is but a particular case of a more general phenomenon, namely the influence on crystallisation of the presence of foreign bodies or "impurities," represented in this case by the gel or colloid matrix. F. S. Beudant had shewn in a fine paper, more than a hundred years ago, that impurities were the chief cause of variation of crystal habit*. Faraday proved that to diffusion in presence of slight impurities, not to actual stratification or alternate deposition, could be ascribed the banded structure of ice,

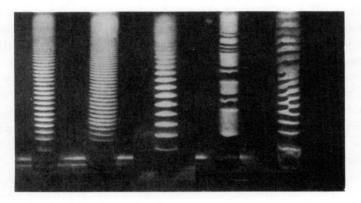

Fig. 304. The Liesegang phenomena. After Emil Hatschek.

of agate or of onyx; and Quincke and Tomlinson added to our scanty knowledge of this remarkable phenomenon†. Ruskin, who knew a great deal about agates, spoke of the perpetual difficulty of distinguishing "between concretionary separation and successive deposition." And Rayleigh shewed how to such a *periodic,* but

* F. S. Beudant, Recherches sur les causes qui peuvent varier les formes crystallines d'une même substance minérale, *Ann. de Chimie*, VIII, pp. 5–52, 1818. See also his Mémoire sur les parties solides des Mollusques, *Mém. du Muséum*, XV, pp. 66–75, 1810.

† Cf. Faraday, On ice of irregular fusibility, *Phil. Trans.* 1858, p. 228; *Researches in Chemistry, etc.*, 1859, p. 374; Canon Moseley, On the veined structure of the ice of glaciers, *Phil. Mag.* (4), XXXIX, p. 241, 1870; R. Weber, in *Poggend. Ann.* CIX, p. 379, 1860; Tyndall, *Forms of Water*, 1872, p. 178; C. Tomlinson, On some effects of small quantities of foreign matter on crystallisation, *Phil. Mag.* (5) XXXI, p. 393, 1891, and other papers. Cf. Liesegang, *Centralbl. f. Mineralogie*, XVI, p. 497, 1911; E. S. Hedges and J. E. Myers, *The problem of physico-chemical periodicity*, London, 1926; W. F. Berg, Crystal growth from solutions, *Proc. R.S.* (A), CLXIV, pp. 79–95, 1938.

unstratified structure all the colours of the opal and the iridescence of ancient glass are alike due.

Besides the tendency to rhythmic action, as manifested in Liesegang's rings, the association of colloid matter with a crystalloid in solution may lead to other well-marked effects. These include *: (1) the total prevention of crystallisation; (2) suppression of certain of the lines of crystal growth; (3) extension of the crystal to abnormal proportions, with a tendency to become compound; (4) a curving or gyrating of the crystal or its parts.

It would seem that, if the supply of material to the growing crystal begin to run short (as may well happen in a colloid medium for lack of convection-currents), then growth will follow only the strongest lines of crystallising force, and will be suppressed or partially suppressed along other axes. The crystal will have a tendency to become filiform, or "fibrous"; and the raphides of our plant-cells, and the needle-like "oxyotes" of sponges, are cases in point. Again, the long slender crystal so formed, pushing its way into new material, may start a new centre of crystallisation: whereby we get the phenomenon known as a "relay," along the principal lines of force and sometimes along subordinate axes as well. This phenomenon is illustrated in the accompanying figure of common salt crystallising in a colloid medium; and it may be that we have here an explanation, or part of an explanation, of the compound siliceous spicules of the Hexactinellid sponges. Lastly, when the crystallising force is nearly equalled by the resistance of the viscous medium, the crystal takes the line of least resistance, with very various results. One of these results would seem to be a gyratory course, giving to the crystal a curious wheel-like shape, as in Fig. 306; and other results are the feathery, fern-like or arborescent shapes so frequently seen in microscopic crystallisation.

To return to Liesegang's rings, the typical appearance of concentric rings upon a plate of gelatine may be modified in various experimental ways. For instance, if our gelatinous medium be placed in a capillary tube immersed in a solution of the precipitating salt, we obtain (Fig. 304) a vertical succession of bands or zones regularly

* Cf. J. H. Bowman, A study in crystallisation, *Journ. Soc. of Chem. Industry*, xxv, p. 143, 1906.

interspaced: the result being very closely comparable to the banded pigmentation which we see in the hair of a rabbit or a rat. In the ordinary plate preparation, the free surface of the gelatine is under different conditions to the layers below and especially to the lowest layer of all in contact with the glass; and so we often obtain a

Fig. 305. Relay-crystals of common salt. After Bowman.

double series of rings, one deep and the other superficial, which by occasional blending or interlacing may produce a netted pattern. Sometimes, when only the inner surface of our capillary tube is covered with a layer of gelatine, there is a tendency for the deposit

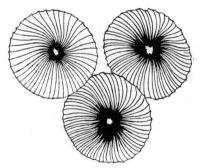

Fig. 306. Wheel-like crystals in a colloid. After Bowman.

to take place in a continuous spiral, rather than in concentric and separate zones. By such means, according to Küster*, various forms of annular, spiral and reticulated thickenings in the vascular tissue of plants may be closely imitated; and he and certain other writers have been inclined to carry the same chemico-physical

* E. Küster, Ueber die Schichtung der Stärkekörner, *Ber. d. botan. Gesellsch.* XXXI, pp. 339–346, 1913; Ueber Zonenbildung in kolloidalen Medien, *Koll. Ztschr.* XIII, pp. 192–194; XIV, pp. 307–319, 1913–14.

phenomenon a very long way, in the explanation of various banded, striped, and other rhythmically successional types of structure or pigmentation.   The striped leaves of many plants (such as *Eulalia japonica*), the striped or clouded colouring of many feathers or of a cat's skin, the patterns of many fishes, such for instance as the brightly coloured tropical Chaetodonts and the like, are all regarded by him as so many instances of "diffusion-figures" closely related to the typical Liesegang phenomenon.   Gebhardt* declares that the banded wings of *Papilio podalirius* are analogous to or even closely imitated in Liesegang's experiments; that the finer markings on the wings of the goatmoth shew a double rhythm, alternately coarse and fine, such as is manifested in certain experimental cases of the same kind; .that the alternate banding of the antennae (for instance in *Sesia spheciformis*), a pigmentation not concurrent with the antennal joints, is explicable in the same way; and that the ocelli on the wings of the Emperor moth are typical illustrations of the common concentric type.   Darwin's well-known disquisition on the ocellar pattern of the feathers of the Argus pheasant, as a result of sexual selection, will occur to the reader's mind, in striking contrast to this or to any other direct physical explanation†.

To turn from the distribution of pigment to more deeply seated structural characters, Leduc has argued, for instance, that the laminar structure of the cornea or the lens is, or may be, a similar phenomenon.   In the lens of the fish's eye, we have a very curious appearance, the consecutive lamellae being roughened or notched by close-set, interlocking sinuosities; and the same appearance, save that it is not quite so regular, is presented in one of Küster's figures as the effect of precipitating a little sodium phosphate in a gelatinous medium.   Biedermann has studied, from the same

* *Verh. d. d. zool. Gesellsch.* p. 179, 1912.

† As a matter of fact, the phenomena associated with the development of an "ocellus" are or may be of great complexity, inasmuch as they involve not only a graded distribution of pigment, but also, in "optical" coloration, a symmetrical distribution of structure or form.   The subject therefore deserves very careful discussion, such as Bateson gives to it (*Variation*, chap. xii).   This, by the way, is one of the very rare cases in which Bateson appears inclined to suggest a purely physical explanation of an organic phenomenon: "The suggestion is strong that the whole series of rings (in *Morpho*) may have been formed by some one central disturbance, somewhat as a series of concentric waves may be formed by the splash of a stone thrown into a pool."   Cf. Darwin, *Descent of Man*, ii, p. 132, 1871.

point of view, the structure and development of the molluscan shell, the problem which Rainey had first attacked more than fifty years before*; and Liesegang himself has applied his results to the formation of pearls, and, as Bechhold has also done, to the development of bone†.

The presence of concentric rings or zones in slow-growing structures is evidently after some fashion a function of the time, and an indication of periodic acceleration or variation of growth; it is apt to be referred, rightly or wrongly, to the seasons of the year, and to be interpreted (with or without confirmation and proof) as a sure mark and measure of the creature's age. This is the case, for instance, with the scales, bones and otoliths of fishes; and a kindred phenomenon in starch-grains has given rise, in like manner, to the belief that they indicate a diurnal and nocturnal periodicity of activity and rest‡ on the part of the cell wherein they grew.

That this is actually the case in growing starch-grains is often if not generally believed, on the authority of Meyer§; but while under certain circumstances a marked alternation of growing and resting periods may occur, and may leave its impress on the structure of the grain, there is now more reason to believe that, apart from such external influences, the internal phenomena of diffusion may, just as in the typical Liesegang experiment, produce the well-known concentric rings. The spherocrystals of inulin, in like manner, shew, like the calcospherites of Harting (Fig. 307), a concentric structure which in all likelihood has had no causative impulse save from within.

The striation, or concentric lamellation, of the scales and otoliths of fishes has been much employed, not as a mere indication, but

---

* Cf. also Sir D. Brewster, On optical properties of mother of pearl, *Phil. Trans.* 1814, p. 397; and J. F. W. Herschel, in *Edin. Phil. Journ.* II, p. 116, 1819.

† W. Biedermann, Ueber die Bedeutung von Kristallisationsprozessen der Skelette wirbelloser Thiere, namentlich der Molluskenschalen, *Z. f. allg. Physiol.* I, p. 154, 1902; Ueber Bau und Entstehung der Molluskenschale, *Jen. Zeitschr.* XXXVI, pp. 1–164, 1902. Cf. also Steinmann, Ueber Schale und Kalksteinbildungen, *Ber. Naturf. Ges. Freiburg i. Br.* IV, 1889; Liesegang, *Naturw. Wochenschr.* 1910. p. 641; *Arch. f. Entw. Mech.* XXXIV, p. 452, 1912; H. Bechhold, *Ztschr. f. phys. Chem.* LII, p. 185, 1905.

‡ Cf. Bütschli, Ueber die Herstellung künstlicher Stärkekörner oder von Sphärokrystallen der Stärke, *Verh. nat. med. Ver. Heidelberg*, V, pp. 457–472, 1896.

§ *Untersuchungen über die Stärkekörner*, Jena, 1905.

as a trustworthy and unmistakeable measure of the fish's age (see *ante*, p. 180). There are some difficulties in the way of accepting this hypothesis, not the least of which is the fact that the otolith-zones, for instance, are extremely well marked even in the case of some fishes which spend their lives in deep water, where temperature and other physical conditions shew little or no appreciable fluctuation with the seasons of the year. There are, on the other hand, phenomena which seem strongly confirmatory of the hypothesis: for instance, the fact (if it be fully established) that in such a fish as the cod, zones of growth, *identical in number*, are found both on

Fig. 307. A sphero-crystal of inulin.

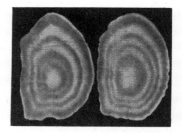

Fig. 308.   Otoliths of plaice, shewing four zones or "age-rings."   After Wallace.

the scales and in the otoliths*.   The subject is as difficult as it is important, but it is at least certain, with the Liesegang phenomenon in view, that we have no right to *assume*, without proof and confirmation, that rhythm and periodicity in structure and growth are necessarily bound up with, and indubitably brought about by, a periodic or seasonal recurrence of particular *external* conditions†.

But while in the ordinary Liesegang phenomenon rhythmic

* Cf. Winge, *Meddel. fra Komm. for Havundersögelse (Fiskeri)*, IV, p. 20, Copenhagen, 1915.

† A. W. Morosow strongly supports the view—uncertain as it seems to be—that the concentric pattern of a fish's scale is due to the Liesegang phenomenon; he produces an "artificial scale," with its "summer and winter rings," by precipitating sodium carbonate and calcium chloride in gelatin: Zur Frage über die Natur des Schuppenwachstums bei Fischen (and in Russian), *Nation. Comm. Agriculture: Rep. Sci. Inst. Fisheries*, I, Moscow, 1924; abstract in Michael Graham's Studies of age-determination in fish, *Rep. Ministry of Agr. and Fisheries, Fishery Investigations*, (2) XI, no. 3, p. 28, 1928.

precipitation depends only on forces intrinsic to the system, and is independent of any corresponding rhythmic changes in external conditions, we have not far to seek for analogous chemico-physical phenomena where rhythmic alternations of structure are produced in close relation to periodic fluctuations of temperature. The banding, or "varving," of Swedish and Irish glacial clays is a remarkable instance. A well-known and a simple case is that of the Stassfurt deposits, where the rock-salt alternates with thin layers of "anhydrite," or (in another series of beds) with "polyhalite*": and where these zones are commonly regarded as marking years, and their alternate bands as due to the seasons. A discussion, however, of this remarkable and significant phenomenon, and of how the chemist explains it, by help of the "phase-rule," in connection with temperature conditions, would lead us far beyond our scope.

We may turn aside to touch, for a single moment, on certain forms and patterns not easy to classify: some of which depend on the molecular structure of a colloid matrix, while others are of a coarser and more mechanical grade. So many organic forms and patterns await explanation that we cannot seek too widely for examples, nor for explanations, of such things. For instance, a drop of dried egg-albumin shews beautiful radial cracks, with cross-lines here and there; and a drop of blood drying on a glass plate shews a complete system of radial fissures, in series after series, sometimes with and sometimes without a clear central space. The general resemblance to the cross-section of a stem, with its pith and its primary and secondary medullary rays, is striking enough to have led some even to look upon a tree as one great complicated but symmetrical colloid mass†. We may compare also the beautiful radiating structure which Bütschli observed long ago around small

---

* The anhydrite is sulphate of lime ($CaSO_4$); the polyhalite is a triple sulphate of lime, magnesia and potash ($2CaSO_4 \cdot MgSO_4 \cdot K_2SO_4 + 2H_2O$).

† Cf. H. Wislicenus, *Ztschr. f. Chemie u. Kolloide*, VI, 1910; A. Lingelsheim, Pflanzenanatomische Strukturbilder in trocknenden Kolloiden, *Arch. f. Entw. Mech.* XLII, pp. 117–125, 1917. Cf. also Liesegang, Trocknungserscheinungen bei Gelen, *Ztschr. f. Ch. u. K.* x, p. 229 *sq.*, 1912; Bütschli, *Verh. n. h. Ver. Heidelberg*, VII, p. 653, 1904. Also (*int. al.*) Norman Stuart, on Spiral growths in silica gel, *Nature*, Oct. 2, 1937, p. 589.

bubbles in chrome-gelatine, and which he used in one of his early (and none too fortunate) speculations on the nature of the nuclear spindle.

We see that the methods by which we attempt to study the chemico-physical characteristics of an inorganic concretion or spicule within the body of an organism soon introduce us to a multitude of phenomena of which our knowledge is extremely scanty, and which we must not attempt to discuss at greater length. As regards our main point, namely the formation of spicules and other elementary skeletal forms, we have seen that some of them may be safely ascribed to precipitation or crystallisation of inorganic materials in ways modified by the presence of albuminous or other colloid substances. The effect of these latter is found to be much greater in the case of some crystallisable bodies than in others. For instance Harting, and Rainey also, found that calcium oxalate was much less affected by a colloid medium than was calcium carbonate; it shewed in their hands no tendency to form rounded concretions or "calcospherites" in presence of a colloid, but continued to crystallise, either normally or with a tendency to form needles or raphides. It is doubtless for this reason that, as we have seen, *crystals* of calcium oxalate are so common in the tissues of plants, while those of other calcium salts are rare; but true calcospherites, or spherocrystals, even of the oxalate are occasionally found, for instance in certain Cacti, and Bütschli* has succeeded in making them artificially in Harting's usual way, that is to say by crystallisation in a colloid medium. If the nature of the salt has a marked specific effect, so also has the gel: silver chromate is thrown down in rings in gelatin but not in agar; replace the silver by lead, and the rings come in agar but not in gelatin; while neither lead nor silver produce them in silicic acid gel.

There link on to such observations as Harting's, and to the statement already quoted that calcareous deposits are associated with the dead residua, or "formed materials," rather than with the living cells of the organism, certain very interesting facts in regard to the *solubility* of salts in colloid media, which go far to account for the presence (apart from the form) of calcareous pre-

---

* Sphärocrystalle von Kalkoxalat bei Kakteen, *Ber. d. d. Bot. Gesellsch.* p. 178, 1885.

cipitates within the organism*. It has been shewn, in the first place, that the presence of albumin has a notable effect on the solubility in a watery solution of calcium salts, increasing the solubility of the phosphate in a marked degree and that of the carbonate in still greater proportion; but the sulphate is only very little more soluble in presence of albumin than in pure water, and the rarity of its occurrence within the organism is accounted for thereby. On the other hand, the bodies derived from the breaking down of the albumins—their "catabolic" products, such as the peptones, etc.—dissolve the calcium salts to a much less degree than albumin itself; and phosphate of lime is scarcely more soluble in them than in water. The probability is, therefore, that the actual precipitation of the calcium salts is not due to the direct action of carbonic acid on a more soluble salt (as was at one time believed); but to catabolic changes in the proteids of the organism, which throw down salts that had been already formed, but had remained hitherto in albuminous solution. The very slight solubility of calcium phosphate under such circumstances accounts for its predominance in mammalian bone†; and, in short, wherever a supply of this salt has been available to the organism.

To sum up, we see that, whether from food or from sea-water, calcium sulphate will tend to pass but little into solution in the albuminoid substances of the body: that calcium carbonate will enter more freely, but a considerable part of it will tend to remain in solution: while calcium phosphate will pass into solution in considerable amount, but will be almost wholly precipitated again as the albumin becomes broken down in the normal process of metabolism. We have still to wait for a similar and equally illuminating study of the solution and precipitation of *silica* in presence of organic colloids.

When carbonate of lime is secreted or precipitated by living organisms, to form bone, shell, egg-shell, coral and what not, its mineralogical form may vary, but the causes which determine it

* W. Pauli u. M. Samec, Ueber Löslichkeitsbeeinflüssung von Elektrolyten durch Eiweisskörper, *Biochem. Zeitschr.* xvii, p. 235, 1910. Some of these results were known much earlier; cf. Fokker in *Pflüger's Archiv*, vii, p. 274, 1873; also Robert Irvine and Sims Woodhead, *op. cit.* p. 347.

† Which, in 1000 parts of ash, contains about 840 parts of phosphate and 76 parts of calcium carbonate.

are all but unknown. It is amorphous in our bones. It has the form of calcite in an oyster, a starfish, a *Gorgonia*, a *Globigerina*; but of aragonite in most molluscs and in all ordinary corals. It is of calcite in a bird's egg, of aragonite in a tortoise's; of the one in *Argonauta*, of the other in *Nautilus*; of the one in an Ammonite, and the other in its *Aptychus*-lid; of the one in *Ostrea*, the other in *Unio*; of the one in the outer and the other in the inner layers of a limpet or a mussel-shell. Physical chemistry has little to say of the formation of these two, of the parts played by temperature, by the presence of sulphate of lime, or of magnesia or of various impurities; it leaves us in the dark as to what brings the one form or the other into being in the organism*.

Organic fibres, animal and vegetable, proteid and non-proteid, hair and wool, silk, cotton and the rest, may be mentioned here in passing: because, as formed material, they have a certain analogy to the spicular formations with which we are concerned. A hair or a wool-fibre may shew upon its surface the scaly or scurfy remnants of the living cells among which its substance was laid down; but the wool itself is by no means living, but is so nearly crystalline as to shew, in an X-ray photograph, the Laue interference-figures well known to physicists. Moreover, the same identical figure is obtained from such diverse sources as human hair, merino-wool and porcupine's quill. But if we stretch the thread, whether of hair or wool, the first Laue diagram changes to another; one crystalline arrangement has shifted over into a new form of molecular equilibrium. We are dealing with a crystalline, or crystal-like, form of *keratin*, the substance of which hoof and horn, nail, scale and feather are made; and this remarkable substance turns out to be a comparatively simple substance after all, with no very high or protein-like molecule †.

From the comparatively small group of inorganic formations which, arising within living organisms, owe their form to precipitation or to crystallisation, that is to say to chemical or other molecular

---

* Cf. Marcel Prenant, Les formes minéralogiques du calcaire chez les êtres vivants, *Biol. Reviews*, ii, pp. 365–393, 1927.

† The study of wool and other fibres has much technical importance, and has gone far during the last few years; cf. W. T. Astbury, in *Phil. Trans.* (A), ccxxx, pp. 75–100, 1931, and other papers.

forces, we shall presently pass to that other and larger group which appears to be conformed in direct relation to the forms and the arrangement of cells or other protoplasmic elements*. The two principles of conformation are both illustrated in the spicular skeletons of the sponges.

In a considerable number but withal a minority of cases, the form of the sponge-spicule may be deemed sufficiently explained on the lines of Harting's and Rainey's experiments, that is to say as the direct result of chemical or physical phenomena associated with the deposition of lime or of silica in presence of colloids†. This is the case, for instance, with various small spicules of a globular or spheroidal form, consisting of amorphous silica, concentrically striated within, and often developing irregular knobs or tiny tubercles over their surfaces. In the aberrant sponge *Astrosclera* ‡, we have, to begin with, rounded, striated discs or globules, which in like manner are nothing more nor less than the calcospherites of Harting's experiments; and as these grow they become closely aggregated together (Fig. 309), and assume an angular, polyhedral form, once more in complete accordance with the results of experiment§. Again, in many monaxonid sponges, we have irregularly shaped, or branched spicules, roughened or tuberculated by secondary superficial deposits, and reminding one of the spicules of the *Alcyonaria*. These also must be looked upon as the simple result of chemical deposition, the form of the deposit being somewhat modified in conformity with the surrounding tissues: just as in the simple experiment the form of the concretionary precipitate is affected by the heterogeneity, visible or invisible, of the matrix. Lastly, the simple needles of amorphous

* Cf. Fr. Dreyer, Die Principien der Gerüstbildung bei Rhizopoden, Spongien und Echinodermen, *Jen. Zeitschr.* XXVI, pp. 204–468, 1892.

† In a very anomalous Australian sponge, described by Professor Dendy (*Nature*, May 18, 1916, p. 253) under the name of *Collosclerophora*, the spicules are "gelatinous," consisting of a gel of colloid silica with a high percentage of water. It is not stated whether an organic colloid is present together with the silica. These gelatinous spicules arise as exudations on the outer surface of cells, and come to lie in intercellular spaces or vesicles.

‡ J. J. Lister, in Willey's *Zoological Results*, pt IV, p. 459, 1900.

§ The peculiar spicules of *Astrosclera* are said to consist of spherules, or calcospherites, of aragonite, spores of a certain red seaweed forming the nuclei or starting-points of the concretions (R. Kirkpatrick, *Proc. R.S.* (B), LXXXIV, p. 579, 1911).

silica which constitute one of the commonest types of spicule call
for little in the way of explanation; they are accretions or deposits
about a linear axis, or fine thread of organic material, just as the
ordinary rounded calcospherite is deposited about some minute
point or centre of crystallisation, and as ordinary crystallisation
may be started by a particle of dust; in some cases they also, like
the others, are apt to be roughened by more irregular secondary

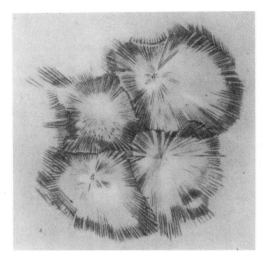

Fig. 309.   Close-packed calcospherites, or so-called "spicules,"
of *Astrosclera*.   After Lister.

deposits, which probably, as in Harting's experiments, assume this
irregular form when material runs short.

Our few foregoing examples, diverse as they are in look and kind,
from the spicules of *Astrosclera* or *Alcyonium* to the otoliths of a
fish, seem all to have their free origin in some larger or smaller
fluid-containing space or cavity of the body: pretty much as
Harting's calcospheres made their appearance in the albuminous
content of a dish.   But we come at last to a much larger class of
spicular and skeletal structures, for whose regular and often complex
forms some other explanation than the intrinsic forces of crystallisation or molecular adhesion is required.   As we enter on this

subject, which is certainly no small nor easy one, it may conduce to simplicity and to brevity if we make a rough classification, by way of forecast, of the conditions we are likely to meet with.

Just as we look upon animals as constituted, some of a great number of cells, others of a single cell or of but few, and just as the shape of the former has no longer a visible relation to the individual shapes of its constituent cells while in the latter it is cell-form which dominates or is actually equivalent to the form of the organism, so shall we find it to be, with more or less exact analogy, in the case of the skeleton. For example, our own skeleton consists of bones, in the formation of each of which a vast number of minute living cellular elements are necessarily concerned; but the form and even the arrangement of these bone-forming cells or corpuscles are monotonously simple, and give no physical explanation of the outward and visible configuration of the bone. It is as part of a far larger field of force—in which we must consider gravity, the action of various muscles, the compressions, tensions and bending moments due to variously distributed loads, the whole interaction of a very complex mechanical system—that we must explain (if we are to explain at all) the configuration of a bone.

In contrast to these massive skeletons we have other skeletal elements whose whole magnitude is commensurate with that of a living cell, or (as comes to very much the same thing) is comparable to the range of action of the molecular forces. Such is the case with the ordinary spicules of a sponge, with the delicate skeleton of a radiolarian, or with the denser and robuster shells of the foraminifera. The effect of *scale*, then, of which we had so much to say in our introductory chapter on Magnitude, is bound to be apparent in the study of skeletal fabrics, and to lead to essential differences between the big and the little, the massive and the minute, in regard to their controlling forces and resultant forms. And if all this be so, and if the range of action of the molecular forces be now the important and fundamental thing, then we may somewhat extend our statement of the case, and include among our directive or constructive influences not only association with the living cellular elements of the body, but also association with any bubbles, drops, vacuoles or vesicles which may be comprised within the bounds of the organism, and which are (as their names and characters connote) of the order of magnitude of which we are speaking.

Proceeding a little farther in our classification, we may conceive each little skeletal element to be associated with, and developed by, a single cell or vesicle, or alternatively a cluster or "system" of consociated cells. In either case there are various possibilities. For instance, the calcified or other skeletal material may tend to overspread the entire outer surface of the cell or cluster of cells, and so tend to assume a configuration comparable to the surface of a fluid drop or aggregation of drops; this, in brief, is the gist and essence of our story of the foraminiferal shell. Another common but very different condition will arise if, in the case of the cell-aggregates, the skeletal material tends to accumulate in the interstices *between* the cells, in the partition-walls which separate them, or in the still more restricted edges, or junctions between these partition-walls; conditions such as these will go a long way to help us to understand many sponge-spicules and an immense variety of radiolarian skeletons. And lastly (for the present), there is a possible and very interesting case of a skeletal element associated with the surface of a cell, not so as to cover it like a shell, but only so as to pursue a course of its own within it, and subject to the restraints imposed by such confinement to a curved and limited surface. With this curious condition we shall deal immediately.

This preliminary and much simplified classification of the lesser skeletal, or micro-skeletal, forms does not pretend (as is evident enough) to completeness. It leaves out of account some conformations and configurations with which we shall attempt to deal, and others which we must perforce omit. But nevertheless it may help to clear or mark our way towards the subjects which this chapter has to consider, and the conditions by which they are at least partially defined.

Among the possible, or conceivable, types of microscopic skeletons let us begin with the case of a spicule, more or less simply linear as far as its *intrinsic* powers of growth are concerned, but which owes its more complicated form to a restraint imposed by the cell to which it is confined, and within whose bounds it is generated.

The conception of a spicule developed under such conditions came from that very great mathematical physicist, G. F. FitzGerald. Many years ago, Sollas pointed out that if a spicule begin to grow in some particular way, presumably under the control or constraint imposed by the organism, it continues to grow by further chemical deposition in the same form or direction even after it has got beyond the boundaries of the organism or its cells. This phenomenon is what we see in, and this imperfect explanation goes so far to account for, the continued growth in straight lines of the long calcareous spines of *Globigerina* or *Hastigerina*, or the similarly radiating but siliceous spicules of many Radiolaria. In physical language, if our crystalline structure has once begun to be laid down in a definite orientation, further additions tend to accrue in a like regular fashion and in an identical direction: corresponding to the phenomenon of so-called "orientirte Adsorption," as described by Lehmann.

In *Globigerina* or in *Acanthocystis* the long needles grow out freely into the surrounding medium, with nothing to impede their rectilinear growth and approximately radiate symmetry. But let us consider some simple cases to illustrate the forms which a spicule will tend to assume when, striving (as it were) to grow straight, it comes under some simple and constant restraint or compulsion.

If we take any two points on a smooth curved surface, such as that of a sphere or spheroid, and imagine a string stretched between them, we obtain what is known in mathematics as a "geodesic" curve. It is the shortest line which can be traced between the two points upon the surface itself, and it has always the same direction upon the surface to which it is confined; the most familiar of all cases, from which the name is derived, is that curve, or "rhumb-line," upon the earth's surface which the navigator learns to follow in the practice of "great-circle sailing," never altering his direction nor departing from his nearest road. Where the surface is spherical, the geodesic is literally a "great circle," a circle, that is to say, whose centre is the centre of the sphere. If instead of a sphere we be dealing with a spheroid, whether prolate or oblate (that is to say a figure of revolution in which an ellipse rotates about its long or its short axis), then the system of geodesics becomes more complicated. For in it the elliptic meridians are all geodesics, and so is the circle of the equator; though the circles of latitude are not so, any more than in the sphere. But a line which crosses the equator at an oblique angle, if it is to be geodesic, will go on so far and then turn back again, winding its way in a continual figure-of-eight curve between two extreme latitudes, as when we wind a ball of wool. To say, as we have done, that the geodesic is the shortest line between two points upon the surface, is as much as to say that it is a *trace* of some particular straight line upon the surface in question; and it follows that, if any linear body be confined to that surface, while retaining a tendency to grow (save only for its confinement to that surface) in a straight line, the resultant form which it will assume will be that of a geodesic.

Let us now imagine a spicule whose natural tendency is to grow into a straight linear element, either by reason of its own molecular anisotropy or because it is deposited about a thread-like axis, and let us suppose that it is confined either within a cell-wall or in adhesion thereto; its line of growth will be a geodesic to the surface of the cell. And if the cell be an imperfect sphere, or a more or less regular ellipsoid, the spicule will tend to grow into one or other of three forms: either a plane curve of nearly circular arc; or, more commonly, a plane curve which is a portion of an ellipse; or, most commonly of all, a curve which is a portion of a spiral in space. In the latter case, the number of turns of the spiral will depend not only on the length of the spicule, but on the relative dimensions of the ellipsoidal cell, as well as on the angle by which the spicule is inclined to the ellipsoid axes; but a very common case will probably be that in which the spicule looks at first sight to be a plane **C**-shaped figure, but is discovered, on more careful inspection, to lie not in one plane but in a more complicated twist. This investigation includes a series of forms which are abundantly represented among actual sponge-spicules, as illustrated in Figs. 310 and 311.

Growth or motion, when confined to some particular curved surface, may appear in various forms and in unexpected places. An amoeba, creeping along the inside or the outside of a glass tube, was found in either case to follow a winding, spiral path: it was really doing its best to go straight—in other words it was following a geodesic or loxodromic path, determined by whatsoever angle of obliquity to the axis of the tube it had chanced to start out upon.

The spiral bands of chlorophyll in *Spirogyra*, set at varying angles of helicoid obliquity, are (I take it) very beautiful examples of continuous growth under the restraint of a cylindrical surface.

Fig. 310.   Sponge and holothurian spicules.

To return to our sponge-spicules.   If the spicule be not restricted to linear growth, but have a tendency to expand, or to branch out from a main axis, we shall obtain a series of more complex figures, all related to the geodesic system of curves.   A notable case will arise where the spicule occupies, in the first instance, the axis of the containing cell, and then, on reaching its boundary, tends to branch or spread outwards.   We shall now get various figures, in some of

Fig. 311.

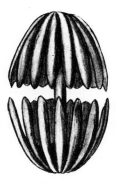

Fig. 312.   An "amphidisc"
of *Hyalonema*.

which the spicule will appear as an axis expanding into a disc or wheel at either end; and in other cases, the terminal disc will be replaced by rays or spokes with a reflex curvature, corresponding to the spherical or ellipsoid curvature of the cell.   Such spicules as these are exceedingly common among various sponges (Fig. 312).

Furthermore, if these mechanical methods of conformation, and others like to these, be the true cause of the shapes which the spicules assume, it is plain that the production of these spicular shapes is not a specific function of the sponge, but that we should expect the same or similar spicules to occur in other organisms, wherever the conditions of inorganic secretion within closed cells were very much the same. As a matter of fact, in the sea-cucumbers, where the formation of intracellular spicules is a characteristic feature of the group, all the principal types of conformation which we have just described can be closely paralleled; indeed, in many cases, the forms of the holothurian spicules are identical and indistinguishable from those of the sponges*. But the holothurian spicules are composed of calcium carbonate while those which we have just described in the case of sponges are siliceous: this being just another proof of the fact that in such cases as these the form of the spicule is not due to its chemical nature or molecular structure, but to the external forces to which it is subjected.

The broad fact that the skeleton is calcareous in certain large groups of animals and calcareous in others is as remarkable as its causes are obscure. I for one have no idea why some sponge-skeletons are of the one and some the other, with never the least admixture of the two; or why the diatoms and radiolarians are all the one, and the molluscs and corals and foraminifera are all the other†.

So much for that small class of sponge-spicules whose forms seem due to the fact that they are developed within, or under the restraint imposed by, the surface of a single cell or vesicle. Such spicules are usually of small size as well as of simple form; and they are greatly outstripped in number, in size, and in supposed importance as guides to zoological classification, by another class of spicules. These are the many and various cases which we explain on the

---

* See for instance the plates in Théel's Monograph of the Challenger *Holothuroidea*; also Sollas's *Tetractinellida*, p. lxi. Cf. also E. Merke, Studien am Skelet der Echinodermen, *Zool. Jahrbücher* (*Abth. f. allgem. Zoologie*), 1916–19.

† The particles of lime and silica tend to bear opposite charges; siliceous organisms seem to flourish in the colder waters as the calcareous certainly do in warmer seas. And such facts, or tendencies, as these may help some day to explain the phenomenon.

assumption that they develop in association (of some sort or another) with the *lines of junction*, or boundary-edges, of contiguous cells. They include the triradiate spicules of the calcareous sponges, the quadriradiate or "tetractinellid" spicules which occur sometimes in the same group but more characteristically in certain siliceous sponges known as the Tetractinellidae, and perhaps (though these last are somewhat harder to understand) the six-rayed spicules of the Hexactinellids. We shall come later on to more complicated skeletons of the same type among the Radiolaria.

The spicules of the calcareous sponges are commonly triradiate, and the three radii are usually inclined to one another at *nearly* equal angles; in certain cases, two of the three rays are nearly in a straight line, and at right angles to the third*. They are not always in a plane, but are often inclined to one another in a trihedral angle, not easy of precise measurement under the microscope. The three rays are often supplemented by a fourth, which is set tetra-hedrally, making *nearly* co-equal angles with the other three. The calcareous spicule consists mainly of carbonate of lime in the form of calcite, with (according to von Ebner) some admixture of soda and magnesia, of sulphates and of water. According to the same writer there is no organic matter in the spicule, either in the form of an axial filament or otherwise, and the appearance of stratifica-tion, often simulating the presence of an axial fibre, is due to "mixed crystallisation" of the various constituents. The spicule is a true crystal, and therefore its existence and its form are *primarily* due to the molecular forces of crystallisation; moreover it is a single crystal and not a group of crystals, as is seen by its behaviour in polarised light. But its axes are not crystalline axes, its angles are variable and indefinite, and its form neither agrees with, nor in any way resembles, any one of the countless, polymorphic forms in which calcite is capable of crystallising. It is as though it were carved out of a solid crystal; it is, in fact, a crystal under restraint, a crystal growing, as it were, in an artificial mould, and this mould is constituted by the surrounding cells or structural vesicles of the sponge.

* For very numerous illustrations of the triradiate and quadriradiate spicules of the calcareous sponges, see (*int. al.*), papers by Dendy (*Q.J.M.S.* xxxv, 1893), Minchin (*P.Z.S.* 1904), Jenkin (*P.Z.S.* 1908), etc.

We have already studied in an elementary way, but enough for our purpose, the manner in which three, four or more cells, or bubbles, meet together under the influence of surface-tension, in configurations geometrically similar to what may be brought about by a uniform distribution of mechanical pressure. And we have seen how surface-energy leads to the *adsorption* of certain chemical substances, first at the corners, then at the edges, lastly in the partition-walls, of such an assemblage of cells. A spicule formed in the interior of such a mass, starting at a corner where four cells meet and extending along the adjacent edges, would then (in theory) have the characteristic form which the geometry of the bee's cell has taught us, of four rays radiating from a point, and set at co-equal angles to one another of 109°, approximately. Precisely such "tetractinellid" spicules are often formed.

But when we confine ourselves to a plane assemblage of cells, or to the outer surface of a mass, we need only deal with the simpler geometry of the hexagon. In such a plane assemblage we find the cells meeting one another in threes; when the cells are uniform in size the partitions are straight lines, and combine to form regular hexagons; but when the cells are unequal, the partitions tend to be curved, and to combine to form other and less regular polygons. Accordingly, a skeletal secretion originating in a layer or surface of cells will begin at the corners and extend to the edges of the cells, and will thus take the form of triradiate spicules, whose rays (in a typical case) will be set at co-equal angles of 120° (Fig. 313, F). This latter condition of inequality will be open to modification in various ways. It will be modified by any inequality in the specific tensions of adjacent cells; as a special case, it will be apt to be greatly modified at the surface of the system, where a spicule happens to be formed in a plane perpendicular to the cell-layer, so that one of its three rays lies between two adjacent cells and the other two are associated with the surface of contact between the cells and the surrounding medium; in such a case (as in the cases considered in connection with the forms of the cells themselves on p. 494), we shall tend to obtain a spicule with two equal angles and one unequal (Fig. 313, A, C); in the last case, the two outer, or superficial rays, will tend to be markedly curved. Again, the equiangular condition will be departed from, and more or less curvature will be imparted to the rays, wherever the cells of the system cease to be uniform in size, and when the hexagonal symmetry of the system is lost accordingly. Lastly, although we speak of the rays as meeting at certain definite angles, this statement applies to their *axes* rather than to the rays themselves. For if the triradiate spicule be developed in the *interspace* between three juxtaposed cells it is obvious that its sides will tend to be concave, because the space between three contiguous equal circles is an equilateral, curvilinear triangle; and even if our

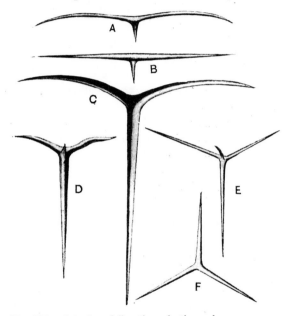

Fig. 313.   Spicules of *Grantia* and other calcareous sponges.
After Haeckel.

spicule be deposited, not in the space between our three cells, but in the mere thickness of an intervening wall, then we may recollect that the several partitions never actually meet at sharp angles, but the angle of contact is always bridged over by an accumulation of material (varying in amount according to its fluidity) whose boundary takes the form of a circular arc, and which constitutes the "bourrelet" of Plateau. In any sample of the triradiate spicules of *Grantia*, or in any series of careful drawings, such as Haeckel's, we shall find all these various configurations severally and completely illustrated.

The tetrahedral, or rather tetractinellid, spicule needs no further explanation in detail (Fig. 313, D, E). For just as a triradiate spicule corresponds to the case of three cells in mutual contact, so does the four-rayed spicule to that of a solid aggregate of four cells: these latter tending to meet one another in a tetrahedral system, shewing four edges, at each of which three facets or partitions meet, their edges being inclined to one another at equal angles of about 109°—the "Maraldi" angle. And even in the case of a single layer, or superficial layer, of cells, if the skeleton originate in connection with all the edges of mutual contact, we shall (in complete and typical cases) have a four-rayed spicule, of which one straight limb will correspond to the 'line of junction between the three cells, and the other three limbs (which will then be curved limbs) will correspond to the three edges where the three cells meet in pairs on the surface of the system.

But if such a physical explanation of the forms of our spicules is to be accepted, we must seek for some physical agency to explain the presence of the solid material just at the junctions or interfaces of the cells, and for the forces by which it is confined to, and moulded to the form of, these intercellular or interfacial contacts. We owe to Dreyer the physical or mechanical theory of spicular conformation which I have just described—a theory which ultimately rests on the form assumed, under surface-tension, by an aggregation of cells or vesicles. But this fundamental point being granted, we have still several possible alternatives by which to explain the details of the phenomenon.

Dreyer, if I understand him aright, was content to assume that the solid material, secreted or excreted by the organism, accumulates in the interstices between the cells, and is there subjected to mechanical pressure or constraint as the cells get crowded together by their own growth and that of the system generally. As far as the general form of the spicules goes such explanation is not inadequate, though under it we might have to renounce some of our assumptions as to what takes place at the surface of the system. But where a few years ago the concept of secretion seemed precise enough, we turn now-a-days to the phenomenon of adsorption as a further stage towards the elucidation of our facts, and here we have a case in point. In the tissues of our sponge, wherever two cells meet, there we have a definite *surface* of contact, and there accordingly we have a manifestation of surface-energy; and the concentration of surface-energy will tend to be a maximum at the *lines* or *edges* whereby such surfaces are conjoined. Of the micro-chemistry of the sponge-cells our ignorance is great; but (without venturing on any hypothesis involving the chemical details of the process) we may safely assert that there is an inherent probability that certain substances will tend to be concentrated and ultimately deposited just in these lines of intercellular contact and conjunction. In other words, adsorptive concentration, under osmotic pressure, at and in the surface-film which bounds contiguous cells, and especially in the *edges* where these films meet and intersect, emerges as an alternative (and, as it seems to me, a highly preferable alternative) to Dreyer's conception of an accumulation under mechanical pressure in the vacant spaces left between one cell and another.

But a purely chemical, or purely molecular, adsorption is not the only form of the hypothesis on which we may rely. For from the purely physical point of view, angles and edges of contact between adjacent cells will be *loci* in the field of distribution of surface-energy, and any material particles whatsoever will tend to undergo a diminution of freedom on entering one of those boundary regions. Let us imagine a couple of soap bubbles in contact with one another; over the surface of each bubble tiny bubbles and droplets glide in every direction; but as soon as these find their way into the groove or re-entrant angle between the two bubbles, there their freedom of movement is so far restrained, and out of that groove they have little tendency, or little freedom, to emerge. A cognate phenomenon is to be witnessed in microscopic sections of steel or other metals. Here, together with its crystalline structure, the metal develops a cellular structure by reason of its lack of homogeneity; for in the molten state one constituent tends to separate out into drops, while the other spreads over these and forms a filmy reticulum between —the *disperse* phase and the *continuous* phase of the colloid chemists. In a polished section we easily observe that the little particles of graphite and other foreign bodies common in the matrix have tended to aggregate themselves in the walls and at the angles of the polygonal cells—this being a direct result of the diminished freedom which they undergo on entering one of these boundary regions. And the same phenomenon is turned to account in the various "separation-processes" in which metallic particles are caught up in the interstices of a froth, that is to say in the walls of the *foam-cells* or *Schaumkammern* *.

It is by a combination of these two principles, chemical adsorption on the one hand and physical quasi-adsorption or concentration of grosser particles on the other, that I conceive the substance of the sponge-spicule to be concentrated and aggregated at the cell boundaries; and the forms of the triradiate and tetractinellid spicules are in precise conformity with this hypothesis. A few general matters, and a few particular cases, remain to be considered. It matters little or not at all for the phenomenon in question, what is the histological nature or "grade" of the vesicular structures on which it depends. In some cases (apart from sponges), they may be no more than little alveoli of an intracellular protoplasmic network, and this would seem to be the case at least in the protozoan *Entosolenia aspera*, within the vesicular protoplasm of whose single cell Möbius has described tiny spicules in the shape of little tetrahedra with sunken or concave sides. It is probably the case also in the small beginnings of Echinoderm spicules, which are likewise intracellular and are of similar shape. Among the sponges we have many varying conditions. In some cases there is reason to believe that the spicule is formed at the boundaries of true cells or histological units; but in the case of the larger triradiate or tetractinellid spicules they far surpass in size the actual "cells." We find them lying, regularly and symmetrically arranged, between the "pore-canals" or "ciliated chambers," and it is in conformity with the shape and arrangement of these large rounded or spheroidal structures that their shape is assumed.

Again, it is not at variance with our hypothesis to find that, in the adult sponge, the larger spicules may greatly outgrow the bounds not only of actual cells but also of the ciliated chambers, and may even appear to project freely from the surface of the sponge. For we have already seen that the spicule is capable of

---

* The crystalline composition of iron was recognised by Hooke in the *Micrographia* (1665); and the cellular or polyhedral structure of the metal was clearly recognised by Réaumur, in his *Art de convertir le fer forgé en acièr*, 1722.

growing, without marked change of form, by further deposition, or crystallisation, of layer upon layer of calcareous molecules, even in an artificial solution; and we are entitled to believe that the same process may be carried on in the tissues of the sponge, without greatly altering the symmetry of the spicule, long after it has established its characteristic but non-crystalline form of a system of slender trihedral or tetrahedral rays.

Neither is it of great importance to our hypothesis whether the rayed spicule necessarily arises as a single structure, or does so from separate minute centres of aggregation. Minchin has shewn that, in some cases at least, the latter is the case; the spicule begins, he tells us, as three tiny rods, separate from one another, each developed in the interspace between two sister-cells, which are themselves the results of the division of one of a little trio of cells; and the little rods meet and fuse together while still very minute, when the whole spicule is only about $\frac{1}{200}$ of a millimetre long. At this stage, it is interesting to learn that the spicule is non-crystalline; but the new accretions of calcareous matter are soon deposited in crystalline form.

This observation threw difficulties in the way of former mechanical theories of the conformation of the spicule, and was quite at variance with Dreyer's theory, according to which the spicule was bound to begin from a central nucleus coinciding with the meeting-place of three contiguous cells, or rather the interspace between them. But the difficulty is removed when we import the concept of adsorption; for by this agency it is natural enough, or conceivable enough, that deposition should go on at separate parts of a common system of surfaces; and if the cells tend to meet one another by their interfaces before these interfaces extend to the angles and so complete the polygonal cell, it is again only natural that the spicule should first arise in the form of separate and detached limbs or rays.

Among the "tetractinellid" sponges, whose spicules are composed of amorphous silica or opal, all or most of the above-described main types of spicule occur, and, as the name of the group implies, the four-rayed, tetrahedral spicules are especially represented. A somewhat frequent type of spicule is one in which one of the four rays is greatly developed, and the other three constitute small prongs diverging at equal angles from the main or axial ray. In all probability, as Dreyer suggests, we have here had to do with a group of four vesicles, of which three were large and co-equal, while

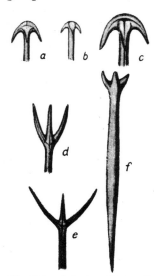

a fourth and very much smaller one lay above and between the other three. In certain cases where we have likewise one large and three much smaller rays, the latter are recurved, as in Fig. 314, *a–c*. This type, save for the constancy of the number of rays and the limitation of the terminal ones to three, and save also for the more important difference that they occur only at one and not at both ends of the long axis, is similar to the type of spicule illustrated in Fig. 312, which we have explained as being probably developed within an oval cell, by whose walls its branches have been cabined and confined. But it is more probable that we have here to do with a spicule developed in the midst of a group of three co-equal and more or less elongated or cylindrical cells or vesicles, the long axial

Fig. 314. Spicules of tetractinellid sponges (after Sollas).
a–e, anatriaenes; *d–f*, pro-
triaenes.

ray corresponding to their common edge or line of contact, and the three short rays having each lain in the surface furrow between two out of the three adjacent cells.

Just as in the case of the little **S**-shaped spicules formed within the bounds of a single cell, so also in the case of the larger tetractinellid types do we find the same configurations reproduced among the holothuroids as we have dealt with in the sponges. The holothurian spicules are a little less neatly formed, a little rougher, than the sponge-spicules, and certain forms occur among the former group which do not present themselves among the latter; but for the most part a community of type is obvious and striking (Fig. 315).

The very peculiar spicules of the holothurian *Synapta*, where a tiny anchor is pivoted or hinged on a perforated plate, are a puzzle indeed; but we may at least solve part of the riddle. How the hinge is formed, I do not know; the anchor gets its shape, perhaps, in some such way as we have supposed the "amphidiscs" of *Hyalo-*

nema to acquire their reflexed spokes, but the perforated plate is more comprehensible. Each plate starts in a little clump of cells in whose boundary-walls calcareous matter is deposited, doubtless by adsorption, the holes in the finished plate thus corresponding to the cells which formed it. Close-packing leads to an arrangement of six cells round a central one, and the normal pattern of the plate displays this hexagonal configuration. The calcareous plate begins as a little rod whose ends fork, and then fork again: in the same inevitable trinodal pattern which includes the "polar furrow" of the embryologists. The anchor had been first formed, and the

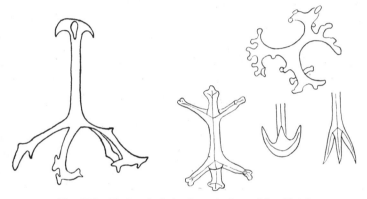

Fig. 315.   Various holothurian spicules.   After Théel.

little plate is added on beneath it. The first spicular rudiment of the plate may lie parallel to the stock of the anchor or it may lie athwart* it. From the physical point of view it would seem to be a mere matter of chance which way the cluster of cells happens to lie; but this difference of direction will cause a certain difference in the symmetry of the resulting plate. It is this very difference which systematic zoologists at one time seized upon to distinguish *S. Buskii* from our two commoner "species." The two latter

* Cf. S. Becher, Nicht-funktionelle Korrelation in der Bildung selbständiger Skeletelemente, *Zool. Jahrbücher (Physiol.)*, xxxi, pp. 1–189, 1912; Hedwiga Wilhelmi, Skeletbildung der füsslosen Holothurien, *ibid.* xxxvii, pp. 493–547, 1920; *Arch. f. Entw. Mech.* xlvi, pp. 210–258, 1920. See also W. Woodland, Studies in spicule-formation, *Q.J.M.S.* xlix, pp. 535–559, 1906; li, pp. 483–509, 1907 and R. Semon, Naturgeschichte der Synaptiden, *Mitth. Zool. St. Neapel*, vii, pp. 272–299, 1886. On the common species of *Synapta*, see Koehler, *Faune de France, Echinodermes*, 1921, pp. 188–9.

(*S. inhaerens* and *S. digitata*) are mainly distinguished from one another by the number of holes in the plate, that is to say, by the average number of cells in the little cluster of which the plate or spicule was formed. In many or perhaps most other holothurians

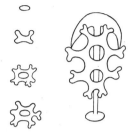

Fig. 316.   Development of anchor-plate in *Synapta*.   After Semon.

the spicules consist of little perforated plates or baskets, developed in the same way, about cells or vesicles more or less close-packed, and therefore more or less symmetrically arranged (Fig. 316).

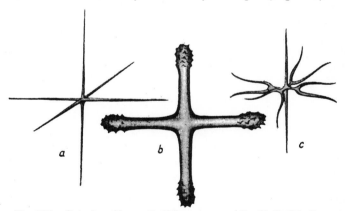

Fig. 317.   Spicules of hexactinellid sponges.   After F. E. Schultze.

The six-rayed siliceous spicules of the hexactinellid sponges, while they are perhaps the most regular and beautifully formed spicules to be found within the entire group, have been found very difficult to explain, and Dreyer has confessed his complete inability to account for their conformation*. But, though it may only be

---

* Cf. Albr. Schwan, Ueber die Funktion des Hexactinellidenskelets, u. seine Vergleichbarkeit mit dem Radiolarienskelet, *Zool. Jb., Abth. allg. Zool. u. Physiol.* xxxiii, pp. 603–616, 1913; cf. V. Hacker, Bericht über d. Tripyleenausbeute d. d. Tiefsee-Exped. *Verh. d. zool. Ges.* 1904.

throwing the difficulty a little further back, we may so far account for them by considering that the cells or vesicles by which they are conformed are not arranged in what is known as "closest packing," but in linear series; so that in their arrangement, and by their mutual compression, we tend to get a pattern not of hexagons but of squares: or, looking to the solid, not of dodecahedra but of cubes or parallelepipeda. This indeed appears to be the case, not with the individual cells (in the histological sense), but with the larger units or vesicles which make up the body of the hexactinellid. And this being so, the spicules formed between the linear, or cubical series of vesicles, will have the same tendency towards a "hexactinellid" shape, corresponding to the angles and adjacent edges of a system of cubes, as in our former case they had to a triradiate or a tetractinellid form, when developed in connection with the angles and edges of a system of hexagons, or a system of rhombic dodecahedra.

However the hexactinellid spicules be arranged (and this is none too easy to determine) in relation to the tissues and chambers of the sponge, it is at least clear that, whether they lie separate or be fused together in a composite skeleton, they effect a symmetrical partitioning of space according to the cubical system, in contrast to that closer packing which is represented and effected by the tetrahedral system[*].

Histologically, the case is illustrated by a well-known phenomenon in embryology. In the segmenting ovum, there is a tendency for the cells to be budded off in linear series; and so they often remain, in rows side by side, at least for a considerable time and during the course of several consecutive cell divisions. Such an arrangement constitutes what the embryologists call the "radial type" of segmentation[†]. But in what is described as the "spiral type" of segmentation, it is stated that, as soon as the first horizontal furrow has divided the cells into an upper and a lower layer, those of "the upper layer are shifted in respect to the lower layer,

---

[*] *Chall. Rep.*, *Hexactinellida*, pls. xvi, liii, lxxvi, lxxxviii.

[†] See, for instance, the figures of the segmenting egg of Synapta (after Selenka), in Korschelt and Heider's *Vergleichende Entwicklungsgeschichte*. On the spiral type of segmentation as a secondary derivative, due to mechanical causes, of the "radial" type of segmentation, see E. B. Wilson, Cell-lineage of *Nereis*, *Journ. Morph.* vi, p. 450, 1892.

by means of a rotation about the vertical axis*." It is, of course, evident that the whole process is merely that which is familiar to physicists as "close packing." It is a very simple case of what Lord Kelvin used to call "a problem in tactics." It is a mere question of the rigidity of the system, of the freedom of movement on the part of its constituent cells, whether or at what stage this tendency to slip into the closest propinquity, or position of *minimum potential*, will be found to manifest itself.

Lastly, a curious case is presented by the so-called "chessman" spicules of *Latrunculia* and of a few other sponges, where the spicular shaft is thickened at regular intervals, and the thickenings grow into whorled and flattened lobes. Dendy suggested that the developing spicule is in a state of vibration (due perhaps to the water-currents of the sponge), and that the whorls correspond to nodes, or *loci* of comparative rest, where the formative cells tend to settle down and do their work undisturbed. The position of the nodes and internodes will depend on many circumstances, on whether the spicule be a fixed rod or a free one, straight or curved, uniform in section or tapering towards either end. In the free bar there should tend, in any case, to be a node in the middle, and two more at definite distances from either end. It so happens that in the forms investigated there are only two whorls, the median and one other; but J. W. Nicholson has calculated the positions of these according to the vibration theory, and the theoretical results are found to agree with those of observation very closely indeed. That one of the whorls should be lacking might seem to imperil the proof; but on the other hand among large numbers of spicules no one was found to have its whorls in a position inconsistent with the theory, and there was the required agreement between the shape of the spicule and the position of the whorls. The absence of a third whorl is explained as due to a lack of the necessary formative cells at that part of the spicule†. The theory is in a way supported by recent work (by R. W. Wood of Baltimore and others) on "supersonic vibrations," showing excessively rapid

* Korschelt and Heider, p. 16.

† A. Dendy and J. W. Nicholson, On the influence of vibration upon the form of certain sponge-spicules, *Proc. R.S.* (B), LXXXIX, pp. 573–587, 1917; A. Dendy, The chessman spicules of the genus *Latrunculia*, etc., *Journ. Quekett Microsc. Club*, XIII, pp. 1–16, 1917.

vibrations in quartz rods, more rapid even than Dendy's hypothesis would seem to require. But on the other hand, it is only to few and even exceptional spicules that the theory would seem to apply.

This question of the origin and causation of the forms of sponge-spicules, with which we have now sought to deal, is all the more important and all the more interesting because it has been discussed time and again, from points of view which are characteristic of very different schools of thought in biology. Haeckel found in the form of the sponge-spicule a typical illustration of his theory of "bio-crystallisation"; he considered that these "biocrystals" represented something midway—*ein Mittelding*—between an inorganic crystal and an organic secretion; that there was a "compromise between the crystallising efforts of the calcium carbonate and the formative activity of the fused cells of the syncytium"; and that the semi-crystalline secretions of calcium carbonate "were utilised by natural selection as 'spicules' for building up a skeleton, and afterwards, by the interaction of adaptation and heredity, became modified in form, and differentiated in a vast variety of ways, in the struggle for existence*." What Haeckel precisely meant by these words is not clear to me.

F. E. Schultze, perceiving that identical forms of spicule were developed whether the material were crystalline or non-crystalline, abandoned all theories based upon crystallisation; he simply saw in the form and arrangement of the spicules something which was "best fitted" for its purpose, that is to say for the support and strengthening of the porous walls of the sponge, and finding clear evidence of "utility" in the specific characters of these skeletal elements, had no difficulty in ascribing them to natural selection.

Sollas and Dreyer, as we have seen, introduced in various ways the conception of physical causation—as indeed Haeckel himself had done in regard to one particular, when he supposed the *position* of the spicules to be due to the constant passage of the water-

---

* "Hierbei nahm der kohlensaure Kalk eine halb-krystallinische Beschaffen-heit an, und gestaltete sich unter Aufnahme von Krystallwasser und in Verbindung mit einer geringen Quantität von organischer Substanz zu jenen individuellen, festen Körpern, welche durch die natürliche Züchtung als *Spicula* zur Skeletbildung benützt, und späterhin durch die Wechselwirkung von Anpassung und Vererbung im Kampfe ums Dasein auf das Vielfältigste umgebildet und differenziert wurden." *Die Kalkschwämme*, I, p. 377, 1872; cf. also pp. 482, 483.

currents; though even here, by the way, if I understand Haeckel aright, he was thinking not of a direct or immediate physical causation, but rather of one manifesting itself through the agency of natural selection*. Sollas laid stress upon the "path of least resistance" as determining the direction of growth; while Dreyer dealt in greater detail with the tensions and pressures to which the growing spicule was exposed, amid the alveolar or vesicular structure which was represented alike by the chambers of the sponge, by the constituent cells, or by the minute structure of the intracellular protoplasm. But neither of these writers, so far as I can discover, was inclined to doubt for a moment the received canon of biology, which sees in such structures as these the characteristics of true organic *species*, the indications of blood-relationship and family likeness, and the evidence by which evolutionary descent throughout geologic time may be deduced or deciphered.

Minchin, in a well-known paper†, took sides with F. E. Schultze, and gave his reasons for dissenting from such mechanical theories as those of Sollas and of Dreyer. For example, after pointing out that all protoplasm contains a number of "granules" or microsomes, contained in an alveolar framework and lodged at the nodes of a reticulum, he argued that these also ought to acquire a form such as the spicules possess, if it were the case that these latter owed their form to their similar or identical position. "If vesicular tension cannot in any other instance cause the granules at the nodes to assume a tetraxon form, why should it do so for the sclerites?" The answer is not far to seek. If the force which the "mechanical" hypothesis has in view were simply that of *mechanical pressure*, as between solid bodies, then indeed we should expect that *any* substances lying between the impinging spheres would tend to assume the quadriradiate or "tetraxon" form; but this conclusion does not follow at all, in so far as it is to *surface-energy* that we ascribe the phenomenon. Here the specific nature of the substance makes all the difference. We cannot argue from one

---

* *Op. cit.* p. 483. "Die geordnete, oft so sehr regelmässige und zierliche Zusammensetzung des Skeletsystems ist zum grössten Theile unmittelbares Product der Wasserströmung; die charakteristische Lagerung der Spicula ist von der constanten Richtung des Wasserstroms hervorgebracht; zum kleinsten Theile ist sie die Folge von Anpassungen an untergeordnete äussere Existenzbedingungen."

† Materials for a Monograph of the Ascones, *Q.J.M.S.* XL, pp. 469–587, 1898.

substance to another; adsorptive attraction shews its effect on one and not on another; and we have no reason to be surprised if we find that the little granules of protoplasmic material, which as they lie bathed in the more fluid protoplasm have (presumably, and as their shape indicates) a strong surface-tension of their own, behave towards the adjacent vesicles in a very different fashion to the incipient aggregations of calcareous or siliceous matter in a colloid medium. "The ontogeny of the spicules," says Professor Minchin, "points clearly to their regular form being a *phylogenetic adaptation, which has become fixed and handed on by heredity, appearing in the ontogeny as a prophetic adaptation.*" And again, "The forms of the spicules are the result of adaptation to the requirements of the sponge as a whole, produced by *the action of natural selection upon variation in every direction.*" It would scarcely be possible to illustrate more briefly and more cogently than by these few words (or the similar words of Haeckel quoted on p. 691), the fundamental difference between the Darwinian conception of the causation and determination of Form, and that which is based on, and characteristic of, the physical sciences.

Last of all, Dendy took a middle course. While admitting that the majority of sponge-spicules are "the outcome of conditions which are in large part purely physical," he still saw in them "a very high taxonomic value," as "indications of phylogenetic history" all on the ground that "it seems impossible to account in any other way for the fact that we can actually arrange the different forms in such well-graduated series." At the same time he believed that "the vast majority of spicule-characters appear to be non-adaptive," "that no one form of spicule has, as a rule, any greater survival-value than another," and that "the natural selection of favourable varieties can have had very little to do with the matter*."

The quest after lines and evidences of descent dominated morphology for many years, and preoccupied the minds of two or three generations of naturalists. We find it easier to see than they did that a graduated or consecutive series of forms may be based on physical causes, that forms mathematically akin may belong to

---

* Cf. A. Dendy, The Tetraxonid sponge-spicule: a study in evolution, *Acta Zoologica*, 1921, pp. 136, 146, etc. Cf. also Bye-products of organic evolution, *Journ. Quekett Microscop. Club*, xii, pp. 65–82, 1913.

organisms biologically remote, and that, in general, mere formal likeness may be a fallacious guide to evolution in time and to relationship by descent and heredity.

If I have dealt comparatively briefly with the inorganic skeletons of sponges, in spite of the interest of the subject from the physical point of view, it has been owing to several reasons. In the first place, though the general trend of the phenomena is clear, it must be admitted that many points are obscure, and could only be discussed at the cost of a long argument. In the second place, the physical theory is too often (as I have shewn) in conflict with the accounts given by embryologists of the development of the spicules, and with the current biological theories which their descriptions embody; it is beyond our scope to deal with such descriptions in detail. Lastly, we find ourselves able to illustrate the same physical principles with greater clearness and greater certitude in another group of animals, namely the Radiolaria.

The group of microscopic organisms known as the Radiolaria is extraordinarily rich in diverse forms or "species." I do not know how many of such species have been described and defined by naturalists, but some fifty years ago the number was said to be over four thousand, arranged in more than seven hundred genera*; of late years there has been a tendency to reduce the number. But apart from the extraordinary multiplicity of forms among the Radiolaria, there are certain features in this multiplicity which arrest our attention. Their distribution in space is curious and vague; many species are found all over the world, or at least every here and there, with no evidence of specific limitations of geographical habitat; some occur in the neighbourhood of the two poles, some are confined to warm and others to cold currents of the ocean. In time their distribution is not less vague: so much so that it has been asserted of them that "from the Cambrian age downwards, the families and even genera appear identical with those now living." Lastly, except perhaps in the case of a few large "colonial forms," we seldom if ever find, as is usual in most animals, a local predominance of one particular species. On the

---

* Haeckel, in his *Challenger Monograph*, p. clxxxviii (1887), estimated the number of known forms at 4314 species, included in 739 genera. Of these, 3508 species were described for the first time in that work.

contrary, in a little pinch of deep-sea mud or of some fossil "radio-larian earth," we shall probably find scores, and it may be even hundreds, of different forms. Moreover, the radiolarian skeletons are of quite extraordinary delicacy and complexity, in spite of their minuteness and the comparative simplicity of the "unicellular" organisms within which they grow; and these complex conforma-tions have a wonderful and unusual appearance of geometric regularity. All these general considerations seem such as to prepare us for some physical hypothesis of causation. The little skeletons remind us of such things as snow-crystals (themselves almost endless in their diversity), rather than of a collection of animals, constructed in accordance with functional needs and distributed in accordance with their fitness for particular situations. Nevertheless, great efforts have been made to attach "a biological meaning" to these elaborate structures, and "to justify the hope that in time their utilitarian character will be more completely recognised*."

As Ernst Haeckel described and figured many hundred "species" of radiolarian skeletons, so have the physicists depicted snow-crystals in several thousand different forms†. These owe their multitudinous variety to symmetrical repetitions of one simple crystalline form—a beautiful illustration of Plato's *One among the Many*, τὸ ἓν παρὰ τὰ πολλά. On the other hand, the radiolarian skeleton rings its endless changes on combinations of certain facets, corners and edges within a filmy and bubbly mass. The broad difference between the two is very plain and instructive.

Kepler studied the snowflake with care and insight, though he said that to care for such a trifle was like Socrates measuring the hop of a flea. The first drawings I know are by Dominic Cassini; and if that great astronomer was content with them they shew how the physical sciences lagged behind astronomy. They date from the time when Maraldi, Cassini's nephew, was studying the bee's cell;

* Cf. Gamble, *Radiolaria* (Lankester's *Treatise on Zoology*), I, p. 131, 1909. Cf. also papers by V. Häcker, in *Jen. Zeitschr.* XXXIX, p. 581, 1905; *Z. f. wiss. Zool.* LXXXIII, p. 336, 1905; *Arch. f. Protistenkunde*, IX, p. 139, 1907; etc.

† See above, p. 411; and see (besides the works quoted there) Kepler, De nive sexangula (1611), *Opera*, ed. Fritsch, VII, pp. 715–730; Erasmus Bartholin, *De figura nivis, Diss.*, Hafniae, 1661; Dom. Cassini, Obs. de la figure de la neige (Abstr.), *Mém. Acad. R. des Sciences* (1666–1699), X, 1730; J. C. Wilcke, Om de naturliga snö-figurers, *K. V. Akad. Handl.* XXII, 1761.

and they shew once more how very rough his measurements of the honeycomb are bound to have been.

Crystals lie outside the province of this book; yet snow-crystals, and all the rest besides, have much to teach us about the variety, the beauty and the very nature of form. To begin with, the snow-crystal is a regular hexagonal plate or thin prism; that is to

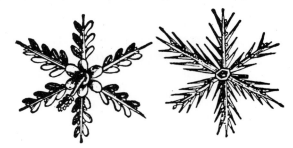

Fig. 318 *a*.   Snow-crystals, or "snow-flowers."   From Dominic Cassini (*c*. 1600).

say, it shews hexagonal faces above and below, with edges set at co-equal angles of 120°.   Ringing her changes on this fundamental form, Nature superadds to the primary hexagon endless combinations of similar plates or prisms, all with identical angles but varying lengths of side;   and she repeats, with an exquisite symmetry,

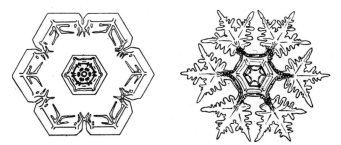

Fig. 318 *b*.   Snow-crystals.   From Bentley and Humphreys, 1931.

about all three axes of the hexagon, whatsoever she may have done for the adornment and elaboration of one.   These snow-crystals seem (as Tutton says) to give visible proof of the space-lattice on which their structure is framed.

The beauty of a snow-crystal depends on its mathematical regularity and symmetry;   but somehow the association of many variants of a single type, all related but no two the same, vastly increases our pleasure and admiration. Such is the peculiar beauty which a Japanese artist sees in a bed of rushes or a clump of bamboos, especially when the wind's ablowing; and such (as we saw before) is the phase-beauty of a flowering spray when it shews every gradation from opening bud to fading flower.

The snow-crystal is further complicated, and its beauty is notably enhanced, by minute occluded bubbles of air or drops of water, whose symmetrical form and arrangement are very curious and not always easy to explain*. Lastly, we are apt to see our snow-crystals after a slight thaw has rounded their sharp edges, and has heightened their beauty by softening their contours.

In the majority of cases, the skeleton of the Radiolaria is composed, like that of so many sponges, of silica; in one large family, the Acantharia, and perhaps in some others, it is made of a very unusual constituent, namely strontium sulphate†. There is no important morphological character in which the shells made of these two constituents differ from one another; and in no case can the chemical properties of these inorganic materials be said to influence the form of the complex skeleton or shell, save only in this general way that, by their hardness, toughness and rigidity, they give rise to a fabric more slender and delicate than we find among calcareous organisms.

A slight exception to this rule is found in the presence of true crystals, which occur within the central capsules of certain Radiolaria, for instance the genus *Collosphaera*‡. Johannes Müller (whose knowledge and insight never fail to astonish us§) remarked

* We may find some suggestive analogies to these occlusions in Emil Hatschek's paper, Gestalt und Orientirung von Gasblasen in Gelen, *Kolloid. Ztschr.* xx, pp. 226–234, 1914.

† Bütschli, Ueber die chemische Natur der Skeletsubstanz der Acantharia, *Zool. Anz.* xxx, p. 784, 1906.

‡ For figures of these crystals see Brandt, *F. u. Fl. d. Golfes von Neapel*, xiii, *Radiolaria*, 1885, pl. v. Cf. Johannes Müller, Ueber die Thalassicollen, etc., *Abh. K. Akad. Wiss. Berlin*, 1858.

§ It is interesting to think of the lesser discoveries or inventions, due to men famous for greater things. Johannes Müller first used the tow-net, and Edward Forbes first borrowed the oyster-man's dredge. When we watch a living polyp under the microscope in its tiny aquarium of a glass-cell, we are doing what John Goodsir was the first to do; and the microtome itself was the invention of that best of laboratory-servants, "old Stirling," Goodsir's right-hand man.

that these were identical in form with crystals of celestine, a sulphate of strontium and barium; and Bütschli's discovery of sulphates of strontium and of barium in kindred forms renders it all but certain that they are actually true crystals of celestine *.

In its typical form, the radiolarian body consists of a spherical mass of protoplasm, around which, and separated from it by some sort of porous "capsule," lies a frothy protoplasm, bubbled up into a multitude of alveoli or vacuoles, filled with a fluid which can scarcely differ much from sea-water†. According to their surface-tension conditions, these vacuoles may appear more or less isolated and spherical, or joined together in a "froth" of polyhedral cells; and in the latter, which is the commoner condition, the cells tend to be of equal size, and the resulting polygonal meshwork beautifully regular. In some cases a large number of such simple individual organisms are associated together, forming a floating colony; and it is probable that many others, with whose scattered skeletons we are alone acquainted, had likewise formed part of a colonial organism.

In contradistinction to the sponges, in which the skeleton always begins as a loose mass of isolated spicules, which only in a few exceptional cases (such as *Euplectella* and *Farrea*) fuse into a continuous network, the characteristic feature of the radiolarians lies in the production of a continuous skeleton, of netted mesh or perforated lacework, sometimes replaced by and oftener associated with minute independent spicules. Before we proceed to treat of the more complex skeletons, we may begin by dealing with those comparatively few simple cases where the skeleton is represented by loose, separate spicules or aciculae, which seem, like the spicules of *Alcyonium*, to be isolated formations or deposits, precipitated in the colloid matrix, with no relation to cellular or vesicular boundaries. These simple acicular spicules occupy a definite position in the organism. Sometimes, as for instance among the fresh-water Heliozoa (e.g. *Raphidiophrys*), they lie on the outer surface of the organism, and not infrequently (when few in number) they tend to

---

* Celestine, or celestite, is $SrSO_4$ with some BaO replacing SrO.

† With the colloid chemists, we may adopt (as Rhumbler has done) the terms *spumoid* or *emulsoid* to denote an agglomeration of fluid-filled vesicles, restricting the name *froth* to such vesicles when filled with air or some other gas.

collect round the bases of the pseudopodia, or around the larger radiating spicules or axial rays in cases where these latter are present. When the spicules are thus localised around some prominent centre, they tend to take up a position of symmetry in regard to it; instead of forming a tangled or felted layer, they come to lie side by side, in a radiating cluster round the focus. In other cases (as for instance in the well-known radiolarian *Aulacantha scolymantha*) the felted layer of aciculae lies at some depth below the surface, forming a sphere concentric with the entire spherical organism. In either case, whether the layer of spicules be deep or be superficial, it tends to mark a "surface of discontinuity," a meeting place either between two distinct layers of protoplasm or between the protoplasm and the water around; and it is evident that, in either case, there are manifestations of surface-energy at the boundary, which cause the spicules to be retained there and to take up their position in its plane. The case is analogous to that of a cirrus cloud, which marks a surface of discontinuity in a stratified atmosphere.

We have, then, to enquire what are the conditions, apart from gravity, which confine an extraneous body to a surface-film; and we may do this very simply, by considering the surface-energy of the entire system. In Fig. 319 we have two fluids in contact with one another (let us call them water and protoplasm), and a body (*b*) which may be immersed in either, or may be restricted to the boundary between. We have here three possible "interfacial contacts," each with its own specific surface-energy per unit of surface area: namely, that between our particle and the water (let us call it α), that between the particle and the protoplasm (β), and

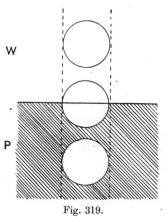

Fig. 319.

that between water and protoplasm (γ). When the body lies in the boundary of the two fluids, let us say half in one and half in the other, the surface-energies concerned are equivalent to $(S/2)\,\alpha + (S/2)\,\beta$; but we must also remember that, by the presence of the particle, a small portion (equal to its sectional area s) of the original contact-surface between water and protoplasm has been obliterated, and with it a proportionate quantity of energy, equivalent to $s\gamma$, has been set free. When, on the other hand, the body lies entirely within one or other fluid, the surface-energies of the system (so far as we are concerned) are equivalent to $S\alpha + s\gamma$, or $S\beta + s\gamma$, as the case may be. Accordingly as α be less or greater than β, the particle will have a tendency to remain immersed in the water or in the protoplasm; but if $(S/2)(\alpha + \beta) - s\gamma$ be less than either $S\alpha$ or $S\beta$, then the condition of minimal potential will be found when the particle lies, as we have said, in the boundary zone, half in one fluid and half in the other; and, if we were to attempt a more general solution of the problem, we should have to deal with possible conditions of equilibrium under which the necessary balance of energies would be attained by the particle rising or sinking in the boundary zone, so as to adjust the relative magnitudes of the surface-areas concerned. This principle may, in certain cases, help us to explain the position even of a *radial* spicule, which is just a case where the surface of the solid spicule is distributed between the fluids with a minimal disturbance, or minimal replacement, of the original surface of contact between the one fluid and the other.

In like manner we may provide for the case (a common and an important one) where the protoplasm "creeps up" the spicule, covering it with a delicate film, and forming catenary curves or festoons between one spicule and another; and a less fluid or more tenacious thread of protoplasm may serve, like a solid spicule, to extend the more fluid film, as we see, for instance, in *Chlamydomyxa* or in *Gromia*. When the spicules are numerous and close-set, the surface-film of protoplasm stretching between them will tend to look like a layer of concave or inverted bubbles; and this honeycombed bubbly surface is sometimes beautifully regular, to judge from a well-known figure of the living *Globigerina**. In *Acanthocystis* we have yet another special case, where the radial spicules plunge only a certain distance into the protoplasm of the cell, being arrested at a boundary-surface between an inner and an outer layer of

* See H. B. Brady's *Challenger Monograph*, pl. lxxvii; and see the figure of *Chlamydomyxa* in Doflein's *Protozoenkunde*, p. 374.

cytoplasm; here we have only to assume that there is a tension at this surface, between the two layers of protoplasm, sufficient to balance the tensions which act directly on the spicule*.

In various Acanthometridae, besides such typical characteristics as the radial symmetry, the concentric layers of protoplasm, and the capillary surfaces in which the outer vacuolated protoplasm is festooned upon the projecting radii, we have another curious feature. On the surface of the protoplasm where it creeps up the sides of the long radial spicules, we find a number of elongated bodies, forming in each case one or several little groups, and lying neatly arranged in parallel bundles. A Russian naturalist, Schewiakoff, whose views have been accepted in the text-books, tells us that these are muscular structures, serving to raise or lower the conical masses of protoplasm about the radial spicules, which latter serve as so many "tent-poles" or masts, on which the protoplasmic membranes are hoisted up; and the little elongated bodies are dignified with various names, such as "myonemes" or "myophriscs," in allusion to their supposed muscular nature†. This explanation is by no means convincing. To begin with, we have precisely similar festoons of protoplasm in a multitude of other cases where the "myonemes" are lacking; from their minute size (0·006–0·012 mm.) and the amount of contraction they are said to be capable of, the myonemes can hardly be very efficient instruments of traction; and further, for them to act (as is alleged) for a specific purpose, namely the "hydrostatic regulation" of the organism giving it power to sink or to swim, would seem to imply a mechanism of action and of coordination not easy to conceive in these minute and simple organisms. The fact is that the whole explanation is unnecessary. Just as the supposed "hauling up" of the protoplasmic festoons may be at once explained by capillary phenomena, so also (in all probability) may the position and arrangement of the little elongated bodies. Whatever the actual nature of these bodies may be, whether they be truly portions of differentiated protoplasm, or whether they be foreign bodies or spicular structures (as bodies occupying a similar position in other cases undoubtedly

---

* Cf. Koltzoff, Zur Frage der Zellgestalt, *Anat. Anzeiger*, XLI, p. 190, 1912.

† *Mém. de l'Acad. des Sci.*, St Pétersbourg, XII, Nr. 10, 1902.

are), we can explain their situation on the surface of the protoplasm, and their arrangement around the radial spicules, all by the principles of capillarity.

This last case is not of the simplest; and I do not forget that my explanation of it, which is wholly theoretical, implies a doubt of Schewiakoff's statements, founded on his personal observation. This I am none too willing to do; but whether it be justly done in this case or not, I hold that it is in principle justifiable to look with suspicion upon all such statements where the observer has obviously left out of account the physical aspect of the phenomenon, and all the opportunities of simple explanation which the consideration of that aspect might afford.

Whether it be applicable to this particular and complex case or no, our general theorem of the localisation and arrestment of solid particles in a surface-film is of great biological significance; for on it depends the power displayed by many little naked protoplasmic organisms of covering themselves with an "agglutinated" shell. Sometimes, as in *Difflugia*, *Astrorhiza* (Fig. 320) and others, this covering consists of sand-grains picked up from the surrounding medium, and sometimes, on the other hand, as in *Quadrula*, it consists of solid particles said to arise as inorganic deposits or concretions within the protoplasm itself, and to find their way outwards to a position of equilibrium in the surface-layer; and in both cases, the mutual capillary attractions between the particles, confined to the boundary-layer but enjoying a certain measure of freedom therein, tends to the orderly arrangement of the particles one with another, and even to the appearance of a regular "pattern" as the result of this arrangement.

The "picking up" by the protoplasmic organism of a solid particle with which "to build its house" (for it is hard to avoid this customary use of figures of speech, misleading though it be) is a physical phenomenon akin to that by which an amoeba "swallows" a particle of food. This latter process has been reproduced or imitated in various pretty experimental ways. For instance, Rhumbler has shewn that if a splinter of glass be covered with shellac and brought near a drop of chloroform suspended in water, the drop takes in the spicule, robs it of its shellac covering,

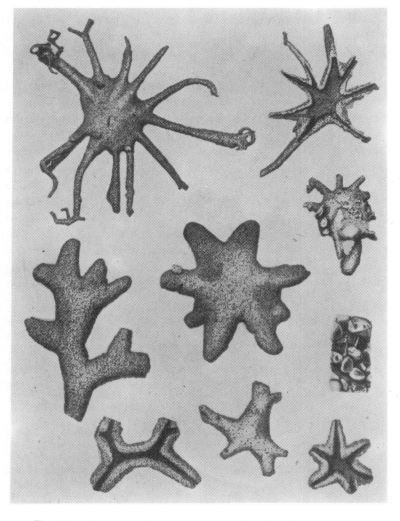

Fig. 320.   Arenaceous Foraminifera; *Astrorhiza limicola* and *arenaria*.
From Brady's *Challenger Monograph*.

and then passes it out again*. In another case a thread of shellac, laid on a drop of chloroform, is drawn in and coiled within it: precisely as we may see a filament of Oscillatoria ingested by an Amoeba, and twisted and coiled within its cell. It is all a question of relative surface-energies, leading to different degrees of "adhesion" between the chloroform and the splinter of glass or its shellac covering. Thus it is that the Amoeba takes in the diatom, dissolves off its proteid covering, and casts out the shell.

Furthermore, as the whole phenomenon depends on a distribution of surface-energy, the amount of which is specific to certain particular substances in contact with one another, we have no difficulty in understanding the *selective action* which is very often a conspicuous feature in the phenomenon†. Just as some caddis-worms make their houses of twigs, and others of shells and again others of stones, so some Rhizopods construct their agglutinated "test" out of stray sponge-spicules, or frustules of diatoms, or again of tiny mud particles or of larger grains of sand. In all these cases, we have to deal with specific surface-energies, and also doubtless with differences in the total available amount of surface-

---

* Rhumbler, Physikalische Analyse von Lebenserscheinungen der Zelle, *Arch. f. Entw. Mech.* VII, p. 250, 1898.

† The whole phenomenon has been described as a "surprising exhibition of constructive and selective activity," and ascribed, in varying phraseology, to intelligence, skill, purpose, psychical activity, or "microscopic mentality": that is to say, to Galen's τεχνικὴ φύσις, or "artistic creativeness" (cf. Brock's *Galen*, 1916, p. xxix); cf. Carpenter, *Mental Physiology*, 1874, p. 41; Norman, Architectural achievements of Little Masons, etc., *Ann. Mag. Nat. Hist.* (5), I, p. 284, 1878; Heron-Allen, Contributions...to the study of the Foraminifera, *Phil. Trans.* (B), CCVI, pp. 227–279, 1915; Theory and phenomena of purpose and intelligence exhibited by the Protozoa, as illustrated by selection and behaviour in the Foraminifera, *Journ. R. Microsc. Soc.* 1915, pp. 547–557; *ibid.*, 1916, pp. 137–140. Sir J. A. Thomson (*New Statesman*, Oct. 23, 1915) describes a certain little foraminifer, whose protoplasmic body is overlaid by a crust of sponge-spicules, as "a psycho-physical individuality, whose experiments in self-expression include a masterly treatment of sponge-spicules, and illustrate that organic skill which came before the dawn of Art." Sir Ray Lankester finds it "not difficult to conceive of the existence of a mechanism in the protoplasm of the Protozoa which selects and rejects building-material, and determines the shapes of the structures built, comparable to that mechanism which is assumed to exist in the nervous system of insects and other animals which 'automatically' go through wonderfully elaborate series of complicated actions." And he agrees with "Darwin and others [who] have attributed the building up of these inherited mechanisms to the age-long action of Natural Selection, and the survival of those individuals possessing qualities or 'tricks' of life-saving value," *Journ. R. Microsc. Soc.* April, 1916, p. 136.

energy in relation to gravity or other extraneous forces. In my early student days, Wyville Thomson used to tell us that certain deep-sea "Difflugias," after constructing a shell out of particles of the black volcanic sand common in parts of the North Atlantic, finished it off with "a clean white collar" of little grains of quartz. Even this phenomenon may be accounted for on surface-tension principles, if we may assume that the surface-energy ratios have tended to change, either with the growth of the protoplasm or by reason of external variation of temperature or the like; we are by no means obliged to attribute even this phenomenon to a manifestation of volition, or taste, or aesthetic skill, on the part of the microscopic organism. Nor, when certain Radiolaria tend more than others to attract into their own substance diatoms and suchlike foreign bodies, is it scientifically correct to speak, as some text-books do, of species "in which diatom-selection has become *a regular habit*." To do so is an exaggerated misuse of anthropomorphic phraseology.

The formation of an "agglutinated" shell is thus seen to be a purely physical phenomenon, and indeed a special case of a more general physical phenomenon which has important consequences in biology. For the shell to assume the solid and permanent character which it acquires in *Difflugia*, we have only to make the further assumption that small quantities of a cementing substance are secreted by the animal, and that this substance flows or creeps by capillary attraction through all the interstices of the little quartz grains, and ends by binding them together. Rhumbler* has shewn us how these agglutinated tests of spicules or of sand-grains can be precisely imitated, and how they are formed with greater or less ease and greater or less rapidity according to the nature of the materials employed, that is to say according to the specific surface-tensions which are involved. If we mix up a little powdered glass with chloroform, and set a drop of the mixture in water, the glass particles gather neatly round the surface of the drop so quickly that the eye cannot follow the operation. If we do the same with oil and fine sand, dropped into 70 per cent. alcohol, a still more

---

* Rhumbler, Beiträge z. Kenntniss 'd. Rhizopoden, I–V, *Z. f. w. Z.* 1891–5; *Das Protoplasma als physikalisches System*, Jena, p. 591, 1914; also in *Arch. f. Entwickelungsmech.* VII, pp. 279–335, 1898; *Biol. Centralbl.* XVIII, 1898; etc.

beautiful artificial Rhizopod-shell is formed, but it takes some three hours to do. Where the action is quick the little test forms as the droplet exudes from the pipette: precisely as in the living *Difflugia* when new protoplasm, laden with solid particles, is being extruded from the mouth of the parent-cell. The experiment can be varied, simply and easily. Instead of a spherical drop a pear-shaped one may easily be formed, so exactly like the common *Difflugia pyriformis* that Rhumbler himself was unable, sometimes, to tell under the microscope the real from the artefact. Again he found that, when the alcohol dissolved the oily substance of the drop and shrinkage took place accordingly, the surface-layer with its solid particles got kinked or folded in—and reproduced in doing so, with startling accuracy, a little shell of common occurrence, known by the generic name of *Lesqueureusia*, or *Difflugia spiralis*. The peculiar shape of this little twisted and bulging shell has been taken to shew that "it had enlarged after its first formation, a very rare occurrence in this group*"; the very opposite is the case. Neither here nor in any allied form does the agglutinated test, once set in order by capillary forces, yield scope for intercalation and enlargement.

At the very time when Rhumbler was thus demonstrating the physical nature of the Difflugian shell, Verworn, a very notable person, was studying the same and kindred organisms from the older standpoint of an incipient psychology†. But as Rhumbler himself admits, Verworn (unlike many another) was doing his best not to over-estimate the appearance of volition, or selective choice, in the little organism's use of materials to construct its dwelling.

This long parenthesis has led us away, for the time being, from the subject of the radiolarian skeleton, and to that subject we must now return. Leaving aside, then, the loose and scattered spicules, which we have sufficiently discussed, the more perfect radiolarian skeletons consist of a continuous and regular structure; and the siliceous (or other inorganic) material of which this framework is composed tends to be deposited in one or other of two ways or in both combined: (1) in long radial spicules, emanating symmetrically from, and usually conjoined at, the centre of the protoplasmic body;

---

* Cf. *Cambridge Natural History*, Protozoa, p. 55.

† Max Verworn, *Psycho-physiologische Protisten-Studien*, Jena, 1889 (219 pp.); Biologische Protisten-Studien, *Z: f. wiss. Z.* L, pp. 445–467, 1890.

(2) in the form of a crust, developed either on the outer surface of the organism or in relation to one, or more of the internal surfaces which separate its concentric layers or its component vesicles. Not infrequently, this superficial skeleton comes to constitute a spherical shell, or a system of concentric spheres.

We have already seen that a great part of the body of the Radiolarian, and especially that outer portion to which Haeckel has given the name of the "calymma," is built up of a mass of "vesicles," forming a sort of stiff froth, and equivalent in the physical though not necessarily in the biological sense to "cells," inasmuch as the little vesicles have their own well-defined boundaries, and their own surface phenomena. In short, all that we have said of cell-surfaces and cell-conformations in our discussion of cells and of tissues will apply in like manner, and under appropriate conditions, to these. In certain cases, even in so common and so simple a one as the vacuolated substance of an *Actinosphaerium*, we may see a close resemblance, or formal analogy, to a cellular or parenchymatous tissue in the close-packed arrangement and consequent configuration of these vesicles, and even at times in a slight membranous hardening of their walls. Leidy has figured * some curious little bodies like small masses of consolidated froth, which seem to be nothing else than the dead and empty husks, or filmy skeletons, of *Actinosphaerium*; and Carnoy † has demonstrated in certain cell-nuclei an all but precisely similar framework, of extreme minuteness and tenuity, formed by adsorption or partial solidification of interstitial matter in a close-packed system of alveoli (Fig. 321). In short, we are again dealing or about to deal with a network or basketwork, whose meshes correspond to the boundary lines between associated cells or vesicles. It is just in those boundary walls or films, still more in their edges or at their corners, that surface-energy will be concentrated and adsorption will be hard at work; and the whole arrangement will follow, or tend to follow, the rules of *areae minimae*—the partition-walls meeting at co-equal angles, three by three in an edge, and their edges meeting four by four in a corner.

Let us suppose the outer surface of our Radiolarian to be covered

---

* J. Leidy, *Fresh-water Rhizopods of North America*, 1879, p. 262, pl. xli, figs. 11, 12.

† Carnoy, *Biologie Cellulaire*, p. 244, fig. 108; cf. Dreyer, *op. cit.* 1892, fig. 185.

by a layer of froth-like vesicles, uniform in size or nearly so. We know that their mutual tensions will *tend* to conform them into the fashion of a honeycomb, or regular meshwork of hexagons, and that the free end of each hexagonal prism will be a little spherical cap. Suppose now that it be at the outer surface of the protoplasm (in contact with the surrounding sea-water) that the siliceous particles have a tendency to be secreted or adsorbed; the distribution of surface-energy will lead them to accumulate in the grooves which separate the vesicles, and the result will be the development

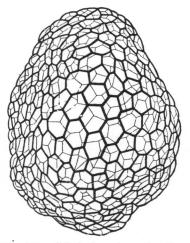
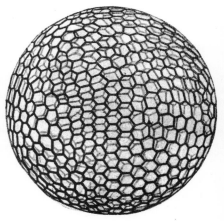

Fig. 321. "*Reticulum plasmatique.*"   Fig. 322.   *Aulonia hexagona* Hkl.
         After Carnoy.

of a delicate sphere composed of tiny rods arranged, or apparently arranged, in a hexagonal network after the fashion of Carnoy's *reticulum plasmatique*, only more solid, and still more neat and regular. Just such a spherical basket, looking like the finest imaginable Chinese ivory ball, is found in the siliceous skeleton of *Aulonia*, another of Haeckel's Radiolaria from the Challenger.

But here a strange thing comes to light. *No system of hexagons can enclose space;* whether the hexagons be equal or unequal, regular or irregular, it is still under all circumstances mathematically impossible. So we learn from Euler: the array of hexagons may be extended as far as you please, and over a surface either plane or curved, but *it never closes in.* Neither our *reticulum plasmatique* nor what seems the very perfection of hexagonal symmetry in *Aulonia* are as we are wont to conceive them; hexagons indeed predominate in both, but a certain number of facets are and must be other than hexagonal. If we look carefully at Carnoy's careful drawing we see that both pentagons and heptagons are shewn in his reticulum, and Haeckel actually states, in his brief description of his *Aulonia hexagona*, that a few square or pentagonal facets are to be found among the hexagons.

Such skeletal conformations are common: and Nature, as in all her handiwork, is quick to ring the changes on the theme. Among

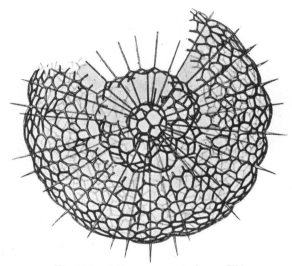

Fig. 323. *Actinomma arcadophorum* Hkl.

its many variants may be found cases (e.g. *Actinomma*) where the vesicles have been less regular in size; and others in which the meshwork has been developed not on an outer surface only but at successive levels, producing a system of concentric spheres. If the siliceous material be not limited to the linear junctions of the cells but spread over a portion of the outer spherical surfaces or caps, then we shall have the condition represented in Fig. 324 (*Ethmosphaera*), where the shell appears perforated by circular instead of hexagonal apertures and the circular pores are set on slight spheroidal eminences; and, interconnected with such types as this, we have others in which the accumulating pellicles of skeletal matter have extended from the edges into the substances of the boundary walls and have so produced a system of films, normal to the surface of the sphere, constituting a very perfect honeycomb, as in *Cenosphaera favosa* and *vesparia**.

In one or two simple forms, such as the fresh-water *Clathrulina*, just such a spherical perforated shell is produced out of some

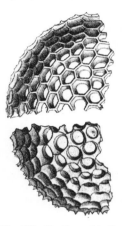

Fig. 324.   *Ethmosphaera conosiphonia* Hkl.

Fig. 325. Portions of shells of two "species" of *Cenosphaera*: upper figure, *C. favosa*; lower, *C. vesparia* Hkl.

organic, acanthin-like substance; and in some examples of *Clathrulina* the chitinous lattice-work of the shell is just as regular and delicate, with the meshes for the most part as beautifully hexagonal as in the siliceous shells of the oceanic Radiolaria. This is only another proof (if proof be needed) that the peculiar form and character of these little skeletons are due not to the material of which they are composed, but to the moulding of that material upon an underlying vesicular structure.

Let us next suppose that another and outer layer of cells or vesicles develops upon some such lattice-work as has just been

---

* In all these latter cases we recognise a relation to, or extension of, the principle of Plateau's *bourrelet*, or van der Mensbrugghe's *masse annulaire*, or Gibbs's ring, of which we have had much to say.

described; and that instead of forming a second hexagonal lattice-work, the skeletal matter tends to be developed normally to the surface of the sphere, that is to say along the *radial* edges where the external vesicles (now compressed into hexagonal prisms) meet one another three by three. The result will be that, if the vesicles be removed, a series of radiating spicules will be left, directed outwards from the angles of the original polyhedron meshwork, all as is seen in Fig. 326. And it may further happen that these radiating skeletal rods branch at their outer ends into divergent rays, forming a triple fork, and corresponding (after the fashion

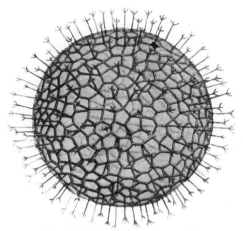

Fig. 326.   *Aulastrum triceros* Hkl..

which we have already described as occurring in certain sponge-spicules) to the superficial furrows between the three adjacent cells; this is, as it were, a halfway stage between simple rods or radial spicules and the full completion of another sphere of latticed hexagons. Another possible case, among many, is when the large, uniform vesicles of the outer protoplasm are replaced by smaller vesicles, piled on one another in concentric layers. In this case the radial rods will no longer be straight, but will be bent zig-zag, with their angles in three vertical planes corresponding to the alternate contacts of the successive layers of cells (Fig. 327).

The solid skeleton is confined, in all these cases, to the boundary-lines, or edges, or grooves between adjacent cells or vesicles, but adsorptive energy may extend throughout the intervening walls. This happens in not a few Radiolaria, and in a certain group called the Nassellaria it produces geometrical forms of peculiar elegance and mathematical beauty.

When Plateau made the wire framework of a regular tetrahedron and dipped it in soap-solution, he obtained in an instant (as we well know) a beautifully symmetrical system of six films, meeting

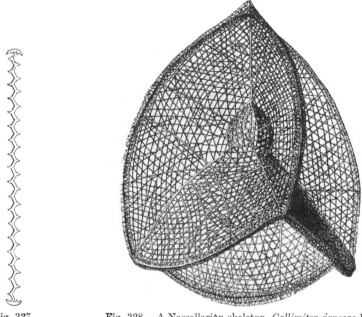

Fig. 327.   Fig. 328.   A Nassellarian skeleton, *Callimitra agnesae* Hkl.
                              (0·15 mm. diameter).

three by three in four edges, and these four edges running from the corners of the figure to its centre of symmetry. Here they meet, two by two, at the Maraldi angle; and the films meet three by three, to form the re-entrant solid angle which we have called a "Maraldi pyramid" in our account of the architecture of the honey-comb. The very same configuration is easily recognised in the minute siliceous skeleton of *Callimitra*. There are two discrepancies, neither of which need raise any difficulty. The figure is not a rectilinear but a *spherical tetrahedron*, such as might be formed by the boundary-edges of a tetrahedral cluster of four co-equal bubbles; and just as Plateau extended his experiment by blowing a small bubble in the centre of his tetrahedral system, so we have a central bubble also here.

This bubble may be of any size*; but its situation (if it be present at all) is always the same, and its shape is always such as to give the Maraldi angles at its own four corners. The tensions

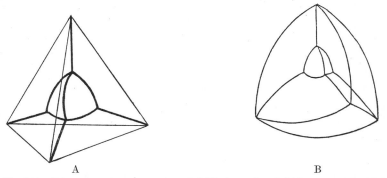

A   B

Fig. 329. Diagrammatic construction of *Callimitra*. A, a bubble suspended within a tetrahedral cage. B, another bubble within a skeleton of the former bubble.

of its own walls, and those of the films by which it is supported or slung, all balance one another. Hence the bubble appears in plane projection as a curvilinear equilateral triangle; and we have only got to convert this plane diagram into the corresponding solid to obtain the spherical tetrahedron we have been seeking to explain (Fig. 329).

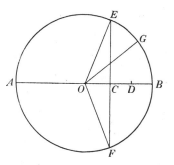

We may make a simplified model (omitting the central bubble) of the tetrahedral skeleton of *Callimitra*, after the fashion of that of the bee's cell (p. 535). Take $OC = CD = DB$, and draw a circle with radius OB and diameter AB. Erect a perpendicular to AB at C, cutting the circle at E, F.

Fig. 330. Geometrical construction of *Callimitra*-skeleton.

AOE, AOF will be (as before) Maraldi angles of 109°; the arcs AE,

* Plateau introduced the central bubble into his cube or tetrahedron by dipping the cage a second time, and so adding an extra face-film; under these circumstances the bubble has a definite magnitude.

AF will be edges of the spherical tetrahedron, and *O* will be the centre of symmetry. Make the angle $FOG = FOA = EOA$. Cut out the circle *EAFG*, and cut through the radius *EO*: fold at *AO*, *FO*, *GO*, and fasten together, using *EOG* for a flap. Make four such sheets, and fasten together back to back. The model will be much improved if little cusps be left at the corners in the cutting out.

The geometry of the little inner tetrahedron is not less simple and elegant. Its six edges and four faces are all equal. The films attaching it to the outer skeleton are all planes. Its faces are spherical, and each has its centre in the opposite corner. The edges are circular arcs, with cosine $\frac{1}{3}$; each is in a plane perpendicular to the chord of the arc opposite, and each has its centre in the middle of that chord. Along each edge the two intersecting spheres meet each other at an angle of 120°*.

This completes the elementary geometry of the figure; but one or two points remain to be considered.

We may notice that the outer edges of the little skeleton are thickened or intensified, and these thickened edges often remain whole or strong while the rest of the surfaces shew signs of imperfection or of breaking away; moreover, the four corners of the tetrahedron are not re-entrant (as in a group of bubbles) but a surplus of material forms a little point or cusp at each corner. In all this there is nothing anomalous, and nothing new. For we have already seen that it is at the margins or edges, and *a fortiori* at the corners, that the surface-energy reaches its maximum—with the double effect of accumulating protoplasmic material in the form of a Gibbs's ring or bourrelet, and of intensifying along the same lines the adsorptive secretion of skeletal matter. In some other tetrahedral systems analogous to *Callimitra*, the whole of the skeletal matter is concentrated along the boundary-edges, and none left to spread over the boundary-planes or interfaces: just as among our spherical Radiolaria it was at the boundary-edges of their many cells or vesicles, and often there alone, that skeletal formation occurred, and gave rise to the spherical skeleton and its meshwork

* For proof, see Lamarle, *op. cit.* pp. 6–8. Lamarle shewed that the sphere can be so divided in seven ways, but of these seven figures the tetrahedron alone is stable. The other six are the cube and the regular dodecahedron; prisms, triangular and pentagonal, with equilateral base and a certain ratio of base to height; and two polyhedra constructed of pentagons and quadrilaterals.

of hexagons. In the beautiful form which Haeckel calls *Archiscenium* the boundary edges disappear, the four edges converging on the median point are intensified, and only three of the six convergent facets are retained; but, much as the two differ in appearance, the geometry of this and of *Callimitra* remain essentially the same.

We learned also from Plateau that, just as a tetrahedral bubble can be inserted within the tetrahedral skeleton or cage, so may a cubical bubble be introduced within a cubical cage; and the edges of the inner cube will be just so curved as to give the Maraldi angles at the corners. We find among Haeckel's Radiolaria one (he calls it *Lithocubus geometricus*) which precisely corresponds to the skeleton

A   B

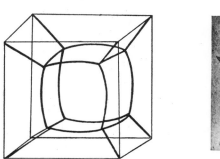

Fig. 331. A, bubble suspended within a cubical cage.
B, *Lithocubus geometricus* Hkl.

of this inner cubical bubble; and the little spokes or spikes which project from the corners are parts of the edges which once joined the corners of the enclosing figure to those of the bubble within (Fig. 331).

Again, if we construct a cage in the form of an equilateral triangular prism, and proceed as before, we shall probably see a vertical edge in the centre of the prism connecting two nodes near either end, in each of which the Maraldi figure is displayed. But if we gradually shorten our prism there comes a point where the two nodes disappear, a plane curvilinear triangle appears horizontally in the middle of the figure, and at each of its three corners four curved edges meet at the familiar angle. Here again we may insert a central bubble, which will now take the form of a curvilinear equilateral triangular prism; and Haeckel's *Prismatium tripodium* repeats this configuration (Fig. 332).

In a framework of two crossed rectangles, we may insert one bubble after another, producing a chain of superposed vesicles whose shapes vary as we alter the relative positions of the rectangular frames. Various species of *Triolampas, Theocyrtis*, etc. are more or less akin to these complicated figures of equilibrium. A very beautiful series of forms may be made by introducing successive bubbles within the film-system formed by a tetrahedron or a

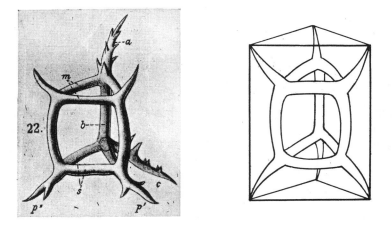

Fig. 332.   *Prismatium tripodium* Hkl.

parallelepipedon. The shape and the curvature of the bubbles and of their suspensory films become extremely beautiful, and we have certain of them reproduced unmistakably in various Nassellarian genera, such as *Podocyrtis* and its allies.

In Fig. 333 we see a curious little skeletal structure or complex spicule, whose conformation is easily accounted for. Isolated spicules such as this form the skeleton in the genus *Dictyocha*, and occur scattered over the spherical surface of the organism (Fig. 334). The basket-shaped spicule has evidently been developed about a cluster of four cells or vesicles, lying in or on the surface of the organism, and therefore arranged, not in the three-dimensional, tetrahedral form of *Callimitra*, but in the manner in which four contiguous cells lying side by side in one plane normally set themselves, like the four cells of a segmenting egg: that is to say with an intervening "polar furrow," whose ends mark the meeting place, at equal angles, of the four cells in groups of three. The little projecting spokes, or spikes, which are set normally to the main basket-work, seem to be uncompleted portions of a larger basket, corresponding to a more numerous aggregation of cells. Similar but more complex forma-

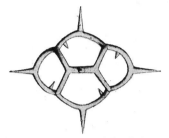

Fig. 333. An isolated portion of the skeleton of *Dityocha*.

tions, all explicable as basket-like frameworks developed around a cluster of cells, and adsorbed or secreted in the grooves common to adjacent cells or bubbles, are found in great variety.

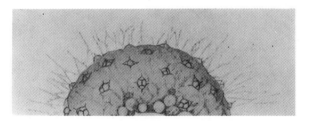

Fig. 334. *Dictyocha stapedia* Hkl.

The *Dictyocha*-spicule, laid down as a siliceous framework in the grooves between a few clustered cells, is too simple and natural to be confined to one group of animals. We have already seen it, as a calcareous spicule, in the holothurian genus *Thyone*, and we may find it again, in many various forms, in the protozoan group known as the Silicoflagellata*. Nothing can better illustrate the physico-mathematical character of these configurations than their

* See (*int. al.*) G. Deflandre, Les Silicoflagellés, etc., *Bull. Soc. Fr. de Microscopie*, i, p. 1, 1932: the figures in which article are mostly drawn from Ehrenberg's *Mikrogeologie*, 1854.

common occurrence in diverse groups of organisms. And the simple fact is, that we seem to know less and less of these things on the biological side, the more we come to understand their physical and

Fig. 335.   Various species of *Distephanus* (Silicoflagellata).
From Deflandre, after Ehrenberg.

mathematical characters. I have lost faith in Haeckel's four thousand "species" of Radiolaria.

In *Callimitra* itself, and elsewhere where the boundary-walls (and not merely their edges) are silicified, the skeletal matter is not

Fig. 336.   Holothurian-spicules.   A, of *Thyone* (Mortensen); B, *Holothuria lactea* (Perrier); C, D, *Holothuria* and *Phyllophorus* (Deichman).

deposited in an even layer, like the waxen walls of a bee's cell, but in a close meshwork of fine curvilinear threads; and the curves seem to form three main series more or less closely related to the three edges of the partition. Sometimes (as may also be seen in our figure) the system is further complicated by a radial series running from the centre towards the free edge of each partition.

As to the former, their arrangement is such as would result if deposition or solidification had proceeded *in waves*, starting independently from each of the three boundary-edges of the little partition-wall, and something of this kind is doubtless what actually happened. We are reminded of the wave-like periodicity of the Liesegang phenomenon, and especially, perhaps, of the criss-cross rings which Liesegang observed in frozen gelatine (*supra*, p. 663). But there may be other explanations. For instance the film, liquid or other, which originally constituted the partition, might conceivably be thrown into *vibrations*, and then (like the dust upon a Chladni plate) minute particles in or on the film would tend to take up position in an orderly way, in relation to the nodal points or lines of the vibrating surface*. Some such hypothetical vibration may (to my thinking) account for the minute and varied and very beautiful patterns upon many diatoms, the resemblance of which patterns to the Chladni figures (in certain of their simpler cases) seems here and there striking and obvious. But I have not attempted to investigate the many special problems suggested by the diatom-skeleton.

The cusps at the four corners of the tetrahedral skeleton are a marked peculiarity of our Nassellarian shell, and we should by no means expect to see them in a skeleton formed at the boundary-edges of a simple tetrahedral pyramid of four bubbles or cells. But when we introduce another bubble into the centre of a system of four, then, as Plateau shewed, the tensions of its walls and of the surrounding partitions so balance one another that it becomes a regular curvilinear tetrahedron, or, as seen in plane projection (Fig. 337), a curvilinear, equilateral triangle, with prominent, not re-entrant angles. A drop of fluid tends to accumulate at each corner where four edges meet, and forms a bourrelet†; it is drawn out in the directions of the four films which impinge upon it, and

---

* Cf. Faraday's beautiful experiments, On the moving groups of particles found on vibrating elastic surfaces, etc., *Phil. Trans.* 1831, p. 299; *Researches*, 1859, pp. 314–358.

† The bourrelet is not only, as Plateau expresses it, a "surface of continuity," but we also recognise that it tends (so far as material is available for its production) to further lessen the free surface-area. On its relation to vapour-pressure and to the stability of foam, see FitzGerald's interesting note in *Nature*, Feb. 1, 1894 (*Works*, p. 309); and on its effect in thinning the soap-bubble to bursting-point, see Willard Gibbs, *Coll. Papers*, I, p. 307 *seq.*

so tends to assume in miniature the very same shape as the tetrahedron to whose corner it is attached.   Out of these bourrelets, then,
the cusps at the four corners of our little skeleton, are formed.

A large and curiously beautiful family of radiolarian (or "polycystine") skeletons look, in a general way, like tiny helmets or
Pickelhauben, with spike above, and three (or sometimes six) curved
lobes, like helmet-straps, below.   We recognise a family likeness,
even a mathematical identity, between this figure and the last, for
both alike are based on a tetrahedral symmetry: the body of the
helmet corresponding to the inner vesicle of *Callimitra*, and the
spike and the three straps to the four edges which ran out from the
inner to the outer tetrahedron.   In the one case an inner vesicle

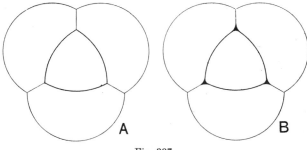

Fig. 337.

is surrounded by a tetrahedral figure whose outer walls, indeed, are
absent, but its edges remain, and so do the walls connecting the
outer and inner vesicles.   In the other case the outer edges are
gone, and so are the filmy partition-walls, save parts which correspond to the four internal edges between them*.   There are apt
to be two slight discrepancies.   The helmet is often of somewhat
complicated form, easily explained as due to the presence of two
superposed bubbles instead of one.   The other apparent anomaly
is that the three helmet-straps are curved, while the corresponding
edges are straight in Plateau's figure of the regular tetrahedron.
But it is a paramount necessity (as we well know) for each set of
four edges in a system of fluid films to meet in a point two and two
at the Maraldi angle;   just as it is necessary for the faces to meet

* Looking through Haeckel's very numerous figures, we see that now and then
something more is left than the mere edges of the partition-walls.

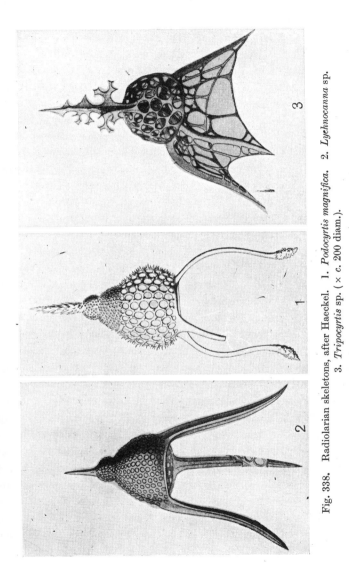

Fig. 338. Radiolarian skeletons, after Haeckel. 1. *Podocyrtis magnifica*. 2. *Lychnocanna* sp. 3. *Tripocyrtis* sp. (× c. 200 diam.).

at 120°; and faces and edges become curved, whenever necessary, in order that these conditions may be fulfilled. The need may arise in various ways. Suppose (as in Fig. 339) that our little central

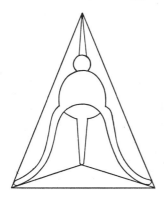

bubble be no longer in the centre of symmetry, but near one corner of the enclosing tetrahedron; the short edge running out from that corner will tend to remain straight (and so form the spike of the helmet); while the other three will each form an S-shaped curve, as a condition of making co-equal angles at their two extremities. An analogous case is figured in ·one of Sir David Brewster's papers where he repeats and amplifies some experiments of Plateau's*.

Fig. 339.   Diagram of one of the He made a tetrahedral cage, and fitted
helmet-shaped radiolaria, e.g. it with three more wires, leading from
Podocyrtis: to shew its tetra- the apex to the middle point of each basal
hedral symmetry.
edge. On dipping this into soap-solution, various complications were seen. At the apex, six films must not meet together, for no more than three surfaces may meet in an edge; intermediate or interstitial films make their appearance, with which we are not greatly concerned. But six films now ascend towards the apex from the base instead of three, and the three which come from the corners have a longer path than the other three which come from the mid-points of the basal edges; they *must* be curved in different degrees in order that all three may make at either end their co-equal angles. And if we now introduce a bubble (or two bubbles) into the interior of the system, we obtain the characteristic form of our helmet-shaped Radiolarian—the spike above, the single or double vesicle of the body, and the straps or lappets with their peculiar and characteristic curvatures.

The little shell is perforated with many rounded holes, and it remains to account for these. They are required, so we are told, for the passage of pseudopodia; well and good. We have referred the *Dictyocha*-spicule and the hexagonal meshwork of *Aulonia* to froth-like associations of vesicles and to adsorption taking place

* On the figures of equilibrium in liquid films, *Trans. R.S.E.* xxiv, 1866.

between; so far so good, again.　But the irregular lacework of the little helmets does not suit this explanation very well, and there may be yet other possibilities.　We have already mentioned Tomlinson's "cohesion figures*."　Experimenting with a great variety of substances, Tomlinson studied the innumerable ways in which drops, jets or floating films "cohere," disrupt or otherwise behave, under the resultant influences of surface-tension, cohesion, viscosity and friction; in one case a film runs out into a wavy or broken edge, in another it gives way here and there, and makes rents or holes in its surface.　I take the little holes in our polycystine skeleton to be cohesion-figures, in Tomlinson's sense of the word— spots where the delicate film has given way and run into holes, and where surface-tension has rounded off the broken edges, and made the rents into rounded apertures.

In the foregoing examples of Radiolaria, the symmetry which the organism displays seems identical with that symmetry of forces which results from the play and interplay of surface-tensions in the whole system: this symmetry being displayed, in one class of cases, in a more or less spherical mass of froth, and in another class in a simpler aggregation of a few, otherwise isolated, vesicles.　In either case skeletons are formed, in great variety, by one and the same kind of surface-action, namely by the adsorptive deposition of silica in walls and edges, corresponding to the manifold surfaces and interfaces of the system.　But among the vast number of known Radiolaria, there are certain forms (especially among the Phaeodaria and Acantharia) which display a no less remarkable symmetry the origin of which is by no means clear, though surface-tension may play a part in its causation.　Even this is doubtful; for the fact that three-way nodes are no longer to be seen at the junctions of the cells suggests that another law than that of minimal areas had been in action here.　They are cases in which (as in some of those already described) the skeleton consists (1) of radiating spicular rods, definite in number and position, and (2) of inter-connecting rods or plates, tangential to the more or less spherical body of the organism, whose form becomes, accordingly, that of a geometric, polyhedral solid.　The great regularity, the numerical

* Cf. *supra*, p. 418

symmetries and the apparent simplicity of these latter forms makes of them a class apart, and suggests problems which have not been solved or even investigated.

The matter is partially illustrated by the accompanying figures (Fig. 340) from Haeckel's *Monograph of the Challenger Radiolaria**. In one of these we see a regular octahedron, in another a regular, or pentagonal, dodecahedron, in a third a regular icosahedron. In all cases the figure appears to be perfectly symmetrical, though neither the triangular facets of the octahedron and icosahedron, nor the pentagonal facets of the dodecahedron, are necessarily plane surfaces. In all of these cases, the radial spicules correspond to the corners of the figure; and they are, accordingly, six in number in the octahedron, twenty in the dodecahedron, and twelve in the icosahedron. If we add to these three figures the regular tetrahedron which we have just been studying, and the cube (which is represented, at least in outline, in the skeleton of the hexactinellid sponges), we have completed the series of the five regular polyhedra known to geometers, the *Platonic bodies†* of the older mathematicians. It is at first sight all the more remarkable that we should here meet with the whole five regular polyhedra, when we remember that, among the vast variety of crystalline forms known among minerals, the regular dodecahedron and icosahedron, simple as they are from the mathematical point of view, never occur. Not only do these latter never occur in crystallography, but (as is explained in textbooks of that science) it has been shewn that they cannot occur, owing to the fact that their indices (or numbers expressing the relation of the faces to the three primary axes) involve an irrational quantity: whereas it is a fundamental law of crystallography, involved in the whole theory of space-partitioning, that "the indices of any and every face of a crystal are small whole numbers‡." At the same time, an imperfect pentagonal dodeca-

---

* Of the many thousand figures in the hundred and forty plates of this beautifully illustrated book, there is scarcely one which does not depict some subtle and elegant *geometrical* configuration.

† They were known long before Plato: Πλάτων δὲ καὶ ἐν τούτοις πυθαγορίζει.

‡ If the equation of any plane face of a crystal be written in the form $hx + ky + lz = 1$, then h, k, l are the indices of which we are speaking. They are the reciprocals of the parameters, or reciprocals of the distances from the origin at which the plane meets the several axes. In the case of the regular or pentagonal dodecahedron these indices are 2, $1 + \sqrt{5}$, 0. Kepler described as follows, briefly

hedron, whose pentagonal sides are non-equilateral, is common among crystals. If we may safely judge from Haeckel's figures, the pentagonal dodecahedron of the Radiolarian (*Circorhegma*) is perfectly regular, and we may rest assured, accordingly, that it is not brought about by principles of space-partitioning similar to those which manifest themselves in the phenomenon of crystallisation. It will be observed that in all these radiolarian polyhedral shells, the surface of each external facet is formed of a minute hexagonal network, whose probable origin, in relation to a vesicular structure, is such as we have already discussed.

In certain allied Radiolaria of the family Acanthometridae (Fig. 341), which have twenty radial spines, the arrangement of these spines is commonly described in a somewhat singular way. The twenty spines are referred to five whorls of four spines each, arranged as parallel circles on the sphere, and corresponding to the equator, the tropics and the polar circles. This rule was laid down by the celebrated Johannes Müller, and has ever since been used and quoted as Müller's law*. But when we come to examine the figure, we find that Müller's law hardly does justice to the facts, and seems to overlook a simpler symmetry. We see in the first place that here, unlike our former cases, the twenty radial spines issue through the facets (and *all* the facets) of the polyhedron, instead of coming from its corners; and that our twenty spines correspond, therefore, not to the corners of a dodecahedron, but to the facets of some sort of an icosahedron. We see, in the next place, that this icosahedron is composed of faces of two kinds, hexagonal and pentagonal; and that the whole figure may be described as a hexagonal prism, whose twelve corners are truncated, and replaced by pentagonal facets. Both hexagons and pentagons appear to be equilateral, but if we

but adequately, the common characteristics of the dodecahedron and icosahedron: "Duo sunt corpora regularia, dodecaedron et icosaedron, quorum illud quinquangulis figuratur expresse, hoc triangulis quidem sed in quinquanguli formam coaptatis. Utriusque horum corporum ipsiusque adeo quinquanguli *structura perfici non potest sine proportione illa, quam hodierni geometrae divinam appellant*" (*De nive sexangula* (1611), Opera, ed. Fritsch, VII, p. 723). Here Kepler was dealing, somewhat after the manner of Sir Thomas Browne, with the mysteries of the quincunx, and also of the hexagon; and was seeking for an explanation of the mysterious or even mystical beauty of the 5-petalled or 3-petalled flower—*pulchritudinis aut proprietatis figurae, quae animam harum plantarum characterisavit.*

* See Johannes Müller, Ueber die Thalassicollen, Polycistinen und Acanthometren des Mittelmeeres, *Abh. d. Akad. Wiss. Berlin*, 1858, pp. 1–62, 11 pl.

try to construct a plane-sided polyhedron of this kind, we find it
to be impossible; for into the angles between the six equatorial

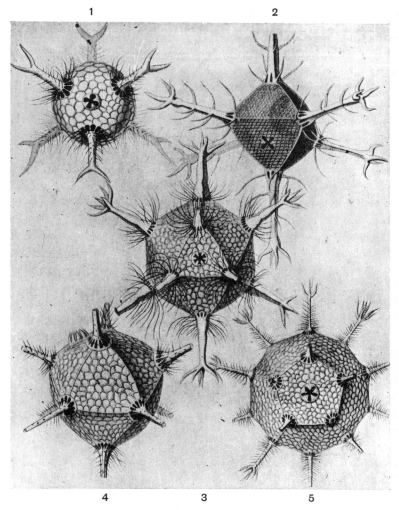

Fig. 340.   Skeletons of various Radiolarians, after Haeckel. 1, *Circoporus sexfurcus*;
  2, *C. octahedrus*; 3, *Circogonia icosahedra*; 4, *Circospathis novena*; 5, *Circorrhegma
  dodecahedra*.

regular hexagons six regular pentagons will not fit. The figure,
however, can be easily constructed if we replace the straight edges

(or some of them) by curves, the plane facets by slightly curved surfaces, or the regular by non-equilateral polygons*.

In some cases, such as Haeckel's *Phatnaspis cristata* (Fig. 342), we have an ellipsoidal body from which the spines emerge in the order described, but which is not obviously divided into facets. In Fig. 234 I have indicated the facets corresponding to the rays, and dividing the surface in the usual symmetrical way.

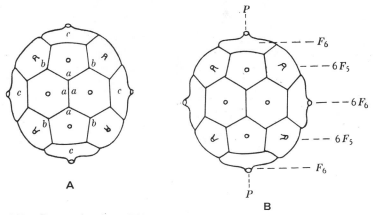

A

B

Fig. 341. *Dorataspis cristata* Hkl. A, viewed according to Müller's law: *a*, four polar plates; *b*, four intermediate or "tropical" plates; *c*, four equatorial plates. B, an alternative description: F_6, two polar and six equatorial hexagonal plates; F_5, two rows of six intermediate pentagonal plates.

About any polyhedron (within or without) we may describe another whose corners correspond to the sides, and whose sides to the corners, of the original figure; or the one configuration may be developed from the other by bevelling off, to a certain definite extent, the corners of the original polyhedron. The two figures, thus reciprocal to one another, form a "conjugate pair," and the principle is known as the "principle of duality" in polyhedra†. Of the regular solids, cube and octahedron, dodecahedron and

* Müller's interpretation was emended by Brandt, and what is known as Brandt's Law, viz. that the symmetry consists of two polar rays and three whorls of six each, coincides so far with the above description: save only that Brandt says plainly that the intermediate whorls stand equidistant between the equator and the poles, i.e. in latitude 45°, which, though not very far wrong, is geometrically inaccurate. But Brandt, if I understand him rightly, did not propose his "law" as a substitute for Müller's, but rather as a second law, applicable to a few special cases.

† First proved by Legendre, *Elém. de Géométrie*, vii, Prop. 25, 1794.

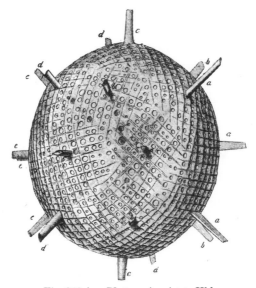

Fig. 342 A.   *Phatnaspis cristata* Hkl.

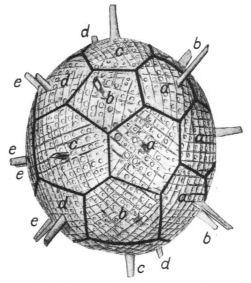

Fig. 342 B.   The same, diagrammatic.

icosahedron, are conjugate pairs; but the first and simplest of all solid figures, the tetrahedron, has no conjugate but itself.

In our little shell of *Dorataspis*, the twenty spicules (as we have seen) spring from and correspond to the twenty facets of the polyhedron, twelve pentagonal and eight hexagonal, meeting at thirty-six corners, in all cases three by three; we may write the formula of the polyhedron, accordingly, as

$$12F_5 + 8F_6 + 36C_3.$$

If we now connect up the twenty spicules, three by three, we shall obtain thirty-six triangles, completely covering the figure; but we shall find that of the twenty corners twelve are surrounded by five, and eight by six triangles. The formula is now

$$36F_3 + 12C_5 + 8C_6,$$

and the two figures are fully reciprocal or conjugate.

I do not know of any radiolarian in which this configuration is to be found; nor does it seem a likely one, owing to the large and variable number of edges which meet in its corners. But we may have polyhedra related to, or derived from, one another in a less full and perfect degree. For instance, letting the twenty spicules of *Dorataspis* again serve as corners for the new figure, let *four facets* meet in each corner; or (which comes to the same thing) let each spicule give off four branches or offshoots, which shall meet their corresponding neighbours, and form the boundary-edges of a new network of facets. The result (Fig. 343) is a symmetrical figure, not geometrically perfect but elegant in its own way, which we recognise in a number of described forms[*]. It shews eight triangular and fourteen rhomboidal facets; and its formula is

$$8F_3 + 14F_4 + 20C_4.$$

Many subsidiary varieties may arise in turn: when, for instance, certain of the little branches fail to meet, or others grow large and widely confluent, always in symmetrical fashion.

We now see how in all such cases as these there is a *double symmetry* involved, that of two superimposed, and conjugate or semi-conjugate, figures. And the ambiguity which attends such descriptions as that which Johannes Müller embodied in his "law"

[*] Cf. W. Mielck, *Acanthometren aus Neu-Pommern*, Diss., Kiel, 1907.

seems due to a failure to recognise this twofold or alternative symmetry.

In all these latter cases it is the arrangement of the axial rods—the "polar symmetry" of the entire organism—which lies at the root of the matter; and which, if only we could account for it, would make it comparatively easy to explain the superficial configuration. But there are no obvious mechanical forces by which we can so explain this peculiar polarity. This at least is evident, that it arises in the central mass of protoplasm, which is the essential living portion of the organism as distinguished from that frothy

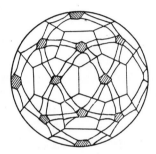

Fig. 343. *Acanthometra* sp. A derivative of the *Dorataspis* figure. After Mielck.

Fig. 344. *Phractaspis prototypus* Hkl.

peripheral mass whose structure has helped us to explain so many phenomena of the superficial or external skeleton. To say that the arrangement depends upon a specific polarisation of the cell is merely to refer the problem to other terms, and to set it aside for future solution. But it is possible that we may learn something about the lines in which *to seek for* such a solution by considering the case of Lehmann's "fluid crystals," and the light which they throw upon the phenomena of molecular aggregation.

The phenomenon of "fluid crystallisation" is found in a number of chemical bodies; it is exhibited at a specific temperature for each substance; and it would seem to be limited to bodies in which there is an elongated, or "long-chain" arrangement of the atoms in the molecule. Such bodies, at the appropriate temperature, tend to aggregate themselves into masses, which are sometimes spherical drops or globules (the so-called "spherulites"), and sometimes have the definite form of needle-like or prismatic crystals. In either case they remain liquid, and are also doubly refractive, polarising light in brilliant colours. Together with them are formed ordinary solid crystals, also with characteristic polarisation, and into such solid crystals all the fluid material ultimately turns. It seems that in these liquid crystals, though the molecules are freely mobile, just as are those of water, they are yet subject to, or endowed with, a "directive force," a force which confers upon them a definite configuration or "polarity," the "Gestaltungskraft" of Lehmann.

Such an hypothesis as this has been gradually extruded from the theories of mathematical crystallography*; and it has come to be understood that the symmetrical conformation of a homogeneous crystalline structure is sufficiently explained by the mere mechanical fitting together of appropriate structural units along the easiest and simplest lines of "close packing": just as a pile of oranges becomes definite, both in outward form and inward structural arrangement, without the play of any *specific* directive force. But while our conceptions of the tactical arrangement of crystalline molecules remain the same as before, and our hypotheses of "modes of packing" or of "space-lattices" remain as useful and as adequate as ever for the definition and explanation of the molecular arrangements, a new conception is introduced when we find something like such space-lattices maintained in what has hitherto been considered the molecular freedom of a liquid field; and Lehmann would persuade us, accordingly, to postulate a specific molecular force, or "Gestaltungskraft" (not unlike Kepler's "facultas formatrix"), to account for the phenomenon†.

Now just as some sort of specific "Gestaltungskraft" had been of old the *deus ex machina* accounting for all crystalline phenomena (*gnara totius geometriae, et in ea exercita*, as Kepler said), and as

---

* Cf. Tutton, *Crystallography*, 1911, p. 932.

† Kepler, if I understand him aright, saw his way to account for the shape of the bee's cell or the pomegranate-seed; and it was for want of any such mechanical explanation, and as little more than a confession of ignorance, that he fell back on a *facultas formatrix* to account for the six rays of the snow-crystal or the five petals of the flower. He was equally ready, unfortunately, to explain, by the same *facultas formatrix in aere*, the appearance of a plague of locusts or a swarm of flies.

such an hypothesis, after being dethroned and repudiated, has now
fought its way back and claims a right to be heard, so it may be
also in biology. We begin by an easy and general assumption of
specific properties, by which each organism assumes its own specific
form; we learn later (as it is the purpose of this book to shew) that
throughout the whole range of organic morphology there are innumerable phenomena of form which are not peculiar to living things,
but which are more or less simple manifestations of ordinary physical
law. But every now and then we come to deep-seated signs of
protoplasmic symmetry or polarisation, which *seem* to lie beyond
the reach of the ordinary physical forces. It by no means follows
that the forces in question are not essentially physical forces, more
obscure and less familiar to us than the rest; and this would seem
to be a great part of the lesson for us to draw from Lehmann's
beautiful discovery. For Lehmann claims to have demonstrated,
in non-living, chemical bodies, the existence of just such a deter-
minant, just such a "Gestaltungskraft," as would be of infinite help
to us if we might postulate it for the explanation (for instance) of
our Radiolarian's axial symmetry. Further than this we cannot
go; such analogy as we seem to see in the Lehmann phenomenon
soon evades us, and refuses to be pressed home. The symmetry
of crystallisation, which Haeckel tried hard to discover and to reveal
in these and other organisms, resolves itself into remote analogies
from which no conclusions can be drawn. Many a beautiful
protozoan form has lent itself to easy physico-mathematical explanation; others, no less simple and no more beautiful prove harder
to explain. That Nature keeps some of her secrets longer than
others—that she tells the secret of the rainbow and hides that of
the northern lights—is a lesson taught me when I was a boy.

A note on Polyhedra.

The theory of Polyhedra, Euler's *doctrina solidorum*, is a branch
of geometry which deals with the more or less regular solids; and
the rudiments of the theory may help us to study certain more or
less symmetrical organic forms. Euler, a contemporary of Linnaeus, is the most celebrated of the many mathematicians who
have carried this subject beyond where Pythagoras, Plato, Euclid
and Archimedes had left it. He drew up a classification of poly-

hedral solids, using a binomial nomenclature based on the number of their corners or vertices, and sides or faces. Thus, for example, he called a figure with eight corners and seven faces *Octogonum heptaedrum*; and the analogy between this and Linnaeus's botanical classification and nomenclature—e.g. *Hexandria trigynia* and the rest—is very close and curious.

A simple theorem, of which Euler was vastly proud and which we still speak of as Euler's Law*, is fundamental to the theory of polyhedra. It tells us that in every polyhedron whatsoever, the faces and corners together outnumber the edges by two†:

$$C - E + F = 2 \qquad \ldots\ldots(1).$$

Another fundamental theorem follows. We know from Euclid that the three angles of a triangle are equal to two right angles; consequently, that in a polygon of C angles, the sum of the angles $= 2(C - 2)$ right angles. And there follows from this—but by no means expectedly—the analogous and extremely simple relation

* Euler, Elementa doctrinae solidorum, *Novi Comment. Acad. Sci. Imp. Petropol.* IV, p. 109 *seq.* (ad annos 1752 et 1753), 1758: "In omni solido hedris planis inclusum, aggregatum ex numero angulorum solidorum et ex numero hedrarum binario excedit numerum acierum." For a proof, see (*int. al.*) De Morgan, article Polyhedron in the *Penny Cyclopaedia.* There is reason to believe that Descartes was acquainted with this theorem between 1672 and 1676; cf. *Foucher de Careil, Œuvres inédites de Descartes*, Paris, II, p. 214. Cf. Baltzer, *Monatsber. Berlin. Akad.* 1861, p. 1043; and de Jonquières, *C.R.* 1890, p. 261. (The student will be struck by the *resemblance* between this formula and the phase rule of Willard Gibbs.)

† If we include, besides the corners, edges and faces (i.e. points, lines and surfaces) the solid figure itself, Euler's Law becomes

$$C - E + F - S = 1.$$

And in this form the theorem extends to n dimensions, as follows:

$$k_0 - k_1 + k_2 - k_3 + k_4 - \ldots = 1.$$

With equal beauty and simplicity, the *simplest* figure in each n-dimensional space is given as follows:

	k_0	k_1	k_2	k_3	k_4 etc.		
$n=0$	1					$=1$	(point)
1	2	-1				$=1$	(line)
2	3	-3	$+1$			$=1$	(triangle)
3	4	-6	$+4$	-1		$=1$	(tetrahedron)
4	5	-10	$+10$	-5	$+1$	$=1$	(pentahedroid)
etc.							

And, in a figure of n-dimensions, the sum of the plane angles $= 2^{(n-1)}(C - 2)$ right angles.

that in a polyhedron of C corners, the sum of the plane angles

$$= 4 (C - 2) \text{ right angles} \qquad \ldots\ldots(2)^*.$$

Hence, if the polyhedron be isogonal, the sum of the plane angles at each corner

$$= \frac{4 (C - 2)}{C} \times 90°, \text{ or } \left(4 - \frac{8}{C}\right) \text{ right angles} \qquad \ldots\ldots(3).$$

The five regular solids, or Platonic bodies—there can be no more —have been known from remote antiquity; they have their corners all alike and their faces all alike, they are isogonal and isohedral. Three of them, the tetrahedron, octahedron and icosahedron, have triangular faces; three of them, the tetrahedron, cube and dodecahedron, have trihedral or three-way corners. One or other of these, triangles or three-way corners, must (as we shall soon see) be present in every polyhedron whatsoever.

The semi-regular solids are regular in one respect or other, but not in both; they are *either* isogonal or isohedral—isohedral, when every face is an identical polygon and isogonal when at every corner the same set of faces is combined. The semi-regular *isogonal* solids, with all their corners alike but with two or more kinds of regular polygons for their faces, are thirteen in number—there can be no more; they were all described by Archimedes, and we call them by his name. One of them, with six square and eight hexagonal facets, derived by truncating the octahedron or the cube, we have found to be of peculiar interest, and it has become familiar to us as, of all homogeneous space-fillers, the one which encloses a given volume within a minimal area of surface. It is the *cubo-octahedron* of Kepler or of Fedorow, the *tetrakaidekahedron* of Kelvin, which latter name we commonly use.

Of semi-regular *isohedral* bodies, with all their sides alike (though no longer regular polygons) and their corners of two kinds or more, only one was known to antiquity; it is the rhombic dodecahedron, which is the crystalline form of the garnet, and appears in part

---

* On this remarkable parallel see Jacob Steiner, *Gesammelte Werke*, I, p. 97. It follows that the sum of the plane angles in a polyhedron, as in a plane polygon, is at once determined by the number of its corners: a result which delighted Euler, and led him to base his primary or generic classification of polyhedra on their corners rather than their sides.

again as a "space-filler" at the base of the bee's cell. A closely
related rhombic icosahedron was known to Kepler; but it was
left to Catalan* to discover, only some seventy-five years ago,
that the isohedral bodies were thirteen in number, and were
precisely comparable with and reciprocal to the Archimedean
solids.

The semi-regular solids, both of Archimedes and of Catalan, are
all, like the Platonic bodies, related to the sphere†, for a circum-
scribing sphere meets all the corners of an Archimedean solid, and
an inscribed sphere touches all the faces of a solid of Catalan; and
while the isogonal bodies can be constructed by various simple
geometrical means, the general method of constructing the thirteen
isohedral bodies is by dividing the sphere into so many similar and
equal areas ‡. It is a matter of spherical trigonometry rather than of
simple geometry, and the problem, for that very reason, remained
long unsolved.

The thirteen Archimedean bodies are derivable from the five
Platonic bodies, in most cases easily, by so truncating their corners
and their edges as to produce new and regularly polygonal faces in
place of the old faces, corners and edges, and the possible number
of faces in the new figure will be easily derived from the edges,
corners and faces of the old. Part of the old faces will remain;
each truncated corner will yield one new face; but each edge may
be truncated, or bevelled, more than once, so as to yield one, two,
or possibly three new faces. In short, if the faces, corners and
edges of a regular solid be F, C, E, those of the Archimedean solids
derivable from it (F_A) will be

$$F_A = F + mC + nE,$$

where $m = 0$ or 1, and $n = 0$, 1, or 2.

From the cube six Archimedean bodies may be derived, from the
dodecahedron six, and from the tetrahedron one.

---

* *Journal de l'école impér. polytechnique*, xli, pp. 1–71, 1865.

† It follows that the Chinese carved and perforated ivory balls, which are
based on regular and symmetrical division of the sphere, can all be referred to one
or another of the Platonic or Archimedean bodies.

‡ As a matter of fact, the Catalan bodies can be formed by adding to the Platonic
bodies, just as (but not so easily as) the Archimedean bodies can be formed by
truncating them.

The derivatives of the cube (with its six sides and eight corners) have the following numbers of sides:

$$F_A = F + C \qquad = 6 + 8 \qquad = 14$$
$$F + C + E = 6 + 8 + 12 = 26$$
$$F + C + 2E = 6 + 8 + 24 = 38.$$

The derivatives of the dodecahedron have, in like manner, 32, 62 or 92 sides; while the tetrahedron yields, by truncation of its four corners, a solid with eight sides.

The growth and form of crystals is a subject alien to our own, yet near enough to attract and tempt us. It is a curious thing (probably traceable to the Index Law of the crystallographer) that the Archimedean or isogonal bodies seldom occur and certainly play no conspicuous part in crystallography, while several of Catalan's isohedral figures are the characteristic forms of well-known minerals*.

Just as we pass from the Platonic to the Archimedean bodies by truncating the corners or edges of the former, so conversely, by producing their faces to a limit we obtain another family of figures —in all cases save the tetrahedron, which admits of no such extension; and the figures of this family are remarkable for the "twinned," or duplicate or multiple appearance which they present. If we extend the faces of an octahedron we get what looks like two tetrahedra, "twinned with" or interpenetrating one another; but there has been no interpenetration in the construction of this twin-like figure, only further accretion upon, and extension of, the facets of the octahedron. Among the higher polyhedra there are many figures which look, in a far more complicated way, like the twinning of simpler but still complicated forms; and these also have been constructed, not by interpenetration, but by the mere superposition of new parts on old.

An elementary, even a very elementary, knowledge of the theory of polyhedra becomes useful to the naturalist in various ways. Among organic structures we often find many-sided boxes (or what may be regarded as such), like the capsular seed-vessels of plants, the skeletons of certain Radiolaria, the shells of the Peridinia, the carapace of a tortoise, and a great many more. Or we may go

* E.g. the triakis, tetrakis and hexakis octahedra of fluor-spar. However the Archimedean tetrakaidekahedron ($6F_4 \, 8F_6$) occurs in alum.

further and treat any cluster of cells, such as a segmenting ovum, as a species of polyhedron and study it from the point of view of Euler's Law and its associated theorems. We should have to include, as the geometer seldom does, the case of two-sided facets—facets with two corners and two curved sides or edges—like the "liths"* of a peeled orange; but the general formula would include these as a matter of course. On the other hand, we need very seldom consider any other than trihedral or three-way corners.

When we limit ourselves to polyhedra with trihedral corners the following formula applies:

$$4f_2 + 3f_3 + 2f_4 + f_5 \pm 0.f_6 - f_7 - 2f_8 - \ldots = 12 \quad \ldots\ldots(4).$$

That this formula applies to the tetrahedron with its four triangles, the cube with its six squares and the dodecahedron with its twelve pentagons, is at once obvious. The now familiar case of our four-celled egg with its polar furrows (Fig. 486, B, etc.) appears in two forms, according as the polar furrows run criss-cross or parallel. In the one case we have a curvilinear tetrahedron, in the other a figure with two two-sided and two four-sided facets; in either case the formula is obviously satisfied.

But the main lesson for us to learn is the broad, general principle that we cannot group as we please any number and sort of polygons into a polyhedron, but that the number and kind of facets in the latter is strictly limited to a narrow range of possibilities. For example, the case of *Aulonia* has already taught us that a polyhedron composed entirely of hexagons is a mathematical impossibility[†]; and the zero-coefficient which defines the number of hexagons in the above formula (4) is the mathematical statement of the fact[‡]. We can state it still more simply by the following corollary, likewise limited to the case of three-way corners:

$$(6 - n)\, F_n = 12.$$

---

* *Lith*, a useful Scottish word for a joint or segment. Cromwell, according to Carlyle, "gar'd kings ken they had a *lith* in their necks."

† That hexagons cannot enclose space, or form a "three-way graph," has been recognised as a significant fact in organic chemistry: where, for instance, it limits, somewhat unexpectedly, the ways in which a closed cyclol, or space-enclosing protein molecule, can be imagined to be built up. Cf. Dorothy Wrinch, in *Proc. R. S.* (A), No. 907, p. 510, 1937.

‡ Euler shewed at the same time the singular fact that no polyhedron can exist with seven edges.

This applies at once to the tetrahedron, the cube and the regular dodecahedron, and at once excludes the possibility of the closed hexagonal network.

We found in *Dorataspis* a closed shell consisting only of hexagons and pentagons; without counting these latter we know, by our formula, that they must be twelve in number, neither more nor less. Lord Kelvin's tetrakaidekahedron consists only of squares and hexagons; the squares are, and must be, six in number.

In a typical Peridinian, such as *Goniodoma*, there are twelve plates, all meeting by three-way nodes or corners; we know, and we have no difficulty in verifying the fact, that the twelve plates are all pentagonal.

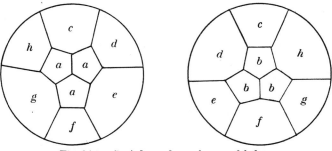

Fig. 345. *Goniodoma*, from above and below.

Without going beyond the elements of our subject we may want to extend our last formula, and remove the restriction to three-way corners under which it lay. We know that a tetrahedron has four triangles and four trihedral corners; that a cube has six squares and eight trihedral corners; an octahedron eight triangles and six four-way corners; an icosahedron twenty triangles and twelve five-way corners. By inspection of these numbers we are led to the following rule, and may establish it as a deduction from Euler's Law:

$$(f_3 + c_3) = 8 + 0\,(f_4 + c_4) + (f_5 + c_5) + 2\,(f_6 + c_6) + \text{etc.} \quad \ldots\ldots(5).$$

This important formula further illustrates the limitations to which all polyhedra are subject; for it shews us, among other things, that (if we neglect the exceptional case of dihedral facets or "liths") every polyhedron *must* possess either triangular faces or trihedral corners, and that these taken together are never less than eight in number.

We may add yet two more formulae, both related to the last, and all derivable, ultimately, from Euler's Law*:

$$3f_3 + 2f_4 + f_5 = 12 + 0 \cdot c_3 + 2c_4 + 4c_5 + \dots + 0 \cdot f_6 + f_7 + 2f_8 \quad \dots \dots (6)$$

and

$$3c_3 + 2c_4 + c_5 = 12 + 0 \cdot f_3 + 2f_4 + 4f_5 + \dots + 0 \cdot c_6 + c_7 + 2c_8 \quad \dots (7).$$

These imply that in every polyhedron the triangular, quadrangular and pentagonal faces (*or* corners) must, taken together and multiplied as above, be at least twelve in number. Therefore no polyhedron can exist which has not a certain number of triangles, squares or pentagons in its composition; and the impossibility of a polyhedron consisting only of hexagons is demonstrated once again.

Formulae (5), (6) and (7) further shew us that not only is a three-way polyhedron of hexagons impossible, but also a four-way polyhedron of quadrangles, or one of six-way corners and triangular facets; all of which become the more obvious when we reflect that the plane angles meeting in each point or node must be, *on the average*, in the first case $3 \times 120°$, in the second $4 \times 90°$, and in the third $6 \times 60°$.

Lastly, having now considered the case of other than trihedral corners, we may learn a simple but very curious relation between the number of faces and corners, arising (like so much else) out of Euler's Law. In a polyhedron whose corners are all n-hedral, $nC = 2E$; therefore (by Euler) $nC/2 + 2 = F + C$; therefore $2F = (n - 2) C + 4$.

Therefore, if

$$\begin{array}{lll} n = 3, & 2F = 4 + & C \\ = 4, & = 4 + 2C \\ = 5, & = 4 + 3C \end{array} \right\} \quad \dots \dots (8).$$

Let us look again at the microscopic skeleton of *Dorataspis* (Fig. 341). We have seen that some of its facets are hexagonal, the rest pentagonal; there happen to be eight of the former, and therefore (as we now know) there *must* be twelve of the latter.

---

* Derivable from Euler together with the formulae for the "edge-counts," viz. $\Sigma n F_n = 2E$, and $\Sigma n C_n = 2E$; which merely mean that each edge separates two faces, and joins two corners.

We know also that, having no triangular facets, the polyhedral skeleton *must* possess trihedral corners; and these, moreover, must (by equation 5) be eight in number, plus the number of pentagons, plus twice the number of hexagons. The total,

$$8 + f_5 + 2f_6 = 8 + 12 + (2 \times 8) = 36,$$

is precisely the number of corners in the figure, all of them trihedral. We also know (from equation 2) that the sum of its plane angles

$$= 4\,(36 - 2) \times 90° = 12{,}240°,$$

which agrees with the sum of the angles of twelve pentagons and eight hexagons. The configuration, then, is a possible one.

So here and elsewhere an apparently infinite variety of form is defined by mathematical laws and theorems, and limited by the properties of space and number. And the whole matter is a running commentary on the cardinal fact that, under such *foedera Naturai* as Lucretius recognised of old, there are things which are possible, and things which are impossible, even to Nature herself.

CHAPTER X

A PARENTHETIC NOTE ON GEODESICS

WE have made use in the last chapter of the mathematical principle of Geodesics (or Geodetics) in order to explain the conformation of a certain class of sponge-spicules; but the principle is of much wider application in morphology, and would seem to deserve attention which it has not yet received. The subject is not an easy one, and if we are to avoid mathematical difficulties we must keep within narrow bounds.

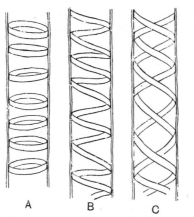

Fig. 346. Annular and spiral thickenings in the walls of plant-cells.

Defining, meanwhile, our geodesic line (as we have already done) as the shortest distance between two points on the surface of a solid of revolution, we find that the cylinder gives us some of the simplest of cases. Here it is plain that the geodesics are of three kinds: (1) a series of annuli around the cylinder, that is to say, a system of circles, in planes parallel to one another and at right angles to the axis of the cylinder (Fig. 346, A); (2) a series of straight lines parallel to the axis; and (3) a series of spiral curves winding round the wall of the cylinder (B, C). These three systems are all of frequent occurrence, and are illustrated in the local thickenings of the wall of the cylindrical cells or vessels of plants.

The spiral, or rather helical, geodesic is particularly common in cylindrical structures, and is beautifully shewn for instance in the spiral coil which stiffens the tracheal tubes of an insect, or the so-called tracheides of a woody stem. A like phenomenon is often witnessed in the splitting of a glass tube. If a crack appear in a test-tube it has a tendency to be prolonged in its own direction, and the more isotropic be the glass the more evenly will the split tend to follow the straight course in which it began. As a result, the crack often continues till our test-tube is split into a continuous spiral ribbon.

One may stretch a tape along a cylinder, but it no sooner swerves to one side than it begins to wind itself around; it is tracing its geodesic*.

In a circular cone, the spiral geodesic falls into closer and closer coils as the cone narrows, till it comes to the end, and then it winds back the same way; and a beautiful geodesic of this kind is exemplified in the sutural line of a spiral shell, such as *Turritella*, or in the striations which run parallel with the spiral suture. On a prolate spheroid, the coils of a spiral geodesic come closer together as they approach the ends of the long axis of the ellipse, and wind back and forward from one pole to the other†. We have a case of this kind in an *Equisetum*-spore, when the integument splits into the spiral "elaters," though the spire is not long enough to shew all its geodesic features in detail.

We begin to see that our first definition of a geodesic requires to be modified; for it is only subject to conditions that it is "the shortest distance between two points on the surface of the solid," and one of the commonest of these restricting conditions is that our geodesic may be constrained to go twice, or many times, round the surface on its way. In short, we may re-define a geodesic, as a curve drawn upon a surface such that, if we take any two *adjacent*

---

* It is not that the geodesic is rectified into a straight line; but that a straight line (the midline of the tape or ribbon) is converted into the geodesic on the given surface.

† In all these cases, $r \cos \alpha = $ a constant, where r is the radius of the circular section, and α the angle at which it is crossed by the geodesic. And the constant is measured by the smallest circle which the geodesic can reach.

points on the curve, the curve gives the shortest distance between them. It often happens, in the geodesic systems which we meet with in morphology, that two opposite spirals or rather helices run separate and distinct from one another, as in Fig. 346, C; and it is also common to find the two interfering with one another, and forming a criss-cross or reticulated arrangement. This indeed is a common source of reticulated patterns.

The microscopic and even ultramicroscopic structure of the cell-wall shews analogous configurations: as in the large cells of the alga *Valonia*, where the wall consists of many lamellae, each composed of parallel fibrillae running in spiral geodesics, and alternating in direction from one lamella to another. Here, and not less clearly in the young parenchyma of seedling oats, it is the long-chain cellulose molecules which follow a spiral course around the cell-wall, right-handed or left-handed as the case may be, and inclined more or less steeply according to the elongation of the cell. But these highly interesting questions of molecular, or micellar, structure lie beyond our scope*.

Among the ciliated infusoria, we have a variety of beautiful geodesic curves in the spiral patterns in which their cilia are arranged; though it is probable enough that in some complicated cases these are not simple geodesics, but developments of curves other than a straight line upon the surface of the organism. In other words, they seem to be instances of "geodesic curvature."

Lastly, an instructive case is furnished by the arrangement of the muscular fibres on the surface of a hollow organ, such as the heart or the stomach. Here we may consider the phenomenon from the point of view of mechanical efficiency, as well as from that of descriptive anatomy. In fact we have a right to expect that the muscular fibres covering such hollow organs will coincide with geodesic lines, in the sense in which we are using the term. For if we imagine a contractile fibre, or an elastic band, to be fixed by its two ends upon a curved surface, it is obvious that its first effort of contraction will tend to expend itself in accommodating

---

* Cf. (e.g.) C. Correns, Innere Struktur einiger Algenmembranen, *Beitr. zur Morphol. u. Physiol. d. Pflanzenzelle*, 1893, p. 260; W. T. Astbury and others, *Proc. R.S.* (B), cix, p. 443, 1932, and other papers; G. van Iterson, jr., *Nature*, cxxxviii, p. 364, 1936; R. D. Preston, *Proc. R.S.* (B), cxxv, p. 772, 1938; etc.

the band to the form of the surface, in "stretching it tight," or in other words in causing it to assume a direction which is the shortest possible line *upon the surface* between the two extremes: and it is only then that further contraction will have the effect of constricting the tube and so exercising pressure on its contents. Thus the muscular fibres, as they wind over the curved surface of an organ, arrange themselves automatically in geodesic curves: in precisely the same manner as we also automatically construct complex systems of geodesics whenever we wind a ball of wool or a spindle of tow, or when the skilful surgeon bandages a limb; indeed the surgeon must fold and crease his bandage if it is not to keep on geodesic lines. It is as a simple, necessary result of geodesic principles that we see those "figures-of-eight" produced, to which, in the case for instance of the heart-muscles, Pettigrew and other anatomists have ascribed peculiar importance. In the case of both heart and stomach we must look upon these organs as developed from a simple cylindrical tube, after the fashion of the glass-blower, as is further discussed on p. 1049 of this book, the modification of the simple cylinder consisting of various degrees of dilatation and of twisting. In the primitive undistorted cylinder, as in an artery or in the intestine, the muscles run in simple geodesic lines, and constitute the circular and longitudinal coats which form (or are said to form) the normal musculature of all tubular organs, or the cylindrical body of a worm. However, we can often recognise, in a small artery for instance, that the so-called circular fibres tend to take a slightly oblique or spiral course; and that the so-called *annular* muscle-fibres are really spirals is an old statement which may very likely be true*. If we consider each muscular fibre as an elastic strand embedded in the elastic membrane which constitutes the wall of the organ, it is evident that, whatever be the distortion suffered by the entire organ, the individual fibre will follow its own course, which will still, in a sense, be geodesic. But if the distortion be considerable, as for instance if the tube

* See A Discourse concerning the Spiral, instead of the supposed Annular, structure of the Fibres of the Intestins; discover'd and shewn by the Learn'd and Inquisitive Dr. William Cole to the Royal Society, *Phil. Trans.* XI, pp. 603–609, 1676. Cf. Eben J. Carey, Studies on the...small intestine, *Anat. Record*, XXI, pp. 189–215, 1921; F. T. Lewis, The spiral trend of intestinal muscle fibres, *Science*, LV, June 30, 1922.

become bent upon itself, or if at some point its walls bulge outwards in a diverticulum or pouch, then the old system of geodesics will only mark the shortest distance between two points more or less approximate to one another, and new systems of geodesics, peculiar to the new surface, will tend to appear, and link up points more remote from one another. This is evidently the case in the human stomach. We still have the systems, or their unobliterated remains, of circular and longitudinal muscles; but we also see two new systems of fibres, both obviously geodesic (or rather, when we look more closely, both parts of one and the same geodesic system), in the form of annuli encircling the pouch or diverticulum at the cardiac end of the stomach, and of oblique fibres taking a spiral course from the neighbourhood of the oesophagus over the sides of the organ.

In the heart we have a similar, but more complicated phenomenon. Its musculature consists, in great part, of the original simple system of circular and longitudinal muscles which enveloped the original arterial tubes, which tubes, after a process of local thickening, expansion, and especially *twisting*, came together to constitute the composite, or double, mammalian heart; and these systems of muscular fibres, geodesic to begin with, remain geodesic (in the sense in which we are using the word) after all the twisting which the primitive cylindrical tube or tubes have undergone. That is to say, these fibres still run their shortest possible course, from start to finish, over the complicated curved surface of the organ; and, as Borelli well understood, it is only because they do so that their contraction, or longitudinal shortening, is able to produce its direct effect in the contraction or systole of the heart*.

As a parenthetic corollary to the case of the spiral pattern upon the wall of a cylindrical cell, we may consider for a moment the spiral line which many small organisms tend to follow in their path

---

* The spiral fibres, or a large portion of them, constitute what Searle called "the rope of the heart" (Todd's *Cyclopaedia*, II, p. 621, 1836). The "twisted sinews of the heart" were known to early anatomists, and have been frequently and elaborately studied: for instance, by Gerdy (*Bull. Fac. Med. Paris*, 1820, pp. 40–148), and by Pettigrew (*Phil. Trans.* 1864), and again by J. B. Macallum (*Johns Hopkins Hospital Report*, IX, 1900) and by Franklin P. Mall (*Amer. Journ. Anat.* XI, 1911).

of locomotion*. A certain physiologist observed that an Amoeba, crawling within a narrow tube, wound its slow way in a spiral course instead of going straight along the tube. The creature was going nowhere in particular, but merely following the direction in which it had begun: in curious illustration of a familiar statement in the "dynamics of a particle," that a particle moving on a surface without constraint will describe geodesic lines.

But it is after a different fashion, and without any constraint to a surface, that the smaller ciliated organisms, such as the ciliate and flagellate infusoria, the rotifers, the swarm-spores of various Protista, and so forth, shew a tendency to pursue a spiral path in their ordinary locomotion. The means of locomotion which they possess in their cilia are at best somewhat primitive and inefficient; they have no apparent means of steering, or modifying their direction; and, if their course tended to swerve ever so little to one side, the result would be to bring them round and round again in an approximately circular path (such as a man astray on the prairie is said to follow), with little or no progress in a definite longitudinal direction. But as a matter of fact, by reason of a more or less unsymmetrical form of the body, all these creatures tend more or less to *rotate* about their long axis while they swim. And this axial rotation, just as in the case of a rifle-bullet, causes their natural swerve, which is always in the same direction as regards their own bodies, to be in a continually changing direction as regards space: in short, to make a spiral course around, and more or less near to, a straight axial line†.

In this short chapter we have touched on phenomena where form repeats itself, and mathematical analogies recur, in very different things and very different orders of magnitude. The spiral muscles of heart or stomach are the mechanical outcome of twists which these tubular organs have undergone in the course of their development, and come, accordingly, under the general category of organic

---

* Cf. Bütschli, "Protozoa," in Bronn's *Thierreich*, II, p. 848, III, p. 1785, etc., 1883–87; Jennings, *Amer. Nat.* XXXV, p. 369, 1901; Pütter, Thigmotaxie bei Protisten, *Arch. f. Anat. u. Phys. (Phys. Abth. Suppl.)*, pp. 243–302, 1900.

† Cf. W. Ludwig, Ueber die Schraubenbahnen niederer Organismen, *Arch. f. vergl. Physiologie*, IX, 1919.

or embryological growth.   But the spiral thickenings in the woody
fibres of a plant are of another order of things, and lie in the region
of molecular phenomena.   The delicate spirals of the cell-wall of
a cotton-hair are based on a complicated cellulose space-lattice,
recalling Nägeli's micellar hypothesis in a new setting; and giving
us a glimpse of organic growth after the very fashion of crystalline
growth, that is to say from the starting-point of molecular structure
and configuration*.

* W. Lawrence Balls, Determiners of cellulose structure as seen in the cell-wall
of cotton-hairs, *Proc. R.S.* (B), xcv, pp. 72–89, 1923, and other papers.   Cf. also
Wilfred Robinson, Microscopical features of mechanical strains in timber, and the
bearing of these on the structure of the cell-wall in plants, *Phil. Trans.* (B), ccx,
pp. 49–82, 1920.

CHAPTER XI

THE EQUIANGULAR SPIRAL

THE very numerous examples of spiral conformation which we meet with in our studies of organic form are peculiarly adapted to mathematical methods of investigation. But ere we begin to study them we must take care to define our terms, and we had better also attempt some rough preliminary classification of the objects with which we shall have to deal.

In general terms, a Spiral is a curve which, starting from a point of origin, continually diminishes in curvature as it recedes from that point; or, in other words, whose *radius of curvature* continually increases. This definition is wide enough to include a number of different curves, but on the other hand it excludes at least one which in popular speech we are apt to confuse with a true spiral. This latter curve is the simple *screw*, or cylindrical *helix*, which curve neither starts from a definite origin nor changes its curvature as it proceeds. The "spiral" thickening of a woody plant-cell, the "spiral" thread within an insect's tracheal tube, or the "spiral" twist and twine of a climbing stem are not, mathematically speaking, *spirals* at all, but *screws* or *helices*. They belong to a distinct, though not very remote, family of curves.

Of true organic spirals we have no lack*. We think at once of horns of ruminants, and of still more exquisitely beautiful molluscan shells—in which (as Pliny says) *magna ludentis Naturae varietas.* Closely related spirals may be traced in the florets of a sunflower; a true spiral, though not, by the way, so easy of investigation, is seen in the outline of a cordiform leaf; and yet again, we can recognise typical though transitory spirals in a lock of hair, in a staple of wool†, in the coil of an elephant's trunk, in the "circling spires"

* A great number of spiral forms, both organic and artificial, are described and beautifully illustrated in Sir T. A. Cook's *Spirals in Nature and Art*, 1903, and *Curves of Life*, 1914.

† On this interesting case see, e.g. J. E. Duerden, in *Science*, May 25, 1934.

of a snake, in the coils of a cuttle-fish's arm, or of a monkey's or
a chameleon's tail.

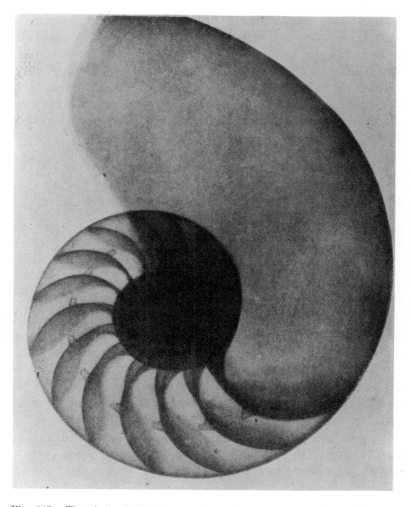

Fig. 347. The shell of *Nautilus pompilius*, from a radiograph: to shew the
equiangular spiral of the shell, together with the arrangement of the internal
septa. From Green and Gardiner, in *Proc. Malacol. Soc.* II, 1897.

Among such forms as these, and the many others which we
might easily add to them, it is obvious that we have to do with
things which, though mathematically similar, are biologically

speaking fundamentally different; and not only are they biologically remote, but they are also physically different, in regard to the causes to which they are severally due. For in the first place, the spiral coil of the elephant's trunk or of the chameleon's tail is, as we have said, but a transitory configuration, and is plainly the result of certain muscular forces acting upon a structure of a definite, and normally an essentially different, form. It is rather a position, or an *attitude*, than a *form*, in the sense in which we have been using this latter term; and, unlike most of the forms which we have been studying, it has little or no direct relation to the phenomenon of growth.

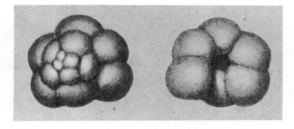

Fig. 348.   A foraminiferal shell (*Pulvinulina*).

Again, there is a difference between such a spiral conformation as is built up by the separate and successive florets in the sunflower, and that which, in the snail or *Nautilus* shell, is apparently a single and indivisible unit. And a similar if not identical difference is apparent between the *Nautilus* shell and the minute shells of the Foraminifera which so closely simulate it: inasmuch as the spiral shells of these latter are composite structures, combined out of successive and separate chambers, while the molluscan shell, though it may (as in *Nautilus*) become secondarily subdivided, has grown as one continuous tube. It follows from all this that there cannot be a physical or dynamical, though there may well be a mathematical *law of growth*, which is common to, and which defines, the spiral form in *Nautilus*, in *Globigerina*, in the ram's horn, and in the inflorescence of the sunflower. Nature at least exhibits in them all "*un reflet des formes rigoureuses qu'étudie la géométrie*.*"

* Haton de la Goupillière, in the introduction to his important study of the *Surfaces Nautiloides*, *Annaes sci. da Acad. Polytechnica do Porto*, Coimbra, III, 1908.

Of the spiral forms which we have now mentioned, every one (with the single exception of the cordate outline of the leaf) is an example of the remarkable curve known as the equiangular or logarithmic spiral. But before we enter upon the mathematics of the equiangular spiral, let us carefully observe that the whole of the organic forms in which it is clearly and permanently exhibited, however different they may be from one another in outward appearance, in nature and in origin, nevertheless all belong, in a certain sense, to one particular class of conformations. In the great majority of cases, when we consider an organism in part or whole, when we look (for instance) at our own hand or foot, or contemplate an insect or a worm, we have no reason (or very little) to consider one part of the existing structure as *older* than another; through and through, the newer particles have been merged and commingled among the old; the outline, such as it is, is due to forces which for the most part are still at work to shape it, and which in shaping it have shaped it as a whole. But the horn, or the snail-shell, is curiously different; for in these the presently existing structure is, so to speak, partly old and partly new. It has been conformed by successive and continuous increments; and each successive stage of growth, starting from the origin, remains as an integral and unchanging portion of the growing structure.

We may go further, and see that horn and shell, though they belong to the living, are in no sense alive *. They are by-products of the animal; they consist of "formed material," as it is sometimes called; their growth is not of their own doing, but comes of living cells beneath them or around. The many structures which display the logarithmic spiral increase, or accumulate, rather than grow. The shell of nautilus or snail, the chambered shell of a foraminifer, the elephant's tusk, the beaver's tooth, the cat's claws or the canary-bird's—all these shew the same simple and very beautiful spiral curve. And all alike consist of stuff secreted or deposited by living cells; all grow, as an edifice grows, by accretion of accumulated

---

* For Oken and Goodsir the logarithmic spiral had a profound significance, for they saw in it a manifestation of life itself. For a like reason Sir Theodore Cook spoke of the *Curves of Life*; and Alfred Lartigues says (in his *Biodynamique générale*, 1930, p. 60): "Nous verrons la Conchyliologie apporter une magnifique contribution à la Stéréodynamique du tourbillon vital." The fact that the spiral is always formed of non-living matter helps to contradict these mystical conceptions.

material; and in all alike the parts once formed remain in being, and are thenceforward incapable of change.

In a slightly different, but closely cognate way, the same is true of the spirally arranged florets of the sunflower. For here again we are regarding serially arranged portions of a composite structure, which portions, similar to one another in form, *differ in age*; and differ also in magnitude in the strict ratio of their age. Somehow or other, in the equiangular spiral the *time-element* always enters in; and to this important fact, full of curious biological as well as mathematical significance, we shall afterwards return.

In the elementary mathematics of a spiral, we speak of the point of origin as the pole (O); a straight line having its extremity in the pole, and revolving about it, is called the radius vector; and a point (P), travelling along the radius vector under definite conditions of velocity, will then describe our spiral curve.

Of several mathematical curves whose form and development may be so conceived, the two most important (and the only two with which we need deal) are those which are known as (1) the equable spiral, or spiral of Archimedes, and (2) the equiangular or logarithmic spiral.

The former may be roughly illustrated by the way a sailor coils a rope upon the deck; as the rope is of uniform thickness, so in the whole spiral coil is each whorl of the same breadth as that which precedes and as that which follows it. Using its ancient definition, we may define it by saying, that "If a straight line revolve uniformly about its extremity, a point which likewise travels uniformly along it will describe the equable spiral*." Or, putting the same thing into our more modern words, "If, while the radius vector revolve uniformly about the pole, a point (P) travel with uniform velocity along it, the curve described will be that called the equable spiral, or spiral of Archimedes." It is plain that the spiral of Archimedes may be compared, but again roughly, to a *cylinder* coiled up. It is plain also that a radius $(r = OP)$, made up of the successive and equal whorls, will increase in *arithmetical* progression: and will equal a certain constant quantity (a) multiplied

---

* Leslie's *Geometry of Curved Lines*, 1821, p. 417. This is practically identical with Archimedes' own definition (ed. Torelli, p. 219); cf. Cantor, *Geschichte der Mathematik*, I, p. 262, 1880.

by the whole number of whorls, (or more strictly speaking) multiplied by the whole angle (θ) through which it has revolved: so that $r = a\theta$. And it is also plain that the radius meets the curve (or its tangent) at an angle which changes slowly but continuously, and which tends towards a right angle as the whorls increase in number and become more and more nearly circular.

But, in contrast to this, in the equiangular spiral of the *Nautilus* or the snail-shell or *Globigerina*, the whorls continually increase in breadth, and do so in a steady and unchanging ratio. Our definition is as follows: "If, instead of travelling with a *uniform* velocity, our point move along the radius vector with a velocity *increasing as its distance from the pole*, then the path described is

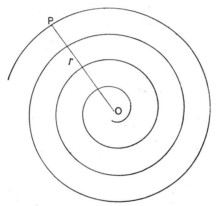

Fig. 349.   The spiral of Archimedes.

called an equiangular spiral." Each whorl which the radius vector intersects will be broader than its predecessor in a definite ratio; the radius vector will increase in length in *geometrical* progression, as it sweeps through successive equal angles; and the equation to the spiral will be $r = a^\theta$. As the spiral of Archimedes, in our example of the coiled rope, might be looked upon as a coiled cylinder, so (but equally roughly) may the equiangular spiral, in the case of the shell, be pictured as a *cone* coiled upon itself; and it is the conical shape of the elephant's trunk or the chameleon's tail which makes them coil into a rough simulacrum of an equiangular spiral.

While the one spiral was known in ancient times, and was investigated if not discovered by Archimedes, the other was first recognised by Descartes, and discussed in the year 1638 in his letters to Mersenne*. Starting with the conception of a growing curve which should cut each radius vector at a constant angle—just as a circle does—Descartes shewed how it would necessarily follow that radii at equal angles to one another at the pole would be in continued proportion; that the same is therefore true of the parts cut off from a common radius vector by successive whorls or convolutions of the spire; and furthermore, that distances measured along the curve from its origin, and intercepted by any radii, as at *B, C,* are proportional to the lengths of these radii, *OB, OC.* It follows that

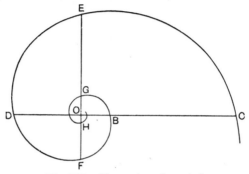

Fig. 350.   The equiangular spiral.

the sectors cut off by successive radii, at equal vectorial angles, are similar to one another in every respect; and it further follows that the figure may be conceived as growing continuously without ever changing its shape the while.

If the whorls increase very slowly, the equiangular spiral will come to look like a spiral of Archimedes. The Nummulite is a case in point. Here we have a large number of whorls, very narrow, very close together, and apparently of equal breadth, which give rise to an appearance similar to that of our coiled rope. And, in a case of this kind, we might actually find that the whorls *were* of equal breadth, being produced (as is apparently the case in the Nummulite) not by any very slow and gradual growth in thickness of a continuous tube, but by a succession of similar cells or chambers laid on, round and round, determined as to their size by constant surface-tension conditions and therefore of unvarying dimensions. The Nummulite must always have a central core, or initial cell, around which the coil is not only wrapped, but out of which it springs; and this initial chamber corresponds to our *a′* in the expression $r = a′ + a\theta \cot \alpha$.

* *Œuvres,* ed. Adam et Tannery, Paris, 1898, p. 360.

The many specific properties of the equiangular spiral are so interrelated to one another that we may choose pretty well any one of them as the basis of our definition, and deduce the others from it either by analytical methods or by elementary geometry. In algebra, when $m^x = n$, x is called the logarithm of n to the base m. Hence, in this instance, the equation $r = a^\theta$ may be written in the form $\log r = \theta \log a$, or $\theta = \log r / \log a$, or (since a is a constant) $\theta = k \log r$*. Which is as much as to say that (as Descartes discovered) the vector angles about the pole are proportional to the logarithms of the successive radii; from which circumstance the alternative name of the "logarithmic spiral" is derived†.

Moreover, for as many properties as the curve exhibits, so many names may it more or less appropriately receive. James Bernoulli called it the logarithmic spiral, as we still often do; P. Nicolas called it the geometrical spiral, because radii at equal polar angles are in geometrical progression; Halley, the proportional spiral, because the parts of a radius cut off by successive whorls are in continued proportion; and lastly, Roger Cotes, going back to Descartes' first description or first definition of all, called it the equiangular spiral‡. We may also recall Newton's remarkable demonstration that, had the force of gravity varied inversely as the *cube* instead of the *square* of the distance, the planets, instead of being bound to their

---

* Instead of $r = a^\theta$, we might write $r = r_0 a^\theta$; in which case r_0 is the value of r for zero value of θ.

† Of the two names for this spiral, equiangular and logarithmic, I used the latter in my first edition, but equiangular spiral seems to be the better name; for the constant angle is its most distinguishing characteristic, and that which leads to its remarkable property of continuous self-similarity. Equiangular spiral is its name in geometry; it is the analyst who derives from its geometrical properties its relation to the logarithm. The mechanical as well as the mathematical properties of this curve are very numerous. A Swedish admiral, in the eighteenth century, shewed an equiangular spiral (of a certain angle) to be the best form for an anchor-fluke (*Sv. Vet. Akad. Hdl.* xv, pp. 1–24, 1796), and in a parrot's beak it has the same efficiency. Macquorn Rankine shewed its advantages in the pitch of a cam or non-circular wheel (*Manual of Mechanics*, 1859, pp. 99–102; cf. R. C. Archibald, *Scripta Mathem.* iii (4), p. 366, 1935).

‡ James Bernoulli, in *Acta Eruditorum*, 1691, p. 282; P. Nicolas, *De novis spiralibus*, Tolosae, 1693, p. 27; E. Halley, *Phil. Trans.* xix, p. 58, 1696; Roger Cotes, *ibid.* 1714, and *Harmonia Mensurarum*, 1722, p. 19. For the further history of the curve see (e.g.) Gomes de Teixeira, *Traité des courbes remarquables*, Coimbre, 1909, pp. 76–86; Gino Loria, *Spezielle algebräische Kurven*, ii, p. 60 *seq.*, 1911; R. C. Archibald (to whom I am much indebted) in *Amer. Mathem. Monthly*, xxv, pp. 189–193, 1918, and in Jay Hambidge's *Dynamic Symmetry*, 1920, pp. 146–157.

ellipses, would have been shot off in spiral orbits from the sun, the equiangular spiral being one case thereof.*

A singular instance of the same spiral is given by the route which certain insects follow towards a candle. Owing to the structure of their compound eyes, these insects do not look straight ahead but make for a light which they see abeam, at a certain angle. As they continually adjust their path to this constant angle, a spiral pathway brings them to their destination at last†.

In mechanical structures, *curvature* is essentially a mechanical phenomenon. It is found in flexible structures as the result of

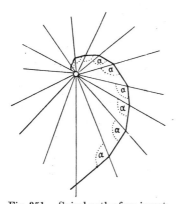

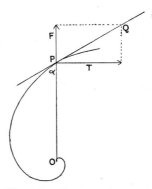

Fig. 351.   Spiral path of an insect, as it draws towards a light. From Wigglesworth (after van Buddenbroek).

Fig. 352.   Dynamical aspect of the equiangular spiral.

bending, or it may be introduced into the construction for the purpose of resisting such a bending-moment. But neither shell nor tooth nor claw are flexible structures; they have not been *bent* into their peculiar curvature, they have *grown* into it.

We may for a moment, however, regard the equiangular or logarithmic spiral of our shell from the dynamical point of view, by looking on *growth* itself as the force concerned. In the growing structure, let growth at any point P be resolved into a force F acting along the line joining P to a pole O, and a force T acting in a direction perpendicular to OP; and let the magnitude of these forces (or of these rates of growth) remain constant. It follows that

* *Principia*, I, 9; II, 15.   On these "Cotes's spirals" see Tait and Steele, p. 147.
† Cf. W. Buddenbroek, *Sitzungsber. Heidelb. Akad.*, 1917; V. H. Wigglesworth, *Insect Physiology*, 1839, p. 167.

the resultant of the forces F and T (as PQ) makes a constant angle with the radius vector. But a constant angle between tangent and radius vector is a fundamental property of the "equiangular" spiral: the very property with which Descartes started his investigation, and that which gives its alternative name to the curve.

In such a spiral, radial growth and growth in the direction of the curve bear a constant ratio to one another. For, if we consider a consecutive radius vector, OP', whose increment as compared with OP is dr, while ds is the small arc PP', then $dr/ds = \cos \alpha = \text{constant}$.

In the growth of a shell, we can conceive no simpler law than this, namely, that it shall widen and lengthen in the same unvarying proportions: and this simplest of laws is that which Nature tends to follow. The shell, like the creature within it, grows in size *but does not change its shape*; and the existence of this constant relativity of growth, or constant similarity of form, is of the essence, and may be made the basis of a definition, of the equiangular spiral*.

Such a definition, though not commonly used by mathematicians, has been occasionally employed; and it is one from which the other properties of the curve can be deduced with great ease and simplicity. In mathematical language it would run as follows: "Any [plane] curve proceeding from a fixed point (which is called the pole), and such that the arc intercepted between any two radii at a given angle to one another is always similar to itself, is called an equiangular, or logarithmic, spiral."

In this definition, we have the most fundamental and "intrinsic" property of the curve, namely the property of continual similarity, and the very property by reason of which it is associated with organic growth in such structures as the horn or the shell. For it is peculiarly characteristic of the spiral shell, for instance, that it does not alter as it grows; each increment is similar to its predecessor, and ·the whole, after every spurt of growth, is just like what it was before. We feel no surprise when the animal which secretes the shell, or any other animal whatsoever, grows by such symmetrical expansion as to preserve its form unchanged; though even there, as we have already seen, the unchanging form denotes a nice balance between the rates of growth in various directions, which is

---

* See an interesting paper by W. A. Whitworth, The equiangular spiral, its chief properties proved geometrically, *Messenger of Mathematics* (1), I, p. 5, 1862. The celebrated Christian Wiener gave an explanation on these lines of the logarithmic spiral of the shell, in his highly original *Grundzüge der Weltordnung*, 1863.

but seldom accurately maintained for long. But the shell retains its unchanging form in spite of its *asymmetrical* growth; it grows at one end only, and so does the horn. And this remarkable property of increasing by *terminal* growth, but nevertheless retaining unchanged the form of the entire figure, is characteristic of the equiangular spiral, and of no other mathematical curve. It well deserves the name, by which James Bernoulli was wont to call it, of *spira mirabilis*.

We may at once illustrate this curious phenomenon by drawing the outline of a little *Nautilus* shell within a big one. We know, or we may see at once, that they are of precisely the same shape; so that, if we look at the little shell through a magnifying glass, it becomes identical with the big one. But we know, on the other hand, that the little *Nautilus* shell grows into the big one, not by growth or magnification in all parts and directions, as when the boy grows into the man, but by growing *at one end only*.

If we should want further proof or illustration of the fact that the spiral shell remains of the same shape while increasing in magnitude by its terminal growth, we may find it by help of our ratio $W : L^3$, which remains constant so long as the shape remains unchanged. Here are weights and measurements of a series of small land-shells (*Clausilia*):*

W (mgm.)	L (mm.)	$\sqrt[3]{W}/L$
50	14·4	2·56
53	15·1	2·49
56	15·2	2·52
56	15·2	2·52
56	15·4	2·44
58	15·5	2·50
61	16·4	2·40
63	16·0	2·49
67	16·0	2·54
69	16·1	2·56
	Mean	2·50

Though of all plane curves, this property of continued similarity is found only in the equiangular spiral, there are many rectilinear figures in which it may be shewn. For instance, it holds good of

* In 100 specimens of *Clausilia* the mean value of $\sqrt[3]{W}/L$ was found to be 2·517, the coefficient of variation 0·092, and the standard deviation 3·6. That is to say, over 90 per cent. grouped themselves about a mean value of 2·5 with a deviation of less than 4 per cent. Cf. C. Petersen, *Das Quotientengesetz*, 1921, p. 55.

any cone; for evidently, in Fig. 353, the little inner cone (represented in its triangular section) may become identical with the larger one either by magnification all round (as in *a*), or by an increment at one end (as in *b*); or for that matter on the rest of its surface, represented by the other two sides, as in *c*. All this is associated with the fact, which we have already noted, that the *Nautilus* shell is but a cone rolled up; that, in other words, the cone is but a particular variety, or "limiting case," of the spiral shell.

This singular property of continued similarity, which we see in the cone, and recognise as characteristic of the logarithmic spiral, would seem, under a more general aspect, to have engaged the particular attention of ancient mathematicians even from the days of Pythagoras, and so, with little doubt, from the still more ancient days of that Egyptian school whence he derived the foundations of

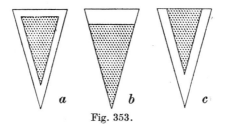

a   *b*   *c*

Fig. 353.

his learning*; and its bearing on our biological problem of the shell, however indirect, is close enough to deserve our very careful consideration.

There are certain things, says Aristotle, which suffer no alteration (save of magnitude) when they grow†. Thus if we add to a square an **L**-shaped portion, shaped like a carpenter's square, the resulting figure is still a square; and the portion which we have so added, with this singular result, is called in Greek a "gnomon."

Euclid extends the term to include the case of any parallelogram‡, whether rectangular or not (Fig. 354); and Hero of Alexandria

---

* I am well aware that the debt of Greek science to Egypt and the East is vigorously denied by many scholars, some of whom go so far as to believe that the Egyptians never had any science, save only some "rough rules of thumb for measuring fields and pyramids" (Burnet's *Greek Philosophy*, 1914, p. 5).

† *Categ.* 14, 15*a*, 30: ἔστι τινὰ αὐξανόμενα ἃ οὐκ ἀλλοιοῦται, οἷον τὸ τετράγωνον, γνώμονος περιτεθέντος, ηὔξηται μὲν ἀλλοιότερον δὲ οὐδὲν γεγένηται.

‡ Euclid (ii, def. 2).

specifically defines a gnomon (as indeed Aristotle had implicitly defined it), as any figure which, being added to any figure whatsoever, leaves the resultant figure similar to the original. Included in this important definition is the case of numbers, considered geometrically; that is to say, the εἰδητικοὶ ἀριθμοί, which can be translated into *form*, by means of rows of dots or other signs (cf. Arist. *Metaph.* 1092 b 12), or in the pattern of a tiled floor: all according to "the mystical way of Pythagoras, and the secret

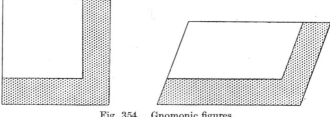

Fig. 354.   Gnomonic figures.

magick of numbers." For instance, the triangular numbers, 1, 3, 6, 10 etc., have the natural numbers for their "differences"; and so the natural numbers may be called their gnomons, because they keep the triangular numbers still triangular. In like manner the square numbers have the successive odd numbers for their gnomons, as follows:

$$0 + 1 = 1^2$$
$$1^2 + 3 = 2^2$$
$$2^2 + 5 = 3^2$$
$$3^2 + 7 = 4^2 \quad \text{etc.}$$

And this gnomonic relation we may illustrate graphically (σχηματο-γραφεῖν) by the dots whose addition keeps the annexed figures perfect squares*:

There are other gnomonic figures more curious still. For example, if we make a rectangle (Fig. 355) such that the two sides are in the

* Cf. Treutlein, *Ztschr. f. Math. u. Phys. (Hist. litt. Abth.)*, xxviii, p. 209, 1883.

ratio of $1 : \sqrt{2}$, it is obvious that, on doubling it, we obtain a similar figure; for $1 : \sqrt{2} :: \sqrt{2} : 2$; and each half of the figure, accordingly, is now a gnomon to the other. Were we to make our paper of such a shape (say, roughly, 10 in. × 7 in.), we might fold and fold it, and the shape of folio, quarto and octavo pages would be all the same. For another elegant example, let us start with a rectangle (A) whose sides are in the proportion of the "divine" or "golden section*" that is to say as $1 : \frac{1}{2}(\sqrt{5} - 1)$, or, approximately, as $1 : 0.618\ldots$. The gnomon to this rectangle is the square (B) erected on its longer side, and so on successively (Fig. 356).

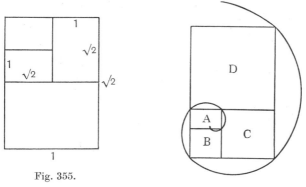

Fig. 355.

Fig. 356.

In any triangle, as Hero of Alexandria tells us, one part is always a gnomon to the other part. For instance, in the triangle ABC (Fig. 357), let us draw BD, so as to make the angle CBD equal to the angle A. Then the part BCD is a triangle similar to the whole triangle ABC, and ABD is a gnomon to BCD. A very elegant case is when the original triangle ABC is an isosceles triangle having one angle of 36°, and the other two angles, therefore, each equal to 72° (Fig. 358). Then, by bisecting one of the angles of the base, we subdivide the large isosceles triangle into two isosceles triangles, of which one is similar to the whole figure and the other is its gnomon†. There is good reason to believe that this triangle was especially studied by the Pythagoreans; for it lies at the root of

* Euclid, II, 11.

† This is the so-called *Dreifachgleichschenkelige Dreieck*; cf. Naber, *op. infra cit.* The ratio 1 : 0·618 is again not hard to find in this construction.

many interesting geometrical constructions, such as the regular
pentagon, and its mystical "pentalpha," and a whole range of other
curious figures beloved of the ancient mathematicians*: culminating

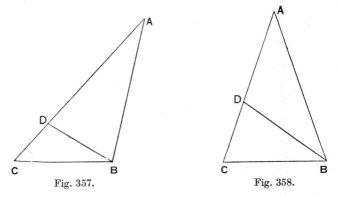

Fig. 357.   Fig. 358.

in the regular, or pentagonal, dodecahedron, which symbolised the
universe itself, and with which Euclidean geometry ends.

If we take any one of these figures, for instance the isosceles
triangle which we have just described, and add to it (or subtract

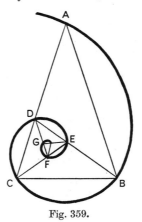

Fig. 359.

from it) in succession a series of gnomons, so converting it into larger
and larger (or smaller and smaller) triangles all similar to the first,
we find that the apices (or other corresponding points) of all these

* See, on the mathematical history of the gnomon, Heath's *Euclid*, I, *passim*,
1908; Zeuthen, *Théorème de Pythagore*, Genève, 1904; also a curious and
interesting book, *Das Theorem des Pythagoras*, by Dr H. A. Naber, Haarlem, 1908.

triangles have their *locus* upon a equiangular spiral: a result which follows directly from that alternative definition of the equiangular spiral which I have quoted from Whitworth (p. 757).

If in this, or any other isosceles triangle, we take corresponding median lines of the successive triangles, by joining C to the mid-point (M) of AB, and D to the mid-point (N) of BC, then the pole of the spiral, or centre of similitude of ABC and BCD, is the point of intersection of CM and $DN*$.

Again, we may build up a series of right-angled triangles, each of which is a gnomon to the preceding figure; and here again, an equiangular spiral is the locus of corresponding points in these successive triangles. And lastly, whensoever we fill up space with a collection of equal and similar figures, as in Figs. 360, 361, there we can always discover a series of equiangular spirals in their successive multiples†.

Once more, then, we may modify our definition, and say that: "Any plane curve proceeding from a fixed point (or pole), and such that the vectorial area of any sector is always a gnomon to the whole preceding figure, is called an equiangular, or logarithmic, spiral." And we may now introduce this new concept and nomenclature into our description of the *Nautilus* shell and other related organic forms, by saying that: (1) if a growing structure be built up of successive parts, similar in form, magnified in geometrical progression, and similarly situated with respect to a centre of similitude, we can always trace through corresponding points a series of equiangular spirals; and (2) it is characteristic of the

---

* I owe this simple but novel construction, like so much else, to Dr G. T. Bennett.

† In each and all of these gnomonic figures we may now recognise a never-ending polygon, with equal angles at its corners, and with its successive sides in geometrical progression; and such a polygon we may look upon as the natural precursor of the equiangular spiral. If we call the exterior or "bending" angle of the polygon β, and the ratio of its sides λ, then the vertices lie on an equiangular spiral of angle α, given by $\log_e \lambda = \beta \cot \alpha$. In the spiral of Fig. 359 the constant angle is thus found to be about 75° 40′, in that of Fig. 355, 77° 40′, and in that of Fig. 356, 72° 50′.

The calculation is as follows. Taking, for example, the successive triangles of Fig. 359, the ratio (λ) of the sides, as $BC : AC$, is that of the golden section, $1:1·618$. The external angle (β), as ADB, is 108°, or in radians 1·885. Then

$$\log 1·618 = 0·209, \text{ from which } \log_e 1·618 = 0·481$$

and
$$\cot \alpha = \frac{\log_e \lambda}{\beta} = \frac{0·481}{1·885} = 0·255 = \cot 75° \ 45′.$$

growth of the horn, of the shell, and of all other organic forms in which an equiangular spiral can be recognised, that *each successive increment of growth is similar, and similarly magnified, and similarly*

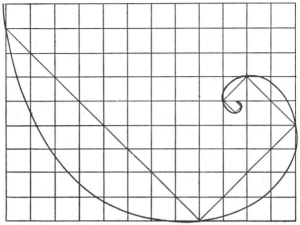

Fig. 360*. Logarithmic spiral derived from corresponding points in a system of squares.

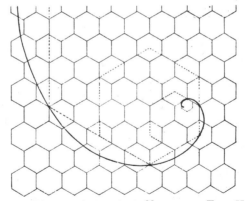

Fig. 361. The same in a system of hexagons. From Naber.

situated to its predecessor, and is in consequence a gnomon to the entire pre-existing structure. Conversely (3) it follows that in the spiral

* This diagram was at fault in my first edition (p. 512), as Dr G. T. Bennett shews me. The curve met its chords at equal angles at either end: whereas it ought to meet the further end at a lesser angle than the other, and ought in consequence to intersect the lines of the coordinate framework. The constant angle of this spiral is about 66° 11′ (tan $\alpha = \pi/2 \log_e 2$).

outline of the shell or of the horn we can always inscribe an endless variety of other gnomonic figures, having no necessary relation, save as a mathematical accident, to the nature or mode of development of the actual structure*. But observe that the gnomons to a square may form increments of any size, and the same is true of the gnomons to a *Haliotis*-shell; but in the higher symmetry, of a chambered *Nautilus*, or of the successive triangles in Fig. 359, growth goes on by a progressive series of gnomons, each one of which is the gnomon to another.

Fig. 362. A shell of *Haliotis*, with two of the many lines of growth, or generating curves, marked out in black: the areas bounded by these lines of growth being in all cases gnomons to the pre-existing shell.

Of these three propositions, the second is of great use and advantage for our easy understanding and simple description of the molluscan shell, and of a great variety of other structures whose mode of growth is analogous, and whose mathematical properties are therefore identical. We see that the successive chambers of a spiral *Nautilus* or of a straight *Orthoceras*, each whorl or part of a whorl of a periwinkle or other gastropod, each new increment of the operculum of a gastropod, each additional increment of an elephant's tusk, or each new chamber of a spiral foraminifer, has its leading characteristic at once described and its form so far explained by the

* For many beautiful geometrical constructions based on the molluscan shell, see S. Colman and C. A. Coan, *Nature's Harmonic Unity* (ch. IX, Conchology), New York, 1912.

simple statement that it constitutes a *gnomon* to the whole previously existing structure. And herein lies the explanation of that "time-element" in the development of organic spirals of which we have spoken already; for it follows as a simple corollary to this theory

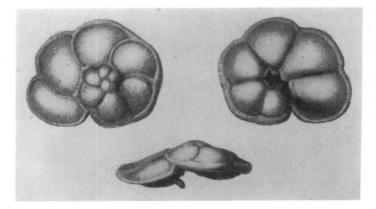

Fig. 363. A spiral foraminifer (*Pulvinulina*), to shew how each successive chamber continues the symmetry of, or constitutes a *gnomon* to, the rest of the structure.

of gnomons that we must never expect to find the logarithmic spiral manifested in a structure whose parts are simultaneously produced, as for instance in the margin of a leaf, or among the many curves that make the contour of a fish. But we most look for it wherever

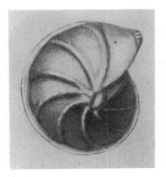

Fig. 364. Another spiral fora-minifer, *Cristellaria*.

the organism retains, and still presents at a single view, the successive phases of preceding growth: the successive magnitudes attained, the successive outlines occupied, as growth pursued the even tenor of its way. And it follows from this that it is in the hard parts of organisms, and not the soft, fleshy, actively growing parts, that this spiral is commonly and characteristically found: not in the fresh mobile tisssue whose form is constrained merely by the active forces of the moment; but in things like shell and tusk, and horn and claw, visibly composed of parts successively and permanently laid down. The shell-less molluscs are never spiral; the snail is spiral but not the slug*. In short, it is the shell which curves the snail, and not the snail which curves the shell. The logarithmic spiral is characteristic, not of the living tissues, but of the dead. And for the same reason it will always or nearly always be accompanied, and adorned, by a pattern formed of "lines of growth," the lasting record of successive stages of form and magnitude†.

The cymose inflorescences of the botanists are analogous in a curious and instructive way to the equiangular spiral.

In Fig. 365 B (which represents the *Cicinnus* of Schimper, or *cyme unipare scorpioide* of Bravais, as seen in the Borage), we begin with a primary shoot from which is given off, at a certain definite angle, a secondary shoot: and from that in turn, on the same side and at the same angle, another shoot, and so on. The deflection, or curvature, is continuous and progressive, for it is caused by no external force but only by causes intrinsic in the system. And the whole system is symmetrical: the angles at which the successive shoots are given off being all equal, and the lengths of the shoots diminishing *in constant ratio*. The result is that the successive shoots, or successive increments of growth, are tangents to a curve, and this curve is a true logarithmic spiral. Or in other words, we may regard each successive shoot as forming, or defining, a gnomon to the preceding structure. While in this simple case the successive shoots are depicted as lying in *a plane*, it may also happen that, in addition to their successive angular divergence from one another within that plane, they also tend to

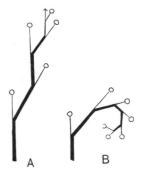

Fig. 365. A, a helicoid; B, a scorpioid cyme.

---

* Note also that *Chiton*, where the pieces of the shell are disconnected, shews no sign of spirality.

† That the invert to an equiangular spiral is identical with the original curve does not concern us in our study of organic form, but it is one of the most beautiful and most singular properties of the curve. It was this which led James Bernoulli, in imitation of Archimedes, to have the logarithmic spiral inscribed upon his tomb; and on John Goodsir's grave near Edinburgh the same symbol is reinscribed. Bernoulli's account of the matter is interesting and remarkable: "Cum autem ob proprietatem tam singularem tamque admirabilem mire mihi placeat spira haec mirabilis, sic ut ejus contemplatione satiari vix nequeam: cogitavi illam ad varias res symbolice repraesentandas non inconcinne adhiberi posse. Quoniam enim semper sibi et eandem spiram gignit, utcunque volvatur, evolvatur, radiet, hinc poterit esse vel sobolis parentibus per omnia similis Emblema: *Simillima Filia Matri*; vel (si rem aeternae veritatis Fidei mysteriis accommodare non est

diverge by successive equal angles *from* that plane of reference; and by this means, there will be superposed upon the equiangular spiral a twist or screw.   And, in the particular case where this latter angle of divergence is just equal to 180°, or two right angles, the successive shoots will once more come to lie in a plane, but they will appear to come off from one another on *alternate* sides, as in Fig. 365 A.   This is the *Schraubel* or *Bostryx* of Schimper, the *cyme unipare hélicoïde* of Bravais.   The equiangular spiral is still latent in it, as in the other; but is concealed from view by the deformation resulting from the helicoid.   Many botanists did not recognise (as the brothers Bravais did) the mathematical significance of the latter case, but were led by the snail-like spiral of the scorpoid cyme to transfer the name "helicoid" to it*.

The spiral curve of the shell is, in a sense, a vector diagram of its own growth; for it shews at each instant of time the direction, radial and tangential, of growth, and the unchanging ratio of velocities in these directions.   Regarding the *actual* velocity of growth in the shell, we know very little by way of experimental measurement; but if we make a certain simple assumption, then we may go a good deal further in our description of the equiangular spiral as it appears in this concrete case.

Let us make the assumption that *similar* increments are added to the shell in *equal* times; that is to say, that the amount of growth in unit time is measured by the areas subtended by equal angles.   Thus, in the outer whorl of a spiral shell a definite area marked out by ridges, tubercles, etc., has very different linear dimensions to the corresponding areas of an inner whorl, but the symmetry of the figure implies that it subtends an equal angle with these; and it is reasonable to suppose that the successive regions, marked out in this way by successive natural boundaries or patterns, are produced in equal intervals of time.

prohibitum) ipsius aeternae generationis Filii, qui Patris veluti Imago, et ab illo ut Lumen a Lumine emanans, eidem ὁμοιούσιος existit, qualiscunque adumbratio.   Aut, si mavis, quia Curva nostra mirabilis in ipsa mutatione semper sibi constantissime manet similis at numero eadem, poterit esse vel fortitudinis et constantiae in adversitatibus, vel etiam Carnis nostrae post varias alterationes et tandem ipsam quoque mortem, ejusdem numero resurrecturae symbolum: adeo quidem, ut si Archimedem imitandi hodiernum consuetudo obtineret, libenter Spiram hanc tumulo meo juberem incidi, cum Epigraphe, *Eadem numero mutata resurget*"; *Acta Eruditorum*, M. Maii, 1692, p. 213.   Cf. L. Isely, Épigraphes tumulaires de mathématiciens, *Bull. Soc. Sci. nat. Neuchâtel.* xxvii, p. 171, 1899.

* The names of these structures have been often confused and misunderstood; cf. S. H. Vines, The history of the scorpioid cyme, *Journ. Bot.* (n.s.), x, pp. 3–9, 1881.

If this be so, the radii measured from the pole to the boundary of the shell will in each case be proportional to the velocity of growth at this point upon the circumference, and at the time when it corresponded with the outer lip, or region of active growth; and while the direction of the radius vector corresponds with the direction of growth in thickness of the animal, so does the tangent to the curve correspond with the direction, for the time being, of the animal's growth in length. The successive radii are a measure of the acceleration of growth, and the spiral curve of the shell itself, if the radius rotate uniformly, is no other than the *hodograph* of the growth of the contained organism*.

So far as we have now gone, we have studied the elementary properties of the equiangular spiral, including its fundamental property of *continued similarity*; and we have accordingly learned that the shell or the horn tends *necessarily* to assume the form of this mathematical figure, because in these structures growth proceeds by successive increments which are always similar in form, similarly situated, and of constant relative magnitude one to another. Our chief objects in enquiring further into the mathematical properties of the equiangular spiral will be: (1) to find means of confirming and verifying the fact that the shell (or other organic curve) is actually an equiangular spiral; (2) to learn how, by the properties of the curve, we may further extend our knowledge or simplify our descriptions of the shell; and (3) to understand the factors by which the characteristic form of any particular equiangular spiral is determined, and so to comprehend the nature of the specific or generic differences between one spiral shell and another.

Of the elementary properties of the equiangular spiral the following are those which we may most easily investigate in the concrete case of the molluscan shell: (1) that the polar radii whose vectorial angles are in arithmetical progression are themselves in geometrical progression; hence (2) that the vectorial angles are proportional to the *logarithms* of the corresponding radii; and (3) that the tangent at any point of an equiangular spiral makes a constant angle (called the *angle of the spiral*) with the polar radius vector.

* The hodograph of a logarithmic spiral (i.e. of a point which lies on a uniformly revolving radius and describes a logarithmic spiral) is likewise a logarithmic spiral: W. Walton, *Collection of Problems in Theoretical Mechanics* (3rd ed.), 1876, p. 296.

The first of these propositions may be written in a simpler form, as follows: radii which form equal angles about the pole of the equiangular spiral are themselves continued proportionals. That is to say, in Fig. 366, when the angle ROQ is equal to the angle QOP, then $OP : OQ :: OQ : OR$.

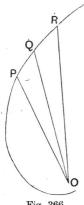

A particular case of this proposition is when the equal angles are each angles of $360°$: that is to say when in each case the radius vector makes a complete revolution, and when, therefore, P, Q and R all lie upon the same radius.

It was by observing with the help of very careful measurement this continued proportionality, that Moseley was enabled to verify his first assumption, based on the general appearance of the shell, that the shell of *Nautilus* was actually an equiangular spiral, and this demonstration he was soon after-wards in a position to generalise by extending it to all spiral Ammonitoid and Gastropod mollusca*. For, taking a median transverse section of a *Nautilus pompilius*, and carefully measuring the successive breadths of the whorls (from the dark line which marks what was originally the outer surface, before it was covered up by fresh deposits on the part of the growing and advancing shell), Moseley found that "the distance of any two of its whorls measured upon a radius vector is one-third that of the two next whorls measured upon the same radius vector†. Thus (in

Fig. 366.

---

* The Rev. H. Moseley, On the geometrical forms of turbinated and discoid shells, *Phil. Trans.* 1838, Pt. I, pp. 351–370. Réaumur, in describing the snail-shell (*Mém. Acad. des Sci.* 1709, p. 378), had a glimpse of the same geometrical law: "Le diamètre de chaque tour de spirale, ou sa plus grande longueur, est à peu près double de celui qui la précède et la moitié de celui qui la suit." Leslie (in his *Geometry of Curved Lines*, 1822, p. 438) compared the "general form and the elegant *septa* of the *Nautilus*" to an equiangular spiral and a series of its involutes.

† It will be observed that here Moseley, speaking as a mathematician and considering the *linear* spiral, speaks of *whorls* when he means the linear boundaries, or lines traced by the revolving radius vector; while the conchologist usually applies the term *whorl* to the whole space between the two boundaries. As con-chologists, therefore, we call the *breadth of a whorl* what Moseley looked upon as the *distance between two consecutive whorls*. But this latter nomenclature Moseley himself often uses. Observe also that Moseley gets a very good approximate result by his measurements "upon a radius vector," although he has to be content with a very rough determination of the pole.

Fig. 367), *ab* is one-third of *bc*, *de* of *ef*, *gh* of *hi*, and *kl* of *lm*. The curve is therefore an equiangular spiral."

The numerical ratio in the case of the *Nautilus* happens to be one of unusual simplicity. Let us take, with Moseley, a somewhat more complicated example.

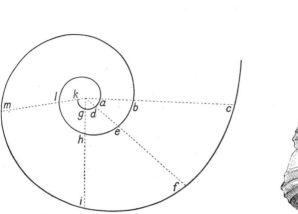

Fig. 367. Spiral of the *Nautilus*.

Fig. 368. *Turritella duplicata* (L.), Moseley's *Turbo duplicatus*. From Chenu. × ½.

From the apex of a large *Turritella* (*Turbo*) *duplicata*** a line was drawn across its whorls, and their widths were measured upon it in succession, beginning with the last but one. The measure-

---

* In the case of "*Turbo*", and all other turbinate shells, we are dealing not with a plane logarithmic spiral, as in *Nautilus*, but with a "gauche" spiral, such that the radius vector no longer revolves in a plane perpendicular to the axis of the system, but is inclined to that axis at some constant angle (β). The figure still preserves its continued similarity, and may be called a logarithmic spiral in space; indeed it is commonly spoken of as a logarithmic spiral *wrapped upon a cone*, its pole coinciding with the apex of the cone. It follows that the distances of successive whorls of the spiral measured on the same straight line passing through the apex of the cone are in geometrical progression, and conversely; just as in the former case. But the ratio between any two consecutive interspaces (i.e. $R_3 - R_2 / R_2 - R_1$) is now equal to $\epsilon^{2\pi \sin \beta \cot \alpha}$, β being the semi-angle of the enveloping cone. (Cf. Moseley, *Phil. Mag.* xxi, p. 300, 1842.)

ments were, as before, made with a fine pair of compasses and a diagonal scale. The sight was assisted by a magnifying glass. In a parallel column to the following admeasurements are the terms of a geometric progression, whose first term is the width of the widest whorl measured, and whose common ratio is 1·1804.

Turritella duplicata

Widths of successive whorls, measured in inches and parts of an inch	Terms of a geometrical progression, whose first term is the width of the widest whorl, and whose common ratio is 1·1804
1·31	1·310
1·12	1·110
0·94	0·940
0·80	0·797
0·67	0·675
0·57	0·572
0·48	0·484
0·41	0·410

The close coincidence between the observed and the calculated figures is very remarkable, and is amply sufficient to justify the conclusion that we are here dealing with a true logarithmic spiral*.

Nevertheless, in order to verify his conclusion still further, and to get partially rid of the inaccuracies due to successive small measurements, Moseley proceeded to investigate the same shell, measuring not single whorls but groups of whorls taken several at a time: making use of the following property of a geometrical progression, that "if μ represent the ratio of the sum of every even number (m) of its terms to the sum of half that number of terms, then the common ratio (r) of the series is represented by the formula

$$r = (\mu - 1)^{\frac{2}{m}}."$$

* Moseley, writing a hundred years ago, uses an obsolete nomenclature which is apt to be very misleading. His *Turbo duplicatus*, of Linnaeus, is now *Turritella duplicata*, the common large Indian *Turritella*, a slender, tapering shell with a very beautiful spiral, about six or seven inches long. But the operculum which he describes as that of *Turbo* does indeed belong to that genus, *sensu stricto*; it is the well-known calcareous operculum or "eyestone" of some such common species as *Turbo petholatus*. *Turritella* has a very different kind of operculum, a thin chitinous disc in the form of a close spiral coil, not nearly filling up the aperture of the shell. Moseley's *Turbo phasianus* is again no true *Turbo*, but is (to judge from his figure) *Phasianella bulimoides* Lam. = *P. australis* (Gmelin); and his *Buccinum subulatum* is *Terebra subulata* (L.).

Accordingly, Moseley made the following measurements, beginning from the second and third whorls respectively:

Width of		Ratio μ
Six whorls	Three whorls	
5·37	2·03	2·645
4·55	1·72	2·645
Four whorls	Two whorls	
4·15	1·74	2·385
3·52	1·47	2·394

"By the ratios of the two first admeasurements, the formula gives

$$r = (1\cdot645)^{\frac{1}{4}} = 1\cdot1804.$$

By the mean of the ratios deduced from the second two admeasurements, it gives

$$r = (1\cdot389)^{\frac{1}{2}} = 1\cdot1806.$$

"It is scarcely possible to imagine a more accurate verification than is deduced from these larger admeasurements, and we may with safety annex to the species *Turbo duplicatus* the characteristic number 1·18."

By similar and equally concordant observations, Moseley found for *Turbo phasianus* the characteristic ratio, 1·75; and for *Buccinum subulatum* that of 1·13.

From the measurements of *Turritella duplicata* (on p. 772), it is perhaps worth while to illustrate the logarithmic statement of the same thing: that is to say, the elementary fact, or corollary, that if the successive radii be in geometric progression, their logarithms will differ from one another by a constant amount.

Turritella duplicata

Widths of successive whorls	Logarithms of do.	Differences of logarithms	Ratios of successive widths
131	2·11727	—	—
112	2·04922	0·06805	1·170
94	1·97313	0·07609	1·191
80	1·90309	0·07004	1·175
67	1·82607	0·07702	1·194
57	1·75587	0·07020	1·175
48	1·68124	0·07463	1·188
41	1·61278	0·06846	1·171
		Mean 0·07207	1·1806

And 0·07207 is the logarithm of 1·1805.

Lastly, we may if we please, in this simple case, reduce the whole matter to arithmetic, and, dividing the width of each whorl by that of the next, see that these quotients are nearly identical, and that their mean value, or common ratio, is precisely that which we have already found.

We may shew, in the same simple fashion, by measurements of *Terebra* (Fig. 397), how the relative widths of successive whorls fall into a geometric progression, the criterion of a logarithmic spiral.

Measurements of a large specimen (15·5 *cm.*) *of* Terebra maculata, *along three several tangents* (a, b, c) *to the whorls.* (*After Chr. Peterson*, 1921.)

a		b		c	
Width (mm.)	Ratio	Width	Ratio	Width	Ratio
25		24·5		23	
	1·25		1·32		1·31
20		18·5		17·5	
	1·33		1·32		1·31
15		14		13·3	
	1·25		1·30		1·36
12		10·75		9·75	
	1·33		1·34		1·34
9		8		7·25	
Mean	1·29		1·32		1·33

Mean ratio, 1·31

The logarithmic spiral is not only very beautifully manifested in the molluscan shell[*], but also, in certain cases, in the little lid or "operculum" by which the entrance to the tubular shell is closed after the animal has withdrawn itself within[†]. In the spiral shell of *Turbo*, for instance, the operculum is a thick calcareous structure, with a beautifully curved outline, which grows by successive increments applied to one portion of its edge, and shews, accordingly, a spiral line of growth upon its surface. The successive increments leave their traces on the surface of the operculum (Fig. 370), which traces have the form of curved lines in *Turbo*, and of straight lines

---

[*] It has even been proposed to use a logarithmic spiral in place of a table of logarithms. Cf. Ant. Favaro, *Statique graphique*, Paris, 1885; Hele-Shaw, in *Brit. Ass. Rep.* 1892, p. 403.

[†] Cf. Fred. Haussay, *Recherches sur l'opercule*, Diss., Paris, 1884.

in (e.g.) *Nerita* (Fig. 371); that is to say, apart from the side constituting the outer edge of the operculum (which side is always and of necessity curved) the successive increments constitute curvilinear

Fig. 369.   Operculum of *Turbo*.

triangles in the one case, and rectilinear triangles in the other. The sides of these triangles are tangents to the spiral line of the

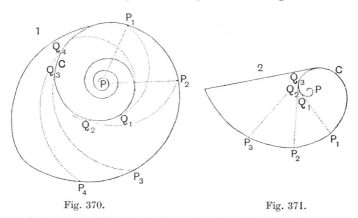

Fig. 370.   Fig. 371.

Figs. 370, 371.   Opercula of *Turbo* and *Nerita*.   After Moseley.

operculum, and may be supposed to generate it by their consecutive intersections.

In a number of such opercula, Moseley measured the breadths

of the successive whorls along a radius vector*, just in the same way as he did with the entire shell in the foregoing cases; and here is one example of his results.

Operculum of Turbo *sp.; breadth (in inches) of successive whorls, measured from the pole*

Distance	Ratio	Distance	Ratio	Distance	Ratio	Distance	Ratio
0·24		0·16		0·2		0·18	
	2·28		2·31		2·30		2·30
0·55		0·37		0·6		0·42	
	2·32		2·30		2·30		2·24
1·28		0·85		1·38		0·94	

The ratio is approximately constant, and this spiral also is, therefore, a logarithmic spiral.

But here comes in a very beautiful illustration of that property of the logarithmic spiral which causes its whole shape to remain unchanged, in spite of its apparently unsymmetrical, or unilateral, mode of growth. For the mouth of the tubular shell, into which the operculum has to fit, is growing or widening on all sides: while the operculum is increasing, not by additions made at the same time all round its margin, but by additions made only on one side of it at each successive stage. One edge of the operculum thus remains unaltered as it advances into its new position, and comes to occupy a new-formed section of the tube, similar to but greater than the last. Nevertheless, the two apposed structures, the chamber and its plug, at all times fit one another to perfection. The mechanical problem (by no means an easy one) is thus solved: "How to shape a tube of a variable section, so that a piston driven along it shall, by one side of its margin, coincide continually with its surface as it advances, provided only that the piston be made at the same time continually to revolve in its own plane."

As Moseley puts it: "That the same edge which fitted a portion of the first less section should be capable of adjustment, so as to fit a portion of the next similar but greater section, supposes a geometrical provision in the curved form of the chamber of great

---

* As the successive increments evidently constitute similar figures, similarly related to the pole (P), it follows that their linear dimensions are to one another as the radii vectores drawn to similar points in them: for instance as PP_1, PP_2, which (in Fig. 370) are radii vectores drawn to the points where they meet the common boundary.

apparent complication and difficulty. But God hath bestowed upon this humble architect the practical skill of a learned geometrician, and he makes this provision with admirable precision in that curvature of the logarithmic spiral which he gives to the section of the shell. This curvature obtaining, he has only to turn his operculum slightly round in its own plane as he advances it into each newly formed portion of his chamber, to adapt one margin of it to a new and larger surface and a different curvature, leaving the space to be filled up by increasing the operculum wholly on the other margin." The fact is that self-similar or gnomonic growth is taking place both in the shell and its operculum; in both of them growth is in reference to a fixed centre, and to a fixed axis through that centre; and in both of them growth proceeds in geometric progression from the centre while rotation takes place in arithmetic progression about the axis. The same architecture which builds the house constructs the door. Moreover, not only are house and door governed by the same law of growth, but, growing together, door and doorway adapt themselves to one another.

The operculum of the gastropods varies from a more or less close-wound spiral, as in *Turritella*, *Trochus* or *Pleurotomaria*, to cases in which accretion takes place, by concentric (or more or less excentric) rings, all round. But these latter cases, so Mr Winckworth tells me, are not very common. *Paludina* and *Ampullaria* come near to having a concentric operculum, and so do some of the Murices, such as *M. tribulus*, and a few Turrids, and the genus *Helicina*; but even these opercula probably begin as spirals, adding on their gnomonic increments at one end or side, and only growing on all sides later on. There would seem to be a truly concentric operculum in the *Siphonium* group of *Vermetus*, where the spiral of the shell itself is lost, or nearly so; but it is usually overgrown with Melobesia, and hard to see.

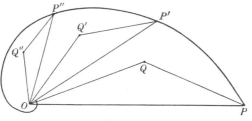

Fig. 372.

One more proposition, an all but self-evident one, we may make passing mention of here: If upon any polar radius vector *OP*, a triangle OPQ be drawn similar to a given triangle, the locus of the vertex Q will be a spiral similar to the original spiral. We may extend this proposition (as given by Whitworth) from the simple case of the triangle to any similar figures whatsoever; and see from it how every spot or ridge or tubercle repeated symmetrically from one radius vector (or one generating curve) to another becomes part of a spiral pattern on the shell.

Viewed in regard to its own fundamental properties and to those of its limiting cases, the equiangular spiral is one of the simplest of all known curves; and the rigid uniformity of the simple laws by which it is developed sufficiently account for its frequent manifestation in the structures built up by the slow and steady growth of organisms.

In order to translate into precise terms the whole form and growth of a spiral shell, we should have to employ a mathematical notation considerably more complicated than any that I have attempted to make use of in this book. But we may at least try to describe in elementary language the general method, and some of the variations, of the mathematical development of the shell. But here it is high time to observe that, while we have been speaking of the *shell* (which is a *surface*) as a logarithmic spiral (which is a *line*), we have been simplifying the case, in a provisional or preparatory way. The logarithmic spiral is but one factor in the case, albeit the chief or dominating one. The problem is one not of plane but of solid geometry, and the solid in question is described by the movement in space of a certain area, or closed curve*.

Let us imagine a closed curve in space, whether circular or elliptical or of some other and more complex specific form, not necessarily in a plane: such a curve as we see before us when we consider the mouth, or terminal orifice, of our tubular shell. Let

* For a more advanced study of the family of surfaces of which the *Nautilus* is a simple case, see M. Haton de la Goupillière (*op. cit.*). The turbinate shells represent a sub-family, which may be called that of the "surfaces cérithioides"; and "surfaces à front générateur" is a short title of the whole family. The form of the generating curve, its rate of expansion, the direction of its advance, and the angle which the generating front makes with the directrix, define, and give a wide extension to, the family. These parameters are all severally to be recognised in the growth of the living object; and they make of a collection of shells an unusually beautiful materialisation of the rigorous definitions of geometry.

the shell. And in not a few cases, as in *Harpa, Dolium perdix*, etc., both alike are conspicuous, ridges and colour-bands intersecting one another in a beautiful isogonal system.

In ordinary gastropods the shell is formed at or near the mantle-edge. Here, near the mantle-border, is a groove lined with a secretory epithelium which produces the horny cuticle or perio stracum of the shell*. A narrow zone of the mantle just behind this secretes lime abundantly, depositing it in a layer below the periostracum; and for some little way back more lime may be secreted, and pigment superadded from appropriate glands. Growth and secretion are periodic rather than continuous. Even in a snail-shell it is easy to see how the shell is built up of narrow annular increments; and many other shells record, in conspicuous colour-patterns, the alternate periods of rest and of activity which their pigment-glands have undergone.

The periodic accelerations and retardations in the growth of a shell are marked in various ways. Often we have nothing more than an increased activity from time to time at or near the mantle-edge—enough to give rise to slight successive ridges, each corresponding to a "generating curve" in the conformation of the shell. But in many other cases, as in *Murex, Ranella* and the like, the mantle-edge has its alternate phases of rest and of turgescence, its outline being plain and even in the one and folded and contorted in the other; and these recurring folds or pleatings of the edge leave their impress in the form of various ridges, ruffles or comb-like rows of spines upon the shell†.

In not a few cases the colour-pattern shews, or seems to shew, how some play of forces has fashioned and transformed the first elementary pattern of pigmentary drops or jets. As the book-binder drops or dusts a little colour on a viscous fluid, and then produces the beautiful streamlines of his marbled papers by stirring

---

* That the shell grows by accretion at the mantle-edge was one of Réaumur's countless discoveries (*Mém. Acad. Roy. des Sc.* 1709, p. 364 *seq.*). It follows that the mathematical "generating curves," as Moseley chose them, correspond to the material increments of the shell.

† The periodic appearance of a ridge, or row of tubercles, or other ornament on the growing shell is illustrated or even exaggerated in the delicate "combs" of *Murex aculeatus*. Here normal growth is interrupted for the time being, the mantle-edge is temporarily folded and reflexed, and shell-substance is poured out into the folds.

and combing the colloid mass, so we may see, in the harp-shells or the volutes, how a few simple spots or lines have been drawn out into analogous wavy patterns by streaming movements during the formation of the shell.

In the complete mathematical formula for any given turbinate shell, we may include, with Moseley, factors for the following elements: (1) for the specific form of a section of the tube, or (as we have called it) the generating curve; (2) for the specific rate of growth of this generating curve; (3) for its inclination to the directrix, or to the axis; (4) for its specific rate of angular rotation about the pole, in a projection perpendicular to the axis; and (5) in turbinate (as opposed to nautiloid) shells, for its rate of screw-translation, parallel to the axis, as measured by the angle between a tangent to the whorls and the axis of the shell*. It seems a complicated affair; but it is only a pathway winding at a steady slope up a conical hill. This uniform gradient is traced by any given point on the generating curve while the vector angle increases in arithmetical progression, and the scale changes in geometrical progression; and a certain *ensemble*, or bunch, of these spiral curves in space constitutes the self-similar surface of the shell.

But after all this is not the only way, neither is it the easiest way, to approach our problem of the turbinate shell. The conchologist turned mathematician is apt to think of the generating curve by which the spiral surface is described as necessarily identical, or coincident, with the mouth or lip of the shell; for this is where growth actually goes on, and where the successive increments of shell-growth are visibly accumulated. But it does not follow that this particular generating curve is chosen for the best from the mathematical point of view; and the mathematician, unconcerned with the physiological side of the case and regardless of the succession of the parts in time, is free to choose any other generating curve which the geometry of the figure may suggest to him. We are following Moseley's example (as is usually done) when we think of no other generating curve but that which takes the form of a

---

* Note that this tangent touches the curve at a series of points, whorl by whorl, instead of at one only. Observe also that we may have various tangent-cones, all centred on the apex of the shell. In an open spiral, like a ram's horn, or a half-open spiral like the shell *Solarium*, we have two cones, one touching the outside, the other the inside of the shell.

frontal plane, outlined by the lip, and *sliding along* the axis while revolving round it; but the geometer takes a better and a simpler way. For, when of two similar figures in space one is derived from the other by a screw-displacement accompanied by change of scale— as in the case of a big whelk and a little whelk—there is a unique (apical) point which suffers no displacement; and if we choose for our generating curve a sectional figure centred on the apical point and passing through the axis of rotation, the whole development of the surface may be simply described as due to a rotation of this generating figure about the axis (z), together with a change of scale with the point 0 as centre of similitude. We need not, and now must not, think of a *slide* or *shear* as part of the operation; the translation along the axis is merely part and parcel of the *magnification* of the new generating curve. It follows that angular rotation in arithmetical progression, combined with change of scale (from 0) in geometrical progression, causes any arbitrary point on the generating curve to trace a path of uniform gradient round a circular cone, or in other words to describe a helico-spiral or gauche equiangular spiral in space. The spiral curve cuts all the straight-line generators of the cone at the same angle; and it further follows that the successive increments are, and the whole figure constantly remains, "self-similar"*.

Apart from the specific form of the generating curve, it is the ratios which happen to exist between the various factors, the ratio for instance between the growth-factor and the rate of angular revolution, which give the endless possibilities of permutation of form. For example, a certain rate of growth in the generating curve, together with a certain rate of vectorial rotation, will give us a spiral shell of which each successive whorl will just touch its predecessor and no more; with a slower growth-factor the whorls will stand asunder, as in a ram's horn; with a quicker growth-factor

---

* The equation to the surface of a turbinate shell is discussed by Moseley both in terms of polar and of rectangular coordinates, and the method of polar coordinates is used also by Haton de la Goupillière; but both accounts are subject to mathematical objection. Dr G. T. Bennett, choosing his generating curve (as described above) in the axial plane from which the vertical angles are measured (the plane $\theta = 0$), would state his equation in cylindrical coordinates, $f(za^\theta, ra^\theta) = 0$: that is to say in terms of z, conjointly with ordinary plane cylindrical coordinates.

each will cut or intersect its predecessor, as in an Ammonite or the majority of gastropods, and so on.

A similar relation of velocities suffices to determine the apical angle of the resulting cone, and give us the difference, for example, between the sharp, pointed cone of *Turritella*, the less acute one of *Fusus* or *Buccinum*, and the obtuse one of *Harpa* or of *Dolium*. In short it is obvious that *all* the differences of form which we observe between one shell and another are referable to matters of *degree*, depending, one and all, upon the relative magnitudes of the various factors in the complex equation to the curve. This is an immensely important thing. To learn that all the multitudinous shapes of shells, in their all but infinite variety, may be reduced to the variant properties of a single simple curve, is a great achievement. It exemplifies very beautifully what Bacon meant in saying that the forms or differences of things are simple and few, and the degrees and coordinations of these make all their variety*. And after such a fashion as this John Goodsir imagined that the naturalist of the future would determine and classify his shells, so that conchology should presently become, like mineralogy, a mathematical science†.

The paper in which, more than a hundred years ago, Canon Moseley‡ gave a simple mathematical account, on lines like these, of the spiral forms of univalve shells, is one of the classics of Natural History. But other students before, and sometimes long before, him had begun to recognise the same simplicity of form and structure. About the year 1818 Reinecke had declared *Nautilus* to be a well-defined geometrical figure, whose chambers followed

* For a discussion of this idea, and of the views of Bacon and of J. S. Mill, see J. M. Keynes, *op. cit.* p. 271.

† On the employment of mathematical modes of investigation in the determination of organic forms; in *Anatomical Memoirs*, II, p. 205, 1868 (posthumous publication).

‡ The Rev. Henry Moseley (1801–1872), of St John's College, Cambridge, Canon of Bristol, Professor of Natural Philosophy in King's College, London, was a man of great and versatile ability. He was father of H. N. Moseley, naturalist on board the *Challenger* and Professor of Zoology in Oxford; and he was grandfather of H. G. J. Moseley (1887–1915)—Moseley of the Moseley numbers—whose death at Gallipoli, long ere his prime, was one of the major tragedies of the Four Years War.

one another in a constant ratio or continued proportion*; and Leopold von Buch and others accepted and even developed the idea.

Long before, Swammerdam had grasped with a deeper insight the root of the whole matter; for, taking a few diverse examples, such as *Helix* and *Spirula*, he shewed that they and all other spiral shells whatsoever were referable to one common type, namely to that of a simple tube, variously curved according to definite mathematical laws; that all manner of ornamentation, in the way of spines, tuberosities, colour-bands and so forth, might be superposed upon them, but the type was one throughout and specific differences were of a geometrical kind. "Omnis enim quae inter eas animadvertitur differentia ex sola nascitur diversitate gyrationum: quibus si insuper externa quaedam adjunguntur ornamenta pinnarum, sinuum, anfractuum, planitierum, eminentiarum, profunditatum, extensionum, impressionum, circumvolutionum, colorumque: ...tunc deinceps facile est, quarumcumque Cochlearum figuras geometricas, curvosque, obliquos atque rectos angulos, ad unicam omnes speciem redigere: ad oblongum videlicet tubulum, qui vario modo curvatus, crispatus, extrorsum et introrsum flexus, ita concrevit†."

Nay more, we may go back yet another hundred years and find Sir Christopher Wren contemplating the architecture of a snail-shell, and finding in it the logarithmic spiral. For Wallis‡, after defining and describing this curve with great care and simplicity, tells us that Wren not only conceived the spiral shell to be a sort of cone or pyramid coiled round a vertical axis, but also saw that on the magnitude of *the angle of the spire* depended the specific form of the shell: "Hanc ipsam curvam...contemplatus est Wrennius noster. Nec tantum curvae longitudinem, partiumque ipsius, et

* J. C. M. Reinecke, *Maris protogaei Nautilos,* etc., Coburg, 1818, p. 17: "In eius forma, quae canalis spiram convoluti formam et proportiones simul subministrat, totius testae forma quoddammodo data est. Restaret solum scire, quota cujusque anfractus pars sequenti inclusa sit, ut testam geometrice construere possimus." Cf. Leopold von Buch, Ueber die Ammoniten in den älteren Gebirgsschichten, *Abh. Berlin. Akad., Phys. Kl.* 1830, pp. 135–158; *Ann. Sc. Nat.* xxviii, pp. 5–43, 1833; cf. Elie de Beaumont, Sur l'enroulement des Ammonites, *Soc. Philom., Pr. verb.* 1841, pp. 45–48.

† *Biblia Naturae sive Historia Insectorum,* Leydae, 1737, p. 152.

‡ Joh. Wallis, *Tractatus duo, de Cycloide,* etc., Oxon., 1659, pp. 107, 108.

magnitudinem adjacentis plani; sed et, ipsius ope, Limacum et Conchiliorum domunculos metitur. Existimat utique, magna veri-similitudine, domunculos hosce non alios esse quam Pyramides convolutas: quarum Axis sit, istiusmodo Spiralis: non quidem in plano jacens, sed sensim in convolutione (circa erectum axim) assurgens: pro variis autem curvae, sive ad rectam circumductam sive ad subjacens planum, angulis, variae Conchiliorum formae enascantur. Atque hac hypothesi, mensurata Pyramide, metitur etiam ea conchiliorum spatia."

For some years after the appearance of Moseley's paper, a number of writers followed in his footsteps, and attempted in various ways to put his conclusions to practical use. For instance, d'Orbigny

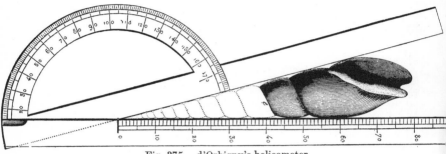

Fig. 375.   d'Orbigny's helicometer.

devised a very simple protractor, which he called a Helicometer*, and which is represented in Fig. 375. By means of this little instrument the apical angle of the turbinate shell was immediately read off, and could then be used as a specific and diagnostic character. By keeping one limb of the protractor parallel to the side of the cone while the other was brought into line with the suture between two adjacent whorls, another specific angle, the "sutural angle," could in like manner be recorded. And, by the linear scale upon the instrument, the relative breadths of the consecutive whorls, and that of the terminal chamber to the rest of the shell, might

* Alcide d'Orbigny, *Bull. de la soc. géol. Fr.* XIII, p. 200, 1842; *Cours élém. de Paléontologie*, II, p. 5, 1851. A somewhat similar instrument was described by Boubée, in *Bull. soc. géol.* I, p. 232, 1831. Naumann's conchyliometer (*Poggend. Ann.* LIV, p. 544, 1845) was an application of the screw-micrometer; it was provided also with a rotating stage for angular measurement. It was adapted for the study of a discoid or ammonitoid shell, while d'Orbigny's instrument was meant for the study of a turbinate shell.

also, though somewhat roughly, be determined. For instance, in *Terebra dimidiata* the apical angle was found to be 13°, the sutural angle 109°, and so forth.

It was at once obvious that, in such a shell as is represented in Figs. 369 and 375 the entire outline (always excepting that of the immediate neighbourhood of the mouth) could be restored from a broken fragment. For if we draw our tangents to the cone, it follows from the symmetry of the figure that we can continue the projection of the sutural line, and so mark off the successive whorls, by simply drawing a series of consecutive parallels, and by then filling into the quadrilaterals so marked off a series of curves similar to one another, and to the whorls which are still intact in the broken shell. But the use of the helicometer soon shewed that it was by no means universally the case that one and the same cone was tangent to all the turbinate whorls; in other words, there was not always one specific apical angle which held good for the entire system. In the great majority of cases, it is true, the same tangent touches all the whorls, and is a straight line. But in others, as in the large *Cerithium nodosum*, such a line is slightly concave to the axis of the shell; and in the short spire of *Dolium*, for instance, the concavity is marked, and the apex of the spire is a distinct cusp. On the other hand, in *Pupa* and *Clausilia* the common tangent is convex to the axis of the shell.

So also is it, as we shall presently see, among the Ammonites: where there are some species in which the ratio of whorl to whorl remains, to all appearance, perfectly constant; others in which it gradually though only slightly increases; and others again in which it slightly and gradually falls away. It is obvious that, among the manifold possibilities of growth, such conditions as these are very easily conceivable. It is much more remarkable that, among these shells, the relative velocities of growth in various dimensions should be as constant as they are than that there should be an occasional departure from perfect regularity. In these latter cases the logarithmic law of growth is only approximately true. The shell is no longer to be represented simply as a cone which has been rolled up, but as a cone which (while rolling up) had grown trumpet-shaped, or conversely whose mouth had narrowed in, and which in longitudinal section is a curvilinear instead of a rectilinear triangle. But all that has happened is that a new factor, usually of small or all but imperceptible magnitude, has been introduced into the case; so that the ratio, $\log r = \theta \log \alpha$, is no longer constant but varies slightly, and in accordance with some simple law.

Some writers, such as Naumann[*] and Grabau, maintained that the molluscan spiral was no true logarithmic spiral, but differed from it specifically, and they gave it the name of *Conchospiral*. They said that the logarithmic spiral originates in a mathematical point, while the molluscan shell starts with a little embryonic shell, or central chamber (the "protoconch" of the conchologists), around which the spiral is subsequently wrapped. But this need not affect the logarithmic law of the shell as a whole; indeed we have already allowed for it by writing our equation in the form $r = ma^\theta$. And Grabau[†], while he clung to Naumann's conchospiral against Moseley's logarithmic spiral, confessed that they were so much alike that ordinary measurements would seldom shew a difference between them.

There would seem, by the way, to be considerable confusion in the books with regard to the so-called "protoconch." In many cases it is a definite structure, of simple form, representing the more or less globular embryonic shell before it began to elongate into its conical or spiral form. But in many cases what is described as the "protoconch" is merely an empty space in the middle of the spiral coil, resulting from the fact that the actual spiral shell must have some magnitude to begin with, and that we cannot follow it down to its vanishing point in infinity. For instance, in the accompanying figure, the large space a is styled the protoconch, but it is the little bulbous or hemispherical chamber within it, at the end of the spire, which is the real beginning of the tubular shell. The form and magnitude of the space a are determined by the "angle of retardation," or ratio of rate of growth between the inner and outer curves of the spiral shell. They

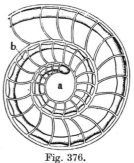

Fig. 376.

---

[*] C. F. Naumann, Beitrag zur Konchyliometrie, *Poggend. Ann.* L, p. 223, 1840; Ueber die Spiralen der Ammoniten, *ibid.* LI, p. 245, 1840; *ibid.* LIV, p. 541, 1845; etc. (See also p. 755.) Cf. also Lehmann, *Die von Seyfriedsche Konchyliensammlung und das Windungsgesetz von einigen Planorben*, Constanz, 1855.

[†] A. H. Grabau, Ueber die Naumannsche Conchospirale, und ihre Bedeutung für die Conchyliometrie, *Inauguraldiss.*, Leipzig, 1872; Ueber die Spiralen der Conchylien, etc., *Leipzig Progr.* No. 502, 1880; cf. *Sb. naturf. Gesellsch.* Leipzig, 1881, pp. 23–32.

are independent of the shape and size of the embryo, and depend only (as we shall see better presently) on the direction and relative rate of growth of the double contour of the shell*.

Now that we have dealt, in a general way, with some of the more obvious properties of the equiangular or logarithmic spiral, let us consider certain of them a little more particularly, keeping in view as our chief object of study the range of variation of the molluscan shell.

There is yet another equation to the logarithmic spiral, very commonly employed, and without the help of which we cannot get far. It is as follows: $r = e^{\theta \cot \alpha}$.

This follows directly from the fact that the angle α (the angle between the radius vector and the tangent to the curve) is constant.

For then,

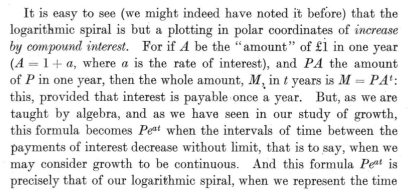

$$\tan \alpha \; (= \tan \phi) = r \, d\theta/dr;$$

therefore $\qquad dr/r = d\theta \cot \alpha,$

and, integrating, $\log r = \theta \cot \alpha,$

or $\qquad r = e^{\theta \cot \alpha}.$

Fig. 377.

It is easy to see (we might indeed have noted it before) that the logarithmic spiral is but a plotting in polar coordinates of *increase by compound interest*. For if A be the "amount" of £1 in one year ($A = 1 + a$, where a is the rate of interest), and PA the amount of P in one year, then the whole amount, M, in t years is $M = PA^t$: this, provided that interest is payable once a year. But, as we are taught by algebra, and as we have seen in our study of growth, this formula becomes Pe^{at} when the intervals of time between the payments of interest decrease without limit, that is to say, when we may consider growth to be continuous. And this formula Pe^{at} is precisely that of our logarithmic spiral, when we represent the time by a vector angle θ, and when for a, the particular rate of interest in the case, we write cot α, the constant measure of growth of the particular spiral.

As we have seen throughout our preliminary discussion, the two most important constants (or "specific characters," as the naturalist would say) in an equiangular or logarithmic spiral are (1) the magnitude of the angle of the spiral, or "constant angle" α, and (2) the rate of increase of the radius vector for any given angle of revolution, θ. But our two magnitudes, that of the constant angle and that of the ratio of the radii or breadths of whorl, are directly related to one another, so that we may determine either of them by measurement and calculate the other.

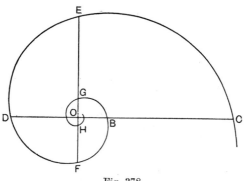

Fig. 378.

In any complete spiral, such as that of *Nautilus*, it is (as we have seen) easy to measure any two radii (r), or the breadths in a radial direction of any two whorls (W). We have then merely to apply the formula

$$\frac{r_{n+1}}{r_n} = e^{\theta \cot \alpha}, \text{ or } \frac{W_{n+1}}{W_n} = e^{\theta \cot \alpha},$$

which we may simply write $r = e^{\theta \cot \alpha}$, etc., when one radius or whorl is regarded, for the purpose of comparison, as equal to unity.

Thus, in Fig. 378, OC/OE, or EF/BD, or DC/EF, being in each case radii, or diameters, at right angles to one another, are all equal to $e^{\frac{\pi}{2} \cot \alpha}$. While in like manner, EO/OF, EG/FH, or GO/HO, all equal $e^{\pi \cot \alpha}$; and BC/BA, or $CO/OB = e^{2\pi \cot \alpha}$.

As soon, then, as we have prepared tables for these values, the determination of the constant angle α in a particular shell becomes a very simple matter.

A complete table would be cumbrous, and it will be sufficient to deal with the simple case of the ratio between the breadths of adjacent, or immediately succeeding, whorls.

Here we have $r = e^{2\pi \cot \alpha}$, or $\log r = \log e \times 2\pi \times \cot \alpha$, from which we obtain the following figures*:

The shape of a nautiloid spiral

Ratio of breadth of each whorl to the next preceding $r/1$	Constant angle α
1·1	89° 8′
1·25	87 58
1·5	86 18
2·0	83 42
2·5	81 42
3·0	80 5
3·5	78 43
4·0	77 34
4·5	76 32
5·0	75 38
10·0	69 53
20·0	64 31
50·0	58 5
100·0	53 46
1000·0	42 17
10,000	34 19
100,000	28 37
1,000,000	24 28
10,000,000	21 18
100,000,000	18 50
1,000,000,000	16 52

We learn several interesting things from this short table. We see, in the first place, that where each whorl is about three times the breadth of its neighbour and predecessor, as is the case in *Nautilus*, the constant angle is in the neighbourhood of 80°; and hence also that, in all the ordinary ammonitoid shells, and in all the typically spiral shells of the gastropods†, the constant angle is also a large one, being very seldom less than 80°, and usually between 80° and 85°. In the next place, we see that with smaller

* It is obvious that the ratios of opposite whorls, or of radii 180° apart, are represented by the square roots of these values; and the ratios of whorls or radii 90° apart, by the square roots of these again.

† For the correction to be applied in the case of the helicoid, or "turbinate" shells, see p. 816.

angles the apparent form of the spiral is greatly altered, and the
very fact of its being a spiral soon ceases to be apparent (Figs. 379,
380). Suppose one whorl to be an inch in breadth, then, if the
angle of the spiral were 80°, the next whorl would (as we have just
seen) be about three inches broad; if it were 70°, the next whorl
would be nearly ten inches, and if it were 60°, the next whorl would
be nearly four feet broad. If the angle were 28°, the next whorl
would be a mile and a half in breadth; and if it were 17°, the next
would be some 15,000 miles broad.

In other words, the spiral shells of gentle curvature, or of small
constant angle, such as *Dentalium* or *Cristellaria*, are true equiangular spirals, just as are those of *Nautilus* or *Rotalia*: from

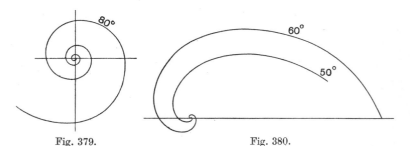

Fig. 379.   Fig. 380.

which they differ only in degree, in the magnitude of an angular
constant. But this diminished magnitude of the angle causes the
spiral to dilate with such immense rapidity that, so to speak,
it never comes round; and so, in such a shell as *Dentalium*, we
never see but a small portion of a single whorl.

We might perhaps be inclined to suppose that, in such a shell as *Dentalium*,
the lack of a visible spiral convolution was only due to our seeing but a small
portion of the curve, at a distance from the pole, and when, therefore, its
curvature had already greatly diminished. That is to say we might suppose
that, however small the angle α, and however rapidly the whorls accordingly
increased, there would nevertheless be a manifest spiral convolution in the
immediate neighbourhood of the pole, as the starting point of the curve.
But it is easy to see that it is not so. It is not that there cease to be convolutions of the spiral round the pole when α is a small angle; on the contrary,
there are infinitely many, mathematically speaking. But as α diminishes,
and cot α increases towards infinity, the ratio between the breadth of one
whorl and the next increases very rapidly. Our table shews us that even
when α is no less than 40°, and our shell still looks strongly curved, one whorl
is a thousandth part of the breadth of the next, and a thousandfold that

of the one before; we cannot expect to see either of them under the materialised conditions of the actual shell. Our shells of small constant angle and gentle curvature, such as *Dentalium*, are accordingly as much as we can ever expect to see of their respective spirals.

The spiral whose constant angle is 45° is both a simple case and a mathematical curiosity; for, since the tangent of 45° is unity, we need merely write $r = e^\theta$; which is as much as to say that the natural logarithms of the radii give us, without more ado, the vector angles. In this spiral the ratio between the breadths of two consecutive whorls becomes $r = e^{2\pi} = e^{2\times 3\cdot 1416}$. Reducing this from Naperian to common logs, we have $\log r = 2\cdot 729$; which tells us (by our tables) that the radius vector is multiplied about $535\frac{1}{2}$ times after a whole polar revolution; it is doubled after turning through a polar angle of less than 40°. Spirals of so low an angle as 45° are common enough in tooth and claw, but rare among molluscan shells; but one or two of the more strongly curved Dentaliums, like *D. elephantinum*, come near the mark. It is not easy to determine the pole, nor to measure the constant angle, in forms like these.

Let us return to the problem of how to ascertain, by direct measurement, the spiral angle of any particular shell. The method already employed is only applicable to complete spirals, that is to say to those in which the angle of the spiral is large, and furthermore it is inapplicable to portions, or broken fragments, of a shell. In the case of the broken fragment, it is plain that the determination of the angle is not merely of theoretic interest, but may be of great practical use to the conchologist as the one and only way by which he may restore the outline of the missing portions. We have a considerable choice of methods, which have been summarised by, and are partly due to, a very careful student of the Cephalopoda, the late Rev. J. F. Blake*.

(1) When an equiangular spiral rolls on a straight line, the pole traces another straight line at an angle to the first equal to the complement of the constant angle of the spiral; for the contact point is the instantaneous centre of the rotational movement, and the line joining it to the pole of the spiral is normal to the roulette path of that point. But the difficulty of determining the pole

---

* On the measurement of the curves formed by Cephalopods and other Mollusks, *Phil. Mag.* (5), VI, pp. 241–263, 1878.

(which is indeed asymptotic) makes this of little use as a method of determining the constant angle. It is, however, a beautiful property of the curve, and all the more interesting that Clerk Maxwell discovered it when he was a boy*.

(2) The following method is useful and easy when we have a portion of a single whorl, such as to shew both its inner and its outer edge. A broken whorl of an Ammonite, a curved shell such as *Dentalium*, or a horn of similar form to the latter, will fall under this head. We have merely to draw a tangent,
GEH, to the outer whorl at any point *E*; then draw to the inner whorl a tangent parallel to *GEH*, touching the curve in some point *F*. The straight line joining the points of contact, *EF*, must evidently pass through the pole: and, accordingly, the angl *GEF* is the angle required. In shells which bear *longitudinal* striae or other ornaments, any pair of these will suffice for our purpose, instead of the actual boundaries of the whorl. But it is obvious that this method will be apt to fail us when the angle α is very small; and when, consequently, the points *E* and *F* are very remote.

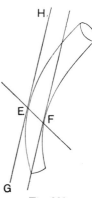

Fig. 381.

(3) In shells (or horns) shewing rings or other *transverse* ornamentation, we may take it that these ornaments are set at

Fig. 382.   An Ammonite, to shew corrugated surface-pattern.

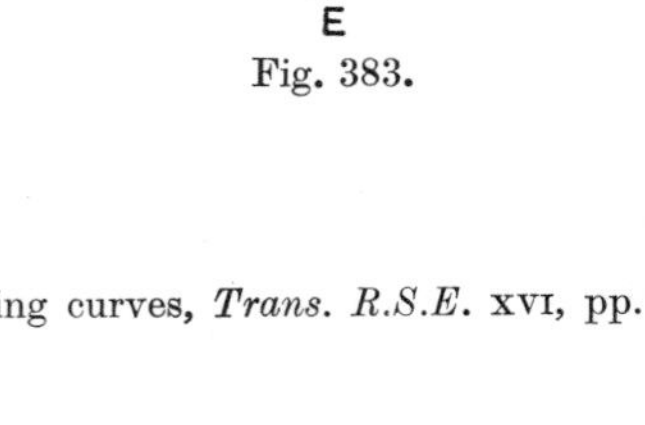

Fig. 383.

* Clerk Maxwell, On the theory of rolling curves, *Trans. R.S.E.* xvi, pp. 519–540, 1849; *Sci. Papers*, i, pp. 4–29.

a constant angle to the spire, and therefore to the radii. The angle (θ) between two of them, as AC, BD, is therefore equal to the angle θ between the polar radii from A and B, or from C and D; and therefore $BD/AC = e^{\theta \cot \alpha}$, which gives us the angle α in terms of known quantities.

(4) If only the outer edge be available, we have the ordinary geometrical problem—given an arc of an equiangular spiral, to find its pole and spiral angle. The methods we may employ depend (i) on determining directly the position of the pole, and (ii) on determining the radius of curvature.

The first method is theoretically simple, but difficult in practice; for it requires great accuracy in determining the points. Let AD, DB be two tangents drawn to the curve. Then a circle drawn through the points A, B, D will pass through the pole O, since the angles OAD, OBE (the supplement of OBD) are equal. The point O may be determined by the intersection of two such circles; and the angle DBO is then the angle, α, required.

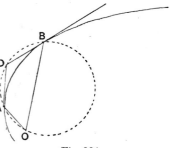

Fig. 384.

Or we may determine graphically, at two points, the radii of curvature $\rho_1 \rho_2$. Then, if s be the length of the arc between them (which may be determined with fair accuracy by rolling the margin of the shell along a ruler),

$$\cot \alpha = (\rho_1 - \rho_2)/s.$$

The following method*, given by Blake, will save actual determination of the radii of curvature.

Measure along a tangent to the curve the distance, AC, at which a certain small offset, CD, is made by the curve; and from another point B, measure the distance at which the curve makes an equal offset. Then, calling the offset μ; the arc AB, s; and AC, BE, respectively x_1, x_2, we have

$$\rho_1 = \frac{x_1^2 + \mu^2}{2\mu}, \text{ approximately,}$$

and
$$\cot \alpha = \frac{x_2^2 - x_1^2}{2\mu s}.$$

Of all these methods by which the mathematical constants, or specific characters, of a given spiral shell may be determined, the

* For an example of this method, see Blake, *loc. cit.* p. 251.

only one of which much use has been made is that which Moseley first employed, namely, the simple method of determining the relative breadths of the whorl at distances separated by some convenient vectorial angle such as 90°, 180°, or 360°. Very elaborate measurements of a number of Ammonites have been made by Naumann*, by Grabau, by Sandberger†, and by Müller, among which we may choose a couple of cases for consideration‡. In the following table I have taken a portion of Grabau's

Ammonites intuslabiatus

Breadth of whorls (180° apart)	Ratio of breadth of successive whorls (360° apart)	The angle (α) as calculated
0·30 mm.	—	— —
0·30	1·333	87° 23′
0·40	1·500	86 19
0·45	1·500	86 19
0·60	1·444	86 39
0·65	1·417	86 49
0·85	1·692	85 13
1·10	1·588	85 47
1·35	1·545	86 2
1·70	1·630	85 33
2·20	1·441	86 40
2·45	1·432	86 43
3·15	1·735	85 0
4·25	1·683	85 16
5·30	1·482	86 25
6·30	1·519	86 12
8·05	1·635	85 32
10·30	1·416	86 50
11·40	1·252	87 57
12·90	—	— —
	Mean	86° 15′

* C. F. Naumann, Ueber die Spiralen von Conchylien, *Abh. k. sächs. Ges.* 1846, pp. 153–196; Ueber die cyclocentrische Conchospirale u. über das Windungsgesetz von *Planorbis corneus, ibid.* I, pp. 171–195, 1849; Spirale von Nautilus u. *Ammonites galeatus, Ber. k. sächs. Ges.* II, p. 26, 1848; Spirale von *Amm. Ramsaueri, ibid.* XVI, p. 21, 1864. Oken, reviewing Naumann's work (in *Isis*, 1847, p. 867) foretold how some day the naturalist and the mathematician would each learn of the other: "Um die Sache zu Vollendung zu bringen wird der Mathematiker Zoolog und Physiolog, und diese Mathematiker werden müssen."

† G. Sandberger, *Clymenia subnautilina, Jahresber. d. Ver. f. Naturk. im Herzogth. Nassau*, 1855, p. 127; Spiralen des *Ammonites Amaltheus, A. Gaytani* und *Goniatites intumescens, Ztschr. d. d. Geolog. Gesellsch.* X, pp. 446–449, 1858. Also Müller, Beitrag zur Konchyliometrie, *Poggend. Ann.* LXXXVI, p. 533, 1850; *ibid.* XC, p. 323, 1853. These two authors upheld the logarithmic law against Naumann and Grabau.

‡ See also Chr. Petersen, *Das Quotientengesetz, eine biologisch-statistische Untersuchung*, 119 pp., Copenhagen, 1921; E. Sporn, Ueber die Gesetzmässigkeit im Baue der Muschelgehaüser, *Arch. f. Entw. Mech.* CVIII, pp. 228–242, 1926.

determinations of the breadth of the whorls in *Ammonites* (*Arcestes*) *intuslabiatus*; these measurements Grabau gives for every 45° of arc, but I have only set forth successive whorls measured along one diameter on both sides of the pole. The ratio between *alternate* measurements is therefore the same ratio as Moseley adopted, namely the ratio of breadth between *contiguous whorls* along a radius vector. I have then added to these observed values the corresponding calculated values of the angle α, as obtained from our usual formula.

There is considerable irregularity in the ratios derived from these measurements, but it will be seen that this irregularity only implies a variation of the angle of the spiral between about 85° and 87°; and the values fluctuate pretty regularly about the mean, which is 86° 15'. Considering the difficulty of measuring the whorls, especially towards the centre, and in particular the difficulty of determining with precise accuracy the position of the pole, it is clear that in such a case as this we are not justified in asserting that the law of the equiangular spiral is departed from.

Ammonites tornatus

Breadth of whorls (180° apart)	Ratio of breadth of successive whorls (360° apart)	The spiral angle (α) as calculated
0·25 mm.	—	—
0·30	1·400	86° 56
0·35	1·667	85 21
0·50	2·000	83 42
0·70	2·000	83 42
1·00	2·000	83 42
1·40	2·100	83 16
2·10	2·179	82 56
3·05	2·238	82 42
4·70	2·492	81 44
7·60	2·574	81 27
12·10	2·546	81 33
19·35	—	— —
	Mean 2·11	83° 22'

In some cases, however, it is undoubtedly departed from. Here for instance is another table from Grabau, shewing the corresponding ratios in an Ammonite of the group of *Arcestes tornatus*. In this case we see a distinct tendency of the ratios to increase as we pass from the centre of the coil outwards, and consequently for the values of the angle α to diminish. The case is comparable to

that of a cone with slightly curving sides: in which, that is to say, there is a slight acceleration of growth in a transverse as compared with the longitudinal direction.

In a tubular spiral, whether plane or helicoid, the consecutive whorls may either be (1) isolated and remote from one another; or (2) they may precisely meet, so that the outer border of one and the inner border of the next just coincide; or (3) they may overlap, the vector plane of each outer whorl cutting that of its immediate predecessor or predecessors.

Looking, as we have done, upon the spiral shell as being essentially a cone rolled up*, it is plain that, for a given spiral angle, intersection or non-intersection of the successive whorls will depend upon *the apical angle* of the original cone. For the wider the cone, the more will its inner border tend to encroach on the preceding whorl. But it is also plain that the greater the apical angle of the cone, and the broader, consequently, the cone itself, the greater difference will there be between the total *lengths* of its inner and outer borders. And, since the inner and outer borders are describing precisely the same spiral about the pole, we may consider the inner border as being *retarded* in growth as compared with the outer, and as being always identical with a smaller and earlier part of the latter.

If λ be the ratio of growth between the outer and the inner curve, then, the outer curve being represented by

$$r = ae^{\theta \cot \alpha},$$

the equation to the inner one will be

$$r' = a\lambda e^{\theta \cot \alpha},$$

or
$$r' = ae^{(\theta - \gamma) \cot \alpha},$$

* To speak of a cone "rolling up," and becoming a nautiloid spiral by doing so, is a rough and non-mathematical description; nor is it easy to see how a cone of wide angle could roll up, and yet remain a cone. But if (i) the centre of a sphere move along a straight line and its radius keep proportional to the distance the centre has moved, the sphere generates as its envelope a circular cone of which the straight line is the axis; and so, similarly, if (ii) the centre of a sphere move along an equiangular spiral and its radius keep proportional to the arc-distance along the spiral back to the pole, the sphere generates as its envelope a self-similar shell-surface, or nautiloid spiral.

and γ may then be called the angle of retardation, to which the inner curve is subject by virtue of its slower rate of growth.

Dispensing with mathematical formulae, the several conditions may be illustrated as follows:

In the diagrams (Fig. 385), $OP_1P_2P_3$, etc. represents a radius, on which P_1, P_2, P_3 are the points attained by the outer border of the tubular shell after as many entire consecutive revolutions. And P_1', P_2', P_3' are the points similarly intersected by the inner border; OP/OP' being always $= \lambda$, which is the ratio of growth, or "cutting-down factor." Then, obviously, (1) when OP_1 is less than

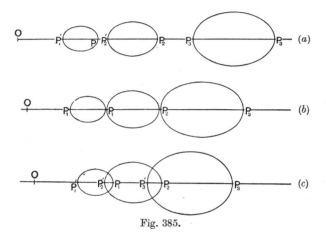

Fig. 385.

OP_2' the whorls will be separated by an interspace (a); (2) when $OP_1 = OP_2'$ they will be in contact (b), and (3) when OP_1 is greater than OP_2' there will be a greater or less extent of overlapping, that is to say of concealment of the surfaces of the earlier by the later whorls (c). And as a further case (4), it is plain that if λ be very large, that is to say if OP_1 be greater, not only than OP_2' but also than OP_3', OP_4', etc., we shall have complete, or all but complete, concealment by the last formed whorl of the whole of its predecessors. This latter condition is completely attained in *Nautilus pompilius*, and approached, though not quite attained, in *N. umbilicatus*; and the difference between these two forms, or "species," is constituted accordingly by a difference in the value of λ. (5) There is also a final case, not easily distinguishable

externally from (4), where P' lies on the opposite side of the radius vector to P, and is therefore imaginary. This final condition is exhibited in *Argonauta*.

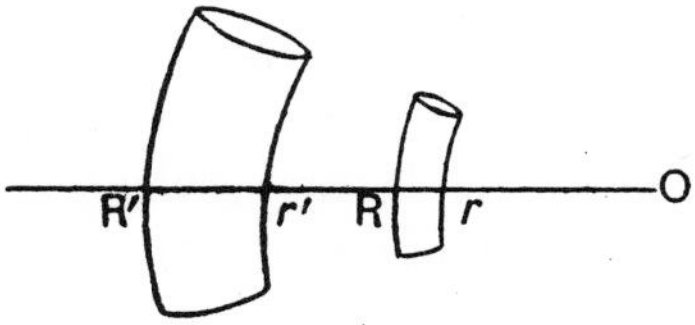

Fig. 386.

The limiting values of λ are easily ascertained.

In Fig. 386 we have portions of two successive whorls, whose corresponding points on the same radius vector (as R and R') are, therefore, at a distance apart corresponding to 2π. Let r and r' refer to the inner, and R, R' to the outer sides of the two whorls. Then, if we consider

$$R = ae^{\theta \cot \alpha},$$

it follows that

$$R' = ae^{(\theta + 2\pi) \cot \alpha},$$

$$r = \lambda ae^{\theta \cot \alpha} = ae^{(\theta - \gamma) \cot \alpha},$$

and

$$r' = \lambda ae^{(\theta + 2\pi) \cot \alpha} = ae^{(\theta + 2\pi - \gamma) \cot \alpha}.$$

Now in the three cases (a, b, c) represented in Fig. 385, it is plain that $r' \gtreqless R$, respectively. That is to say,

$$\lambda ae^{(\theta + 2\pi) \cot \alpha} \gtreqless ae^{\theta \cot \alpha},$$

and

$$\lambda e^{2\pi \cot \alpha} \lesseqgtr 1.$$

The case in which $\lambda e^{2\pi \cot \alpha} = 1$, or $- \log \lambda = 2\pi \cot \alpha \log e$, is the case represented in Fig. 385, b: that is to say, the particular case, for each value of α, where the consecutive whorls just touch, without interspace or overlap. For such cases, then, we may tabulate the values of λ as follows:

Constant angle α of spiral	Ratio (λ) of rate of growth of inner border of tube, as compared with that of the outer border
89°	0·896
88	0·803
87	0·720
86	0·645
85	0·577
80	0·330
75	0·234
70	0·1016
65	0·0534

We see, accordingly, that in plane spirals whose constant angle lies, say, between 65° and 70°, we can only obtain contact between

consecutive whorls if the rate of growth of the inner border of the
tube be a small fraction—a tenth or a twentieth—of that of the
outer border. In spirals whose con-
stant angle is 80°, contact is attained
when the respective rates of growth
are, approximately, as 3 to 1; while
in spirals of constant angle from about
85° to 89°,contact is attained when the
rates of growth are in the ratio of from
about $\frac{3}{5}$ to $\frac{9}{10}$.

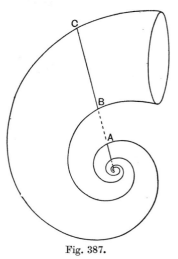

Fig. 387.

If on the other hand we have, for
any given value of α, a value of λ
greater or less than the value given
in the above table, then we have,
respectively, the conditions of separation or of overlap which are exemplified
in Fig. 385, a and c. And, just as we
have constructed this table for the
particular case of simple contact, so we could construct similar tables
for various degrees of separation, or of overlap.

For instance, a case which admits of simple solution is that in
which the interspace between the whorls is everywhere a mean proportional between the breadths of the whorls themselves (Fig. 387).
In this case, let us call $OA = R$, $OC = R_1$, and $OB = r$. We then
have

$$R_1 = OA = ae^{(\theta \cot \alpha)},$$

$$R_2 = OC = ae^{(\theta + 2\pi) \cot \alpha},$$

$$R_1 R_2 = ae^{2(\theta + \pi) \cot \alpha} = r^2 *.$$

And $$r^2 = (1/\lambda)^2 . \epsilon^{2\theta \cot \alpha},$$

whence, equating, $$1/\lambda = e^{\pi \cot \alpha}.$$

---

* It has been pointed out to me that it does not follow at once and obviously
that, because the interspace AB is a mean proportional between the breadths of
the adjacent whorls, therefore the whole distance OB is a mean proportional
between OA and OC. This is a corollary which requires to be proved; but the
proof is easy.

The corresponding values of λ are as follows:

Constant angle (α)	Ratio (λ) of rates of growth of outer and inner border, such as to produce a spiral with interspaces between the whorls, the breadth of which interspaces is a mean proportional between the breadths of the whorls themselves
90°	1·00 (imaginary)
89	0·95
88	0·89
87	0·85
86	0·81
85	0·76
80	0·57
75	0·43
70	0·32
65	0·23
60	0·18
55	0·13
50	0·090
45	0·063
40	0·042
35	0·026
30	0·016

As regards the angle of retardation, γ, in the formula

$$r' = \lambda e^{\theta \cot \alpha}, \quad \text{or} \quad r' = e^{(\theta - \gamma) \cot \alpha},$$

and in the case

$$r' = e^{(2\pi - \gamma) \cot \alpha}, \quad \text{or} \quad -\log \lambda = (2\pi - \gamma) \cot \alpha,$$

it is evident that when $\gamma = 2\pi$, that will mean that $\lambda = 1$. In other words, the outer and inner borders of the tube are identical, and the tube is constituted by one continuous line.

When λ is a very small fraction, that is to say when the rates of growth of the two borders of the tube are very diverse, then γ will tend towards infinity—tend that is to say towards a condition in which the inner border of the tube never grows at all. This condition is not infrequently approached in nature. I take it that *Cypraea* is such a case. But the nearly parallel-sided cone of *Dentalium*, or the widely separated whorls of *Lituites*, are cases where λ nearly approaches unity in the one case, and is still large in the other, γ being correspondingly small; while we can easily find cases where γ is very large, and λ is a small fraction, for instance in *Haliotis*, in *Calyptraea*, or in *Gryphaea*.

For the purposes of the morphologist, then, the main result of this last general investigation is to shew that all the various types of "open" and "closed" spirals, all the various degrees of separation or overlap of the successive whorls, are simply the outward expression of a varying ratio in the *rate of growth* of the outer as compared with the inner border of the tubular shell.

The foregoing problem of contact, or intersection, of successive whorls is a very simple one in the case of the discoid shell but a more complex one in the turbinate. For in the discoid shell contact will evidently take place when the retardation of the inner as compared with the outer whorl is just 360°, and the shape of the whorls need not be considered.

As the angle of retardation diminishes from 360°, the whorls stand further and further apart in an open coil; as it increases beyond 360°, they overlap more and more; and when the angle of retardation is infinite, that is to say when the true inner edge of the whorl does not grow at all, then the shell is said to be completely involute. Of this latter condition we have a striking example in *Argonauta*, and one a little more obscure in *Nautilus pompilius*.

In the turbinate shell the problem of contact is twofold, for we have to deal with the possibilities of contact on the *same* side of the axis (which is what we have dealt with in the discoid) and also with the new possibility of contact or intersection on the *opposite* side; it is this latter case which will determine the presence or absence of an open *umbilicus*. It is further obvious that, in the case of the turbinate, the question of contact or no contact will depend on the shape of the generating curve; and if we take the simple case where this generating curve may be considered as an ellipse, then contact will be found to depend on the angle which the major axis of this ellipse makes with the axis of the shell. The question becomes a complicated one, and the student will find it treated in Blake's paper already referred to.

When one whorl overlaps another, so that the generating curve cuts its predecessor (at a distance of 2π) on the same radius vector, the locus of intersection will follow a spiral line upon the shell, which is called the "suture" by conchologists. It is one of that *ensemble* of spiral lines in space of which, as we have seen, the whole shell may be conceived to be constituted; and we might call it a "contact-spiral," or "spiral of intersection." In discoid shells, such as an *Ammonite* or a *Planorbis*, or in *Nautilus umbilicatus*, there are obviously two such contact-spirals, one on each side of the shell, that is to say one on each side of a plane perpendicular to the axis.   In turbinate shells such a condition is also possible, but is somewhat rare.   We have it for instance in *Solarium perspectivum*, where the one contact-spiral is visible on the exterior of

Fig. 388.   *Solarium perspectivum.*

the shell, and the other lies internally, winding round the open cone of the umbilicus*; but this second contact-spiral is usually imaginary, or concealed within the whorls of the turbinated shell.

Fig. 390.   *Scalaria pretiosa* L.; the wentletrap. From Cooke's *Spirals.*

Fig. 389.   *Haliotis tuberculata* L.; the ormer, or ear shell.

Again, in *Haliotis*, one of the contact-spirals is non-existent, because of the extreme obliquity of the plane of the generating curve.   In

* A beautiful construction: *stupendum Naturae artificium*, Linnaeus.

Scalaria pretiosa and in *Spirula** there is no contact-spiral, because the growth of the generating curve has been too slow in comparison with the vector rotation of its plane. In *Argonauta* and in *Cypraea* there is no contact-spiral, because the growth of the generating curve has been too quick. Nor, of course, is there any contact-spiral in *Patella* or in *Dentalium*, because the angle α is too small

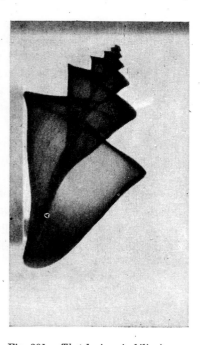

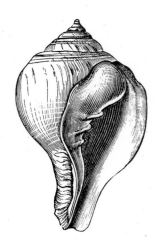

Fig. 392. *Turbinella napus* Lam.; an Indian chank-shell. From Chenu.

Fig. 391. *Thatcheria mirabilis* Angas; from a radiograph by Dr A. Müller.

ever to give us a complete revolution of the spire. *Thatcheria mirabilis* is a peculiar and beautiful shell, in which the outline of the lip is sharply triangular, instead of being a smooth curve: with the result that the apex of the triangle forms a conspicuous "generating spiral", which winds round the shell and is more conspicuous than the suture itself.

In the great majority of helicoid or turbinate shells the innermost

---

* "It [*Spirula*] is curved so as its roundness is kept, and the Parts do not touch one another": R. Hooke, *Posthumous Works*, 1745, p. 284.

or axial portions of the whorls tend to form a solid axis or "columella"; and to this is attached the columellar muscle which on the one hand withdraws the animal within its shell, and on the other hand provides the controlling force or trammel, by which (in the gastropod) the growing shell is kept in its spiral course. This muscle is apt to leave a winding groove upon the columella (Fig. 373); now and then the muscle is split into strands or bundles, and then it leaves parallel grooves with ridges or pleats between, and the number of these folds or pleats may vary with the species, as in the Volutes, or even with race or locality. Thus, among the curiosities of conchology, the chank-shells on the Trincomali coast have four columellar folds or ridges; but all those from Tranquebar, just north of Adam's Bridge, have only three (Fig. 392)*.

The various forms of straight or spiral shells among the Cephalopods, which we have seen to be capable of complete definition by the help of elementary mathematics, have received a very complicated descriptive nomenclature from the palaeontologists. For instance, the straight cones are spoken of as *orthoceracones* or *bactriticones*, the loosely coiled forms as *gyroceracones* or *mimoceracones*, the more closely coiled shells, in which one whorl overlaps the other, as *nautilicones* or *ammoniticones*, and so forth. In such a series of forms the palaeontologist sees undoubted and unquestioned evidence of ancestral descent. For instance we read in Zittel's *Palaeontology*†: "The bactriticone obviously represents the primitive or primary radical of the Ammonoidea, and the mimoceracone the next or secondary radical of this order"; while precisely the opposite conclusion was drawn by Owen, who supposed that the straight chambered shells of such fossil Cephalopods as *Orthoceras* had been produced by the gradual unwinding of a coiled nautiloid shell‡. *The mathematical study of the forms of shells lends no support to these*

---

* Cf. R. Winckworth, *Proc. Malacol. Soc.* xxiii, p. 345, 1939.

† English edition, 1900, p. 537. The chapter is revised by Professor Alpheus Hyatt, to whom the nomenclature is largely due. For a more copious terminology, see Hyatt, *Phylogeny of an Acquired Characteristic*, 1894, p. 422 *seq.* Cf. also L. F. Spath, The evolution of the Cephalopoda, *Biol. Reviews*, viii, pp. 418–462, 1933.

‡ This latter conclusion is adopted by Willey, *Zoological Results*, 1902, p. 747. Cf. also Graham Kerr, on *Spirula*: *Dana Reports*, No. 8, Copenhagen, 1931.

*or any suchlike phylogenetic hypotheses**. If we have two shells in which the constant angle of the spire be respectively 80° and 60°, that fact in itself does not at all justify an assertion that the one is more primitive, more ancient, or more "ancestral" than the other. Nor, if we find a third in which the angle happens to be 70°, does that fact entitle us to say that this shell is intermediate between the other two, in time, or in blood relationship, or in any other sense whatsoever save only the strictly formal and mathematical one. For it is evident that, though these particular arithmetical constants manifest themselves in visible and recognisable differences of form, yet they are not necessarily more deep-seated or significant than are those which manifest themselves only in difference of magnitude; and the student of phylogeny scarcely ventures to draw conclusions as to the relative antiquity of two allied organisms on the ground that one happens to be bigger or less, or longer or shorter, than the other.

At the same time, while it is obviously unsafe to rest conclusions upon such features as these, unless they be strongly supported and corroborated in other ways—for the simple reason that there is unlimited room for *coincidence*, or separate and independent attainment of this or that magnitude or numerical ratio—yet on the other hand it is certain that, in particular cases, the evolution of a race has actually involved gradual increase or decrease in some one or more numerical factors, magnitude itself included— that is to say increase or decrease in some one or more of the actual and relative velocities of growth. When we do meet with a clear and unmistakable series of such progressive magnitudes or ratios, manifesting themselves in a progressive series of "allied" forms, then we have the phenomenon of "*orthogenesis.*" For orthogenesis is simply that phenomenon of continuous lines or series of form (and also of functional or physiological capacity),

---

* Phylogenetic speculation, fifty years ago the chief preoccupation of the biologist, has had its caustic critics. Cf. (*int. al.*) Rhumbler, in *Arch. f. Entw. Mech.* vii, p. 104, 1898: "Phylogenetische Speculationen...werden immer auf Anklang bei den Fachgenossen rechnen dürfen, sofern nicht ein anderer Fachgenosse auf demselben Gebiet mit gleicher Kenntniss der Dinge und mit gleicher Scharfsinn zufällig zu einer anderen Theorie gekommen ist....Die Richtigkeit 'guter' phylogenetischer Schlüsse lässt sich im schlimmsten Fälle anzweifeln, aber direkt widerlegen lässt sich in der Regel nicht."

which was the foundation of the Theory of Evolution, alike to Lamarck and to Darwin and Wallace; and which we see to exist whatever be our ideas of the "origin of species," or of the nature and origin of "functional adaptations." And to my mind, the mathematical (as distinguished from the purely physical) study of morphology bids fair to help us to recognise this phenomenon of orthogenesis in many cases where it is not at once patent to the' eye; and, on the other hand, to warn us in many other cases that even strong and apparently complex resemblances in form may be capable of arising independently, and may sometimes signify no more than the equally accidental numerical coincidences which are manifested in identity of length or weight or any other simple magnitudes.

I have already referred to the fact that, while in general a very great and remarkable regularity of form is characteristic of the molluscan shell, yet that complete regularity is apt to be departed from. We have clear cases of such a departure in *Pupa, Clausilia* and various *Bulimi*, where the spire is not conical, but its sides are curved and narrow in.

The following measurements of three specimens of *Clausilia* shew a gradual change in the ratio to one another of successive whorls, or in other words a marked departure from the logarithmic law:

Clausilia lamellosa.   (From Chr. Petersen*.)

	Width of successive whorls (mm.)				Ratios, or "quotients" of successive whorls			
	I	II	III		I	II	III	Mean
a	2·42	2·51	2·49	a/b	1·43	1·45	1·42	1·44
b	1·69	1·72	1·75	. b/c	1·36	1·33	1·31	1·33
c	1·24	1·30	1·33	c/d	1·21	1·29	1·23	1·24
d	1·02	1·00	1·08	d/e	1·22	1·20	1·26	1·23
e	0·83	0·83	0·86					

In many ammonites, where the helicoid factor does not enter into the case, we have a clear illustration of how gradual and marked

* From Chr. Petersen, *Das Quotientengesetz*, p. 36. After making a careful statistical study of 1000 Clausilias, Peterson found the following mean ratios of the successive whorls, a/b, b/c, etc.: 1·37, 1·33, 1·27, 1·24, 1·22, 1·19.

changes in the spiral angle may be detected even in ammonites which present nothing abnormal to the eye. But let us suppose that the spiral angle increases somewhat rapidly; we shall then get a spiral with gradually narrowing whorls, which condition is characteristic of *Oekotraustes*, a subgenus of *Ammonites*. If on the other hand, the angle α gradually diminishes, and even falls away to zero, we shall have the spiral curve opening out, as it does in *Scaphites*, *Ancyloceras*

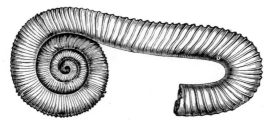

Fig. 393. An ammonitoid shell (*Macroscaphites*) to shew change of curvature.

and *Lituites*, until the spiral coil is replaced by a spiral curve so gentle as to seem all but straight. Lastly, there are a few cases, such as *Bellerophon expansus* and some *Goniatites*, where the outer spiral does not perceptibly change, but the whorls become more "embracing" or the whole shell more involute. Here it is the angle of retardation, the ratio of growth between the outer and inner parts of the whorl, which undergoes a gradual change.

In order to understand the relation of a close-coiled shell to its straighter congeners, to compare (for example) an *Ammonite* with an *Orthoceras*, it is necessary to estimate the length of the right cone which has, so to speak, been coiled up into the spiral shell. Our problem is, to find the length of a plane equiangular spiral, in terms of the radius and the constant angle α. Then, if OP be a radius vector, OQ a line of reference perpendicular to OP, and PQ a tangent to the curve, PQ, or sec α, is equal in length to the spiral arc OP. In other words, the arc measured from the pole is equal to the polar tangent*. And this is practically obvious: for

* Descartes made this discovery, and records it in a letter to Mersenne, 1638. The equiangular spiral was thus the first transcendental curve to be "rectified."

$PP'/PR' = ds/dr = \sec \alpha$, and therefore $\sec \alpha = s/r$, or the ratio of arc to radius vector.

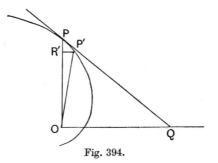

Fig. 394.

Accordingly, the ratio of l, the total length, to r, the radius vector up to which the total length is to be measured, is expressed by a simple table of secants; as follows:

α	l/r	α	l/r
5°	1·004	87°	19·1
10	1·015	88	28·7
20	1·064	89	57·3
30	1·165	89° 10′	68·8
40	1·305	20	85·9
50	1·56	30	114·6
60	2·0	40	171·9
70	2·9	50	343·8
75	3·9	55	687·5
80	5·8	59	3437·7
85	11·5	90	Infinite
86	14·3		

Putting the same table inversely, so as to shew the total length in terms of the radius, we have as follows:

Total length (in terms of the radius)	Constant angle
2	60°
3	70 31′
4	75 32
5	78 28
10	84 16
20	87 8
30	88 6
40	88 34
50	88 51
100	89 26
1000	89 56′ 36″
10,000	89 59 30

Accordingly, we see that (1), when the constant angle of the spiral is small, the shell (or for that matter the tooth, or horn or claw) is scarcely to be distinguished from a straight cone or cylinder; and this remains pretty much the case for a considerable increase of angle, say from 0° to 20° or more; (2) for a considerably greater increase of the constant angle, say to 50° or more, the shell would still only have the appearance of a gentle curve; (3) the characteristic close coils of the Nautilus or Ammonite would be typically represented only when the constant angle lies within a few degrees on either side of about 80°. The coiled up spiral of a Nautilus, with a constant angle of about 80°, is about six times the length of its radius vector, or rather more than three times its own diameter; while that of an Ammonite, with a constant angle of, say, from 85° to 88°, is from about six to fifteen times as long as its own diameter. And (4) as we approach an angle of 90° (at which point the spiral vanishes in a circle), the length of the coil increases with enormous rapidity. Our spiral would soon assume the appearance of the close coils of a Nummulite, and the successive increments of breadth in the successive whorls would become inappreciable to the eye.

The geometrical form of the shell involves many other beautiful properties, of great interest to the mathematician but which it is not possible to reduce to such simple expressions as we have been content to use. For instance, we may obtain an equation which shall express completely the surface of any shell, in terms of polar or of rectangular coordinates (as has been done by Moseley and by Blake), or in Hamiltonian vector notation*. It is likewise possible (though of little interest to the naturalist) to determine the area of a conchoidal surface or the volume of a conchoidal solid, and to find the centre of gravity of either surface or solid†. And Blake has further shewn, with considerable elaboration, how we may deal with the symmetrical distortion due to pressure which fossil shells are often found to have undergone, and how we may reconstitute by calculation their original undistorted form—a problem which, were the available methods only a little easier, would be

---

* Cf. H. W. L. Hime's *Outlines of Quaternions*, 1894, pp. 171–173.

† See Moseley, *op. cit.* p. 361 *seq.* Also, for more complete and elaborate treatment, Haton de la Goupillière, *op. cit.* 1908, pp. 5–46, 69–204.

very helpful to the palaeontologist; for, as Blake himself has shewn, it is easy to mistake a symmetrically distorted specimen of (for instance) an Ammonite for a new and distinct species of the same genus. But it is evident that to deal fully with the mathematical problems contained in, or suggested by, the spiral shell, would require a whole treatise, rather than a single chapter of this elementary book. Let us then, leaving mathematics aside, attempt to summarise, and perhaps to extend, what has been said about the general possibilities of form in this class of organisms.

The univalve shell: a summary

The surface of any shell, whether discoid or turbinate, may be imagined to be generated by the revolution about a fixed axis of a closed curve, which, remaining always geometrically similar to itself, increases its dimensions continually: and, since the scale of the figure increases in geometrical progression while the angle of rotation increases in arithmetical, and the centre of similitude remains fixed, the curve traced in space by corresponding points in the generating curve is, in all such cases, an equiangular spiral. In discoid shells, the generating figure revolves in a plane perpendicular to the axis, as in the Nautilus, the Argonaut and the Ammonite. In turbinate shells, it follows a skew path with respect to the axis of revolution, and the curve in space generated by any given point makes a constant angle to the axis of the enveloping cone, and partakes, therefore, of the character of a helix, as well as of a logarithmic spiral; it may be strictly entitled a helico-spiral. Such turbinate or helico-spiral shells include the snail, the periwinkle and all the common typical Gastropods.

When the envelope of the shell is a right cone—and it is seldom far from being so—then our helico-spiral is a loxodromic curve, and is obviously identical with a projection, parallel with the axis, of the logarithmic spiral of the base. As this spiral cuts all radii at a constant angle, so its orthogonal projection on the surface intersects all generatrices, and consequently all parallel circles, under a constant angle: this being the definition of a loxodromic curve on a surface of revolution. Guido Grandi describes this curve for the first time in a letter to Ceva, printed at the end of his *Demonstratio theorematum Hugenianorum circa...logarithmicam lineam*, 1701 *.

The generating figure may be taken as any section of the shell, whether parallel, normal, or otherwise inclined to the axis. It is very commonly assumed to be identical with the mouth of the shell; in which case it is sometimes a plane curve of simple form; in other and more numerous cases, it becomes complicated in form and its boundaries do not lie in one plane: but in such cases as these we may replace it by its "trace," on a plane at some definite angle to the direction of growth, for instance by its form as it appears in a section through the axis of the helicoid shell. The generating curve is of very various shapes. It is circular in *Scalaria* or *Cyclostoma*, and in *Spirula*; it may be considered as a segment of a circle in *Natica* or in *Planorbis*. It is triangular in *Conus* or *Thatcheria*, and rhomboidal in *Solarium* or *Potamides*. It is very commonly more or less elliptical: the long axis of the ellipse being parallel to the axis of the shell in *Oliva* and *Cypraea*; all but perpendicular to it in many Trochi; and oblique to it in many well-marked cases, such as *Stomatella, Lamellaria, Sigaretus haliotoides* (Fig. 396) and *Haliotis*. In *Nautilus pompilius* it is approximately a semi-ellipse, and in *N. umbilicatus* rather more than a semi-ellipse, the long axis lying in both cases perpendicular to the axis of the shell*. Its form is seldom open to easy mathematical expression, save when it is an actual circle or

Fig. 395.   Section of a spiral univalve, *Triton corrugatus* Lam. From Woodward.

spiral. Paul Serret (*Th. nouv...des lignes à double courbure*, 1860, p. 101) called it "*hélice cylindroconique*"; Haton de la Goupillière calls it a "*cônhélice*." It has also been studied by (*int. al.*) Tissot, *Nouv. ann. de mathém.* 1852; G. Pirondini, *Mathesis*, XIX, pp. 153–8, 1899; etc.

* In *Nautilus*, the "hood" has somewhat different dimensions in the two sexes, and these differences are impressed upon the shell, that is to say upon its "generating curve." The latter constitutes a somewhat broader ellipse in the male than in the female. But this difference is not to be detected in the young; in other words, the form of the generating curve perceptibly alters with advancing age. Somewhat similar differences in the shells of Ammonites were long ago suspected, by d'Orbigny, to be due to sexual differences. (Cf. Willey, *Natural Science*, VI, p. 411, 1895; *Zoological Results*, 1902, p. 742.)

ellipse; but an exception to this rule may be found in certain Ammonites, forming the group "Cordati," where (as Blake points out) the curve is very nearly represented by a cardioid, whose equation is $r = a\,(1 + \cos\theta)$.

When the generating curves of successive whorls cut one another, the line of intersection forms the conspicuous helico-spiral or loxodromic curve called the *suture* by conchologists.

The generating curve may grow slowly or quickly; its growth-factor is very slow in *Dentalium* or *Turritella*, very rapid in *Nerita*, or *Pileopsis*, or *Haliotis* or the Limpet. It may contain the axis in its plane, as in *Nautilus*; it may be parallel to the axis, as in the majority of Gastropods; or it may be inclined to the axis, as it is in a very marked degree in *Haliotis*. In fact, in *Haliotis* the generating

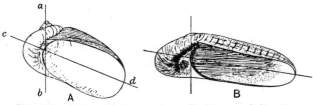

Fig. 396.   A, *Lamellaria perspicua*; B, *Sigaretus haliotoides*.
After Woodward.

curve is so oblique to the axis of the shell that the latter appears to grow by additions to one margin only (cf. Fig. 362), as in the case of the opercula of *Turbo* and *Nerita* referred to on p. 775; and this is what Moseley supposed it to do.

The general appearance of the entire shell is determined (apart from the form of its generating curve) by the magnitude of three angles; and these in turn are determined, as has been sufficiently explained, by the ratios of certain velocities of growth. These angles are (1) the constant angle of the equiangular spiral (α); (2) in turbinate shells, the enveloping angle of the cone, or (taking half that angle) the angle (β) which a tangent to the whorls makes with the axis of the shell; and (3) an angle called the "angle of retardation" (γ), which expresses the retardation in growth of the inner as compared with the outer part of each whorl, and therefore measures the extent to which one whorl overlaps, or the extent to which it is separated from, another.

The spiral angle (α) is very small in a limpet, where it is usually taken as $= 0°$; but it is evidently of a significant amount, though obscured by the shortness of the tubular shell. In *Dentalium* it is still small, but sufficient to give the appearance of a regular curve; it amounts here probably to about 30° to 40°. In *Haliotis* it is from about 70° to 75°; in *Nautilus* about 80°; and it lies between 80° and 85° or even more, in the majority of Gastropods*.

The case of *Fissurella* is curious. Here we have, apparently, a conical shell with no trace of spiral curvature, or (in other words) with a spiral angle which approximates to 0°; but in the minute embryonic shell (as in that of the limpet) a spiral convolution is distinctly to be seen. It would seem, then, that what we have to do with here is an unusually large growth-factor in the generating curve, causing the shell to dilate into a cone of very wide angle, the apical portion of which has become lost or absorbed, and the remaining part of which is too short to show clearly its intrinsic curvature. In the closely allied *Emarginula*, there is likewise a well-marked spiral in the embryo, which however is still manifested in the curvature of the adult, nearly conical, shell. In both cases we have to do with a very wide-angled cone, and with a high retardation-factor for its inner, or posterior, border. The series is continued, from the apparently simple cone to the complete spiral, through such forms as *Calyptraea*.

The angle α, as we have seen, is not always, nor rigorously, a constant angle. In some Ammonites it may increase with age, the whorls becoming closer and closer; in others it may decrease rapidly and even fall to zero, the coiled shell then straightening out, as in *Lituites* and similar forms. It diminishes somewhat, also, in many Orthocerata, which are slightly curved in youth but straight in age. It tends to increase notably in some common land-shells, the *Pupae* and *Bulimi*; and it decreases in *Succinea*.

Directly related to the angle α is the ratio which subsists between the breadths of successive whorls. The following table gives a few

---

* What is sometimes called, as by Leslie, the *angle of deflection* is the complement of what we have called the *spiral angle* (α), or obliquity of the spiral. When the angle of deflection is 6° 17′ 41″, or the spiral angle 83° 42′ 19″, the radiants, or breadths of successive whorls, are doubled at each entire circuit.

illustrations of this ratio in particular cases, in addition to those which we have already studied.

Ratio of breadth of consecutive whorls

Pointed Turbinates		Obtuse Turbinates and Discoids	
Telescopium fuscum ...	1·14	*Conus virgo*	1·25
Terebra subulata	1·16	‡*Clymenia laevigata*	1·33
*Turritella terebellata ...	1·18	*Conus litteratus*	1·40
*Turritella imbricata	1·20	*Conus betulinus*	1·43
Cerithium palustre	1·22	‡*Clymenia arietina*	1·50
Turritella duplicata	1·23	‡*Goniatites bifer*	1·50
Melanopsis terebralis ...	1·23	**Helix nemoralis*	1·50
Cerithium nodulosum ...	1·24	**Solarium perspectivum* ...	1·50
*Turritella carinata	1·25	*Solarium trochleare* ...	1·62
Terebra crenulata	1·25	*Solarium magnificum* ...	1·75
Terebra maculata (Fig. 397)	1·25	**Natica aperta*	2·00
*Cerithium lignitarum ...	1·26	*Euomphalus pentangulatus*	2·00
Terebra dimidiata	1·28	*Planorbis corneus*	2·00
Cerithium sulcatum	1·32	*Solaropsis pellis-serpentis* ...	2·00
Fusus longissimus	1·34	*Dolium zonatum*	2·10
*Pleurotomaria conoidea ...	1·34	‡*Goniatites carinatus* ...	2·50
Trochus niloticus (Fig. 398)	1·41	**Natica glaucina*	3·00
Mitra episcopalis	1·43	*Nautilus pompilius*	3·00
Fusus antiquus	1·50	*Haliotis excavatus*	4·20
Scalaria pretiosa	1·56	*Haliotis parvus*	6·00
Fusus colosseus	1·71	*Delphinula atrata*	6·00
Phasianella australis ...	1·80	*Haliotis rugoso-plicata* ...	9·30
Helicostyla polychroa ...	2·00	*Haliotis viridis*	10·00

Those marked * from Naumann; ‡ from Müller; the rest from Macalister†.

In the case of turbinate shells, we must take into account the angle β, in order to determine the spiral angle α from the ratio of the breadths of consecutive whorls; for the short table given on p. 791 is only applicable to discoid shells, in which the angle β is an angle of 90°. Our formula, as mentioned on p. 771, now becomes

$$R = \epsilon^{2\pi \sin \beta \cot \alpha}.$$

For this formula I have worked out the following table.

† Alex. Macalister, Observations on the mode of growth of discoid and turbinated shells, *Proc. R.S.* xviii, pp. 529–532, 1870; *Ann. Mag. N.H.* (6), iv, p 160, 1870. Cf. also his Law of Symmetry as exemplified in animal form, *Journ. R. Dublin Soc.* 1869, p. 327.

Table shewing values of the spiral angle α corresponding to certain ratios of breadth of successive whorls of the shell, for various values of the apical semi-angle β

Ratio R/1	β = 5°	10°	15°	20°	30°	40°	50°	60°	70°	80°	90°
1·1	80° 8'	85° 0'	86° 44'	87° 28'	88° 16'	88° 39'	88° 52'	89° 0'	89° 4'	89° 7'	89° 8'
1·25	67 51	78 27	82 11	84 5	85 56	86 50	87 21	87 39	87 50	87 56	87 58
1·5	53 30	69 37	76 0	79 21	82 39	84 16	85 13	85 44	86 4	86 15	86 18
2·0	38 20	57 35	66 55	73 11	77 34	80 16	81 52	82 45	83 18	83 37	83 42
2·5	30 53	50 0	60 35	67 0	73 45	77 13	79 19	80 26	81 11	81 35	81 42
3·0	26 32	44 50	56 0	63 0	70 45	74 45	77 17	78 35	79 28	79 56	80 5
3·5	23 37	41 5	52 25	59 50	68 15	72 45	75 35	77 2	78 1	78 33	78 43
4·0	21 35	38 10	49 35	57 15	66 10	71 3	74 9	75 42	76 47	77 22	77 34
4·5	20 0	36 0	47 15	55 5	64 25	69 35	72 54	74 33	75 43	76 20	76 35
5·0	18 45	34 10	45 20	53 15	62 55	68 15	71 48	73 31	74 45	75 25	75 38
10·0	13 25	25 20	35 15	43 5	53 45	60 20	64 57	67 4	68 42	69 35	69 53
20·0	10 25	20 0	28 30	35 45	46 25	53 25	58 52	61 10	63 6	64 10	64 31
50·0	8 0	15 35	22 35	28 50	38 45	45 55	52 1	54 18	56 28	57 42	58 6
100·0	6 50	13 20	19 30	25 5	34 20	41 15	47 35	49 45	52 3	53 20	53 46

From this table, by interpolation, we may easily fill in the approximate values of α, as soon as we have determined the apical angle β and measured the ratio R; as follows:

	R	β	α
Turritella sp.	1·12	7°	81°
Cerithium nodulosum ...	1·24	15	82
Conus virgo	1·25	70	88
Mitra episcopalis... ...	1·43	16	78
Scalaria pretiosa	1·56	26	81
Phasianella australis ...	1·80	26	80
Solarium perspectivum ...	1·50	53	85
Natica aperta	2·00	70	83
Planorbis corneus ...	2·00	90	84
Euomphalus pentangulatus	2·00	90	84

We see from this that shells so different in appearance as *Cerithium*, *Solarium*, *Natica* and *Planorbis* differ very little indeed in the magnitude of the spiral angle α, that is to say in the relative velocities of radial and tangential growth. It is upon the angle β that the difference in their form mainly depends.

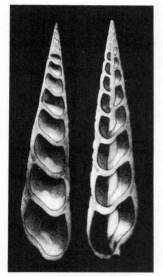

Fig. 397.   *Terebra maculata* L.

The angle, or rather semi-angle (β), of the tangent cone may be taken as 90° in the discoid shells, such as *Nautilus* and *Planorbis:* It is still a large angle, of 70° or 75°, in *Conus* or in *Cymba*, somewhat less in *Cassis*, *Harpa*, *Dolium* or *Natica*; it is about 50° to 55° in the various species of *Solarium*, about 35° in the typical *Trochi*, such as *T. niloticus* or *T. zizyphinus*, and about 25° or 26° in *Scalaria pretiosa* and *Phasianella bulloides*; it becomes a very acute angle, of 15°, 10°, or even less, in *Eulima*, *Turritella* or *Cerithium*. The British species of 'Fusus' form a series in which the apical angle ranges from about 28° in *F. antiquus*, through *F. Norvegicus*, *F. berniciensis*, *F. Turtoni*, *F. Islandicus*, to about 17° in *F. gracilis*. It varies much among the Cones; and the costly *Conus gloria-maris*, one of the great treasures of the conchologist, differs from its congeners in no important particular save in the somewhat "produced" spire, that is to say in the comparatively low value of the angle β.

A variation with advancing age of β is common, but (as Blake points out) it is often not to be distinguished or disentangled from an alteration of α. Whether alone, or combined with a change in α, we find it in all those many gastropods whose whorls cannot all be touched by the same enveloping cone, and whose spire is accordingly described as *concave* or *convex*. The former condition, as we have

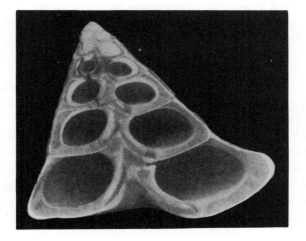

Fig. 398. *Trochus niloticus* L.

it in *Cerithium*, and in the cusp-like spire of *Cassis*, *Dolium* and some Cones, is much the commoner of the two*.

In the vast majority of spiral univalves the shell winds to the right, or turns clockwise, as we look along it in the direction in which the animal crawls and puts out its head. The thread of a carpenter's screw (except in China) runs the same way, and we call it a "right-handed screw." Save that it takes a right-handed movement to

---

* Many measurements of the linear dimensions of univalve shells have been made of late years, and studied by statistical methods in order to detect local races and other instances of variation and variability. But conchological statisticians seem to be content with some arbitrary linear ratio as a measure of "squatness" or the reverse; and the measurements chosen give little or no help towards the determination either of the apical or of the spiral angle. Cf. (e.g.) A. E. Boycott, Conchometry, *Proc. Malacol. Soc.* XVII, p. 8, 1928; C. Price-Jones, *ibid.* XIX, p. 146, 1930; etc. See also G. Duncker, Methode der Variations-Statistik, *Arch. f. Entw. Mech.* VIII, pp. 112–183, 1899.

drive in a "right-handed" screw, the terms right-handed and left-handed are purely conventional; and the mathematicians and the naturalists, unfortunately, use them in opposite ways. Thus the mathematicians call the snail-shell or the joiner's screw *leiotropic*; and Listing for one has much to say about lack of precision or even confusion on the part of the conchologists and the botanists, from Linnaeus downwards, in their attempts to deal with right-handed and left-handed spirals or screws*. The convolvulus twines to the right, the hop to the left; vine-tendrils are said to be mostly right-handed. At any rate, Clerk Maxwell spoke of *hop-spirals* and *vine-spirals*, trying to avoid the confusion or ambiguity of left and right. Some climbing plants are one and some the other; and the architect shews little preference, but builds his spiral staircases or twisted columns either way. But in all these, shells and all, the spiral runs *one way*; it is *isotropic*, while the fir-cone shews spirals running both ways at once, and we call them *heterotropic*, or *diadromic*.

When we find a "reversed shell," a whelk or a snail winding the wrong way, we describe it mathematically by the simple statement that the apical angle (β) has changed sign. Such left-handed shells occur as a well-known but rare abnormality; and the men who handle snails in the Paris market or whelks in Billingsgate keep a sharp look-out for them. In rare instances they become common. While left-handed whelks (*Buccinum* or *Neptunea*) are very rare nowadays, it was otherwise in the epoch of the Red Crag; for *Neptunea* was then extremely common, but right-handed specimens were as rare as left-handed are today. In the beautiful genus *Ampullaria*, or apple-snails, which inhabit tropical and sub-tropical rivers, there is unusual diversity; for the spire turns to the right in some species, and to the left in others, and again some are flat or "discoid," with no spire at all; and there are plenty of half-way stages, with right and left-handed spires of varying steepness or acuteness†; in short, within the limits of this singular genus the apical angle (β) may vary from about $\pm 35°$ to $\pm 125°$. But we need not imagine that the direction of growth actually changes over from right-handed to left-handed; it is enough to suppose

---

* See Listing's *Topologie*, p. 36; and cf. Clerk Maxwell's *Electricity and Magnetism*, I, p. 24.

† See figures in Arnold Lang's *Comparative Anatomy* (English translation), II, p. 161, 1902.

that the skew movement along the axis has changed its direction. For if I take a roll of tape and push the core out to one side or to the other, or if I keep the centre of the roll fixed and push the rim to the one side or to the other, I thereby convert the flat roll into a hollow cone, or (in other words) a plane into a gauche spiral. Whether we push one way or other, whether the spiral coil be plane or gauche, positively or negatively deformed, it remains right-handed or left-handed as the case may be; but it does change its direction as soon as we *turn it upside down*, or as soon as the animal does so in assuming its natural attitude. The linear spirals within and without the cone may change places but must remain congruent with one another; for they are merely the two edges of the ribbon, and as such are inseparable and identical twins. But of the shell itself we may reasonably say that a right-handed has given place to a left-handed spiral. Of these, the one is a mirror-image of the other; and the passing from one to the other through the plane of symmetry (which has no "handedness") is an operation which Listing called *perversion*. The flat or discoid apple-snails are like our roll of tape, which can be *converted* into a conical spire and *perverted* in one direction or the other; and in this genus, by a rare exception, it seems wellnigh as easy to depart one way as the other from the plane of symmetry. But why, in the general run of shells, all the world over, in the past and in the present, one direction of twist is so overwhelmingly commoner than the other, no man knows.

The phenomenon of reversal, or "sinistrality," has an interest of its own from the side of development and heredity. For careful study of certain pond-snails has shewn that dextral and sinistral varieties appear, not one by one, but by whole broods of the one sort or the other; a discovery which goes some way to account for the predominant left-handedness of *Fusus ambiguus* in the Red Crag. The right-handed, or ordinary form, is found to be "dominant" to the other; but the Mendelian heredity is of a curious and complicated kind. For the direction of the twist appears to be predetermined in the germ even prior to its fertilisation; and a left-handed pond-snail will produce a brood of left-handed young even when fertilised by a normal, or right-handed, individual[*].

* See A. E. Boycott and others, Abnormal forms of *Limnaea peregra*...and their inheritance, *Phil. Trans.* (B), ccxxix, p. 51, 1930; and other papers.

The angle of retardation (γ) is very small in *Dentalium* and *Patella*; it is very large in *Haliotis*; it becomes infinite in *Argonauta* and in *Cypraea*. Connected with the angle of retardation are the various possibilities of contact or separation, in various degrees, between adjacent whorls in the discoid shell, and between both adjacent and opposite whorls in the turbinate. But with these phenomena we have already dealt sufficiently.

The beautiful shell of the paper-nautilus (*Argonauta argo* L.) differs in sundry ways both from the Nautilus and from ordinary univalves. Only the female Argonaut possesses it; it is not attached to its owner, but is (so to speak) worn loose; it is rather a temporary cradle for the young than a true shell or bodily covering; and it is not secreted in the usual way, but is plastered on from the outside by two of the eight arms of the little Octopus to which it belongs. The shell shews a single whorl, or but little more; and the spiral is hard to measure, for this reason. It has been supposed by some to obey a law other than the logarithmic spiral. For my part I have made no special study of it, nor has any one else, to my knowledge, of recent years; but the simple fact that it conserves its shape as it grows, or that each increment is a gnomon to the rest, is enough to shew that this delicate and beautiful shell is mathematically, though not morphologically, homologous with all the others.

Of bivalve shells

Hitherto we have dealt only with univalve shells, and it is in these that all the mathematical problems connected with the spiral, or helico-spiral, configuration are best illustrated. But the case of the bivalve shell, whether of the lamellibranch or the brachiopod, presents no essential difference, save only that we have here to do with two conjugate spirals, whose two axes have a definite relation to one another, and some independent freedom of rotatory movement relatively to one another.

The bivalve or lamellibranch mollusca are very different creatures from the rest. The univalves or gastropods, like their cousins the cephalopods, go about their business and get their living in an ordinary way; but the bivalves are unintelligent, "acephalous" animals, and imbibe the invisible plankton-food which ciliary currents bring automatically to their mouths. There is something

Fig. 399. *Argonauta argo* L. The paper-nautilus.   From Cooke's *Spirals in Nature and Art.*

to be said for withdrawing them, as brachiopods and others have been withdrawn, from Cuvier's great class of the Mollusca. But whether bivalves and univalves be near relations or no is not the question. Both of them secrete a shell, and in both the shell grows by the successive addition of similar parts, gnomon after gnomon; so that in both the equiangular spiral makes, and is bound to make, its appearance. There is a mathematical analogy between the two; but it has no more bearing on zoological classification than has the still closer likeness between Nautilus and the nautiloid Foraminifera.

The generating curve is particularly well seen in the bivalve, where it simply constitutes what we call "the outline of the shell." It is for the most part a plane curve, but not always; for there are forms such as *Hippopus, Tridacna* and many Cockles, or *Rhynchonella* and *Spirifer* among the Brachiopods, in which the edges of the two valves interlock, and others, such as *Pholas, Mya*, etc., where they gape asunder. In such cases as these the generating curves, though not plane, are still conjugate, having a similar relation, but of opposite sign, to a median plane of reference or of projection. There are a few exceptional cases, e.g. *Arca (Parallelepipedon) tortuosa*, where there is no median plane of symmetry, but the generating curve, and therefore the outline of the shell itself, is a tortuous curve in three dimensions.

A great variety of form is exhibited among the bivalves by these generating curves. In many cases the curve or outline is all but circular, as in *Anomia, Sphaerium, Artemis, Isocardia*; it is nearly semicircular in *Argiope*; it is approximately elliptical in *Anodon, Lutraria, Orthis*; it may be called semi-elliptical in *Spirifer*; it is a nearly rectilinear triangle in *Lithocardium*, and a curvilinear triangle in *Mactra*. Many apparently diverse but more or less related forms may be shewn to be deformations of a common type, by a simple application of the mathematical theory of "transformations," which we shall have to study in a later chapter. In such a series as is furnished, for instance, by *Gervillea, Perna, Avicula, Modiola, Mytilus*, etc., a "simple shear" accounts for most, if not all, of the apparent differences.

Upon the surface of the bivalve shell we usually see with great clearness the "lines of growth" which represent the successive margins of the shell, or in other words the successive positions assumed during growth by the growing generating curve; and we have a good illustration, accordingly, of how it is characteristic of the generating curve that it should constantly increase, while never altering its geometric similarity.

Underlying these lines of growth, which are so characteristic of a molluscan shell (and of not a few other organic formations), there is, then, a law of growth which we may attempt to enquire into and which may be illustrated in various ways. The simplest cases are those in which we can study the lines of growth on a more or less flattened shell, such as the one valve of an oyster, a *Pecten* or a *Tellina*, or some such bivalve mollusc. Here around an origin, the so-called "umbo" of the shell, we have a series of curves, sometimes nearly circular, sometimes elliptical, often asymmetrical; and such curves are obviously not "concentric," though we are often apt to call them so, but have a common centre of similitude. This arrangement may be illustrated by various analogies. We might for instance compare it to a series of waves, radiating outwards from a point, through a medium which offered a resistance increasing, with the angle of divergence, according to some simple law. We may find another and perhaps a simpler illustration as follows:

In a simple and beautiful theorem, Galileo shewed that, if we imagine a number of inclined planes, or gutters, sloping downwards (in a vertical plane) at various angles from a common starting-point, and if we imagine a number of balls rolling each down its own gutter under the influence of gravity (and without hindrance from friction), then, at any given instant, the locus of all these moving bodies is a circle passing through the point of origin. For the acceleration along any one of the sloping paths, for instance AB (Fig. 400), is such that

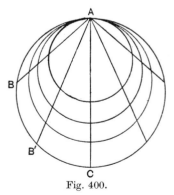

Fig. 400.

$$AB = \tfrac{1}{2}g \cos \theta . t^2$$
$$= \tfrac{1}{2}g . AB/AC . t^2.$$

Therefore $\qquad t^2 = 2/g . AC.$

That is to say, all the balls reach the circumference of the circle at the same moment as the ball which drops vertically from A to C.

Where, then, as often happens, the generating curve of the shell is approximately a circle passing through the point of origin, we may consider the acceleration of growth along various radiants to be governed by a simple mathematical law, closely akin to that simple law of acceleration which governs the movements of a falling body. And, *mutatis mutandis*, a similar definite law underlies the cases where the generating curve is continually elliptical, or where it assumes some more complex, but still regular and constant form.

It is easy to extend the proposition to the particular case where the lines of growth may be considered elliptical. In such a case we have $x^2/a^2 + y^2/b^2 = 1$, where a and b are the major and minor axes of the ellipse.

Or, changing the origin to the vertex of the figure,

$$\frac{x^2}{a^2} - \frac{2x}{a} + \frac{y^2}{b^2} = 0, \quad \text{giving} \quad \frac{(x-a)^2}{a^2} + \frac{y^2}{b^2} = 1.$$

Then, transferring to polar coordinates, where $r.\cos\theta = x$, $r.\sin\theta = y$, we have

$$\frac{r.\cos^2\theta}{a^2} - \frac{2\cos\theta}{a} + \frac{r.\sin^2\theta}{b^2} = 0,$$

which is equivalent to

$$r = \frac{2ab^2\cos\theta}{b^2\cos^2\theta + a^2\sin^2\theta},$$

or, simplifying, by eliminating the sine-function,

$$r = \frac{2ab^2\cos\theta}{(b^2 - a^2)\cos^2\theta + a^2}.$$

Obviously, in the case when $a = b$, this gives us the circular system which we have already considered. For other values, or ratios, of a and b, and for all values of θ, we can easily construct a table, of which the following is a sample:

Chords of an ellipse, whose major and minor axes (a, b)
are in certain given ratios

θ	a/b = 1/3	1/2	2/3	1/1	3/2	2/1	3/1
0°	1·0	1·0	1·0	1·0	1·0	1·0	1·0
10	1·01	1·01	1·002	0·985	0·948	0·902	0·793
20	1·05	1·03	1·005	0·940	0·820	0·695	0·485
30	1·115	1·065	1·005	0·866	0·666	0·495	0·289
40	1·21	1·11	0·995	0·766	0·505	0·342	0·178
50	1·34	1·145	0·952	0·643	0·372	0·232	0·113
60	1·50	1·142	0·857	0·500	0·258	0·152	0·071
70	1·59	1·015	0·670	0·342	0·163	0·092	0·042
80	1·235	0·635	0·375	0·174	0·078	0·045	0·020
90	0·0	0·0	0·0	0·0	0·0	0·0	0·0

The ellipses which we then draw, from the values given in the table, are such as are shewn in Fig. 401 for the ratio $a/b = \frac{3}{1}$, and in Fig. 402 for the ratio $a/b = \frac{1}{2}$; these are fair approximations to the actual outlines, and to the actual arrangement of the lines of growth, in such forms as *Solecurtus* or *Cultellus*, and in *Tellina* or *Psammobia*. It is not difficult to introduce a constant into our equation to meet the case of a shell which is somewhat unsymmetrical on either side of the median axis. It is a somewhat more troublesome matter, however, to bring these configurations into relation with a "law of growth," as was so easily done in the case of the circular figure: in other words, to formulate a law of acceleration according to which

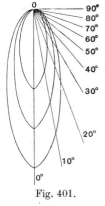

Fig. 401.

points starting from the origin *O*, and moving along radial lines, would all lie, at any future epoch, on an ellipse passing through *O*; and this calculation we need not enter into.

All that we are immediately concerned with is the simple fact that where a velocity, such as our rate of growth, varies with its direction—varies that is to say as a function of the angular divergence from a certain axis—then, in a certain simple case, we get lines of growth laid down as a system of coaxial circles, and, in somewhat less simple cases, we obtain a system of ellipses or of other more complicated coaxial figures, which may or may not be symmetrical on either side of the axis. Among our bivalve mollusca we shall find the lines of growth to be approximately circular in, for instance, *Anomia*; in *Lima* (e.g. *L. subauriculata*) we have a system of nearly symmetrical ellipses with the vertical axis about twice the transverse; in *Solen pellucidus*, we have again a system of lines of growth which are not far from being symmetrical ellipses,

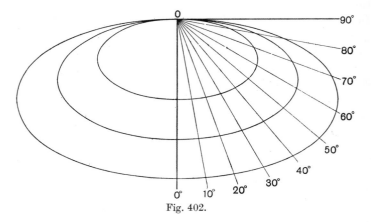

Fig. 402.

in which however the transverse is between three and four times as great as the vertical axis. In the great majority of cases, we have a similar phenomenon with the further complication of slight, but occasionally very considerable, lateral asymmetry.

In the above account of the mathematical form of the bivalve shell, we have supposed, for simplicity's sake, that the pole or origin of the system is at a point where all the successive curves touch one another. But such an arrangement is neither theoretically probable, nor is it actually the case; for it would mean that in a certain direction growth fell, not merely to a minimum, but to zero. As a matter of fact, the centre of the system (the "umbo" of the conchologists) lies not at the edge of the system, but very near to it; in other words, there is a certain amount of growth all round. But to take account of this condition would involve more troublesome mathematics, and it is obvious that the foregoing illustrations are a sufficiently near approximation to the actual case.

In certain little Crustacea (of the genus *Estheria*) the carapace takes the form of a bivalve shell, closely simulating that of a lamellibranchiate mollusc, and bearing lines of growth in all respects analogous to or even identical with those of the latter. The explanation is very curious and interesting. In ordinary Crustacea the carapace, like the rest of the chitinised and calcified integument, is shed off in successive moults, and is restored again as a whole. But in *Estheria* (and one or two other small crustacea) the moult is incomplete: the old carapace is retained, and the new, growing up underneath it, adheres to it like a lining, and projects beyond its edge: so that in course of time the margins of successive old carapaces appear as "lines of growth" upon the surface of the shell. In this mode of formation, then (but not in the usual one), we obtain a structure which "is partly old and partly new," and whose successive increments are all similar, similarly situated, and enlarged

Fig. 403. *Hemicardium inversum* Lam. From Chenu.

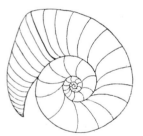

Fig. 404. *Caprinella adversa.* After Woodward.

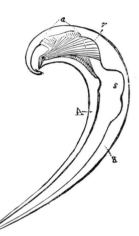

Fig. 405. Section of *Productus* (*Strophonema*) sp. From Woods.

in a continued progression. We have, in short, all the conditions appropriate and necessary for the development of a logarithmic spiral; and this logarithmic spiral (though it is one of small angle) gives its own character to the structure, and causes the little carapace to partake of the characteristic conformation of the molluscan shell.

Among the bivalves the spiral angle (α) is very small in the flattened shells, such as *Orthis*, *Lingula* or *Anomia*. It is larger, as a rule, in the Lamellibranchs than in the Brachiopods, but in the latter it is of considerable magnitude among the Pentameri.

Among the Lamellibranchs it is largest in such forms as *Isocardia* and *Diceras*, and in the very curious genus *Caprinella*; in all of these last-named genera its magnitude leads to the production of a spiral shell of several whorls, precisely as in the univalves. The angle is usually equal, but of opposite sign, in the two valves of the Lamellibranch, and usually of opposite sign but unequal in the two valves of the Brachiopod. It is very unequal in many Ostreidae, and especially in such forms as *Gryphaea*, or in *Caprinella*, which is a kind of exaggerated *Gryphaea*; in the cretaceous genus *Requienia*, the two valves of the shell closely resemble a turbinate gastropod with its flat calcified operculum. Occasionally it is of the same sign in both valves (that is to say, both valves curve the same way) as we see sometimes in *Anomia*, and better in *Productus* or *Strophonema*.

It will be observed, and it may not be difficult to explain, that the more the bivalve shell curves in the one direction the more it curves in the other; each valve tends to be spheroidal, or ellipsoidal, rather than cylindroidal. The cylindroidal form occurs, exceptionally, in *Solen*. But *Pecten, Gryphaea, Terebratula* are all cases of bivalve shells where one valve is flat and the other curved from *side to side*; and the flat valve tends to remain flat in the longitudinal direction also, while the curved valve grows into its logarithmic spiral.

In the genus *Gryphaea*, an oyster-like bivalve from the Jurassic, the creature lay on its side with its left valve downward, as oysters and scallops also do; and this valve adhered to the ground while the animal was young. The upper valve stays flat, and looks like a mere operculum; but the lower or deep valve grows into a more or less pronounced spiral. So is it also in the neighbouring genus *Pecten*, where *P. Jacobaeus* has its under-valve much deeper and more curved than, say, *P. opercularis*; but *Gryphaea incurva* is more spirally curved than any of these, and *G. arcuata* has a spiral angle very near to that of *Nautilus* itself. In both the spiral is a typical equiangular one, built up of a succession of gnomonic increments, which in turn depend on a constant ratio between the expansion of a generating figure and its rotation about a centre of similitude. *Rate of growth* is at the root of the whole matter. Now *Gryphaea*, like some Ammonites of which we spoke before, is

one of those cases in which not only does the form of the shell
vary, but geologists recognise, now and then, a *trend*, or progressive
sequence of variation, from one stratum or one "horizon" to
another. In short, *as time goes on*, we seem to see the shell growing
thicker or wider, or more and more spirally curved, before our
eyes. What meaning shall we give, what importance should we
assign, to these changes, and what sort or grade of evolution do
they imply? Some hold that these palaeontological features are
"strictly comparable with those on which the geneticist bases his
factorial studies"; and that as such they may shew "linkage of
characters," as when "in the evolution of *Gryphaea* the area of
attachment retrogresses as the arching progresses"*. These are
debatable matters. But in so far as the changes depend on mere
gradations of magnitude, they lead indeed to variety but fall short
of the full concept of evolution. For to quote Aristotle once again
(though we need not go to Aristotle to learn it): "some things shew
increase but suffer no alteration; because increase is one thing and
alteration is another."

The so-called "spiral arms" of *Spirifer* and many other Brachiopods are not difficult to explain. They begin as a single structure,
in the form of a loop of shelly substance,
attached to the dorsal valve of the shell,
in the neighbourhood of the hinge, and
forming a skeletal support for two ciliate
and tentaculate arms. These grow to a
considerable length, coiling up within the
shell that they may do so. In *Terebratula*
the loop remains short and simple, and is
merely flattened and distorted somewhat
by the restraining pressure of the ventral
valve; but in *Spirifer, Atrypa, Athyris* and
many more it forms a watchspring coil on
either side, corresponding to the close-
coiled arms of which it was the support and skeleton. In these
curious and characteristic structures we see no sign of progressive

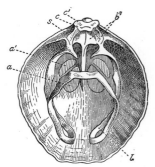

Fig. 406. Skeletal loop of *Terebratula*. From Woods.

* H. H. Swinnerton, Unit characters in fossils, *Biol. Reviews*, VII, pp. 321–335,
1932; cf. A. E. Truman, *Geol. Mag.* LIX, p. 258, LXI, p. 358, 1922–24.

growth, no successional increments, no "gnomons," no self-similarity in the figure. In short it has nothing to do with a logarithmic or equiangular spiral, but is a mere twist, or tapering helix, and it points now one way, now another. The cases in which the helicoid spires point towards, or point away from, the middle line are ascribed, in zoological classification, to particular "families" of Brachiopods, the former condition defining (or helping to define) the Atrypidae and the latter the Spiriferidae

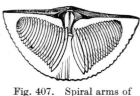

Fig. 407.   Spiral arms of
Spirifer.

Fig. 408.   Inwardly directed
spiral arms of *Atrypa.*

and Athyridae. It is obvious that the incipient curvature of the arms, and consequently the form and direction of the spirals, will be influenced by the surrounding pressures, and these in turn by the general shape of the shell. We shall expect, accordingly, to find the long outwardly directed spirals associated with shells which are transversely elongated, as *Spirifer* is; while the more rounded *Atrypa* will tend to the opposite condition. In a few cases, as in *Cyrtina* or *Reticularia*, where the shell is comparatively narrow but long, and where the uncoiled basal support of the arms is long also, the coils into which the latter grow are turned backwards, in the direction where there is most room for them. And in the few cases where the shell is very considerably flattened, the spirals (if they find room to grow at all) will be constrained to do so in a discoid or nearly discoid fashion, and this is actually the case in such flattened forms as *Koninckina* or *Thecidium.*

The shells of Pteropods

While mathematically speaking we are entitled to look upon the bivalve shell of the Lamellibranch as consisting of two distinct elements, each comparable to the entire shell of the univalve, we

have no biological grounds for such a statement; for the shell arises
from a single embryonic origin, and afterwards becomes split into
portions which constitute the two separate valves. We can perhaps
throw some indirect light upon this phenomenon, and upon several
other phenomena connected with shell-growth, by a consideration
of the simple conical or tubular shells of the Pteropods. The shells
of the latter are in few cases suitable for simple mathematical
investigation, but nevertheless they are of very considerable interest
in connection with our general problem. The morphology of the
Pteropods is by no means well understood, and in speaking of them

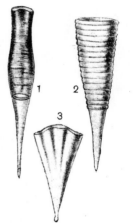

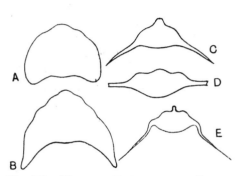

Fig. 409.   Pteropod shells:
   (1) *Cuvierina   columnella*;
   (2)   *Cleodora   chierchiae*;
   (3) *C. pygmaea*. After Boas.

Fig. 410.   Diagrammatic transverse sections, or
outlines of the mouth, in certain Pteropod shells:
A, B, *Cleodora australis*; C, *C. pyramidalis*;
D, *C. balantium*; E, *C. cuspidata*. After Boas.

I will assume that there are still grounds for believing (in spite of
Boas' and Pelseneer's arguments) that they are directly related to,
or may at least be directly compared with, the Cephalopoda*.

The simplest shells among the Pteropods have the form of a tube,
more or less cylindrical (*Cuvierina*), more often conical (*Creseis,
Clio*); and this tubular shell (as we have already had occasion to
remark, on p. 416), frequently tends, when it is very small and
delicate, to assume the character of an unduloid. (In such a case
it is more than likely that the tiny shell, or that portion of it which

* We need not assume a *close* relationship, nor indeed any more than such a
one as permits us to compare the shell of a *Nautilus* with that of a Gastropod.

constitutes the unduloid, has not grown by successive increments or "rings of growth," but has developed as a whole.) A thickened "rib" is often, perhaps generally, present on the dorsal side of the little conical shell. In a few cases (*Limacina, Peraclis*) the tube becomes spirally coiled, in a normal equiangular spiral or helico-spiral.

In certain cases (e.g. *Cleodora, Hyalaea*) the tube or cone is curiously modified. In the first place, its cross-section, originally circular or nearly so, becomes flattened or compressed dorsoventrally; and

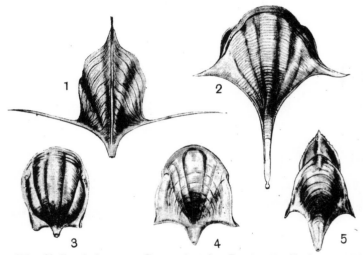

Fig. 411. Shells of thecosome Pteropods (after Boas). (1) *Cleodora cuspidata*; (2) *Hyalaea trispinosa*; (3) *H. globulosa*; (4) *H. uncinata*; (5) *H. inflexa*.

the angle, or rather edge, where dorsal and ventral walls meet, becomes more and more drawn out into a ridge or keel. Along the free margin, both of the dorsal and the ventral portion of the shell, growth proceeds with a regularly varying velocity, so that these margins, or lips, of the shell become regularly curved or markedly sinuous. At the same time, growth in a transverse direction proceeds with an acceleration which manifests itself in a curvature of the sides, replacing the straight borders of the original cone. In other words, the cross-section of the cone, or what we have been calling the generating curve, increases its dimensions more rapidly than its distance from the pole.

In the above figures, for instance in that of *Cleodora cuspidata*, the markings of the shell which represent the successive edges of the lip at former stages of growth furnish us at once with a "graph"

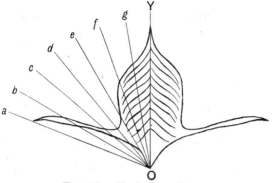

Fig. 412.   *Cleodora cuspidata.*

of the varying velocities of growth as measured, radially, from the apex.   We can reveal more clearly the nature of these variations in the following way, which is simply tantamount to converting our radial into rectangular coordinates.   Neglecting curvature (if any)

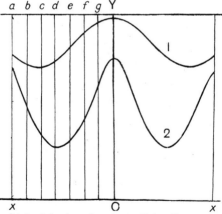

Fig. 413.   Curves obtained by transforming radial ordinates, as in Fig. 412, into vertical equidistant ordinates.   1, *Hyalaea trispinosa*; 2, *Cleodora cuspidata.*

of the sides and treating the shell (for simplicity's sake) as a right cone, we lay off equal angles from the apex *O*, along the radii *Oa*, *Ob*, etc.   If we then plot, as vertical equidistant ordinates, the magnitudes Oa, Ob ... OY, and again on to Oa', we obtain a diagram such as follows in Fig. 413: by help of which we not only see more clearly the way in which the growth-rate varies from point to point, but we also recognise better than before the nature of the law which governs this variation in the different species.

Furthermore, the young shell having become differentiated into a dorsal and a ventral part, marked off from one another by a lateral edge or keel, and the inequality of growth being such as to cause each portion to increase most rapidly in the median line, it follows that the entire shell will appear to have been split into a dorsal and a ventral plate, both connected with, and projecting from, what remains of the original undivided cone. Putting the same thing in other words, we may say that the generating figure, which

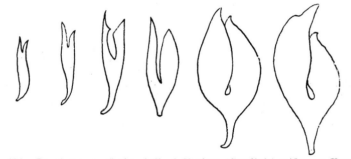

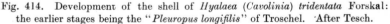

Fig. 414. Development of the shell of *Hyalaea (Cavolinia) tridentata* Forskal: the earlier stages being the "*Pleuropus longifilis*" of Troschel. After Tesch.

lay at first in a plane perpendicular to the axis of the cone, has now, by unequal growth, been sharply bent or folded, so as to lie approximately in two planes, parallel to the anterior and posterior faces of the cone. We have only to imagine the apical connecting portion to be further reduced, and finally to disappear or rupture, and we should have a *bivalve shell* developed out of the original simple cone.

In its outer and growing portion, the shell of our Pteropod now consists of two parts which, though still connected together at the apex, may be treated as growing practically independently. The shell is no longer a simple tube, or simple cone, in which regular inequalities of growth will lead to the development of a spiral; and this for the simple reason that we have now two opposite maxima of growth, instead of a maximum on the one side and a minimum on the other side of our tubular shell. As a matter of fact, the dorsal and the ventral plate tend to curve in opposite directions, towards the middle line, the dorsal curving ventrally and the ventral curving towards the dorsal side.

In the case of the Lamellibranch or the Brachiopod, it is quite possible for both valves to grow into more or less pronounced spirals, for the simple reason that they are *hinged* upon one another; and each growing edge, instead of being brought to a standstill by the growth of its opposite neighbour, is free to move out of the way, by the rotation about the hinge of the plane in which it lies.

But where there is no such hinge, as in the Pteropod, the dorsal and ventral halves of the shell (or dorsal and ventral valves, if we

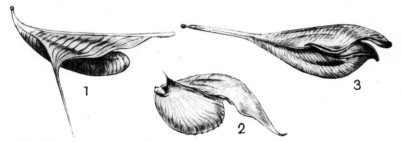

Fig. 415.   Pteropod shells, from the side: (1) *Cleodora cuspidata*; ·(2) *Hyalaea longirostris*; (3) *H. trispinosa.* After Boas.

may call them so) would soon interfere with one another's progress if they curved towards one another (as they do in a cockle), and the development of a pair of conjugate spirals would become impossible. Nevertheless, there is obviously, in both dorsal and ventral valve, a *tendency* to the development of a spiral curve, that of the ventral valve being more marked than that of the larger and overlapping dorsal one, exactly as in the two unequal valves of *Terebratula*. In many cases (e.g. *Cleodora cuspidata*), the dorsal valve or plate, strengthened and stiffened by its midrib, is nearly straight, while the curvature of the other is well displayed. But the case will be materially altered and simplified if growth be arrested or retarded in either half of the shell. Suppose for instance that the dorsal valve grew so slowly that after a while, in comparison with the other, we might speak of it as being absent altogether:

or suppose that it merely became so reduced in relative size as to form no impediment to the continued growth of the ventral one; the latter would continue to grow in the direction of its natural curvature, and would end by forming a complete and coiled logarithmic spiral. It would be precisely analogous to the spiral shell of *Nautilus*, and, in regard to its ventral position, concave towards the dorsal side, it would even deserve to be called directly homologous with it. Suppose, on the other hand, that the ventral valve were to be greatly reduced, and even to disappear, the dorsal valve would then pursue its unopposed growth; and, were it to be markedly curved, it would come to form a logarithmic spiral, concave towards the ventral side, as is the case in the shell of *Spirula**. Were the dorsal valve to be destitute of any marked curvature (or in other words, to have but a low spiral angle), it would form a simple plate, as in the shells of *Sepia* or *Loligo*. Indeed, in the shells of these latter, and especially in that of *Sepia*, we seem to recognise a manifest resemblance to the dorsal plate of the Pteropod shell, as we have it (e.g.) in *Cleodora* or *Hyalaea*; the little "rostrum" of *Sepia* is but the apex of the primitive cone, and the rounded anterior extremity has grown according to a law precisely such as that which has produced the curved margin of the dorsal valve in the Pteropod. The ventral portion of the original cone is nearly, but not wholly, wanting; it is represented by the so-called posterior wall of the "siphuncular space." In many decapod cuttle-fishes also (e.g. *Todarodes*, *Illex*, etc.) we still see at the posterior end of the "pen" a vestige of the primitive cone, whose dorsal margin only has continued to grow; and the same phenomenon, on an exaggerated scale, is represented in the *Belemnites*.

It is not at all impossible that we may explain on the same lines the development of the curious "operculum" of the Ammonites. This consists of a single horny plate (*Anaptychus*), or of a thicker, more calcified plate divided into two symmetrical halves (*Aptychi*), often found inside the terminal chamber of the Ammonite, and occasionally to be seen lying *in situ*, as an operculum which partially closes the mouth of the shell; this structure is known to exist even

---

* Cf. Owen, "These shells [*Nautilus* and Ammonites] are revolutely spiral or coiled over the back of the animal, not involute like *Spirula*": *Palaeontology*, 1861, p. 97; cf. *Memoir on the Pearly Nautilus*, 1832; also *P.Z.S.* 1878, p. 955.

in connection with the early embryonic shell. In form the Anaptychus, or the pair of conjoined Aptychi, shew an upper and a lower border, the latter strongly convex, the former sometimes slightly concave, sometimes slightly convex, and usually shewing a median projection or slightly developed rostrum. From this rostral border the curves of growth start, and course round parallel to, finally constituting, the convex border. It is this convex border which fits into the free margin of the mouth of the Ammonite's shell, while the other is applied to and overlaps the preceding whorl of the spire. Now this relationship is precisely what we should expect, were we to imagine as our starting-point a shell similar to that of *Hyalaea*: in which however the dorsal part of the split cone had become separate from the ventral half, had remained flat, and had grown comparatively slowly, while at the same time it kept slipping forward over the growing and coiling spire into which the ventral half of the original shell develops*. In short, I think there is reason to believe, or at least to suspect, that we have in the shell and Aptychus of the Ammonites, two portions of a once united structure; of which other Cephalopods retain not both parts but only one or other, one as the ventrally situated shell of *Nautilus*, the other as the dorsally placed shell for example of *Sepia* or of *Spirula*.

In the case of the bivalve shells of the Lamellibranchs or of the Brachiopods, we have to deal with a phenomenon precisely analogous to the split and flattened cone of our Pteropods, save only that the primitive cone has been split into two portions, not incompletely, as in the Pteropod (*Hyalaea*), but completely, so as to form two separate valves. Though somewhat greater freedom is given to growth now that the two valves are separate and hinged, yet still the two valves oppose and hamper one another, so that in the longitudinal direction each is capable of only a moderate curvature. This curvature, as we have seen, is recognisable as an equiangular spiral, but only now and then does the growth of the spiral continue so far as to develop successive coils: as it does in a few symmetrical forms such as *Isocardia cor*; and as it does still more conspicuously in a few others, such as *Gryphaea* and *Caprinella*, where one of the

---

* The case of *Terebratula* or of *Gryphaea* would be closely analogous, if the smaller valve were less closely connected and co-articulated with the larger.

two valves is stunted, and the growth of the other is (relatively speaking) unopposed.

Of septa

Before we leave the subject of the molluscan shell, we have still another problem to deal with, in regard to the form and arrangement of the septa which divide up the tubular shell into chambers, in the Nautilus, the Ammonite and their allies.

The existence of septa in a nautiloid shell may probably be accounted for as follows. We have seen that it is a property of a cone that, while growing by increments at one end only, it conserves its original shape: therefore the animal within, which (though growing by a different law) also conserves its shape, will continue to fill the shell if it actually fills it to begin with: as does a snail or other Gastropod. But suppose that our mollusc fills a part only of a conical shell (as it does in the case of *Nautilus*); then, unless it alter its shape, it must move upward as it grows in the growing cone, until it comes to occupy a space similar in form to that which it occupied before: just, indeed, as a little ball drops far down into the cone, but a big one must stay farther up. Then, when the animal after a period of growth has moved farther up in the shell, the mantle-surface continues or resumes its secretory activity, and that portion which had been in contact with the former septum secretes a septum anew. In short, at any given epoch, the creature is not secreting a tube and a septum by separate operations, but is secreting a shelly case about its rounded body, of which case one part appears as the continuation of the tube, and the other part, merging with it by indistinguishable boundaries, appears as the septum*.

The various forms assumed by the septa in spiral shells† present us with a number of problems of great beauty, simple in their essence, but whose full investigation would soon lead us into difficult mathematics.

---

* "It has been suggested, and I think in some quarters adopted as a dogma, that the formation of successive septa [in *Nautilus*] is correlated with the recurrence of reproductive periods. This is not the case, since, according to my observations, propagation only takes place after the last septum is formed"; Willey, *Zoological Results*, 1902, p. 746.

† Cf. Henry Woodward, On the structure of camerated shells, *Pop. Sci. Rev.* xi, pp. 113–120, 1872.

We do not know how these septa are laid down in an Ammonite, but in the Nautilus the essential facts are clear*. The septum begins as a very thin cuticular membrane (composed of a substance called conchyolin), which is secreted by the skin, or mantle-surface, of the animal; and upon this membrane nacreous matter is gradually laid down on the mantle-side (that is to say between the animal's body and the cuticular membrane which has been thrown off from it), so that the membrane remains as a thin pellicle over the *hinder* surface of the septum, and so that, to begin with, the membranous septum is moulded on the flexible and elastic surface of the animal, within which the fluids of the body must exercise a uniform, or nearly uniform pressure.

Let us think, then, of the septa as they would appear in their uncalcified condition, formed of, or at least superposed upon, an elastic membrane. They must follow the general law, applicable to all elastic membranes under uniform pressure, that the tension varies inversely as the radius of curvature; and we come back once more to our old equation of Laplace and Plateau, that

$$P = T \left(\frac{1}{r} + \frac{1}{r'} \right).$$

Moreover, since the cavity below the septum is practically closed, and is filled either with air or with water, P will be constant over the whole area of the septum. And further, we must assume, at least to begin with, that the membrane constituting the incipient septum is homogeneous or isotropic.

Let us take first the case of a straight cone, of circular section, more or less like an *Orthoceras*; and let us suppose that the septum is attached to the shell in a plane perpendicular to its axis. The septum itself must then obviously be spherical. Moreover the extent of the spherical surface is constant, and easily determined. For obviously, in Fig. 417, the angle LCL' equals the supplement of the angle (LOL') of the cone; that is to say, the circle of contact subtends an angle at the centre of the spherical surface, which is constant, and which is equal to $\pi - 2\beta$. The case is not excluded where, owing to an asymmetry of tensions, the septum meets the

* See Willey, *op. cit.*, p. 749. Cf. also Bather, Shell-growth in Cephalopoda, *Ann. Mag. N.H.* (6), I, pp. 298–310, 1888; *ibid.* pp. 421–427, and other papers by Blake, Riefstahl, etc. quoted therein.

side walls of the cone at other than a right angle, as in Fig. 416; and here, while the septa still remain portions of spheres, the geometrical construction for the position of their centres is equally easy.

If, on the other hand, the attachment of the septum to the inner walls of the cone be in a plane oblique to the axis, then the outline of the septum will be an ellipse, but its surface will still be spheroidal. If

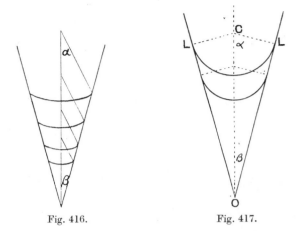

Fig. 416.   Fig. 417.

the attachment of the septum be not in one plane, but forms a sinuous line of contact with the cone, then the septum will be a saddle-shaped surface, of great complexity and beauty. In all cases, provided only that the membrane be isotropic, the form assumed will be precisely that of a soap-bubble under similar conditions of attachment: that is to say, it will be (with the usual limitations or conditions) a surface of minimal area, and of constant mean curvature.

If our cone be no longer straight, but curved, then the septa will by symmetrically deformed in consequence. A beautiful and interesting case is afforded us by *Nautilus* itself. Here the outline of the septum, referred to a plane, is approximately bounded by two elliptic curves, similar and similarly situated, whose areas are to one another in a definite ratio, namely as

$$\frac{A_1}{A_2} = \frac{r_1 r'_1}{r_2 r'_2} = \epsilon^{-4\pi \cot \alpha},$$

and a similar ratio exists in Ammonites and all other close-whorled spirals, in which however we cannot always make the simple

assumption of elliptical form. In a median section of *Nautilus*, we
see each septum forming a tangent to the inner and to the outer
wall, just as it did in a section of the straight *Orthoceras*; but the

Fig. 418.   Section of *Nautilus*, shewing the contour of the septa
in the median plane.

curvatures in the neighbourhood of these two points of contact are
not identical, for they now vary inversely as the radii, drawn from
the pole of the spiral shell. The contour of the septum in this
median plane is a spiral curve—the conformal spiral transformation
of the spherical septum of the rectilinear Orthoceratite.

But while the outline of the septum in median section is simple
and easy to determine, the curved surface of the septum in its
entirety is a very complicated matter, even in *Nautilus* which is
one of the simplest of actual cases. For, in the first place, since
the form of the septum, as seen in median section, is that of a
logarithmic spiral, and as therefore its curvature is constantly

Fig. 419. Cast of the interior of *Nautilus*: to shew the contours of
the septa at their junction with the shell-wall.

altering, it follows that, in successive *transverse* sections, the curvature is also constantly altering. But in the case of *Nautilus*, there
are other aspects of the phenomenon, which we can illustrate, but
only in part, in the following simple manner. Let us imagine a pack
of cards, in which we have cut out of each card a similar concave
arc of a logarithmic spiral, such as we actually see in the median
section of the septum of a *Nautilus*. Then, while we hold the cards
together, foursquare, in the ordinary position of the pack, we have
a simple "ruled" surface, which in any longitudinal section has the

form of a logarithmic spiral but in any transverse section is a straight horizontal line. If we shear or slide the cards upon one another, thrusting the middle cards of the pack forward in advance of the others, till the one end of the pack is a convex, and the other a concave, ellipse, the cut edges which combine to represent our septum will now form a curved surface of much greater complexity; and this is part, but not by any means all, of the deformation produced as a direct consequence of the form in *Nautilus* of the section of the tube within which the septum has to lie. The complex curvature of the surface will be manifested in a sinuous outline of the edge, or line of attachment of the septum to the tube, and will vary according to the configuration of the latter. In the case of *Nautilus*, it is easy to shew empirically (though not perhaps easy to demonstrate mathematically), that the sinuous or saddle-shaped contour of the "suture" (or line of attachment of the septum to the tube) is such as can be precisely accounted for in this manner; and we may find other forms, such as *Ceratites*, where the septal outline is only a little more sinuous, and still precisely analogous to that of *Nautilus*. It is also easy to see that, when the section of the tube (or "generating curve") is more complicated in form, when it is flattened, grooved, or otherwise ornamented, the curvature of the septum and the outline of its sutural attachment will become very complicated indeed*; but it will be comparatively simple in the case of the first few sutures of the young shell, laid down before any over-lapping of whorls has taken place, and this comparative simplicity of the first-formed sutures is a marked feature among Ammonites†.

---

* The "lobes" and "saddles" which arise in this manner, and on whose arrangement the modern classification of the nautiloid and ammonitoid shells largely depends, were first recognised and named by Leopold von Buch, *Ann. Sci. Nat.* xxvii, xxviii, 1829.

† Blake has remarked upon the fact (*op. cit.* p. 248) that in some Cyrtocerata we may have a curved shell in which the ornaments approximately run at a constant angular distance from the pole, while the septa approximate to a radial direction; and that "thus one law of growth is illustrated by the inside, and another by the outside." In this there is nothing at which we need wonder. It is merely a case where the generating curve is set very obliquely to the axis of the shell; but where the septa, which have no necessary relation to the *mouth* of the shell, take their places, as usual, at a certain definite angle to the *walls* of the tube. This relation of the septa to the walls of the tube arises after the tube itself is fully formed, and the obliquity of growth of the open end of the tube has no relation to the matter.

We have other sources of complication, besides those which are
at once introduced by the sectional form of the tube.   For instance,
the siphuncle, or little inner tube which perforates the septa, exercises
a certain amount of tension, sometimes evidently considerable, upon
the latter: which tension is made manifest in *Spirula* (and slightly
so even in *Nautilus*) by a dip in the septal floor where it meets the
siphuncle.   We can no longer, then, consider each septum as an
isotropic surface under uniform pressure; and there may be other
structural modifications, or inequalities, in that portion of the

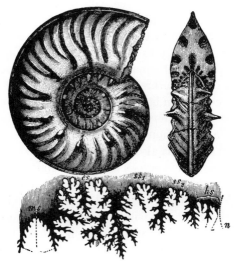

Fig. 420.   *Ammonites Sowerbyi.*   From Zittel.

animal's body with which the septum is in contact, and by which
it is conformed.   It is hardly likely, for all these reasons, that we
shall ever attain to a full and particular explanation of the septal
surfaces and their sutural outlines throughout the whole range of
Cephalopod shells;   but in general terms, the problem is probably
not beyond the reach of mathematical analysis.   The problem might
be approached experimentally, after the manner of Plateau's experiments, by bending a wire into the complicated form of the suture-line,
and studying the form of the liquid film which constitutes the
corresponding surface *minimae areae*.

In certain Ammonites the septal outline is further complicated
in another way.   Superposed upon the usual sinuous outline, with

its "lobes" and "saddles," we have here a minutely ramified, or arborescent outline, in which all the branches terminate in wavy, more or less circular arcs—looking just like the "landscape marble" from the Bristol Rhaetic. We have no difficulty in recognising in this a surface-tension phenomenon. The figures are precisely such as we can imitate (for instance) by pouring a few drops of milk upon a greasy plate, or of oil upon an alkaline solution*; they are what Charles Tomlinson called "cohesion figures."

Fig. 421. Suture-line of a Triassic Ammonite (*Pinacoceras*). From Zittel.

We must not forget that while the nautilus and the ammonite resemble one another, and are mathematically identical in their spiral curves, they are really very different things. The one is an external, the other an internal shell. The nautilus occupies the large terminal chamber of the many-chambered shell, and "Still as the spiral grew, He left the past year's dwelling for the new." But even the largest ammonites never contained the body of the animal, but lay hidden, as *Spirula* does, deep within the substance of the mantle. How the complicated septa and septal outlines of the ammonites are produced I do not know†.

We have very far from exhausted, we have perhaps little more than begun, the study of the logarithmic spiral and the associated curves which find exemplification in the multitudinous diversities of molluscan shells. But, with a closing word or two, we must now bring this chapter to an end.

---

* "The Fimbriae, or Edges, appeared on the Surface like the Outlines of some curious Foliage. This, upon Examination of them, I found to proceed from the Fulness of the Edges of the Diaphragms, whereby the Edges were waved or plaited somewhat in the manner of a Ruff" (R. Hooke, *op. cit.*).

† In certain rare cases the complicated sutural pattern of an ammonite is found *upside down*, but unchanged otherwise. Cf. Otto Haas, A case of inversion of suture lines in *Hysteroceras, Amer. Jl. of Sci.* ccxxxix, p. 661, 1941.

In the spiral shell we have a problem, or a phenomenon, of growth, immensely simplified by the fact that each successive increment is no sooner formed than it is fixed irrevocably, instead of remaining in a state of flux and sharing in the further changes which the organism undergoes. In such a structure, then, we have certain primary phenomena of growth manifested in their original simplicity, undisturbed by secondary and conflicting phenomena. What actually *grows* is merely the lip of an orifice, where there is produced a ring of solid material, whose form we have discussed under the name of the generating curve; and this generating curve grows in magnitude without alteration of its form. Besides its increase in areal magnitude, the growing curve has certain strictly limited degrees of freedom, which define its motions in space. And, though we may know nothing whatsoever about the actual velocities of any of these motions, we do know that they are so correlated together that their *relative* velocities remain constant, and accordingly the form and symmetry of the whole system remain in general unchanged.

But there is a vast range of possibilities in regard to every one of these factors: the generating curve may be of various forms, and even when of simple form, such as an ellipse, its axes may be set at various angles to the system; the plane also in which it lies may vary, almost indefinitely, in its angle relatively to that of any plane of reference in the system; and in the several velocities of growth, of rotation and of translation, and therefore in the ratios between all these, we have again a vast range of possibilities. We have then a certain definite type, or group of forms, mathematically isomorphous, but presenting infinite diversities of outward appearance: which diversities, as Swammerdam said, *ex sola nascuntur diversitate gyrationum*; and which accordingly are seen to have their origin in differences of rate, or of magnitude, and so to be, essentially, neither more nor less than *differences of degree*.

In nature, we find these forms presenting themselves with but little relation to the character of the creature by which they are produced. Spiral forms of certain particular kinds are common to Gastropods and to Cephalopods, and to diverse families of each; while outside the class of molluscs altogether, among the Foraminifera and among the worms (as in *Spirorbis*, *Spirographis*, and in the *Dentalium*-like shell of *Ditrupa*), we again meet with similar and corresponding spirals.

Again, we find the same forms, or forms which (save for external ornament) are mathematically identical, repeating themselves in all periods of the world's geological history; and we see them mixed up, one with another, irrespective of climate or local conditions, in the depths and on the shores of every sea. It is hard indeed (to my mind) to see in such a case as this where Natural Selection necessarily enters in, or to admit that it has had any share whatsoever in the production of these varied conformations. Unless indeed we use the term Natural Selection in a sense so wide as to deprive it of any purely biological significance; and so recognise as a sort of natural selection whatsoever nexus of causes suffices to differentiate between the likely and the unlikely, the scarce and the frequent, the easy and the hard: and leads accordingly, under the peculiar conditions, limitations and restraints which we call "ordinary circumstances," one type of crystal, one form of cloud, one chemical compound, to be of frequent occurrence and another to be rare*.

* Cf. Bacon, *Advancement of Learning*, Bk. II (p. 254): "Doth any give the reason, why some things in nature are so common and in so great mass, and others so rare and in so small quantity?"

CHAPTER XII

THE SPIRAL SHELLS OF THE FORAMINIFERA

W E have already dealt in a few simple cases with the shells of the
Foraminifera*; and we have seen that wherever the shell is but a
single unit or single chamber, its form may be explained in general
by the laws of surface-tension: the argument (or assumption) being
that the little mass of protoplasm which makes the simple shell
behaves as a *fluid drop*, the form of which is perpetuated when the
protoplasm acquires its solid covering. Thus the spherical Orbulinae
and the flask-shaped Lagenae represent drops in equilibrium, under
various conditions of freedom or constraint; while the irregular,
amoeboid body of Astrorhiza is a manifestation not of equilibrium,
but of a varying and fluctuating distribution of surface energy.
When the foraminiferal shell becomes multilocular, the same general
principles continue to hold; the growing protoplasm increases drop
by drop, and each successive drop has its particular phenomena of
surface energy, manifested at its fluid surface, and tending to confer
upon it a certain place in the system and a certain shape of its own.
It is characteristic and even diagnostic of this particular group
of Protozoa (1) that development proceeds by a well-marked alternation of rest and of activity—of activity during which the protoplasm
increases, and of rest during which the shell is formed; (2) that the
shell is formed at the outer surface of the protoplasmic organism,
and tends to constitute a continuous or all but continuous covering;
and it follows (3) from these two factors taken together that each
successive increment is added on outside of and distinct from its
predecessors, that the successive parts or chambers of the shell are
of different and successive ages, so that one part of the shell is always
relatively new, and the rest old in various grades of seniority.

The forms which we set together in the sister-group of Radiolaria
are very differently characterised. Here the cells or vesicles of
which each little composite organism is made up are but little

* Cf. pp. 420, 702, etc.

separated, and in no way walled off, from one another; the hard
skeletal matter tends to be deposited in the form of isolated spicules
or of little connected rods or plates, at the angles, the edges or the
interfaces of the vesicles; the cells or vesicles form a coordinated
and cotemporaneous rather than a successive series. In a word,
the whole quasi-fluid protoplasmic body may be likened to a little
mass of froth or foam: that is to say, to an aggregation of simultaneously formèd drops or bubbles, whose physical properties and
geometrical relations are very different from those of a system of

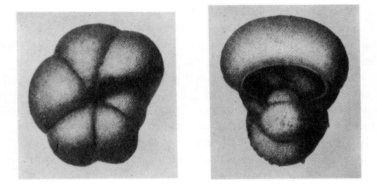

Fig. 422.   *Hastigerina* sp.; to shew the "mouth."

drops or bubbles which are formed one after another, each solidifying
before the next is formed.

With the actual origin or mode of development of the foraminiferal
shell we are now but little concerned. The main factor is the
adsorption, and subsequent precipitation at the surface of the
organism, of calcium carbonate—the shell so formed being interrupted
by pores or by some larger interspace or "mouth" (Fig. 422), which
interruptions we may doubtless interpret as being due to unequal
distributions of surface energy. In many cases the fluid protoplasm
"picks up" sand-grains and other foreign particles, after a fashion
which we have already described (p. 702); and it cements these
together with more or less of calcareous material. The calcareous
shell is a crystalline structure, and the micro-crystals of calcium
carbonate are so set that their little prisms radiate outwards in each
chamber through the thickness of the wall—which symmetry is

subject to corresponding modification when the spherical chambers are more or less symmetrically deformed *.

In various ways the rounded drop-like shells of the Foraminifera, both simple and compound, have been artificially imitated. Thus, if small globules of mercury be immersed in water in which a little chromic acid is allowed to dissolve, as the little beads of quicksilver become slowly covered with a crystalline coat of mercuric chromate they assume various forms reminiscent of the monothalamic Foraminifera. The mercuric chromate has a higher atomic volume than the mercury which it replaces, and therefore the fluid contents of the drop are under pressure, which increases with the thickness of the pellicle; hence at some weak spot in the latter the contents will presently burst forth, so forming a mouth to the little shell. Sometimes a long thread is formed, just as in *Rhabdammina linearis*; and sometimes unduloid swellings make their appearance on such a thread, just as in *R. discreta*. And again, by appropriate modifications of the experimental conditions, it is possible (as Rhumbler has shewn) to build up a chambered shell †.

In a few forms, such as *Globigerina* and its close allies, the shell is beset during life with excessively long and delicate calcareous spines or needles. It is only in oceanic forms that these are present, because only when poised in water can such delicate structures endure; in dead shells, such as we are much more familiar with, every trace of them is broken and rubbed away. The growth of these long needles may be partly explained (as we have already said on p. 675) by the phenomenon which Lehmann calls *orientirte Adsorption*—the tendency for a crystalline structure to grow by accretion, not necessarily in the outward form of a "crystal," but continuing in any direction or orientation which has once been impressed upon it: in this case the spicular growth is in direct continuation of the radial symmetry of the micro-crystalline

---

* In a few cases, according to Awerinzew and Rhumbler, where the chambers are added on in concentric series, as in Orbitolites, we have the crystalline structure arranged radially in the radial walls but tangentially in the concentric ones: whereby we tend to obtain, on a minute scale, a system of orthogonal trajectories, comparable to that which we shall presently study in connection with the structure of bone. Cf. S. Awerinzew, Kalkschale der Rhizopoden, *Z. f. w. Z.* LXXIV, pp. 478–490, 1903.

† L. Rhumbler, Die Doppelschalen von Orbitolites und anderer Foraminiferen, etc., *Arch. f. Protistenkunde*, I, pp. 193–296, 1902; and other papers. Also *Die Foraminiferen der Planktonexpedition*, I, pp. 50–56, 1911.

elements of the shell-wall. But the calcareous needles are secreted
in, or *by*, no less long and delicate pseudopodia or "filopodia,"
and much has been learned since this book was written of the
molecular, or micellar, orientation of the protoplasm in such
filamentous structures; it is known that the long pseudopodia of
the Foraminifera are doubly refractive, and it follows that their
molecules are anisotropically arranged*. Whether the slender form
and asymmetrical structure of calcareous rod and protoplasmic
thread be independent phenomena, or merely two aspects of one
and the same phenomenon, is a hard question, and not one for us
to discuss. Nor can we profitably discuss (much as we should like
to know) how far these patterns of molecular structure in threads,
films and surface-pellicles affect the "fluidity" of the substance,
and conflict with the capillary forces which influence its outward
form. But we may safely say that the effects of surface-tension
on cell-form have been so plainly seen all through this book that
any counter-effects due to protoplasmic asymmetry must be
phenomena of a second order, and inconspicuous on the whole.
Over the whole surface of the shell of *Globigerina* the radiating
spicules tend to occur in a hexagonal pattern, symmetrically
grouped around the pores which perforate the shell. Rhumbler
has suggested that this arrangement is due to diffusion-currents,
forming little eddies about the base of the pseudopodia issuing from
the pores: the idea being borrowed from Bénard, to whom is due
the discovery of this type or order of vortices†. In one of Bénard's
experiments a thin layer of paraffin is strewn with particles of
graphite, then warmed to melting, whereupon each little solid granule
becomes the centre of a vortex; by the interaction of these vortices
the particles tend to be repelled to equal distances from one another,
and in the end they are found to be arranged in a hexagonal pattern‡.

* Cf. W. J. Schmidt, *Die Bausteine des Tierkörpers in polarisiertem Lichte*,
Bonn, 1924; Ueber den Feinbau der Filopodien; insb. ihre Doppelbrechung bei
Miliola, *Protoplasma*, xxvii, p. 587, 1937; also D. L. Mackinnon, Optical pro-
perties of contractile organs in Heliozoa, *Jl. Physiol.* xxxviii, p. 254, 1909;
R. O. Herzog, Lineare u. laminäre Feinstrukturen, *Kolloidzschr.* lxi, p. 280, 1932.
See, for discussion and bibliography, L. E. Picken, *op. cit.*

† H. Bénard, Les tourbillons cellulaires, *Ann. de Chimie* (8), xxiv, 1901. Cf.
also the pattern of cilia on an Infusorian, as figured by Bütschli in Bronn's
Protozoa, iii, p. 1281, 1887.

‡ A similar hexagonal pattern is obtained by the mutual repulsion of floating
magnets in Mr R. W. Wood's experiments, *Phil. Mag.* xlvi, pp. 162–164, 1898.

The analogy is plain between this experiment and those diffusion experiments by which Leduc produces his beautiful hexagonal systems of artificial cells, with which we have dealt in a previous chapter. But let us come back to the shell itself, and consider particularly its spiral form. That the shell in the Foraminifera should tend towards a spiral form need not surprise us; for we have learned that one of the fundamental conditions of the production of a concrete spiral is just precisely what we have here, namely the development of a structure by means of successive graded increments superadded to its exterior, which then form part, successively, of a permanent and rigid structure. This condition is obviously forthcoming in the foraminiferal, but not at all in the radiolarian, shell. Our second fundamental condition of the production of a logarithmic spiral is that each successive increment shall be so posited and so conformed that its addition to the system leaves the form of the whole system unchanged. We have now to enquire into this latter condition; and to determine whether the successive increments, or successive chambers, of the foraminiferal shell actually constitute *gnomons* to the entire structure.

It is obvious enough that the spiral shells of the Foraminifera closely resemble true logarithmic spirals. Indeed so precisely do the minute shells of many Foraminifera repeat or simulate the spiral shells of *Nautilus* and its allies that to the naturalists of the early nineteenth century they were known as the *Céphalopodes microscopiques**, until Dujardin shewed that their little bodies comprised no complex anatomy of organs, but consisted merely of that slime-like organic matter which he taught us to call "sarcode," and which we learned afterwards from Schwann to speak of as "protoplasm."

One striking difference, however, is apparent between the shell of *Nautilus* and the little nautiloid or rotaline shells of the Foraminifera: namely that the septa in these latter, and in all other chambered Foraminifera, are convex outwards (Fig. 423), whereas they are concave outwards in *Nautilus* (Fig. 347) and in the rest of the chambered molluscan shells. The reason is perfectly simple.

* Cf. Alc. d'Orbigny, Tableau méthodique de la classe des Céphalopodes, *Ann. des Sci. Nat.* (1), VII, pp. 245–315, 1826; Félix Dujardin, Observations nouvelles sur les prétendus Céphalopodes microscopiques, *ibid.* (2), III, pp. 108, 109, 312–315, 1835; Recherches sur les organismes inférieurs, *ibid.* IV, pp. 343–377, 1835; etc.

In both cases the curvature of the septum was determined before it became rigid, and at a time when it had the properties 'of a fluid film or an elastic membrane. In both cases the actual curvature is determined by the tensions of the membrane and the pressures to which it was exposed. Now it is obvious that the extrinsic pressure which the tension of the membrane has to withstand is on opposite sides in the two cases. In *Nautilus*, the pressure to be resisted is that produced by the growing body of the animal, lying to the *outer side* of the septum, in the outer, wider portion of the tubular shell. In the Foraminifer the septum at the time of its formation was no septum at all; it was but a portion of the convex

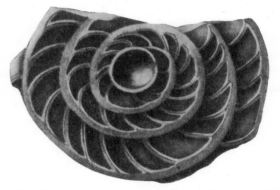

Fig. 423. *Nummulina antiquior* R. and V. After V. von Möller.

surface of a drop—that portion namely which afterwards became overlapped and enclosed by the succeeding drop; and the curvature of the septum is concave towards the pressure to be resisted, which latter is *inside* the septum, being simply the hydrostatic pressure of the fluid contents of the drop. The one septum is, speaking generally, the reverse of the other; the organism, so to speak, is outside the one and inside the other; and in both cases alike, the septum tends to assume the form of a surface of minimal area, as permitted, or as defined, by all the circumstances of the case.

The logarithmic spiral is easily recognisable in typical cases* (and

* It is obvious that the actual *outline* of a foraminiferal, just as of a molluscan shell, may depart widely from a logarithmic spiral. When we say here, for short, that the shell *is* a logarithmic spiral, we merely mean that it is essentially related to one: that it can be inscribed in such a spiral, or that corresponding points (such, for instance, as the centres of gravity of successive chambers, or the extremities of successive septa) will be found to lie upon such a spiral.

especially where the spire makes more than one visible revolution about the pole), by its fundamental property of continued similarity: that is to say, by reason of the fact that the big many-chambered shell is of just the same shape as the smaller and younger shell— which phenomenon is as apparent and even obvious in the nautiloid Foraminifera, as in *Nautilus* itself: but nevertheless the nature of the curve must be verified by careful measurement, just as Moseley determined or verified it in his original study of *Nautilus* (cf. p. 770). This has accordingly been done, by various writers: and in the first instance by Valerian von Möller, in an elaborate study of *Fusulina*— a palaeozoic genus whose little shells have built up vast tracts of carboniferous limestone in European Russia *.

In this genus a growing surface of protoplasm may be conceived as wrapping round and round a small initial chamber, in such a way as to produce a fusiform or ellipsoidal shell—a transverse section of which reveals the close-wound spiral coil. The following are measurements of the successive whorls in a couple of species of this genus:—

	F. cylindrica Fischer		*F. Böcki* v. Möller	
		Breadth (in millimetres)		
Whorl	Observed	Calculated	Observed	Calculated
I	0·132	—	0·079	—
II	0·195	0·198	0·120	0·119
III	0·300	0·297	0·180	0·179
IV	0·449	0·445	0·264	0·267
V	—	—	0·396	0·401

In both cases the successive whorls are very nearly in the ratio of 1 : 1·5; and on this ratio the calculated values are based.

Here is another of von Möller's series of measurements of *F. cylindrica*, the measurements being those of opposite whorls—that is to say of whorls 180° apart:

Breadth (mm.)	0·096	0·117	0·144	0·176	0·216	0·264	0·323	0·395
Log. of do.	0·982	0·068	0·158	0·246	0·334	0·422	0·509	0·597
Diff. of logs.	—	0·086	0·090	0·088	0·088	0·088	0·087	0·088

The mean logarithmic difference is here 0·088, = log 1·225; or the mean difference of alternate logs (corresponding to a vector angle of 2π, i.e. to consecutive measurements along the *same* radius) is 0·176, = log 1·5, the same value as before. And this ratio of 1·5 between the breadths of successive whorls corresponds (as we see

* V. von Möller, Die spiral-gewundenen Foraminifera des rüssischen Kohlenkalks, *Mém. de l'Acad. Imp. Sci.*, St Pétersbourg (7), xxv, 1878.

by our table on p. 791) to 'a constant angle of about 86°, or just such a spiral as we commonly meet with in the Ammonites (cf. p. 796).

In *Fusulina*, and in some few other Foraminifera (cf. Fig. 424, A), the spire seems to wind evenly on, with little or no external sign of the successive periods of growth, or successive chambers of the shell. The septa which mark off the chambers, and correspond to retardations or cessations in the periodicity of growth, are still to be found in sections of the shell of *Fusulina*, but they are somewhat irregular and comparatively inconspicuous; the measurements we have just spoken of are taken without reference to the segments or chambers, but only with reference to the whorls, or in other words with direct reference to the vectorial angle.

| A | B |

Fig. 424.   A, *Cornuspira foliacea* Phil.; B, *Operculina complanata* Defr.

The linear dimensions of successive chambers have been measured in a number of cases. Van Iterson* has done so in various Miliolinidae, with such results as the following:

Triloculina rotunda d'Orb.

No. of chamber ...		1	2	3	4	5	6	7	8	9	10
Breadth of chamber in μ		—	34	45	61	84	114	142	182	246	319
Breadth of chamber in μ, calculated	...	—	34	45	60	79	105	140	187	243	319

* G. van Iterson, *Mathem. u. mikrosk.-anat. Studien über Blattstellungen, nebst Betrachtungen über den Schalenbau der Miliolinen*, 331 pp., Jena, 1907.

Here the mean ratio of breadth of consecutive chambers may be taken as 1·323 (that is to say, the eighth root of 319/34); and the calculated values, as given above, are based on this determination. Again, Rhumbler has measured the linear dimensions of a number of rotaline forms, for instance *Pulvinulina menardi* (Fig. 363): in which common species he finds the mean linear ratio of consecutive chambers to be about 1·187. In both cases, and especially in the latter, the ratio is not strictly constant from chamber to chamber, but is subject to a small secondary fluctuation*.

When the linear dimensions of successive chambers are in continued proportion, then, in order that the whole shell may constitute a logarithmic spiral, it is necessary that the several chambers should subtend equal angles of revolution at the pole. In the case of the Miliolidae this is obviously the case (Fig. 425); for in this family the chambers lie in two rows (*Biloculina*), or three rows (*Triloculina*), or in some other small number of series: so that the angles subtended by them are large, simple fractions of the circular arc, such as 180° or 120°. In many of the nautiloid forms, such as *Cyclammina* (Fig. 426), the angles subtended, though of less magnitude, are still remarkably constant, as we may see by Fig. 427; where the angle subtended by each chamber is made equal to 20°, and this diagrammatic figure is not perceptibly different from the other. In some cases the subtended angle is less constant; and in these it would be necessary to equate the several linear dimensions with the corresponding vector angles, according to our equation $r = e^{\theta \cot \alpha}$. It is probable that, by so taking account of variations of θ, such variations of r as (according to Rhumbler's measurements) *Pulvinulina* and other genera appear to shew, would be found to diminish or even to disappear.

The law of increase by which each chamber bears a constant ratio of magnitude to the next may be looked upon as a simple

* Hans Przibram asserts that the linear ratio of successive chambers tends in many Foraminifera to approximate to 1·26, which $= \sqrt[3]{2}$; in other words, that the volumes of successive chambers tend to double. This Przibram would bring into relation with another law, viz. that insects and other arthropods tend to moult, or to metamorphose, just when they double their weights, or increase their linear dimensions in the ratio of 1 : $\sqrt[3]{2}$. (Die Kammerprogression der Foraminiferen als Parallele zur Häutungsprogression der Mantiden, *Arch. f. Entw. Mech.* xxxiv, p. 680, 1813.) Neither rule seems to me to be well grounded (see above, p. 165).

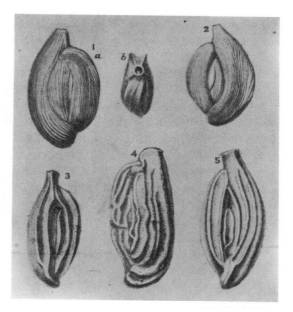

Fig. 425.   1, 2, *Miliolina pulchella* d'Orb.;   3–5, *M. linnaeana* d'Orb
After Brady.

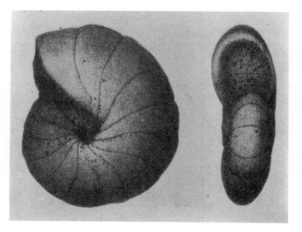

Fig. 426.   *Cyclammina cancellata* Brady.

consequence of the structural uniformity or homogeneity of the organism; we have merely to suppose (as this uniformity would naturally lead us to do) that the rate of increase is at each instant proportional to the whole existing mass. For if V_0, V_1, etc. be

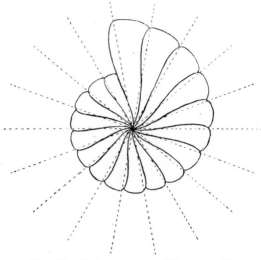

Fig. 427. *Cyclammina* sp. (Diagrammatic.)

the volumes of the successive chambers, let V_1 bear a constant proportion to V_0, so that $V_1 = qV_0$, and let V_2 bear the same proportion to the whole pre-existing volume: then

$$V_2 = q\,(V_0 + V_1) = q\,(V_0 + qV_0) = qV_0\,(1 + q) \quad \text{and} \quad V_2/V_1 = 1 + q.$$

This ratio of $1/(1 + q)$ is easily shewn to be the constant ratio running through the whole series, from chamber to chamber; and if this ratio of volumes be constant, so also are the ratios of corresponding surfaces, and of corresponding linear dimensions, provided always that the successive increments, or successive chambers, are similar in form.

We have still to discuss the similarity of form and the symmetry of position which characterise the successive chambers, and which, together with the law of continued proportionality of size, are the distinctive characters and the indispensable conditions of a series of "gnomons."

The minute size of the foraminiferal shell or at least of each

successive increment thereof, taken in connection with the fluid or
semi-fluid nature of the protoplasmic substance, is enough to suggest
that the molecular forces, and especially the force of surface-tension,
must exercise a controlling influence over the form of the whole
structure; and this suggestion, or belief, is already implied in our
statement that each successive increment of growing protoplasm
constitutes a separate *drop*. These "drops," partially concealed by
their successors, but still shewing in part their rounded outlines,
are easily recognisable in the various foraminiferal shells which are
illustrated in this chapter.

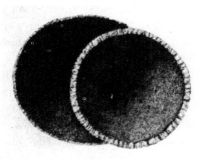

Fig. 428.   *Orbulina universa* d'Orb.

The accompanying figure represents, to begin with, the spherical
shell characteristic of the common, floating, oceanic *Orbulina*.   In
the specimen illustrated, a second chamber, superadded to the first,
has arisen as a drop of protoplasm which exuded through the pores
of the first chamber, accumulated on its surface, and spread over
the latter till it came to rest in a position of equilibrium.   We may
take it that this position of equilibrium is determined, at least in
the first instance, by the "law of the constant angle," which holds,
or tends to hold, in all cases where the free surface of a given liquid
is in contact with a given solid, in presence of another liquid or a gas.
The corresponding equations are precisely the same as those which
we have used in discussing the form of a drop (on p. 466); though
some slight modification must be made in our definitions, inasmuch
as the consideration of surface-*tension* is no longer appropriate at
the solid surfaces, and the concept of surface-*energy* must take its

place. Be that as it may, it is enough for us to observe that, in such a case as ours, when a given fluid (namely protoplasm) is in surface contact with a solid (viz. a calcareous shell), in presence of another fluid (sea-water), then the angle of contact, or angle by which the common surface (or interface) of the two liquids abuts against the solid wall, tends to be constant: and that being so, the drop will have a certain definite form, depending (*inter alia*) on the form of the surface with which it is in contact. After a period of rest, during which the surface of our second drop becomes rigid by calcification, a new period of growth will recur and a new drop of protoplasm be accumulated. Circumstances remaining the same, this new drop will meet the solid surface of the shell at the same angle as did the former one; and, the other forces at work on the system remaining the same, the form of the whole drop, or chamber, will be the same as before.

According to Rhumbler, this "law of the constant angle" is the fundamental principle in the mechanical conformation of the foraminiferal shell, and provides for the symmetry of form as well as of position in each succeeding drop of protoplasm: which form and position, once acquired, become rigid and fixed with the onset of calcification. But Rhumbler's explanation brings with it its own difficulties. It is by no means easy of verification, for on the very complicated curved surfaces of the shell it seems to me extraordinarily difficult to measure, or even to recognise, the actual angle of contact: of which angle of contact, by the way, but little is known, save only in the particular case where one of the three bodies is air, as when a surface of water is exposed to air and in contact with glass. It is easy moreover to see that in many of our Foraminifera the angle of contact, though it may be constant in homologous positions from chamber to chamber, is by no means constant at all points along the boundary of each chamber. In *Cristellaria*, for instance (Fig. 429), it would seem to be (and Rhumbler asserts that it actually is) about 90° on the outer side and only about 50° on the inner side of each septal partition; in *Pulvinulina* (Fig. 363), according to Rhumbler, the angles adjacent to the mouth are of 90°, and the opposite angles are of 60°, in each chamber. For these and other similar discrepancies Rhumbler would account by simply invoking the heterogeneity of the protoplasmic drop: that is to say, by assuming that the protoplasm has a different composition and different properties (including a very different distribution of surface-energy), at points near to and remote from the mouth of the shell. Whether the differences in angle of contact be as great as Rhumbler takes them to be, whether marked heterogeneities of the protoplasm occur, and whether these be enough to account for the differences of angle, I cannot tell. But it seems to me that we had better rest content with a general statement, and that Rhumbler has taken too precise and narrow a view.

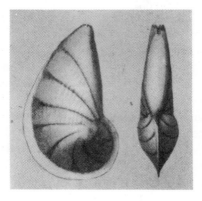

Fig. 429. *Cristellaria reniformis* d'Orb

In the molecular growth of a crystal, although we must of necessity assume that each molecule settles down in a position of minimum potential energy, we find it very hard indeed to explain precisely, even in simple cases and after all the labours of modern crystallographers, why this or that position is actually a place of minimum potential. In the case of our little Foraminifer (just as in the case of the crystal), let us then be content to assert that each drop or bead of protoplasm takes up a position of minimum potential energy, in relation to all the circumstances of the case; and let us not attempt, in the present state of our knowledge, to define that position of minimum potential by reference to angle of contact or any other particular condition of equilibrium. In most cases the whole exposed surface, on some portion of which the drop must come to rest, is an extremely complicated one, and the forces involved constitute a system which, in its entirety, is more complicated still; but from the symmetry of the case and the continuity of the whole phenomenon, we are entitled to believe that the conditions are just the same, or very nearly the same, time after time, from one chamber to another: as the one chamber is conformed so will the next tend to be, and as the one is situated relatively to the system so will its successor tend to be situated in turn. The physical law of minimum potential (including also the law of minimal area) is all that we need in order to explain, *in general terms*, the continued similarity of one chamber to another; and the physiological law of growth, by which a continued proportionality of size tends to run through the series of successive chambers, impresses the form of a logarithmic spiral upon this series of similar increments.

In each particular case the nature of the logarithmic spiral, as defined by its constant angle, will be chiefly determined by the rate of growth; that is to say by the particular ratio in which each new chamber exceeds its predecessor in magnitude. But shells having the same constant angle (α) may still differ from one another in many ways—in the general form and relative position of the chambers, in their extent of overlap, and hence in the actual contour and appearance of the shell; and these variations must correspond to particular distributions of energy within the system, which is governed as a whole by the law of minimum potential.

Our problem, then, becomes reduced to that of investigating the possible configurations which may be derived from the successive symmetrical apposition of similar bodies whose magnitudes are in continued proportion; and it is obvious, mathematically speaking, that the various possible arrangements all come under the head of the logarithmic spiral, together with the limiting cases which it includes. Since the difference between one such form and another depends upon the numerical value of certain coefficients of magnitude, it is plain that any one must tend to pass into any other by small and continuous gradations; in other words, that a *classification* of these forms must (like any classification whatsoever of logarithmic spirals or of any other mathematical curves) be theoretic or "artificial." But we may easily make such an artificial classification, and shall probably find it to agree, more or less, with the usual methods of classification recognised by biological students of the Foraminifera.

Firstly we have the typically spiral shells, which occur in great variety, and which (for our present purpose) we need hardly describe further. We may merely notice how in certain cases, for instance *Globigerina*, the individual chambers are little removed from spheres; in other words, the area of contact between the adjacent chambers is small. In such forms as *Cyclammina* and *Pulvinulina*, on the other hand, each chamber is greatly overlapped by its successor, and the spherical form of each is lost in a marked asymmetry. Furthermore, in *Globigerina* and some others we have a tendency to the development of a gauche spiral in space, as in so many of our univalve molluscan shells. The mathematical problem of how a shell should grow, under the assumptions which we have made, would probably find its most general statement in such a case as that of *Globigerina*, where the whole organism lives and grows freely poised in a medium whose density is little different from its own.

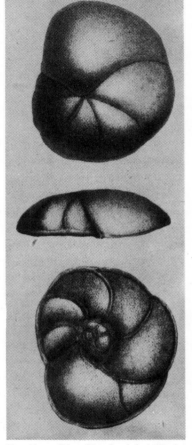

The majority of spiral forms, on the other hand, are plane or discoid spirals, and we may take it that in these cases some force has exercised a controlling influence, so as to keep all the chambers in a plane. This is especially the case in forms like *Rotalia* or *Discorbina* (Fig. 430), where the organism lives

Fig. 430.   *Discorbina bertheloti* d'Orb.

attached to a rock or a frond of sea-weed; for here (just as in the case of the coiled tubes which little worms such as *Serpula* and *Spirorbis* make, under similar conditions) the spiral disc is itself asymmetrical, its whorls being markedly flattened on their attached surfaces.

We may also conceive, among other conditions, the very curious case in which the protoplasm may entirely overspread the surface of the shell without reaching a position of equilibrium; in which case a new shell will be formed *enclosing* the old one, whether the old one be in the form of a single, solitary chamber, or have already attained to the form of a chambered or spiral shell. This is precisely what often happens in the case of *Orbulina*, when within the spherical shell we find a small, but perfectly formed, spiral "*Globigerina**."

The various Miliolidae (Fig. 425) only differ from the typical spiral, or rotaline forms, in the large angle subtended by each chamber, and the consequent abruptness of their inclination to each other. In these cases the *outward* appearance of a spiral tends to be lost; and it behoves us to recollect, all the more, that our spiral curve is not necessarily identical with the *outline* of the shell, but is always a line drawn through corresponding *points* in the successive chambers of the latter.

We reach a limiting case of the logarithmic spiral when the chambers are arranged in a straight line; and the eye will tend to associate with this limiting case the much more numerous forms in which the spiral angle is small, and the shell only exhibits a gentle curve with no succession of enveloping whorls. This constitutes the Nodosarian type (Fig. 134, p. 421); and here again, we must postulate some force which has tended to keep the chambers in a rectilinear series: such for instance as gravity, acting on a system of "hanging drops."

In *Textularia* and its allies (Fig. 431) we have a precise parallel to the helicoid cyme of the botanists (cf. p. 767): that is to say we have a screw translation, perpendicular to the plane of the underlying logarithmic spiral. In other words, in tracing a genetic spiral through the whole succession of chambers, we do so by a continuous vector rotation through successive angles of 180° (or 120° in some cases), while the pole moves along an axis perpendicular to the original plane of the spiral.

Another type is furnished by the "cyclic" shells of the Orbitolitidae, where small and numerous chambers tend to be added

---

* Cf. G. Schacko, Ueber *Globigerina*-Einschluss bei *Orbulina*, *Wiegmann's Archiv*, XLIX, p. 428, 1883; Brady, *Chall. Rep.* 1884, p. 607.

on round and round the system, so building up a circular flattened
disc.   This again we perceive to be, mathematically, a limiting case
of the logarithmic spiral; the spiral has become wellnigh a circle and
the constant angle is wellnigh 90°.

Lastly there are a certain number of Foraminifera in which,
without more ado, we may simply say that the arrangement of the
chambers is irregular, neither the law of constant ratio of magnitude
nor that of constant form being obeyed.   The chambers are heaped
pell-mell upon one another, and such forms are known to naturalists
as the Acervularidae.

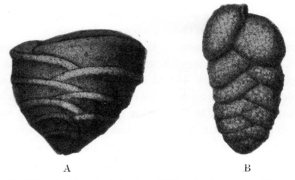

A   B

Fig. 431.   A, *Textularia trochus* d'Orb.   B, *T. concava* Karrer.

While in these last we have an extreme lack of regularity, we
must not exaggerate the regularity or constancy which the more
ordinary forms display.   We may think it hard to believe that the
simple causes, or simple laws, which we have described should operate,
and operate again and again, in millions of individuals to produce
the same delicate and complex conformations.   But we are taking
a good deal for granted if we assert that they do so, and in particular
we are assuming, with very little proof, the "constancy of species"
in this group of animals.   Just as Verworn has shewn that the
typical *Amoeba proteus*, when a trace of alkali is added to the water
in which it lives, tends, by alteration of surface tensions, to protrude
the more delicate pseudopodia characteristic of *A. radiosa*—and
again when the water is rendered a little more alkaline, to turn
apparently into the so-called *A. limax*—so it is evident that a very
slight modification in the surface-energies concerned might tend

to turn one so-called species into another among the Foraminifera. To what extent this process actually occurs, we do not know. But that this, or something of the kind, does actually occur we can scarcely doubt. For example in the genus *Peneroplis*, the first portion of the shell consists of a series of chambers arranged in a spiral or nautiloid series; but as age advances the spiral is apt to be modified in various ways*. Sometimes the successive chambers grow rapidly broader, the whole shell becoming fan-shaped. Sometimes the chambers become narrower, till they no longer enfold the earlier chambers but only come in contact each with its immediate predecessor: the result being that the shell straightens out, and (taking into account the earlier spiral portion) may be described as crozier-shaped. Between these extremes of shape, and in regard to other variations of thickness or thinness, roughness or smoothness, and so on, there are innumerable gradations passing one into another and intermixed without regard to geographical distribution:— "wherever Peneroplides abound this wide variation exists, and nothing can be more easy than to pick out a number of striking specimens and give to each a distinctive name, but *in no other way can they be divided into 'species.'*†" Some writers have wondered at the peculiar variability of this particular shell‡; but for all we know of the life-history of the Foraminifera, it may well be that a great number of the other forms which we distinguish as separate species and even genera are no more than temporary manifestations of the same variability§.

---

* Cf. H. B. Brady, *Challenger Rep.*, *Foraminifera*, 1884, p. 203, pl. XIII.

† Brady, *op. cit.* p. 206; Batsch, one of the earliest writers on Foraminifera, had already noticed that this whole series of ear-shaped and crozier-shaped shells was filled in by gradational forms; *Conchylien des Seesandes*, 1791, p. 4, pl. VI, fig. 15 a–f. See also, in particular, Dreyer, *Peneroplis; eine Studie zur biologischen Morphologie und zur Speciesfrage*, Leipzig, 1898; also Eimer und Fickert, Artbildung und Verwandschaft bei den Foraminiferen, *Tübinger zool. Arbeiten*, III, p. 35, 1899.

‡ Doflein, *Protozoenkunde*, 1911, p. 263: "Was diese Art veranlässt in dieser Weise gelegentlich zu variiren, ist vorläufig noch ganz räthselhaft."

§ In the case of *Globigerina*, some fourteen species (out of a very much larger number of described forms) were allowed by Brady (in 1884) to be distinct; and this list has been, I believe, rather added to than diminished. But these so-called species depend for the most part on slight differences of degree, differences in the angle of the spiral, in the ratio of magnitude of the segments, or in their area of contact one with another. Moreover with the exception of one or two "dwarf"

Conclusion

If we can comprehend and interpret on some such lines as these the form and mode of growth of the foraminiferal shell we may also begin to understand two striking features of the group, on the one hand the large number of diverse types or families which exist and the large number of species and varieties within each, and on the other the persistence of forms which in many cases seem to have undergone little change or none at all from the Cretaceous or even from earlier periods to the present day. In few other groups, perhaps only among the Radiolaria, do we seem to possess so nearly complete a picture of all possible transitions between form and form, and of the whole branching system of the evolutionary tree: as though little or nothing of it had ever perished, and the whole web of life, past and present, were as complete as ever. It leads one to imagine that these shells have grown according to laws so simple, so much in harmony with their material, with their environment, and with all the forces internal and external to which they are exposed, that none is better than another and none fitter or less fit to survive. It invites one also to contemplate the possibility of the lines of possible variation being here so narrow and determinate that identical forms may have come independently into being again and again.

While we can trace in the most complete and beautiful manner the passage of one form into another among these little shells, and ascribe them all at last (if we please) to a series which starts with the simple sphere of *Orbulina* or with the amoeboid body of *Astrorhiza*, the question stares us in the face whether this be an "evolution" which we have any right to correlate with historic *time*. The mathematician can trace one conic section into another, and "evolve" for example, through innumerable graded ellipses, the circle from the straight line: which tracing of continuous steps is a true "evolution," though time has no part therein. It was after this fashion that Hegel, and for that matter Aristotle himself, was an

---

forms, said to be limited to Arctic and Antarctic waters, there is no principle of geographical distribution to be discerned amongst them. A species found fossil in New Britain turns up in the North Atlantic; a species described from the West Indies is rediscovered at the ice-barrier of the Antarctic.

evolutionist—to whom evolution was a mental concept, involving order and continuity in thought but not an actual sequence of events in time. Such a conception of evolution is not easy for the modern biologist to grasp, and is harder still to appreciate. And so it is that even those who, like Dreyer* and like Rhumbler, study the foraminiferal shell as a physical system, who recognise that its whole plan and mode of growth is closely akin to the phenomena exhibited by fluid drops under particular conditions, and who explain the conformation of the shell by help of the same physical principles and mathematical laws—yet all the while abate no jot or tittle of the ordinary postulates of modern biology, nor doubt the validity and universal applicability of the concepts of Darwinian evolution. For these writers the *biogenetisches Grundgesetz* remains impregnable. The Foraminifera remain for them a great family tree, whose actual pedigree is traceable to the remotest ages; in which historical evolution has coincided with progressive change; and in which structural fitness for a particular function (or functions) has exercised its selective action and ensured "the survival of the fittest." By successive stages of historic evolution we are supposed to pass from the irregular *Astrorhiza* to a *Rhabdammina* with its more concentrated disc; to the forms of the same genus which consist of but a single tube with central chamber; to those where this chamber is more and more distinctly segmented; so to the typical many-chambered Nodosariae; and from these, by another definite advance and later evolution to the spiral Trochamminae. After this fashion, throughout the whole varied series of the Foraminifera, Dreyer and Rhumbler (following Neumayr) recognise so many successions of related forms, one passing into another and standing towards it in a definite relationship of ancestry or descent. Each evolution of form, from simpler to more complex, is deemed to have been attended by an advantage to the organism, an enhancement of its chances of survival or perpetuation; hence the historically older forms are on the whole structurally the simpler; or conversely, the simpler forms, such as the simple sphere, were the first to come into being in primeval seas; and finally, the gradual development and increasing complication of the individual

* F. Dreyer, Prinzipien der Gerüstbildung bei Rhizopoden, etc., *Jen. Zeitschr.* XXVI, pp. 204–468, 1892.

within its own lifetime is held to be at least a partial recapitulation of the unknown history of its race and dynasty*.

We encounter many difficulties when we try to extend such concepts as these to the Foraminifera. We are led for instance to assert, as Rhumbler does, that the increasing complexity of the shell, and of the manner in which one chamber is fitted on another, makes for advantage; and the particular advantage on which Rhumbler rests his argument is *strength*. Increase of strength, *die Festigkeitssteigerung*, is according to him the guiding principle in foraminiferal evolution, and marks the historic stages of their development in geologic time. But in days gone by I used to see the beach of a little Connemara bay bestrewn with millions upon millions of foraminiferal shells, simple Lagenae, less simple Nodosariae, more complex Rotaliae: all drifted by wave and gentle current from their sea-cradle to their sandy grave: all lying bleached and dead: one more delicate than another, but all (or vast multitudes of them) perfect and unbroken. And so I am not inclined to believe that niceties of form affect the case very much: nor in general that foraminiferal life involves a struggle for existence wherein breakage is a danger to be averted, and strength an advantage to be ensured†.

In the course of the same argument Rhumbler remarks that Foraminifera are absent from the coarse sands and gravels‡, as Williamson indeed had observed many years ago: so averting, or at least escaping, the dangers of concussion. But this is after all

---

* A difficulty arises in the case of forms (like *Peneroplis*) where the young shell appears to be more complex than the old, the first-formed portion being closely coiled while the later additions become straight and simple: "die biformen Arten verhalten sich, kurz gesagt, gerade umgekehrt als man nach dem biogenetischen Grundgesetz erwarten sollte," Rhumbler, *op. cit.* p. 33, etc.

† "Das Festigkeitsprinzip als *Movens* der Weiterentwicklung ist zu interessant und für die Aufstellung meines Systems zu wichtig um die Frage unerörtert zu lassen, warum diese Bevorzügung der Festigkeit stattgefunden hat. Meiner Ansicht nach lautet die Antwort auf diese Frage einfach, weil die Foraminiferen meistens unter Verhältnissen leben, die ihre Schalen in hohem Grade der Gefahr des Zerbrechens aussetzen; es muss also eine fortwahrende Auslese des Festeren stattfinden," Rhumbler, *op. cit.* p. 22.

‡ "Die Foraminiferen kiesige oder grobsandige Gebiete des Meeresbodens *nicht lieben*, u.s.w.": where the last two words have no particular meaning, save only that (as M. Aurelius says) "of things that use to be, we say commonly that they love to be."

a very simple matter of mechanical analysis. The coarseness or fineness of the sediment on the sea-bottom is a measure of the current: where the current is strong the larger stones are washed clean, where there is perfect stillness the finest mud settles down; and the light, fragile shells of the Foraminifera find their appropriate place, like every other graded sediment, in this spontaneous order of levigation.

The theorem of Organic Evolution is one thing; the problem of deciphering the lines of evolution, the order of phylogeny, the degrees of relationship and consanguinity, is quite another. Among the higher organisms we arrive at conclusions regarding these things by weighing much circumstantial evidence, by dealing with the resultant of many variations, and by considering the probability or improbability of many coincidences of cause and effect; but even then our conclusions are at best uncertain, our judgments are continually open to revision and subject to appeal, and all the proof and confirmation we can ever have is that which comes from the direct, but fragmentary evidence of palaeontology *.

But in so far as forms can be shewn to depend on the play of physical forces, and the variations of form to be directly due to simple quantitative variations in these, just so far are we thrown back on our guard before the biological conception of consanguinity, and compelled to revise the vague canons which connect classification with phylogeny.

The physicist explains in terms of the properties of matter, and classifies according to a mathematical analysis, all the drops and forms of drops and associations of drops, all the kinds of froth and foam, which he may discover among inanimate things; and his task ends there. But when such forms, such conformations and configurations, occur among *living* things, then at once the biologist introduces his concepts of heredity, of historical evolution, of succession in time, of recapitulation of remote ancestry in individual growth, of common origin (unless contradicted by direct evidence) of similar forms remotely separated by geographic space or geologic time, of fitness for a function, of adaptation to an environment, of higher and lower, of "better" and "worse." This is the fundamental

---

* In regard to the Foraminifera, "die Palaeontologie lässt uns leider an Anfang der Stammesgeschichte fast gänzlich im Stiche," Rhumbler, *op. cit.* p. 14.

difference between the "explanations" of the physicist and those of the biologist.

In the order of physical and mathematical complexity there is no question of the sequence of historic time. The forces that bring about the sphere, the cylinder or the ellipsoid are the same yesterday and to-morrow. A snow-crystal is the same to-day as when the first snows fell. The physical forces which mould the forms of *Orbulina*, of *Astrorhiza*, of *Lagena* or of *Nodosaria* to-day were still the same, and for aught we have reason to believe the physical conditions under which they worked were not appreciably different, in that yesterday which we call the Cretaceous epoch; or, for aught we know, throughout all that duration of time which is marked, but not measured, by the geological record.

In a word, the minuteness of our organism brings its conformation as a whole within the range of the molecular forces; the laws of its growth and form appear to lie on simple lines; what Bergson calls* the "ideal kinship" is plain and certain, but the "material affiliation" is problematic and obscure; and, in the end and upshot, it seems to me by no means certain that the biologist's usual mode of reasoning is appropriate to the case, or that the concept of continuous historical evolution must necessarily, or may safely and legitimately, be employed.

That things not only alter but improve is an article of faith, and the boldest of evolutionary conceptions. How far it be true were very hard to say; but I for one imagine that a pterodactyl flew no less well than does an albatross, and that Old Red Sandstone fishes swam as well and easily as the fishes of our own seas.

---

* The evolutionist theory, as Bergson puts it, "consists above all in establishing relations of ideal kinship, and in maintaining that wherever there is this relation of, so to speak, *logical* affiliation between forms, *there is also a relation of chronological succession between the species in which these forms are materialised*" (*Creative Evolution*, 1911, p. 26). Cf. *supra*. p. 412.

CHAPTER XIII

THE SHAPES OF HORNS, AND OF TEETH OR TUSKS: WITH A NOTE ON TORSION

W E have had so much to say on the subject of shell-spirals that we must deal briefly with the analogous problems which are presented by the horns of sheep, goats, antelopes and other horned quadrupeds; and all the more, because these horn-spirals are on the whole less symmetrical, less easy of measurement than those of the shell, and in other ways also are less easy of investigation. Let us dispense altogether in this case with mathematics; and be content with a very simple account of the configuration of a horn.

There are three types of horn which deserve separate consideration: firstly, the horn of the rhinoceros; secondly, the horns of the sheep, the goat, the ox or the antelope, that is to say, of the so-called hollow-horned ruminants; and thirdly, the solid bony horns, or "antlers," which are characteristic of the deer.

The horn of the rhinoceros presents no difficulty. It is physiologically equivalent to a mass of consolidated hairs, and, like ordinary hair, it consists of non-living or "formed" material, continually added to by the living tissues at its base. In section the horn is elliptical, with the long axis fore-and-aft, or in some species nearly circular. Its longitudinal growth proceeds with a maximum velocity anteriorly, and a minimum posteriorly; and the ratio of these velocities being constant, the horn curves into the form of a logarithmic spiral in the manner that we have already studied. The spiral is of small angle, but in the longer-horned species, such as the great white rhinoceros (*Ceratorhinus*), the spiral curvature is distinctly recognised. As the horn occupies a median position on the head—a position, that is to say, of symmetry in respect to the field of force on either side—there is no tendency towards a lateral twist, and the horn accordingly develops as a *plane* logarithmic spiral. When two median horns coexist, the hinder one is much the smaller of the two: which is as much as to say that the force, or rate, of growth diminishes as we pass backwards, just as it does within the limits of the single horn. And accordingly, while both horns have *essentially* the same shape, the spiral curvature is less manifest in the second one, by the mere reason of its shortness.

The paired horns of the ordinary hollow-horned ruminants, such as the sheep or the goat, grow under conditions which are in some respects similar, but which differ in other and important respects from the conditions under which the horn grows in the rhinoceros. As regards its structure, the entire horn now consists of a bony core with a covering of skin; the inner, or dermal, layer of the latter is richly supplied with nutrient blood-vessels, while the outer layer, or epidermis, develops the fibrous or chitinous material, chemically and morphologically akin to a mass of cemented or consolidated hairs, which constitutes the "sheath" of the horn. A zone of active growth at the base of the horn keeps adding to this sheath, ring by ring, and the specific form of this annular zone may be taken as the "generating curve" of the horn*. Each horn no longer lies, as it does in the rhinoceros, in the plane of symmetry of the animal of which it forms a part; and the limited field of force concerned in the genesis and growth of the horn is bound, accordingly, to be more or less laterally asymmetrical. But the two horns are in symmetry one with another; they form "conjugate" spirals, one being the "mirror-image" of the other. Just as in the hairy coat of the animal each hair, on either side of the median "parting," tends to have a certain definite direction of its own, inclined away from the median axial plane of the whole system, so is it both with the bony core of the horn and with the consolidated mass of hairs or hair-like substance which constitutes its sheath; the primary axis of the horn is more or less inclined to, and may even be nearly perpendicular to, the axial plane of the animal.

The growth of the horny sheath is not continuous, but more or less definitely periodic: sometimes, as in the sheep, this periodicity is particularly well-marked, and causes the horny sheath to be com-

---

* In this chapter we keep to Moseley's way of regarding the equiangular spiral in space, of shell or horn, as generated by a certain figure which (*a*) grows, (*b*) revolves about an axis, and (*c*) is translated along or parallel to the said axis, all at certain appropriate and specific velocities. This method is simple, and even adequate, from the naturalist's point of view; but not so, or much less so, from the mathematician's, as we have found in the last chapter (p. 782).

posed of a series of all but separate rings, which are supposed to be formed year by year, and so to record the age of the animal*.

Just as Moseley sought for the true generating curve in the orifice, or "lip," of the molluscan shell, so we begin by assuming that in the spiral horn the generating curve corresponds to the lip or margin of one of the horny rings or annuli. This annular margin, or boundary of the ring, is usually a sinuous curve, not lying in a plane, but such as would form the boundary of an anticlastic surface of great complexity: to the meaning and origin of which phenomenon we shall return presently. But, as we have already seen in the case of the molluscan shell, the complexities of the lip itself, or of the corresponding lines of growth upon the shell, need

Fig. 432.   The Argali sheep; *Ovis Ammon.*   From Cook's
Spirals in Nature and Art.

not concern us in our study of the development of the spiral: inasmuch as we may substitute for these actual boundary lines, their "trace," or projection on a plane perpendicular to the axis—in other words the simple outline of a transverse section of the whorl. In the horn, this transverse section is often circular or nearly so, as in the oxen and many antelopes: it now and then becomes of

* Cf. R. S. Hindekoper, *On the Age of the Domestic Animals,* Philadelphia and London, 1891, p. 173. In the case of the ram's horn, the assumption that the rings are annual is probably justified. In cattle they are much less conspicuous, but are sometimes well-marked in the cow; and in Sweden they are then called "calf-rings," from a belief that they record the number of offspring. That is to say, the growth of the horn is supposed to be retarded during gestation, and to be accelerated after parturition, when superfluous nourishment seeks a new outlet. (Cf. Lönnberg, *P.Z.S.* 1900, p. 689.)

somewhat complicated polygonal outline, as in a highland ram; but in many antelopes, and in most of the sheep, the outline is that of an isosceles or sometimes nearly equilateral triangle, a form which is typically displayed, for instance, in *Ovis Ammon*. The horn in this latter case is a trihedral prism, whose three faces are (1) an upper, or frontal face, in continuation of the plane of the frontal bone; (2) an outer, or orbital, starting from the upper margin of the orbit; and (3) an inner, or nuchal, abutting on the parietal bone*. Along these three faces, and their corresponding angles or edges, we can trace in the fibrous substance of the horn a series of homologous spirals, such as we have called in a preceding chapter the "*ensemble* of generating spirals" which define or constitute the surface.

The case of the horn differs in ways of its own from that of the molluscan shell. For one thing, the horn is always tubular— its generating curve is actually, as well as theoretically, a closed curve; there is no such thing as "involution," or the wrapping of one whorl within another, or successive intersection of the generating curve. Again, while the calcareous substance of the shell is laid down once for all, fixed and immovable, there is reason to believe that the young horn has, to begin with, a certain measure of flexibility, a certain freedom, even though it be slight, to bend or fold or wrinkle. And this being so, while it is no harder in the horn than in the shell to recognise the general field of force or general direction of growth, the actual conditions are somewhat more complex.

In some few cases, of which the male musk ox is one of the most notable, the horn is not developed in a continuous spiral curve. It changes its shape as growth proceeds; and this, as we have seen, is enough to show that it does not constitute a logarithmic spiral. The reason is that the bony exostoses, or horn-cores, about which the horny sheath is shaped and moulded, neither grow continuously nor even remain of constant size after attaining their full growth. But as the horns grow heavy the bony core is bent downwards by their weight, and so guides the growth of the horn in a new direction. Moreover as age advances, the core is further weakened and to a great extent absorbed: and the horny sheath or horn proper,

* Cf. Sir V. Brooke, On the large sheep of the Thian Shan, *P.Z.S.* 1875, p. 511.

deprived of its support, continues to grow, but in a flattened curve very different from its original spiral *. The chamois is a somewhat analogous case. Here the terminal, or oldest, part of the horn is curved; it tends to assume a spiral form, though from its comparative shortness it seems merely to be bent into a hook. But later on the bony core within, as it grows and strengthens, stiffens the horn and guides it into a straighter course or form. The same phenomenon of change of curvature, manifesting itself at the time when, or the place where, the horn is freed from the support of the internal core, is seen in a good many other antelopes (such as the

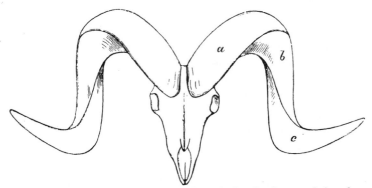

Fig. 433.   Diagram of ram's horns.   *a*, frontal; *b*, orbital; *c*, nuchal surface.
After Sir Vincent Brooke, from *P.Z.S.*

hartebeest) and in many buffaloes; and the cases where it is most manifest appear to be those where the bony core is relatively short, or relatively weak. All these illustrate the cardinal difference between the growth of the horn and that of the bone below: the one dead, the other alive; the one adding and retaining its successive increments, the other mobile, plastic, and in continual flux throughout.

But in the great majority of horns we have no difficulty in recognising a continuous logarithmic spiral, nor in correlating it with an unequal rate of growth (parallel to the axis) on two opposite sides of the horn, the inequality maintaining a constant ratio as long as growth proceeds. In certain antelopes, such as the gemsbok, the spiral angle is very small, or in other words the horn

* Cf. E. Lönnberg, On the structure of the musk ox, *P.Z.S.* 1900, pp. 686–718.

is very nearly straight; in other species of the same genus *Oryx*, such as the Beisa antelope and the Leucoryx, a gentle curve (not unlike though generally less than that of a *Dentalium* shell) is evident; and the spiral angle, according to the few measurements I have made, is found to measure from about 20° to nearly 40°. In some of the large wild goats, such as the Scinde wild goat, we have

Fig. 434.   Head of Arabian wild goat, *Capra sinaitica.*
After Sclater, from *P.Z.S.*

a beautiful logarithmic spiral, with a constant angle of rather less than 70°; and we may easily arrange a series of forms, such for example as the Siberian ibex, the moufflon, *Ovis Ammon*, etc., and ending with the long-horned Highland ram: in which, as we pass from one to another, we recognise precisely homologous spirals with an increasing angular constant, the spiral angle being, for instance, about 75° or rather less in *Ovis Ammon*, and in the Highland ram a very little more.   We have already seen that in the neighbourhood of 70° or 80° a small change of angle makes a marked difference in the appearance of the spire; and we know also that the actual length of the horn makes a very striking difference, for the spiral becomes especially conspicuous to the eye when horn or shell is long enough to shew several whorls, or at least a considerable part of one entire convolution.

Even in the simplest cases, such as the wild goats, the spiral is never a plane but always a *gauche* spiral: in greater or less degree there is always superposed upon the plane logarithmic spiral a helical spiral in space. Sometimes the latter is scarcely apparent, for the horn (though long, as in the said wild goats) is not nearly long enough to shew a complete convolution: at other times, as in the ram, and still better in many antelopes such as the koodoo, the corkscrew curve of the horn becomes its most characteristic feature. So we may study, as in the molluscan shell, the helicoid component of the spire—in other words the variation in what we have called (on p. 816) the angle β. This factor it is which, more than the constant angle of the logarithmic spiral, imparts a characteristic appearance to the various species of sheep, for instance to the various closely allied species of Asiatic wild sheep, or Argali. In all of these the constant angle of the logarithmic spiral is very much the same, but the enveloping angle of the cone differs greatly. Thus the long drawn out horns of *Ovis Poli*, four feet or more from tip to tip, differ conspicuously from those of *Ovis Ammon* or *O. hodgsoni*, in which a very similar logarithmic spiral is wound (as it were) round a much blunter cone.

Let us continue to dispense with mathematics, for the mathematical treatment of a gauche spiral is never very simple, and let us deal with the matter by experiment. We have seen that the generating curve, or transverse section, of a typical ram's horn is triangular in form. Measuring (along the curve of the horn) the length of the three edges of the trihedral structure in a specimen of *Ovis Ammon*, and calling them respectively the outer, inner, and hinder edges (from their position at the base of the horn, relatively to the skull), I find the outer edge to measure 80 cm., the inner 74 cm., and the posterior 45 cm.; let us say that, roughly, they are in the ratio of 9 : 8 : 5. Then, if we make a number of little cardboard triangles, equip each with three little legs (I make them of cork), whose relative lengths are as 9 : 8 : 5, and pile them up and stick them all together, we straightway build up a curve of double curvature precisely analogous to the ram's horn: except only that, in this first approximation, we have not allowed for the gradual increment (or decrement) of the triangular surfaces, that is to say, for the *tapering* of the horn due to the magnification of the generating curve.

In this case then, and in most other trihedral or three-sided horns, one of the three components, or three unequal velocities of growth, is of relatively small magnitude, but the other two are nearly equal one to the other; it would involve but little change for these latter to become precisely equal; and again but little to turn the balance of inequality the other way. But the immediate consequence of this altered ratio of growth would be that the horn would appear to wind the other way, as it does in the antelopes, and also in certain goats, e.g. the markhor, *Capra falconeri*.

For these two opposite directions of twist Dr Wherry has suggested a convenient nomenclature. When the horn winds so that we follow it from base to apex in the direction of the hands of a watch, it is customary to call it a "left-handed" spiral. Such a spiral we have in the horn on the left-hand side of a ram's head. Accordingly, Dr Wherry calls the condition *homonymous*, where, as in the sheep, a right-handed spiral is on the right side of the head, and a left-handed spiral on the left side; while he calls the opposite condition *heteronymous*, as we have it in the antelopes, where the right-handed twist is on the left side of the head, and the left-handed twist on the right-hand side. Among the goats, we may have either condition. Thus the domestic and most of the wild goats agree with the sheep; but in the markhor the twisted horns are heteronymous, as in the antelopes. The difference, as we have seen, is easily explained; and (very much as in the case of our opposite spirals in the apple-snail, referred to on p. 820) it has no very deep importance

Summarised then in a very few words, the argument by which we account for the spiral conformation of the horn is as follows: The horn elongates by dint of continual growth within a narrow zone, or annulus, at its base. If the rate of growth be identical on all sides of this zone, the horn will grow straight; if it be greater on one side than on the other, the horn will become curved; and it probably *will* be greater on one side than on the other, because each single horn occupies an unsymmetrical field with reference to the plane of symmetry of the animal. If the maximal and minimal velocities of growth be precisely at opposite sides of the zone of growth, the resultant spiral will be a plane spiral; but if they be not precisely or diametrically opposite, then the spiral will be a gauche spiral in space; and it is by no means likely that the maximum and minimum *will* occur at precisely opposite ends of a diameter, for no such plane of symmetry is manifested in the field of force to which the growing annulus corresponds or appertains.

Now we must carefully remember that the rates of growth of which we are here speaking are the net rates of longitudinal increment, in which increment the activity of the living cells in the zone of growth at the base of the horn is only one (though it is the fundamental) factor. In other words, if the horny sheath were continually being added to with equal rapidity all round its zone of active growth,

Fig. 435.   Marco Polo's sheep: *Ovis Poli*.   From Cook.

but at the same time had its elongation more retarded on one side than the other (prior to its complete solidification) by varying degrees of adhesion or membranous attachment to the bony core within, then the net result would be a spiral curve precisely such as would have arisen from initial inequalities in the rate of growth itself. It seems probable that this is an important factor, and sometimes even the chief factor in the case. The same phenomenon of attachment to the bony core, and the consequent friction or retardation with which the sheath slides over its surface, will lead to various subsidiary phenomena: among others to the presence of transverse folds or corrugations upon the horn, and to their unequal distribution upon its several faces or edges. And while it is perfectly true that nearly all the characters of the horn can be accounted for by unequal velocities

of longitudinal growth upon its different sides, it is also plain that the
actual field of force is a very complicated one indeed. For example,
we can easily see (at least in the great majority of cases) that the
direction of growth of the horny fibres of the sheath is by no means
parallel to the axis of the core within; accordingly these fibres will
tend to wind in a system of helicoid curves around the core, and not
only this helicoid twist but any other tendency to spiral curvature
on the part of the sheath will tend to be opposed or modified by the
resistance of the core within. On the other hand living bone is a
very plastic structure, and yields easily though slowly to any forces
tending to its deformation; and so, to a considerable extent, the
bony core itself will tend to be modelled by the curvature which the

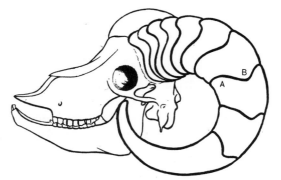

Fig. 436.   Head of *Ovis Ammon*, shewing St Venant's curves.

growing sheath assumes, and the final result will be determined by
an equilibrium between these two systems.

While it is not very safe, perhaps, to lay down any general rule
as to what horns are more and what are less spirally curved, I think
it may be said that, on the whole, the thicker the horn the greater
is its spiral curvature. It is the slender horns, of such forms as
the Beisa antelope, which are gently curved, and it is the robust
horns of goats or of sheep in which the curvature is more pronounced.
Other things being the same, this is what we should expect to find;
for it is where the transverse section of the horn is large that we may
expect to find the more marked differences in the intensity of the
field of force, whether of active growth or of retardation, on opposite
sides or in different sectors thereof.

But there is yet another and a very remarkable phenomenon which we may discern in the growth of a horn when it takes the form of a curve of double curvature, namely, an effect of torsional strain; and this it is which gives rise to the sinuous "lines of growth," or sinuous boundaries of the separate horny rings, of which we have already spoken. It is not at first sight obvious that a mechanical strain of torsion is necessarily involved in the growth of the horn. In our experimental illustration (p. 880), we built up a twisted coil of separate elements, and no torsional strain attended the development of the system. So would it be if the horny sheath grew by successive annular increments, free save for their relation to one another and having no attachment to the solid core within. But as a matter of fact there is such an attachment, by subcutaneous connective tissue, to the bony core; and accordingly a torsional strain will be set up in the growing horny sheath, again provided that the forces of growth therein be directed more or less obliquely to the axis of the core; for a "couple" is thus introduced, giving rise to a strain which the sheath would not experience were it free (so to speak) to slip along, impelled only by the pressure of its own growth from below. And furthermore, the successive small increments of the growing horn (that is to say, of the horny sheath) are not instantaneously converted from living to solid and rigid substance; but there is an intermediate stage, probably long-continued, during which the new-formed horny substance in the neighbourhood of the zone of active growth is still plastic and capable of deformation.

Now we know, from the celebrated experiments of St Venant*, that in the torsion of an elastic body, other than a cylinder of circular section, a very remarkable state of strain is introduced. If the body be thus cylindrical (whether solid or hollow), then a twist leaves each circular section unchanged, in dimensions and in figure. But in all other cases, such as an elliptic rod or a prism of any particular sectional form, forces are introduced which act parallel to the axis of the structure, and which warp each section into a complex "anticlastic" surface. Thus in the case of a triangular and

---

* St Venant, De la torsion des prismes, avec des considérations sur leur flexion, etc., Mém. des Savants Étrangers, Paris, XIV, pp. 233–560, 1856. Karl Pearson dedicated part of his History of the Theory of Elasticity to the memory of this ingenious and original man. For a modern account of the subject see Love's Elasticity (2nd ed.), chap. XIV.

equilateral prism, such as is shewn in section in Fig. 437 A, if the part
of the rod represented in the section be twisted by a force acting
in the direction of the arrow, then the originally plane section will
be warped as indicated in the diagram—where the full contour-lines
represent elevation above, and the dotted lines represent depression
below, the original level. On the external surface of the prism,
then, contour-lines which were originally. parallel and horizontal
will be found warped into sinuous curves, such that, on each of the
three faces, the curve will be convex upwards on one half, and
concave upwards on the other half of the face. The ram's horn,
and still better that of *Ovis Ammon*, is comparable to such a prism,

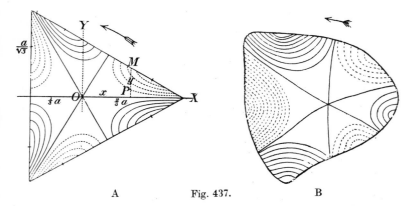

A   Fig. 437.   B

save that in section it is not quite equilateral, and that its three
faces are not plane. The warping is therefore not precisely identical
on the three faces of the horn; but, in the general distribution of
the curves, it is in complete accordance with theory*. Similar
anticlastic curves are well seen in many antelopes; but they are
conspicuous by their absence in the *cylindrical* horns of oxen.

The better to illustrate this phenomenon, the nature of which is
indeed obvious enough from a superficial examination of the horn,
I made a plaster cast of one of the horny rings in a horn of *Ovis
Ammon*, so as to get an accurate pattern of its sinuous edge: and
then, filling the mould up with wet clay, I modelled an anticlastic

* The case of a thin conical shell under torsion is more complicated than either
that of the cylinder or of a prismatic rod; and the more tapering horns doubtless
deserve further study from this point of view. Cf. R. V. Southwell, On the torsion
of conical shells, *Proc. R.S.* (A), CLXIII, pp. 337–355, 1937.

surface, such as to correspond as nearly as possible with the sinuous outline*. Finally, after making a plaster cast of this sectional surface, I drew its contour-lines (as shewn in Fig. 437 B) with the help of a simple form of spherometer. It will be seen that in great part this diagram is precisely similar to St Venant's diagram of the cross-section of a twisted triangular prism; and this is especially the case in the neighbourhood of the sharp angle of our prismatic section. That in parts the diagram is somewhat asymmetrical is not to be wondered at: and (apart from inaccuracies due to the somewhat rough means by which it was made) this asymmetry can be sufficiently accounted for by anisotropy of the material, by inequalities in thickness of different parts of the horny sheath, and especially (I think) by unequal distributions of rigidity due to the presence of the smaller corrugations of the horn. It is on account of these minor corrugations that in such horns as the Highland ram's, where they are strongly marked, the main St Venant effect is not nearly so well shewn as in smoother horns, such as those of *Ovis Ammon* and its congeners†.

The distribution of forces which manifest themselves in the growth and configuration of a horn is no simple nor merely superficial matter. One thing is coordinated with another; the direction of the axis of the horn, the form of its sectional boundary, the specific rates of growth in the mean spiral and at various parts of its periphery—all these play their parts, controlled in turn by the supply of nutriment which the character of the adjacent tissues and the distribution of the blood-vessels combine to determine. To suppose that this or that size or shape of horn has been produced or altered, acquired or lost, *by Natural Selection*, whensoever one type rather than another proved serviceable for defence or attack or any other purpose, is an hypothesis harder to define and to substantiate than some imagine it to be.

There are still one or two small matters to speak of before we leave these spiral horns. It is the way of sportsmen to keep record of big game by measuring the length along the curve of the horn and the span from tip to tip. Now if we study such measurements (as

---

* This is not difficult to do, with considerable accuracy, if the clay be kept well wetted or semi-fluid, and the smoothing be done with a large wet brush.

† The curves are well shewn in most of Sir V. Brooke's figures of the various species of Argali, in the paper quoted above, on p. 877.

they may be found in Mr Rowland Ward's book *), we shall soon
see that the two measurements do not tally one with the other:
but that a pair of horns, the longer when measured along the curve,
may be the shorter from tip to tip, and *vice versa*. We might set
this down to mere variability of form, but the true reason is simpler
still. If the axes of the two horns stood straight out, at right angles
to the median plane, then growth in length and in width of span
would go on together. But if the two horns diverge at any lesser
angle, then as the horns grow their spiral curvature will tend to
bring their tips nearer and farther apart alternately.

There is one last, but not least curious property to be seen in
a ram's horns. However large and heavy the horns may be—and
in *Ovis Poli* 50 or 60 lb. is no unusual weight for the pair to grow to—
the ram carries them with grace and ease, and they neither endanger
his poise nor encumber his movements. The reason is that head
and horns are very perfectly *balanced*, in such a way that no bending
moment tends to turn the head up or down about its fulcrum in
the atlas vertebra; if one puts two fingers into the foramen magnum
one may lift up the heavy skull, and find it hang in perfect equilibrium. Moreover, the horns go on growing, but this equipoise is
never lost nor changed; for the centre of gravity of the logarithmic
spiral remains constant. There are other cases where heavy horns,
well balanced as they doubtless are, yet visibly affect the set and
balance of the head. The stag carries his head higher than a horse,
and an Indian buffalo tilts his muzzle higher than a cow.

A further note upon torsion

The phenomenon of torsion, to which we have been thus introduced, opens up many wide questions in connection with form. Some
of the associated phenomena are admirably illustrated in the case
of climbing plants; but we can only deal with these still more briefly
and parenthetically. The subject has been elaborately dealt with
not only in Darwin's books†, but also by a great number of
earlier and later writers. In "twining" plants, which constitute
the greater number of "climbers," the essential phenomenon is a
tendency of the growing shoot to revolve about a vertical axis—

---

* *Records of Big Game*, 9th edition, 1928.
† *Climbing Plants*, 1865 (2nd ed. 1875); *Power of Movement in Plants*, 1880.

a tendency long ago discussed by de Candolle and investigated by Palm, H. von Mohl and Dutrochet*. This tendency to revolution—circumvolution, as Darwin calls it, revolving nutation, as Sachs puts it—is very closely comparable to the process by which an antelope's horn (such as the koodoo's) acquires its spiral twist, and is due, in like manner, to inequalities in the rate of growth of the growing stem: with this difference between the two, that in the antelope's horn the zone of active growth is confined to the base of the horn, while in the climbing stem the same phenomenon is at work throughout the whole length of the growing structure. This growth is in the main due to "turgescence," that is to the extension, or elongation, of ready-formed cells through the imbibition of water; it is a phenomenon due to osmotic pressure. The particular stimulus to which these movements (that is to say, these inequalities of growth) have been ascribed can hardly be discussed here; but it was hotly debated fifty years ago and for many years thereafter, the point at issue being no other than whether direct physical causation, or the Darwinian concept of fitness or adaptation, should be invoked as an "explanation" of biological phenomena. The old *Naturphilosophie* had been inclined to look for spirals everywhere, and to attribute them to very simple causes: "Man wird nicht gross irren" (said Oken †) "wenn man sagt, alle Pflanzen entstehen als Spirale, und zwar weil sie feststehen und ein End gegen die Sonne kehren, die täglich einen Spiralgang um sie macht, u.s.w." When de Candolle saw a shoot curve under the influence of light (by heliotropism, as we are told to call it), he was content to regard the curvature as the result of different rates of growth on one side or other of the shoot, and these in turn as the direct result of differences of illumination. But by the Darwins, father and son, and by Sachs and by the Würzburg school, the curvature was ascribed to "irritability," a "stimulus" on one side of the shoot being followed by a "motor-reaction" on the other. The curvature was thus taken to be a "response" to external stimuli (such as light and gravity); and stimulus and response were supposed to have

* Palm, *Ueber das Winden der Pflanzen*, 1827; H. von Mohl, *Bau und Winden der Ranken*, etc., 1827; R. H. J. Dutrochet, Sur la volubilité des tiges de certains végétaux, et sur la cause de ce phénomène, *Ann. Sc. Nat.* (*Bot.*), II, pp. 156–167, 1844, and other papers.

† *Isis*, I, p. 222, 1817.

evolved together in the course of ages, to bring about something more and more fitted for survival in the struggle for existence. They were, in short, of the nature of *acquired habits*, rather than *physical phenomena*. But there was no gainsaying the fact that the immediate cause of curvature was inequality of growth on opposite sides *.

A simple stem growing upright in the dark, or in uniformly diffused light, would be in a position of equilibrium to a field of force radially symmetrical about its vertical axis. But this complete radial symmetry will not often occur; and the radial anomalies may be such as arise intrinsically from structural peculiarities in the stem itself, or externally to it by reason of unequal illumination or through various other localised forces. The essential fact, so far as we are concerned, is that in twining plants we have a very marked tendency to inequalities in longitudinal growth on different aspects of the stem—a tendency which is but an exaggerated manifestation of one which is more or less present, under certain conditions, in all plants whatsoever. Just as in the case of the ruminants' horns so we find here that this inequality may be, so to speak, positive or negative, the maximum lying to the one side or the other of the twining stem; and so it comes to pass that some climbers twine to the one side and some to the other: the hop and the honeysuckle following the sun, and the field-convolvulus twining in the reverse direction; there are also some, like the woody nightshade (*Solanum Dulcamara*), which twine indifferently either way.

Together with this circumnutatory movement, there is very generally to be seen an actual *torsion* of the twining stem—a twist, that is to say, about its own axis; and Mohl made the curious observation, confirmed by Darwin, that when a stem twines around a smooth cylindrical stick the torsion does not take place, save "only in that degree which follows as a mechanical necessity from the spiral winding": but that stems which had climbed around a rough stick were all more or less, and generally much, twisted. Here Darwin did not refrain from introducing that teleological argument which pervades his whole train of reasoning: "The stem," he says, "probably gains rigidity by being twisted (on the same

---

* On the whole controversy, see F. F. Blackman's obituary notice of Francis Darwin in *Proc. R.S.* (B), cx, 1932.

principle that a much twisted rope is stiffer than a slackly twisted one), and is thus indirectly benefited so as to be able to pass over inequalities in its spiral ascent, and to carry its own weight when allowed to revolve freely." The mechanical explanation would appear to be very simple, and such as to render the teleological hypothesis unnecessary. In the case of the roughened support, there is a temporary adhesion or "clinging" between it and the growing stem which twines around it; and a system of forces is thus set up, producing a "couple," just as it was in the case of the ram's or antelope's horn through direct adhesion of the bony core to the surrounding sheath. The twist is the direct result of this couple, and it disappears when the support is so smooth that no such force comes to be exerted.

Another important class of climbers includes the so-called "leaf-climbers." In these, some portion of the leaf, generally the petiole, sometimes (as in the fumitory) the elongated midrib, curls round a support; and a phenomenon of like nature occurs in many, though not all, of the so-called "tendril-bearers." Except that a different part of the plant, leaf or tendril instead of stem, is concerned in the twining process, the phenomenon here is strictly analogous to our former case; but in the resulting helix there is, as a rule, this obvious difference, that, while the twining stem, for instance of the hop, makes a slow revolution about its support, the typical leaf-climber makes a close, firm coil: the axis of the latter is nearly perpendicular and parallel to the axis of its support, while in the twining stem the angle between the two axes is comparatively small. Mathematically speaking, the difference merely amounts to this, that the component in the direction of the vertical axis is large in the one case, and the corresponding component is small, if not absent, in the other; in other words, we have in the climbing stem a considerable vertical component, due to its own tendency to grow in height, while this longitudinal or vertical extension of the whole system is not apparent, or little apparent, in the other cases. But from the fact that the twining stem tends to run obliquely to its support, and the coiling petiole of the leaf-climber tends to run transversely to the axis of its support, there immediately follows this marked difference, that the phenomenon of *torsion*, so manifest in the former case, will be absent in the latter.

There is one other phenomenon which meets us in the twining and twisted stem, and which is doubtless illustrated also, though not so well, in the antelope's horn; it is a phenomenon which forms the subject of a second chapter of St Venant's researches on the effects of torsional strain in elastic bodies. We have already seen how one effect of torsion, in for instance a prism, is to produce strains parallel to the axis, elevating parts and depressing other parts of each transverse section. But in addition to this, the same torsion has the effect of materially altering the form of the section itself, as we may easily see by twisting a square or oblong piece of india-rubber. If we start with a cylinder, such as a round piece of catapult india-rubber, and twist it on its own long axis, we have already seen that it suffers no other distortion; it still remains a cylinder, that is to say, it is still in section everywhere circular. But if it be of any other shape than cylindrical the case is different, for now the sectional shape tends to alter under the strain of torsion. Thus, if our rod be elliptical in section to begin with, it will, under torsion, become a more elongated ellipse; if it be square, its angles will become more prominent and its sides will curve inwards, till at length the square assumes the appearance of a four-pointed star with rounded angles. Furthermore, looking at the results of this process of modification, we find experimentally that the resultant figures are more easily twisted, less resistant to torsion, than were those from which we evolved them; and this is a very curious physical or mathematical fact. So a cylinder, which is especially resistant to torsion, is very easily bent or flexed; while projecting ribs or angles, such as an engineer makes in a bar or pillar of iron for the purpose of increasing its resistance to *bending*, actually make it much weaker than before (for the same amount of metal per unit length) in the way of resistance to *torsion*.

In the hop itself, and in a very considerable number of other twining and twisting stems, the ribbed or channelled form of the stem is a conspicuous feature. We may safely take it, (1) that such stems are especially susceptible of torsion; and (2) that the effect of torsion will be to intensify any such peculiarities of sectional outline which they may possess, though not to initiate them in an originally cylindrical structure. In the leaf-climbers the case does not present itself, for there, as we have seen, torsion itself is not, or is very slightly, manifested. There are very distinct traces of the phenomenon in the horns of certain antelopes, but the reason why it is not a more conspicuous feature of the antelope's horn or of the ram's is apparently a very simple one: namely, that the presence of the bony core within tends to check that deformation which is perpendicular, while it permits that which is parallel, to the axis of the horn.

Of deer's antlers

But let us return to our subject of the shapes of horns, and consider briefly our last class of these structures, namely the bony antlers of the elk and deer*. The problems which these present to us are very different from those which we have had to do with in the antelope or the sheep.

With regard to its structure, it is plain that the bony antler corresponds, upon the whole, to the bony core of the antelope's horn; while in place of the hard horny sheath of the latter, we have the soft "velvet," which every season covers the new growing antler, and protects the large nutrient blood-vessels by help of which the antler grows†. The main difference lies in the fact that in the one case the bony core, imprisoned within its sheath, is rendered incapable of branching and incapable also of lateral expansion, and the whole horn is only permitted to grow in length while retaining a sectional contour that is identical with (or but little altered from) that which it possesses at its growing base: but in the antler on the other hand no such restraint is imposed, and the living, growing fabric of bone is free to expand into a broad flat plate over which the blood-vessels run. In the immediate neighbourhood of the main blood-vessels growth will be most active, in the interspaces between it may wholly fail: with the result that we may have great notches cut out of the flattened plate, or may at length find it reduced to the

---

* For an elaborate study of antlers, see A. Rörig, *Arch. f. Entw. Mech.* x, pp. 525–644, 1900; xi, pp. 65–148, 225–309, 1901; C. Hoffmann, *Zur Morphologie der rezenten Hirsche*, 75 pp., 23 pls., 1901; also Sir Victor Brooke, On the classification of the Cervidae, *P.Z.S,* 1878, pp. 883–928. For a discussion of the development of horns and antlers, see H. Gadow, *P.Z.S.* 1902, pp. 206–222, and works quoted therein.

† Cf. L. Rhumbler, Ueber die Abhängigkeit des Geweihwachstums der Hirsche, speziell des Edelhirsches, vom Verlauf der Blutgefässe im Kalbengeweih, *Zeitschr. f. Forst. und Jagdwesen,* 1911, pp. 295–314.

form of a simple branching structure. The main point is that the "horn" is essentially an *axial rod*, while the "antler" is essentially an outspread *surface**. In other words, the whole configuration of an antler is more easily understood by conceiving it as a plate or a surface, more and more notched and scolloped till but a slender skeleton remains, than to look upon it the other way, namely as an axial stem (or beam) giving off branches (or tines), the interspaces between which latter may sometimes fill up to form a continuous surface.

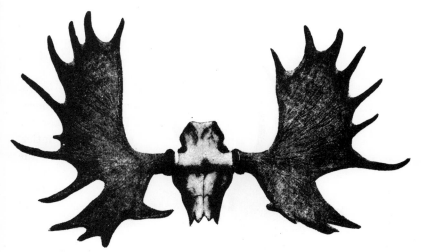

Fig. 438.   Antlers of Swedish elk.   After Lönnberg, from *P.Z.S.*

In a sense it matters very little whether we regard the broad plate-like antlers of the elk or the slender branching antlers of the stag as the more primitive type; for we are not concerned here with questions of hypothetical phylogeny, and even from the mathematical point of view it makes little or no difference whether we describe the plate as constituted by the interconnection of branches, or the branches as derived by the notching or incision

---

* The fact that in one very small deer, the little South American Coassus, the antler is reduced to a simple short spike, does not preclude the general distinction which I have drawn. In Coassus we have the beginnings of an antler, which has not yet manifested its tendency to expand; and in the many allied species of the American genus Cariacus, we find the expansion manifested in various simple modes of ramification or bifurcation.

of a plate. The important point for us is to recognise that
(save for occasional slight irregularities) the branching system in
the one *conforms* essentially to the curved plate or surface which we
see plainly in the other. In short the arrangement of the branches
is more or less comparable to that of the veins in a leaf, or to that of
the blood-vessels as they course over the curved surface of an organ.
It is a process of ramification, not, like that of a tree, in various
planes, but strictly limited to a single surface. And just as the
veins within a leaf are not necessarily confined (as they happen to

Fig. 439.   Head and antlers of the Indian swamp-deer (*Cervus Duvauceli*).
After Lydekker, from *P.Z.S.*

be in most ordinary leaves) to a *plane* surface, but, as in the petal
of a tulip or the capsule of a poppy, may have to run their course
within a curved surface, so does the analogy of the leaf lead us
directly to the mode of branching which is characteristic of the antler.
The surface to which the branches of the antler tend to be confined
is a more or less spheroidal, or occasionally an ellipsoidal one; and
furthermore, when we inspect any well-developed pair of antlers,
such as those of a red deer, a sambur or a wapiti, we have no difficulty
in seeing that the two antlers make up between them *a single surface*,

and constitute a symmetrical figure, each half being the mirror-image of the other. It is what the ghillies call the "cup of the antler".

To put the case in another way, a pair of antlers (apart from occasional slight irregularities) tends to constitute a figure such that we could conceive an elastic sheet stretched over or round the entire system, and to form one continuous and even surface; and not only would the surface curvature be on the whole smooth and even, but the boundary of the surface would also tend to be an even curve: that is to say the tips of all the tines would approximately have their locus in a continuous curve.

It follows from this that if we want to make a simple model of a set of antlers, we shall be very greatly helped by taking some appropriate spheroidal surface as our groundwork or scaffolding. The best form of surface is a matter for trial and investigation in each particular case; but even in a sphere, by selecting appropriate areas thereof, we can obtain sufficient varieties of surface to meet all ordinary cases. With merely a bit of sculptor's clay or plasticine, we should be put hard to it to model the horns of a wapiti or a reindeer: but if we start with an orange (or a round florence flask) and lay our little tapered rolls of plasticine upon it, in simple natural curves, it is surprising to see how quickly and successfully we can imitate one type of antler after another. In either case, we shall be struck by the fact that our model may vary in its mode of branching within very considerable limits, and yet look perfectly natural; for the same wide range of variation is characteristic of the natural antlers themselves. As Sir V. Brooke says (*op. cit.* p. 892), "No two antlers are ever exactly alike; and the variation to which the antlers are subject is so great that in the absence of a large series they would be held to be indicative of several distinct species*." But all these many variations lie within a limited range, for they are all subject to our general rule that the entire structure is essentially confined to a single curved surface. A sheet of stiff paper makes an even simpler model. Fold it in two; cut a deer's head out of the double sheet, and leave a large oval where the antlers are to be; cut a few notches in this oval leaf, for the spaces between the tines (Fig. 440). The likeness to a pair of antlers seems remote to begin

---

* Cf. also the immense range of variation in elks' horns, as described by Lönnberg, *P.Z.S.* II, pp. 352–360, 1902.

with; but it is wonderfully improved as we separate the two antlers and give a twist to each, turning antler, tines and all, into the appropriate curved or twisted surface.

It is probable that in the curvatures both of the beam and of its tines, in the angles by which these latter meet the beam, and in the contours of the entire system, there are involved many elegant mathematical problems with which we cannot attempt to deal. Nor must we attempt meanwhile to enquire into the physical meaning or origin of these phenomena, for as yet the clue seems to be lacking and we should only heap one hypothesis upon another. That there is a complete contrast of mathematical properties between the horn and the antler is the main lesson with which, in the meantime, we must rest content.

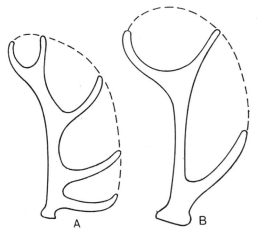

Fig. 440.   Diagrams of antlers, before twisting into shape.
A, Red-deer; B, Swamp-deer.

Of teeth, and of beak and claw

In a fashion similar to that manifested in the shell or the horn, we find the equiangular spiral to be implicit in a great many other organic structures where the phenomena of growth proceed in a similar way: that is to say, where about an axis there is some asymmetry leading to unequal rates of longitudinal growth, and where the structure is of such a kind that each new increment is added on as a permanent and unchanging part of the entire conformation. Nail and claw, beak and tooth, all come under this category. The logarithmic spiral *always* tends to manifest itself in such structures as these, though it usually only attracts our attention in elongated structures, where (that is to say) the radius vector has described a considerable angle. When the canary-bird's claws grow long from lack of use, or when the incisor tooth of a rabbit or a rat grows long by reason of disease or of injury of the opponent tooth against which it was wont to bite*, we know that the tooth or claw tends to grow into a spiral curve, and we speak of it as a malformation†. But there has been no fundamental change of form, only an abnormal increase in length; ·the elongated tooth or claw has the selfsame curvature which it had when it was short, but the spiral becomes more and more manifest the longer it grows. It is only natural, but nevertheless it is curious to see, how closely a rabbit's abnormally overgrown teeth come to resemble the tusks of swine or elephants, of which the normal state is one of hypertrophy. A curiously analogous case is that of the New Zealand Huia bird, in which the beak of the male is comparatively short and straight, while that of the female is long and curved; it is easy to see that there is a slight but identical curve also in the beak of the male, and that the beak of the female shews nothing but an extension or prolongation of the same. In the case of the more curved beaks, such as those of an eagle or a parrot, we may, if we please, determine the constant angle of the logarithmic spiral, just as we have done in the case of the *Nautilus* shell; and here again, as the bird grows older or the beak longer, the spiral nature of the curve becomes more and more apparent, as in the hooked beak of an old eagle, or in the great beak of a hyacinthine macaw.

Let us glance at one or two instances to illustrate the spiral curvature of teeth.

* Cf. John Hunter, *Natural History of the Human Teeth* (3rd ed.), 1808, p. 110: "Where a tooth has lost its opposite, it will in time become really so much longer than the rest as the others grow shorter by abrasion". Cf. James Murie, Notes on some diseased dental conditions in animals, *Tr. Odontol. Soc.* 1867–8, pp. 37–69, 257–298. We now know that a *Coenurus*-cyst in a rabbit's masseter muscle may twist the jaw sideways, so that the incisors fail to meet, and grow accordingly: H. A. Baylis, *Trans. R. Soc. Trop. Medicine*, xxxiii, p. 4, 1939.

† See Professor W. C. McIntosh's paper on "Abnormal teeth in certain mammals, especially in the rabbit," *Trans. R.S.E.* lvi, pp. 333–407, for a large collection of instances admirably illustrated.

A dentist knows that every tooth has a curvature of its own, and that in pulling the tooth he must follow the direction of the curve; but in an ordinary tooth this curvature is scarcely visible, and is least so when the diameter of the tooth is large compared with its length. In simple, more or less conical teeth, such as those of the dolphin, and in the more or less similarly shaped canines and incisors of mammals in general, the curvature of the tooth is particularly well seen. We see it in the little teeth of a hedgehog, and in the canines of a dog or a cat it is very obvious indeed. When the great canine of the carnivore becomes still further enlarged or elongated, as in *Machairodus*, it grows into the strongly curved sabre-tooth of that extinct tiger; and the boar's canine grows into the spiral tusk of wart-hog or babirussa. In rodents, it is the incisors which undergo elongation; their rate of growth differs, though but slightly, on the two sides of the axis, and by summation of these slight differences in the rapid growth of the tooth an unmistakable logarithmic spiral is gradually built up; we see it admirably in the beaver, or in the great ground-rat *Geomys*. The elephant is a similar case, save that the tooth or tusk remains, owing to comparative lack of wear, in a more perfect condition. In the rodent (save only in those abnormal cases mentioned on the last page) the tip, or first-formed part of the tooth wears away as fast as it is added to from behind; and in the grown animal, all those portions of the tooth near to the pole of the logarithmic spiral have long disappeared. In the elephant, on the other hand, we see, practically speaking, the whole unworn tooth, from point to root; and its actual tip nearly coincides with the pole of the spiral. If we assume (as with no great inaccuracy we may do) that the tip actually coincides with the pole, then we may very easily construct the continuous spiral of which the existing tusk constitutes a part; and by so doing, we see the short, gently curved tusk of our ordinary elephant growing gradually into the spiral tusk of the mammoth. No doubt, just as in the case of our molluscan shells, we have a tendency to variation, both individual and specific, in the constant angle of the spiral; some elephants, and some species of elephant, undoubtedly have a higher spiral angle than others. But in most cases, the angle would seem to be such that a spiral configuration would become very manifest indeed if only the tusk pursued its steady growth, unchanged otherwise in form, till it attained the dimensions which we meet with in the mammoth. In a species such as *Mastodon angustidens*, or *M. arvernensis*, the specific angle is low and the tusk comparatively straight; but the American mastodons and the existing species of elephant have tusks which do not differ appreciably, except in size, from the great spiral tusks of the mammoth, though from their comparative shortness the spiral is little developed and only appears to the eye as a gentle curve. Wherever the tooth is very long indeed, as in the mammoth or the beaver, the effect of some slight and all but inevitable lateral asymmetry in the rate of growth begins to shew itself: in other words, the spiral is seen to lie not absolutely in a plane, but to be a gauche curve, like a twisted horn. We see this condition very well in the huge canine tusks of the babirussa; it is a conspicuous feature in the mammoth, and it is more or less perceptible in any large tusk of the ordinary elephants.

The simplest of mammalian teeth are, like those of reptiles, conical buds which spring by single roots from a common origin: much as the pinnules of a compound leaf spring from a common petiole. A dolphin's teeth are typical of what Cope* called a *haplodont* dentition; a sloth's (whether degenerate or no) are no further advanced; canines remain unaltered throughout the mammalia, and incisors vary little save for some flattening due to crowding in a foreshortened jaw. Like the leaf and its pinnules, the tooth-germ buds and branches in endless ways; and we have no criterion of comparison (nor any right to expect it) between the individual cusps of a dog's, an elephant's and a horse's teeth, any more than between the several pinnules, cusps or leaflets of a rose, a maple and a horse-chestnut. The tooth-buds remain apart or coalesce in various numbers and degrees; and crowding, abrasion and mechanical pressure play a large part in the final arrangement and conformation†.

The dolphin's teeth, used only for prehension, do not impinge on one another, and stay sharp accordingly; those of the carnivores

* E. D. Cope, On the homologies and origin of the types of molar teeth in mammalian dentition, *Pr. Ac. N. S. Philad.* xxv, p. 371, 1873; *Journ. Ac. N. S. Philad.* (c), viii, pp. 71–89, 1874.

† Cf. J. A. Ryder, Mechanical genesis of tooth forms, *Pr. Ac. N. S. Philad.* 1878, pp. 45–80.

interlock, rather than meet and oppose*; in herbivorous animals the molars grind one against another, and wear their crowns away. The teeth of ungulates have been studied with especial care by the palaeontologists on the basis of Cope's well-known tritubercular theory, and one is greatly daring who ventures to deal with them in a different way†. The case is neither plain nor easy. We are accustomed to speak of a "tooth" as a single unit, however complicated it may be; but we may err in doing so, and we encounter other difficulties in studying teeth whose crowns are worn away, and in interpreting the "patterns" which successive stages of wear and tear expose.

The elephant's molar is manifestly composite. We see on its worn surface a long succession of "enamel ridges," each marking a narrow ring or island, lying transversely, filled with dentine, surrounded by interstitial cement, and with a root or roots of its own. The molars develop one after another during the animal's lifetime; and each consists, to begin with, of so many separate island-elements, not yet cemented together nor worn down, each with its own roots, its own covering of enamel and its own transversely cuspidate crown. These are true dental units, the primitive individual "teeth", corresponding to the still simpler teeth of the dolphin; and they illustrate, and go far to confirm the view that the molar tooth is formed, both here and elsewhere, by "concrescence"‡. These *rudimenta dentium*, as old Patrick Blair called them, or *denticules* as Owen did, soon fuse together, and begin to wear down as soon as the great composite tooth rolls forward and emerges from the gum. As each denticule begins to wear away, it first appears as a transverse row of separate rings, the so-called *columns*, which represent the cusps of the original crown and vary in size, number and proportion with the species.

---

* This is precisely what Aristotle means when he describes the dog's teeth as *carcharodont*, or sharklike, i.e. interlocking—καρχαρόδοντα γάρ ἐστιν ὅσα ἐπαλλάττει τὰς ὀδόντας τὰς ὀξεῖς, H.A., ii, 501 a 18.

† See E. D. Cope, *loc. cit.* and H. F. Osborn, *passim.* Cf. also W. K. Gregory, A half century of trituberculy, *Proc. Am. Phil. Soc.* LXXIII, pp. 169–317, 1934, who says that "even the most complex molar patterns of the Ungulates are referable to the trituberculate type, in strict accord with the steps postulated by Cope and Osborn."

‡ A view held by Gaudry, Giebel, Kükental and others, but stoutly opposed by Cope and Osborn, who see in the molar tooth a single unit, complicated by "differentiation".

These columns soon fuse and vanish as the cusps wear down; and each denticule now appears as a continuous ring of enamel, within

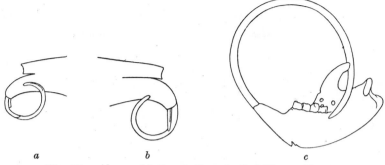

a   *b*   *c*

Fig. 441.   Abnormal incisor teeth: *a, b*, of rabbit; *c*, of beaver.
After McIntosh.

which the dentine is exposed and around which the cement accumulates*.   A single great molar is made up of nearly a dozen of these

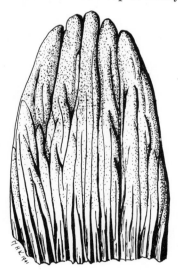

Fig. 442.   A dental unit, or element of the composite molar tooth, of an Indian elephant.   It consists of five "columns", terminating in yet unworn "cusps".

---

* See Blair's *Osteographia elephantina*, 1713, Tab. III, 19; also the figures in F. van Gaver's Étude de la tête d'un jeune Éléphant d'Asie, *Ann. Mus. Marseille*, XX, 1925.   Cf. also L. Bolk, Zur Ontogenie des Elefantengebisses, *Odontologische Studien*, III, Leipzig, 1919.

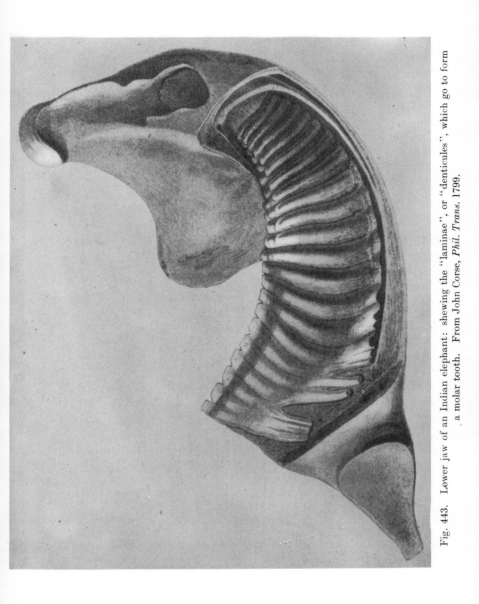

Fig. 443.   Lower jaw of an Indian elephant: shewing the "laminae", or "denticules", which go to form a molar tooth.   From John Corse, *Phil. Trans.* 1799.

elements in the African, and of twice as many (twice as much
flattened or compressed) in the Indian elephant; in the mastodon
they are much fewer and much larger, and their great tuberculated
crowns never wholly wear away. In an old but admirable paper
on the Indian elephant*, Mr John Corse says: "The number of
teeth of which a grinder is composed varies from four to twenty-
three, according as the elephant advances in years; so that a
grinder, or case of teeth, in full-grown elephants, is more than
sufficient to fill one side of the mouth.... The same number of
laminae generally fills the jaw of a young or of an old elephant;
and from three till fifty years there are from ten to twelve teeth or
laminae in use, in each side of either jaw, for the mastication of
the food."

The molar teeth of a mouse, a hare or a capybara are likewise
composite structures; they shew, precisely after the fashion of the
elephant, successive narrow annular islands of enamel, with dentine
within and cement between, all in varying degrees of independence
or coalescence †.

The molars of a hippopotamus are composite but to a less degree;
his upper molars have each two pair of roots, the last molar one
root more. A block of dentine lying transversely to the jaw, with
a pair of roots below and a pair of enamel-covered cusps above, is the
unit of dentition, and is analogous to the young toothlet of the
elephant.

In the horse and its kind the teeth are long and deeply sunk in
the jaw, very much as in a rabbit or hare. Their length is made
up not of root but of elongated crown, in which the deep valleys
between the once high cusps are filled or flooded with cement; and
these long crowns are soon worn down to an all but level surface,

* J. Corse, Observations on the different species of Asiatic elephant, and their
mode of dentition, *Phil. Trans.* 1799, pp. 205–236; and cf. Owen's *Comp. Anat.*
III, p. 361.

† The elephant (in my opinion) shews its likeness or affinity to the rodents
throughout its whole anatomy, the metacromial process of its scapula being one
conspicuous indication. *Hyrax* and *Elephas* are two isolated forms lying near the
common origin of ungulates and rodents; the one lying rather to the ungulate
side, the other to the rodent side, of the vague and indefinable border-line. On
the relation of the rodent's dentition to the elephant's (a view strongly opposed
by Dr W. K. Gregory), see M. Friant, Contribution à l'étude...des dents jugales
chez les Mammifères, *Bull. Mus. Hist. Nat.* I, pp. 1–132, 1933.

in which the enamel-layer which covered the hills and lined the valleys is seen in sectional contour. A horse's incisor is the simplest case. On its worn surface we see an inner ring of enamel concentric with the enamel of the outer edge or surface of the tooth; cement fills up the inner ring, and dentine the space between. The tip of the tooth has sunk down, or been tucked in, till it forms a cement-filled lake on the top of the hill; the lake narrows in, and at last vanishes as the horse grows old and the tooth wears down; in the "aged" horse we see the "mark" no more. To recognise this lake or pit in the simple contours of the young incisor is an easy matter; but in the abraded molar the enamel-layer which once covered all its ups and downs forms a contour-line, or "curve of level," of great complexity. This contour-line alters as the levels change, and varies from one tooth to the next and from one year to another, so long as wear and tear continue. The geographer reads the lie of the land, with all its ups and downs, from a many-contoured map*, but the worn tooth shews us only one level and one contour at a time; we must eke out its scanty evidence by older and younger teeth in other phases or degrees of wear. The "pattern" of a horse's molar tooth is indeed so closely akin to a map-maker's contours that some of the terms he uses may be useful to us. He speaks, for instance, of *ridge-lines* and *course-lines*, *lignes de faîte* and *lignes de thalweg;* of a *gap*, or lowland way between two hills, in contrast to a *col* or *saddle* at the summit of a mountain-pass; or of a *gorge*, which is a narrow steep-sided valley; or a *scarp*, which is a long steep-faced hillside. We must take care all the while to see which side of our contour-line is positive or negative—on which side the ground slopes up and on which down. In our tooth we find that every enamel-contour has dentine on the one side and more or less cement on the other; the dentine belongs to the closed interior of the tooth itself, and on the other side of the enamel-line are spaces open to the world.

In a horse's molar we see the sinuous contours of two small lakes, remains of the two valleys which lay between the three transverse ridges of the compound tooth; and outside the enamel-edges of

---

* Contour-lines or *horizontals*, as some geographers prefer to call them, were invented by Buache, in 1752. These are discussed by Cayley, On contours and slope-lines, *Phil. Mag.* xviii, pp. 294–8, 1859; and by Clerk Maxwell, On hills and dales, *ibid.* xl, pp. 421–7, 1870.

these contoured lakes is the dentinal substance of the tooth, surrounded again by the outer covering of enamel. The space between
the outer and the inner contours is narrowed in each case at a
certain point, suggesting a "col"; while it broadens out at other
places, suggesting the former sites of cusps or hills. In neighbouring sections (B, C) we rise above the level of the cols, find a
way open to the valleys, and see the separate transverse mountain-
ranges (or lophs) of which the tooth is composed. The general plan

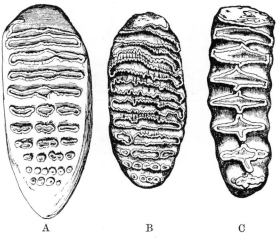

A　　　　　　　B　　　　　　　C

Fig. 444. Upper molar teeth of elephants. A, *E. meridionalis* (Pliocene), largest
of elephants. B, Indian, and C, African elephants.

of this tooth is characteristic of the Perissodactyles; but the varying
steepness of the hills, the depth of the valleys, and the amount of
abrasion or erosion, lead to an infinite variety of patterns, varying
with the species, with the age of the animal, and with the order
of succession of the particular tooth; and so rendering it (as Osborn
says) "one of the most difficult objects to define and describe in
the whole field of vertebrate palaeontology*."

In *Elasmotherium* the hillsides are ridged and channelled, and
their contours folded or sinuous accordingly. In *Rhinoceros* broad
gaps replace the narrow cols, and certain jutting crags figure on

* H. F. Osborn, Equidae of the Oligocene, etc., *Mem. Amer. Mus. of N.H.* (n.s.)
II, p. 3, 1918.

the contour-lines as the so-called *crochet* and *anticrochet*. In *Anchitherium*, erosion goes no farther than the summits of the several cusps or hill-tops.

These, to my thinking, are the few and simple lines on which we may study the architecture of the Perissodactyle tooth. But to say how far we may rely on the innumerable minor differences of pattern

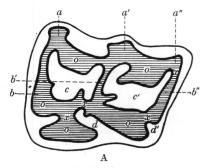

A

Fig. 445 A. Third upper molar of a horse. *a*, the *ectoloph*, with its three *styles*, separated by two *indents* (Owen); *b*, three transverse ridges, the *protoloph*, *mesoloph* and *metaloph* (Osborn); *c*, *c'*, two lakes, valleys or *fossettes*; *d*, *d'*, what Owen calls the *entries of the valleys*; *x*, *x'*, *cols*, where a less worn tooth would shew open roads or *passes*; *o*, *o*, *cusps* or *conules*, the sites of worn-down hills or hillocks.

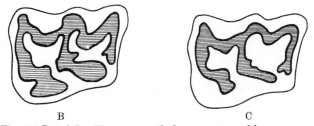

B　　　　　　　　　　　　　　C

Fig. 445 B and C. The same tooth, but younger and less worn down than A. Diagrammatic.

as evidence of blood-relationship and evolutionary descent is quite another story, and deserves much more anxious consideration*.

* In the vast literature of mammalian dentition the following are conspicuous: R. Owen, *Odontography*, 1845; L. Rütimeyer, Zur Kenntniss der fossilen Pferde, und zu einer vergl. Odontographie der Hufthiere, *Verh. Naturf. Ges. Basel*, III, 1963; W. Leche, Zur Entwicklungsgeschichte des Zahnsystems der Säugetiere, *Bibl. Zool.* 1894–5, 160 pp.; E. D. Cope, On the trituberculate type of molar tooth in the Mammalia, *Proc. Amer. Philos. Soc.* XXI, pp. 324–326, 1885; W. K. Gregory, A half-century of trituberculy, *ibid.* LXXIII, pp. 161–317, 1934.

The "horn" or tusk of the narwhal is a very remarkable and a very anomalous thing. It is the only tooth in the creature's head to come to maturity; it grows to an immense and apparently

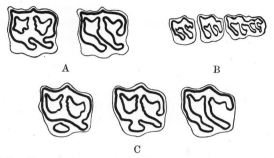

Fig. 446.　Enamel patterns (diagrammatic) of certain fossil Equidae.
A, *Protohippus*; B, *Hyohippus*; C, *Neohippus*.

unwieldy size, say to eight or even nine feet long; it never curves nor bends, but grows as straight as straight can be—a very singular and exceptional thing; it looks as though it were twisted, but really

Fig. 447.　Enamel pattern (diagrammatic) of the upper molar teeth of *Rhinoceros*.
The back-tooth (to the right-hand side) is the least worn, and its contour-line lies at the highest level.

carries on its straight axis a *screw* of several contiguous low-pitched threads; and (last and most anomalous thing of all) when, as happens now and then, two tusks are developed instead of one, one on either side, these two do not form a *conjugate* or symmetrical pair, they are not mirror-images of one another, but are *identical screws, with both threads running the same way**.

* The male narwhal carries the horn, the female being tuskless; but the whalers say that the rare two-horned specimens are all females. A famous two-horned skull in the Hamburg Museum is known to have belonged to a pregnant female. It was brought home in 1684, and is one of the oldest museum specimens in the world; the tusks measure 242 and 236 cm. During my thirty years' close acquaintance with the Dundee whalers, only four two-horned narwhals passed through my hands. Bateson (*Problems of Genetics*, 1913, p. 44) makes the curious remark that "the Narwhal's tusks, in being both twisted in the same direction, are highly anomalous, and are *comparable with pairs of twins*."

All ordinary teeth, as we have seen, have their own natural curvature, less or more, which becomes more manifest and conspicuous the longer they grow. We cannot suppose that the field of force (internal and external) in which the narwhal's tusk develops is so simple and uniform as to allow it to grow in perfect symmetry, year after year, without the least bias or intention toward either side; we must rather suppose that the resistances which the growing tusk encounters average out and cancel one another, and leave no one-sided resultant. The long, straight, tapering tooth is commonly said to have a "spiral twist," but there is no twist at all; the ivory is straight-grained and uniform, through and through. The tusk, in short, is a straight, right-handed, low-pitched screw or helix, with several threads; which threads, in the form of alternate grooves and ridges, wind evenly and continuously from one end of the tusk to the other, *even extending to its root*, deep-set in the socket or alveolus of the upper jaw.

How this composite spiral thread is formed is quite unknown. We have just seen that it is not due to any twisting of the dentinal axis of the tooth. That it is uniform and unbroken from end to end shews that the tooth somehow fashions it as a whole; and that it extends deep down within the alveolus is enough to shew that it is not impressed or graven on the tooth by any external agency. We note, as a minor feature, that the several grooves or ridges which constitute the composite thread have their individual or accidental differences; a broader or a narrower groove continues unchanged and recognisable from one end of the tooth to the other; in other words, whatever makes each ridge or groove goes on acting in the selfsame way, as long as growth goes on. A *screw* is made, in general, by compounding a translatory with a rotatory motion, and by bringing the latter into relation with the mould or matrix by which the thread is fashioned or imposed; and I cannot see how to avoid believing that the narwhal's tooth must revol e in like manner, very slowly on its longitudinal axis, all the while it grows—however strange, anomalous and hard to imagine such a mode of growth may be. We know that the tooth grows throughout life in its longitudinal direction, the open root and "permanent pulp" accounting for this; and only by a simultaneous and equally continuous rotation (so far as I can see) can we account for the perfect

straightness of the tusk, for the grooving or "rifling" of the surface
accompanied by no internal twist, for the extension of that rifling
to the alveolar portion of the tusk within the jaw, and for the fact
that the several associate grooves and ridges preserve their individual
character as they pass along and wind their way around. A very
slow rotation is all we need demand—say four or five complete
revolutions of the tusk in the whole course of a lifetime.

The progress of a whale or dolphin through the water may be
explained as the reaction to a wave which is caused to run from
head to tail, the creature moving through the water somewhat
slower than the wave travels. The same is true, so far, of a fish;
but the wave tends to be in one plane in the fish, the dorsal and
ventral fins helping to keep it so; while in the dolphin it may be
said to be "circularly polarised," or resoluble into two oscillations in
planes normal to one another, and caused by tail and tail-end swishing
around in circular orbits which alter in phase from one transverse
section to another. Just as in the case of a screw-propeller, or as
in a torpedo (where it is specially corrected or compensated), this
mode of action entails a certain waste of energy; it comes of the
development of a "harmful moment," which tends to rotate the
body about its axis, and to *screw* the animal along its course. A
slight left-handed curvature of the dolphin's tail goes some little
way towards correcting this tendency. M. Shuleikin's study of the
kinematics of the dolphin*—a fine piece of work both on its experimental and its theoretical side—shows the dolphin to be a better
swimmer than the fish, inasmuch as its speed of progression comes
nearer to the velocity of the wave which is propagated along its
body; the so-called "step," or fraction of the body-length travelled
in a single period, is found to be about 0·7 in the dolphin, against
0·57 in a fast-swimming fish (tunny or mackerel).

Shuleikin makes the curious remark that the asymmetry of the
skull (discernible in all Cetacea), which in the dolphin shews a screw-
twist with a pitch about equal to the length of the body, acts as
a compensatory check to the screw-component in the creature's
movement of progression, and that "the till now obscure purpose

* Wassilev Shuleikin, Kinematics of a dolphin (Russian), *Bull. Acad. Sci. U.R.S.S.*
(*Cl. sci. math. et phys.*), 1935, pp. 651–671; also, Dynamics, external and internal,
of a fish, *ibid.* 1934, pp. 1151–1186. On the latter subject see James Gray,
Croonian Lecture, 1940, and other papers.

of the skull's asymmetry" is accordingly explained. I should put
this differently, and suggest that this counter-spirality of the skull
is the direct *result* of the spiral component in locomotion. It implies,
I take it, a lagging and incomplete response in the fore-part of the
body to the rotatory impulse of the parts behind: or, in the plain
words of the engineer, a *torque of inertia.*

This tendency, dimly seen in the dolphin's skull, is clearly demonstrated in the narwhal's "horn," and gives a complete explanation
of its many singularities. The narwhal and its horn are joined
together, and move together as one piece—nearly, but not quite!
Stiff, straight and heavy, the great tusk has its centre of inertia
well ahead of the animal, and far from the driving impulse of its
tail. At each powerful stroke of the tail the creature not only darts
forward, but twists or slews all of a sudden to one side; and the
heavy horn, held only by its root, responds (so to speak) with
difficulty. For at its slender base the "couple", by which it has
to follow the twisting of the body, works at no small disadvantage.
A "torque of inertia" is bound to manifest itself. The horn does
not twist round in perfect synchronism with the animal; but the
animal (so to speak) goes slowly, slowly, little by little, round its
own horn! The play of motion, the lag, between head and horn
is slight indeed; but it is repeated with every stroke of the tail.
It is felt just at the growing root, the permanent pulp, of the tooth;
and it puts a strain, or exercises a torque, at the very seat, and during
the very process, of calcification.

Suppose that at every sweep of the tail there be a lag of no more
than a fifth part of a second of arc* between the rotation of the
tusk and of the body, that small amount would amply suffice to
account, on a rough estimate of the age and of the activity of the
animal, for as many turns of the screw as a fair-sized tusk is found
to exhibit.

According to this explanation, or hypothesis, the slow rotation
of the tusk corrects all tendency to flexure or curvature in one
direction or another; the grooves and ridges which constitute the
"thread" of the screw are the result of irregularities or inequalities
within the alveolus, which "rifle" the tusk as it grows; and the

* Or say a hundred-thousandth part of the angle subtended by a minute on the
clock.

identity of direction in the two horns of a pair is at once accounted for.

Beautiful as the spiral pattern of the tusk is, it obviously falls short, in regularity and elegance, of what we find, for instance, in a long tapering *Terebra* or *Turritella*, or any other spiral gasteropod shell. In the narwhal we have, as we suppose, only a *general* and never a precise agreement between rate of torsion and rate of growth; for these two velocities—of translation and rotation—are separate and independent, and their resultant keeps fairly steady but no more. In the snail-shell, on the other hand, actual tissue-growth is the common cause of both longitudinal and torsional displacements, and the resultant spiral is very perfect and regular.

Before we leave the teeth, let us note that their extreme tightness in their sockets is a remarkable thing. A thin "periodontal membrane," less than 0·25 mm. thick, fills up the space between tooth and socket; and this membrane, elastic, homogeneous and incompressible, is analogous to the thin layer of viscous liquid dealt with in modern theories of lubrication. The equilibrium of the system, the tightness of the fit, the displacement of the tooth under given forces, and the conditions of stress and strain in the membrane, are all open to mathematical treatment; distributions of pressure can be assigned to the tooth, a centre of rotation can be found, a critical load can be approximately determined, and the pressures calculated at various points. If the membrane thickens, the tooth loosens; ·its freedom of movement or range of displacement varies with the cube of the thickness of the membrane, and is at most exceedingly small*.

* J. L. Synge, The tightness of the teeth, etc., *Phil. Trans.* (A), ccxxxi, pp. 435–477, 1933.

CHAPTER XIV

ON LEAF-ÁRRANGEMENT, OR PHYLLOTAXIS

THE beautiful configurations produced by the orderly arrangement of leaves or florets on a stem have long been an object of admiration and curiosity; and not the least curious feature of the case is the limited, even the small number of possible arrangements which we observe and recognise. Leonardo da Vinci would seem, as Sir Theodore Cook tells us, to have been the first to record his thoughts upon this subject; but the old Greek and Egyptian geometers are not likely to have left unstudied or unobserved the spiral traces of the leaves upon a palm-stem, or the spiral order of the petals of a lotus or the florets in a sunflower. For so, as old Nehemiah Grew says, "from the contemplation of Plants, men might first be invited to Mathematical Enquirys*."

The spiral leaf-order has been regarded by many learned botanists as involving a fundamental law of growth, of the deepest and most far-reaching importance; while others, such as Sachs, have looked upon the whole doctrine of "phyllotaxis" as "a sort of geometrical or arithmetical playing with ideas," and "the spiral theory as a mode of view gratuitously introduced into the plant." Sachs even went so far as to declare this doctrine to be "in direct opposition to scientific investigation, and based upon the idealism of the Naturphilosophie"—the mystical biology of Oken and his school.

The essential facts of the case are not difficult to understand; but the theories built upon them are so varied, so conflicting, and sometimes so obscure, that we must not attempt to submit them to detailed analysis and criticism. There are said to be two chief ways by which we may approach the question, according to whether we regard as the more fundamental and typical, one or other of two chief modes in which the phenomenon presents itself. That is to say, we may hold that the phenomenon is displayed in its essential

---

* N. Grew, *The Anatomy of Plants,* 1682, p. 152.

simplicity by the corkscrew spirals, or helices, which mark the position of the leaves on a cylindrical stem or tapering fir-cone; or, on

Fig. 448. A giant sunflower, *Helianthus maximus*. From H. A. Naber, after M. Brocard.

the other hand, we may be more attracted by, and may regard as of greater importance, the spirals traced by the curving rows of florets in the discoidal inflorescence of a sunflower. Whether one way or

the other be the better, or even whether one be not positively correct
and the other radically wrong, has been vehemently debated; but as
a matter of fact they are, both mathematically and biologically,
inseparable and even identical phenomena. For the face of the
sunflower is but a shortened stem, and the curves upon its

Fig. 449   A cauliflower, its composite inflorescence shewing spiral patterns
of the first and second order.

surfaces are but the projection on a plane of a more elongated
inflorescence.

We speak, as botanists are wont to do, of these spirals of sunflower,
cauliflower and the rest as logarithmic spirals, but not without
hesitation. They doubtless resemble the logarithmic or equiangular
spiral, but different spirals may look much alike; and these are
ill-suited to the careful admeasurement and rigorous verification

which Moseley gave to the spirals of his molluscan shells *. But in the sunflower, to judge by the eye, the spirals remain self-similar as they grow; each fresh increment forms, or seems to form, a *gnomon* to what went before; each new floret falls into line as part of a continuous and self-similar curve: and this goes a long way to justify our use of the familiar term logarithmic, or equiangular spiral. But the leaf-arrangement or the inflorescence are far less simple than the shell. The shell grew as one continuous and indivisible whole; its tip is the oldest part, it remains the smallest part, and the spiral tube expands continuously as it goes on. But each floret of the sunflower has its own separate and individual growth; the oldest is also the largest, and the youngest is the least; and as younger and younger florets are added on, the spiral advances in the direction of its own focus, or its own little end. And the conditions may be less simple still in other cases, as in the fir-cone itself.

The spiral tesselation of the fir-cone was carefully studied in the middle of the eighteenth century by the celebrated Bonnet, with the help of Calandrini the mathematician. Memoirs published about 1835, by Schimper and Braun, greatly amplified Bonnet's investigations, and introduced a nomenclature which still holds its own in botanical textbooks. Naumann and the brothers Bravais are among those who continued the investigation in the years immediately following, and Hofmeister, in 1868, gave an admirable account and summary of the work of these and many other writers †.

* Thus Dr A. H. Church, in his *Interpretation of Phyllotaxis Phenomena*, 1920, p. 3, begins by saying that "angular measurements on actual plant-specimens... can never hope to come within a range of accuracy admitting of an error of less than half a degree, while precise mathematical theory soon begins to tabulate minutes and seconds."

† Besides papers referred to below, and many others quoted in Sachs's *Botany* and elsewhere, the following are important: Alex. Braun, Vergl. Untersuchung über die Ordnung der Schuppen an den Tannenzapfen, etc., *Nova Acta Acad. Car. Leop.* xv, pp. 199–401, 1831; C. F. Schimper's Vorträge über die Möglichkeit eines wissenschaftlichen Verständnisses der Blattstellung, etc., *Flora*, xviii, pp. 145–191, 737–756, 1835; C. F. Schimper, Geometrische Anordnung der um eine Achse peripherischen Blattgebilde, *Verhandl. Schweiz. Ges.* 1836, pp. 113–117; L. and A. Bravais, Essai sur la disposition des feuilles curvisériées, *Ann. Sci. Nat.* (2), vii, pp. 42–110, 1837; Sur la disposition symétrique des inflorescences, *ibid.* pp. 193–221, 291–348, viii, pp. 11–42, 1838; Sur la disposition générale des feuilles rectisériées, *ibid.* xii, pp. 5–41, 65–77, 1839; *Mémoire sur la disposition géométrique des feuilles et des inflorescences*, Paris, 1838; Zeising, *Normalverhältniss*

The surface of a pine-cone shews a crowded assemblage of woody scales, close-packed and pressed together in such a way that each has a quadrangular, rhomboidal form*. Each scale forms part of, and marks the intersection of, two linear series; these run upwards in a spiral course, one in one direction and one in the other, and are called accordingly *diadromous* spirals. In the little cones of the Scotch Fir (*Pinus silvestris*), the whose assemblage of scales may be looked on as forming five linear series, or spiral bands, running side by side the one way, or as eight such series running the other. But these two sets are far from being all the spirals which we can trace upon the cone. Sometimes the packing is closer still, especially if the cone be long and slender. Then each scale tends to come in contact with six others, and so to become roughly hexagonal; we recognise a third spiral series besides the other two, and this new series is found tò consist of thirteen rows. But let us disregard for the moment this perplexing phenomenon of a cone composed of so many series of scales, five, eight or thirteen in number as we happen to look at them; and try to find a single series in which every scale takes part. We are in no way limited to the fir-cone, which is a somewhat special case; but may consider, in a very general way, the case of any leafy stem.

Starting from some given level and proceeding upwards, let us mark the position of some one leaf (*A*) upon the cylindrical stem.

der chemischen und morphologischen Proportionen, Leipzig, 1856; C. F. Naumann, Ueber den Quincunx als Gesetz der Blattstellung bei Sigillaria, etc., *Neues Jahrb. f. Miner.* 1842, pp. 410–417; T. Lestiboudois, *Phyllotaxie anatomique*, Paris, 1848; G. Henslow, Phyllotaxis, or the arrangement of leaves according to mathematical laws, *Jl. Victoria Inst.* VI, pp. 129–140, 1873; On the origin of the prevailing systems of Phyllotaxis, *Tr. Linn. Soc. (Bot.)*, I, pp. 37–45, 1880. J. Wiesner, Bemerkungen über rationale und irrationale Divergenzen, *Flora*, LVIII, pp. 113–115, 139–143, 1875; H. Airy, On leaf arrangement, *Proc. R.S.* XXI, p. 176, 1873; S. Schwendener, *Mechanische Theorie der Blattstellungen*, Leipzig, 1878; F. Delpino, *Causa meccanica della filotasse quincunciale*, Genova, 1880; *Teoria generale di Filotasse, ibid.* 1883; S. Günther, Das mathematische Grundgesetz im Bau des Pflanzenkörpers, *Kosmos*, IV, pp. 270–284, 1879; F. Ludwig, Wichtige Abschnitte aus der mathematischen Botanik, *Zeitschr. f. mathem. u. naturw. Unterricht*, XIV, p. 161, 1883; Weiteres über Fibonacci-Kurven und die numerische Variation der gesammten Blüthenstände der Kompositen, *Botan. Cblt.* LXVIII, p. 1, 1896; Alex. Dickson, Phyllotaxis of *Lepidodendron* and *Knossia, Jl. Bot.* IX, p. 166, 1871. For a historical account of the earlier literature, see Casimir de Candolle's *Considérations générales sur l'étude de la phyllotaxie*, Genève, 1881.

* Cf. *supra*, p. 515.

Another, and a younger leaf (B) will be found standing at a certain distance *around* the stem, and a certain distance *along* the stem, from the first. The former distance may be expressed as a fractional "divergence" (such as two-fifths of the circumference of the stem) as the botanists describe it, or by an "angle of azimuth" (such as $\phi = 144°$) as the mathematician would be more likely to state it. The position of B relatively to A may be determined, not only by this angle ϕ, in the horizontal plane, but also by an angle of slope (θ), or merely by linear distance from its basal plane; for the height of B above the level of A, in comparison with the diameter of the cylinder, will obviously make a great difference in the appearance of the whole system. But this matter botanical students have not concerned themselves with; in other words, their studies have been limited (or mainly limited) to the relation of the leaves to one another in *azimuth*—in other words, to the angle ϕ and its multiples.

Whatever relation we have found between A and B, let precisely the same relation subsist between B and C: and so on. Let the growth of the system, that is to say, be continuous and uniform; it is then evident that we have the elementary conditions for the development of a simple cylindrical helix; and this "primary helix" or "genetic spiral" we can now trace, winding round and round the stem, through A, B, C, etc. But if we can trace such a helix through A, B, C, it follows from the symmetry of the system, that we have only to join A to some other leaf to trace another spiral helix, such, for instance, as A, C, E, etc.; parallel to which will run another and similar one, namely in this case B, D, F, etc. And these spirals will run in the opposite direction to the spiral ABC*.

In short, the existence of one helical arrangement of points implies and involves the existence of another and then another helical pattern, just as, in the pattern of a wall-paper, our eye travels from one linear series to another.

A modification of the helical system will be introduced when, instead of the leaves appearing, or standing, in singular succession, we get two or more appearing simultaneously upon the same level. If there be two such, then we shall have two generating spirals

---

* For the spiral *ACE* to be different from *ABC*, the angle of divergence, or angle of azimuth for one step, must exceed 90°, so that the nearer way from A to C is backwards; otherwise the spiral *ACE* is *ABCDE*, or *ABC* over again.

precisely equivalent to one another; and we may call them A, B, C, etc., and A', B', C', and so on. These are the cases which we call "whorled" leaves; or in the simplest case, where the whorl consists of two opposite leaves only, we call them "decussate."

Among the phenomena of phyllotaxis, two points in particular have been found difficult of explanation, and have aroused discussion. These are (1), the presence of the logarithmic spirals such as we have already spoken of in the sunflower; and (2) the fact that, as regards the number of the helical or spiral rows, certain numerical coincidences are apt to recur again and again, to the exclusion of others, and so to become characteristic features of the phenomenon. As to the first of these, we have seen that the curves resemble, and sometimes closely resemble, the logarithmic spiral; but that they are, strictly speaking, logarithmic is neither proved nor capable of proof. That they appear as spiral curves (whether equable or logarithmic) is then a mere matter of mathematical "deformation." The stem which we have begun to speak of as a cylinder is not strictly so, inasmuch as it tapers off towards its summit. The curve which winds evenly around this stem is, accordingly, not a true helix, for that term is confined to the curve which winds evenly around the *cylinder*: it is a curve in space which (like the spiral curve we have studied in our turbinate shells) partakes of the characters of a helix and of a spiral, and which is in fact a spiral with its pole drawn out of its original plane by a force acting in the direction of the axis. If we imagine a tapering cylinder, or cone, projected by vertical projection on a plane, it becomes a circular disc; and a helix described about the cone becomes in the disc a spiral described about a pole which corresponds to the apex of our cone. In like manner we may project an identical spiral in space upon such surfaces as (for instance) a portion of a sphere or of an ellipsoid; and in all these cases we preserve the spiral configuration, which is the more clearly brought into view the more we reduce the vertical component by which it was accompanied. The converse is equally true, and equally obvious, namely that any spiral traced upon a circular disc or spheroidal surface will be transformed into a corresponding spiral helix when the plane or spheroidal disc is extended into an elongated cone approximating to a cylinder. This mathematical conception is translated, in botany, into actual fact. The fir-cone may be looked upon as a cylindrical axis contracted at both ends, until it becomes approximately an ellipsoidal solid of revolution, generated about the long axis of the ellipse; and the semi-ellipsoidal capitulum of the teasel, the more or less hemispherical one of the thistle, and the flattened but still convex one of the sunflower, are all beautiful and successive deformations of what is typically a long, conical, and all but cylindrical stem. On the other hand, every stem as it grows out into its long cylindrical shape is but a deformation of the little spheroidal or ellipsoidal or conical surface which was its forerunner in the bud.

This identity of the helical spirals around the stem with spirals projected on a plane was clearly recognised by Hofmeister, who was accustomed to represent his diagrams of leaf-arrangement either in one way or the other, though not in a strictly geometrical projection*.

According to Mr A. H. Church†, who has dealt carefully and elaborately with the whole question of phyllotaxis, the spirals such as we see in the disc of the sunflower have a far greater importance and a far deeper meaning than this brief treatment of mine would accord to them: and Sir Theodore Cook, in his book on the *Curves of Life*, adopted and helped to expound and popularise Mr Church's investigations.

Mr Church, regarding the problem as one of "uniform growth," easily arrives at the conclusion that, *if* this growth can be conceived as taking place symmetrically about a central point or "pole," the uniform growth would then manifest itself in logarithmic spirals, including of course the limiting cases of the circle and straight line. With this statement I have little fault to find; it is in essence identical with much that I have said in a previous chapter. But other statements of Mr Church's, and many theories woven about them by Sir T. Cook and himself, I am less able to follow. Mr Church tells us that the essential phenomenon in the sunflower disc is a series of orthogonally intersecting logarithmic spirals. Unless I wholly misapprehend Mr Church's meaning, I should say that this

* *Allgemeine Morphologie der Gewächse*, 1868, p. 442, etc.

† *Relation of Phyllotaxis to Mechanical Laws*, Oxford, 1901–1903; cf. *Ann. Bot.* xv, p. 481, 1901.

is very far from essential. The spirals intersect isogonally, but orthogonal intersection would be only one particular case, and in all probability a very infrequent one, in the intersection of logarithmic spirals developed about a common pole. Again on the analogy of the hydrodynamic lines of force in certain vortex movements, and of similar lines of force in certain magnetic phenomena, Mr Church proceeds to argue that the energies of life follow lines comparable to those of electric energy, and that the logarithmic spirals of the sunflower are, so to speak, lines of equipotential*. And Sir T. Cook remarks that this "theory, if correct, would be fundamental for all forms of growth, though it would be more easily observed in plant construction than in animals." But the physical analogies are remote, and the deductions I am not able to follow.

Mr Church sees in phyllotaxis an organic mystery, a something for which we are unable to suggest any precise cause: a phenomenon which is to be referred, somehow, to waves of growth emanating from a centre, but on the other hand not to be explained by the division of an apical cell, or any other histological factor. As Sir T. Cook puts it, "at the growing point of a plant where the new members are being formed, there is simply *nothing to see.*"

But it is impossible to deal satisfactorily, in brief space, either with Mr Church's theories, or my own objections to them†. Let it suffice to say that I, for my part, see no subtle mystery in the matter, other than what lies in the steady production of similar growing parts, similarly situated, at similar successive intervals of time. If such be the case, then we are bound to have in consequence

---

* "The proposition is that the genetic spiral is a logarithmic spiral, homologous with the line of current-flow in a spiral vortex; and that in such a system the action of orthogonal forces will be mapped out by other orthogonally intersecting logarithmic spirals—the 'parastichies'"; Church, *op. cit.* I, p. 42.

† Mr Church's whole theory, if it be not based upon, is interwoven with, Sachs's theory of the orthogonal intersection of cell-walls, and the elaborate theories of the symmetry of a growing point or apical cell which are connected therewith. According to Mr Church, "the law of the orthogonal intersection of cell-walls at a growing apex may be taken as generally accepted" (p. 32); but I have taken a very different view of Sachs's law, in the eighth chapter of the present book. With regard to his own and Sachs's hypotheses, Mr Church makes the following curious remark (p. 42): "Nor are the hypotheses here put forward more imaginative than that of the paraboloid apex of Sachs which remains incapable of proof, or his construction for the apical cell of *Pteris* which does not satisfy the evidence of his own drawings."

a series of symmetrical patterns, whose nature will depend upon the form of the entire surface. If the surface be that of a cylinder, we shall have a system, or systems, of spiral helices: if it be a plane with an infinitely distant focus, such as we obtain by "unwrapping" our cylindrical surface, we shall have straight lines; if it be a plane containing the focus within itself, or if it be any other symmetrical surface containing the focus, then we shall have a system of logarithmic spirals. The appearance of these spirals is sometimes spoken of as a "subjective" phenomenon, but the description is inaccurate: it is a purely mathematical phenomenon, an inseparable secondary result of other arrangements which we, for the time being, regard as primary. When the bricklayer builds a factory chimney, he lays his bricks in a certain steady, orderly way, with no thought of the spiral patterns to which this orderly sequence inevitably leads, and which spiral patterns are by no means "subjective." The designer of a wall-paper not only has no intention of producing a pattern of criss-cross lines, but on the contrary he does his best to avoid them; nevertheless, so long as his design is a symmetrical one, the criss-cross intersections inevitably come. And as the train carries us past an orchard we see not one single symmetrical configuration, but a multiplicity of collineations among the trees.

Let us, however, leave this discussion, and return to the facts of the case.

Our second question, which relates to the numerical coincidences so familiar to all students of phyllotaxis, is not to be set and answered in a word.

Let us, for simplicity's sake, avoid consideration of simultaneous or whorled leaf origins, and consider only the more frequent cases where a single "genetic spiral" can be traced throughout the entire system.

It is seldom that this primary, genetic spiral catches the eye, for the leaves which immediately succeed one another in this genetic order are usually far apart on the circumference of the stem, and it is only in close-packed arrangements that the eye readily apprehends the continuous series. Accordingly in such a case as a fir-cone, for instance, it is certain of the secondary spirals or "parastichies" which catch the eye; and among fir-cones, we can easily count these, and we find them to be on the whole very constant in number, according to the species.

Thus in many cones, such as those of the Norway spruce, we can trace five rows of scales winding steeply up the cone in one direction, and three rows winding less steeply the other way; in certain other species, such as the common larch, the normal number is eight rows in the one direction and five in the other; while in the American larch we have again three in the one direction and five in the other. It not seldom happens that two arrangements grade into one another on different parts of one and the same cone. Among other cases in which such spiral series are readily visible we have, for instance, the crowded leaves of the stone-crops and mesembryanthemums, and (as we have said) the crowded florets of the composites. Among these we may find plenty of examples in which the numbers of the serial rows are similar to those of the fir-cones; but in some cases, as in the daisy and others of the smaller composites, we shall be able to trace thirteen rows in one direction and twenty-one in the other, or perhaps twenty-one and thirty-four; while in a great big sunflower we may find (in one and the same species) thirty-four and fifty-five, fifty-five and eighty-nine, or even as many as eighty-nine and one hundred and forty-four. On the other hand, in an ordinary "pentamerous" flower, such as a ranunculus, we may be able to trace, in the arrangement of its sepals, petals and stamens, shorter spiral series, three in one direction and two in the other; and the scales on the little cone of a Cypress shew the same numerical simplicity. It will be at once observed that these arrangements manifest themselves in connection with very different things, in the orderly interspacing of single leaves and of entire florets, and among all kinds of leaf-like structures, foliage-leaves, bracts, cone-scales, and the various parts or members of the flower. Again we must be careful to note that, while the above numerical characters are by much the most common, so much so as to be deemed "normal," many other combinations are known to occur.

The arrangement, as we have seen, is apt to vary when the entire structure varies greatly in size, as in the disc of the sunflower. It is also subject to less regular variation within one and the same species, as can always be discovered when we examine a sufficiently large sample of fir-cones. For instance, out of 505 cones of the

Norway spruce, Beal* found 92 per cent. in which the spirals were in five and eight rows; in 6 per cent. the rows were four and seven, and in 4 per cent. they were four and six. In each case they were nearly equally divided as regards direction; for instance, of the 467 cones shewing the five-eight arrangement, the five-series ran in right-handed spirals in 224 cases, and in left-handed spirals in 243. Omitting the "abnormal" cases, such as we have seen to occur in a small percentage of our cones of the spruce, the arrangements which we have just mentioned may be set forth as follows (the fractional number used being simply an abbreviated symbol for the number of associated helices or parastichies which we can count running in the opposite directions): 2/3, 3/5, 5/8, 8/13, 13/21, 21/34, 34/55, 55/89, 89/144. Now these numbers form a very interesting series, which happens to have a number of curious mathematical properties†. We see, for instance, that the denominator of each

* *Amer. Naturalist*, vii, p. 449, 1873.

† This celebrated series corresponds to the continued fraction $\dfrac{1}{1 + \dfrac{1}{1 + \ldots}}$ etc., and converges to 1·618..., the numerical equivalent of the *sectio divina*, or "*Golden Mean.*" The series of numbers, 1, 1, 2, 3, 5, 8, ..., of which each is the sum of the preceding two, was used by Leonardo of Pisa (*c.* 1170–1250), nicknamed *Fi Bonacci*, or *filius bonassi*, in his *Liber Abbaci*, a work dedicated to *magister meus, summus philosophus*, Michael Scot (*Scritti*, i, pp. 283–284, 1857). This learned man was educated in Morocco, where his father was clerk or dragoman to Pisan merchants; and he is said to have been the first to bring the Arabic numerals, or "*novem figurae Indorum*," into Europe. The Fibonacci numbers were first so-called by Eduard Lucas, *Bollettino di Bibliogr. e Storia dei Sci. Matem. e Fis.* x, p. 129, 1877. The general expression for the series

$$u_n = \frac{1}{\sqrt{5}} \left\{ \left(\frac{1 + \sqrt{5}}{2} \right)^n - \left(\frac{1 - \sqrt{5}}{2} \right)^n \right\},$$

was known to Euler and to Daniel Bernoulli (*Comm. Acad. Sci. Imp. Petropol.* 1732, p. 90), and was rediscovered by Binet, *C.R.* xviii, p. 563, 1843; xix, p. 939, 1844) and by Lamé, *ibid.* xix, p. 867, after whom it is sometimes called Lamé's series. But the Greeks were familiar with the series 2, 3: 5, 7 : 12, 17, etc.; which converges to $\sqrt{2}$, as the other does to the Golden Mean; and so closely related are the two series, that it seems impossible that the Greeks could have known the one and remained ignorant of the other. (See a paper of mine, on "Excess and Defect, etc.," in *Mind*, xxxviii, No. 149, 1928.)

The Fibonacci (or Lamé) series was well known to Kepler, who, in his paper *De nive sexangula* (1611, cf. *supra*, p. 695), discussed it in connection with the form of the dodecahedron and icosahedron, and with the ternary or quinary symmetry of the flower. (Cf. F. Ludwig, Kepler über das Vorkommen der Fibonaccireihe im Pflanzenreich, *Bot. Centralbl.* lxviii, p. 7, 1896.) Professor William Allman, Professor of Botany in Dublin (father of the historian of Greek geometry), fraction is the numerator of the next; and further, that each successive numerator, or denominator, is the sum of the preceding two. Our immediate problem, then, is to determine, if possible, how these numerical coincidences come about, and why these particular numbers should be so commonly met with as to be considered "normal" and characteristic features of the general phenomenon of phyllotaxis. The following account is based on a short paper by Professor P. G. Tait*. Of the two following diagrams, Fig. 450 represents the general case, and Fig. 451 a particular one, for the sake of possibly greater simplicity. Both diagrams represent a portion of a branch, or fir-cone, regarded as cylindrical, and unwrapped to form a plane surface. *A*, *a*, at the two ends of the base-line, represent the same initial leaf or scale; *O* is a leaf which can be reached from *A* by *m* steps

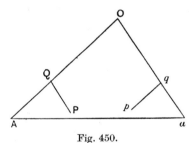

Fig. 450.

in a right-hand spiral (developed into the straight line *AO*), and by *n* steps from *a* in a left-handed spiral *aO*. Now it is obvious in our fir-cone, that we can include *all* the scales upon the cone by taking so many spirals in the one direction, and again include them all by so

speculating on the same facts, put forward the curious suggestion that the cellular tissue of the dicotyledons, or exogens, would be found to consist of dodecahedra, and that of the monocotyledons or endogens of icosahedra (*On the mathematical connection between the parts of vegetables*: abstract of a Memoir read before the Royal Society in the year 1811 (privately printed, n.d.). Cf. De Candolle, *Organogenie végétale*, I, p. 534. See also C. E. Wasteels, Over de Fibonaccigetalen, 3de Natuur. *Congres, Antwerpen*, 1899, pp. 25–37; R. C. Archibald, in Jay Hambidge's *Dynamic Symmetry*, 1920, pp. 146–157; and, on the many mathematical properties of the series, L. E. Dickson, *Theory of Numbers*, I, pp. 393–411, 1919.

Of these famous and fascinating numbers a mathematical friend writes to me: "All the romance of continued fractions, linear recurrence relations, surd approximations to integers and the rest, lies in them, and they are a source of endless curiosity. How interesting it is to see them striving to attain the unattainable, the golden ratio, for instance; and this is only one of hundreds of such relations."

* *Proc. R.S.E.* VII, p. 391, 1872.

many in the other. Accordingly, in our diagrammatic construction, the spirals AO and aO must, and always can, be so taken that m spirals parallel to aO, and n spirals parallel to AO, shall separately include all the leaves upon the stem or cone.

If m and n have a common factor, l, it can easily be shewn that the arrangement is composite, and that there are l fundamental, or genetic spirals, and l leaves (including A) which are situated

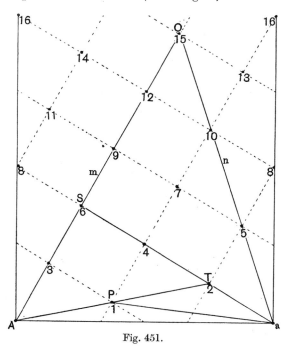

Fig. 451.

exactly on the line Aa. That is to say, we have here a *whorled* arrangement, which we have agreed to leave unconsidered in favour of the simpler case. We restrict ourselves, accordingly, to the cases where there is but one genetic spiral, and when *therefore* m and n are prime to one another.

Our fundamental, or genetic, spiral, as we have seen, is that which passes from A (or a) to the leaf which is situated nearest to the base-line Aa. The fundamental spiral will thus be right-handed (A, P, etc.) if P, which is nearer to A than to a, be this leaf—left-handed if it be p. That is to say, we make it a convention that we shall always, for our fundamental spiral, run round the system, from one leaf to the next, *by the shortest way.*

Now it is obvious, from the symmetry of the figure (as further shewn in Fig. 451), that, besides the spirals running along AO and aO, we have a series running *from the steps on aO* to the steps on AO. In other words we can find a leaf (S) upon AO, which, like the leaf O, is reached directly by a spiral series from A and from a, such that aS includes n steps, and AS (being part of the old spiral line AO) now includes $m - n$ steps. And, since m and n are prime to one another (for otherwise the system would have been a composite or whorled one), it is evident that we can continue this process of convergence until we come down to a 1, 1 arrangement, that is to say to a leaf which is reached by a single step, in opposite directions from A and from a, which leaf is therefore the first leaf, next to A, of the fundamental or generating spiral.

If our original lines along AO and aO contain, for instance, 13 and 8 steps respectively (i.e. $m = 13$, $n = 8$), then our next series, observable in the same cone, will be 8 and $(13 - 8)$ or 5; the next 5 and $(8 - 5)$ or 3; the next 3, 2; and the next 2, 1; leading to the ultimate condition of 1, 1. These are the very series which we have found to be common, or normal; and so far as our investigation has yet gone, it has proved to us that, if one of these exists, it entails, *ipso facto*, the presence of the rest.

In following down our series, according to the above construction, we have seen that at every step we have changed direction, the longer and the shorter sides of our triangle changing places every time. Let us stop for a moment, when we come to the 1, 2 series, or AT, aT of Fig. 451. It is obvious that there is nothing to prevent us making a new 1, 3 series if we please, by continuing the generating spiral through three leaves, and connecting the leaf so reached directly with our initial one. But in the case represented in Fig. 451, it is obvious that these two series (A, 1, 2, 3, etc., and a, 3, 6, etc.) will be running in the same direction; i.e. they will both be right-handed, or both left-handed spirals. The simple meaning of this is that the third leaf of the generating spiral was distant from our initial leaf by *more than the circumference* of the cylindrical stem; in other words, that there were more than two, but *less than three* leaves in a single turn of the fundamental spiral.

Less than two there can obviously never be. When there are exactly two, we have the simplest of all possible arrangements, namely that in which the leaves are placed alternately on opposite sides of the stem. When there are more than two, but less than three, we have the elementary condition for the production of the series which we have been considering, namely 1, 2; 2, 3; 3, 5, etc. To put the latter part of this argument in more precise language, let us say that: If, in our descending series, we come to steps 1 and t, where t is determined by the condition that 1 and $t + 1$ would give spirals both right-handed, or both left-handed; it follows that there are less than $t + 1$ leaves in a single turn of the fundamental spiral. And, determined in this manner, it is found in the great majority of cases, in fir-cones, and a host of other examples of phyllotaxis, that $t = 2$. In other words, in the great majority of cases, we have what corresponds to an arrangement next in order of simplicity to the simplest case of all: next, that is to say, to the arrangement which consists of opposite and alternate leaves.

"These simple considerations," as Tait says, "explain completely the so-called mysterious appearance of terms of the recurring series 1, 2, 3, 5, 8, 13, etc.* The other natural series, usually but misleadingly represented by convergents to an infinitely extended continuous fraction, are easily explained, as above, by taking $t = 3$, 4, 5, etc., etc." Many examples of these latter series have been recorded, as more or less rare abnormalities, by Dickson† and other writers.

We have now learned, among other elementary facts, that wherever any one system of spiral steps is present, certain others invariably and of necessity accompany it, and are definitely related to it. In any diagram, such as Fig. 451, in which we represent our leaf-arrangement by means of uniform and regularly interspaced dots, we can draw one series of spirals after another, and one as easily

---

* The necessary existence of these recurring spirals is also proved, in a somewhat different way, by Leslie Ellis, On the theory of vegetable spirals, in *Mathematical and other Writings*, 1863, pp. 358–372. Leslie Ellis, Whewell's brother-in-law, was a man of great originality. He is best remembered, perhaps, for his views on the Theory of Probabilities (cf. J. M. Keynes, *Treatise on Probabilities*, 1921, p. 92), and for his association with Stebbing as editor of Bacon.

† *Proc. R.S.E.* VII, p. 397, 1872; *Trans. R.S.E.* XXVI, pp. 505–520, 1872.

as another. In a fire-cone one particular series, or rather two conjugate series, are always conspicuous, but the related series may be sought and found with little difficulty. The spruce-fir is commonly said to have a phyllotaxis of 8/13; but we may count still

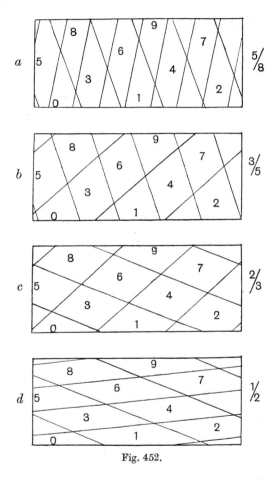

Fig. 452.

steeper and nearly vertical rows of scales to the number of 13/21; and if we take pains to number all the scales consecutively, we may find the lower series, 5/8, 3/5, and even 1/2, with ease and certainty.

The phenomenon is illustrated by Fig. 452, *a–d*. The ground-plan of all these diagrams is identically the same. The generating spiral in each case represents a divergence of 3/8, or 135° of azimuth; and the points succeed one another at the same successional distances parallel to the axis. The rectangular outlines, which correspond to the exposed surface of the leaves or cone-scales, are of equal area, and of equal number. Nevertheless the appearances presented by these diagrams are very different; for in one the eye catches a 5/8 arrangement, in another a 3/5; and so on, down to an arrangement of 1/1. The mathematical side of this very curious phenomenon I have not attempted to investigate. But it is quite obvious that, in a system within which various spirals are implicitly contained, the conspicuousness of one set or another does not depend upon angular divergence. It depends on the relative proportions in length and breadth of the leaves themselves; or, more strictly speaking, on the ratio of the diagonals of the rhomboidal figure by which each leaf-area is circumscribed. When, as in the fir-cone, the scales by mutual compression conform to these rhomboidal outlines, their inclined edges at once guide the eye in the direction of some one particular spiral; and we shall not fail to notice that in such cases the usual result is to give us arrangements corresponding to the middle diagrams in Fig. 452, which are the configurations in which the quadrilateral outlines approach most nearly to a rectangular form, and give us accordingly the least possible ratio (under the given conditions) of sectional boundary-wall to surface area.

The manner in which one system of spirals may be caused to slide, so to speak, into another, has been ingeniously demonstrated by Schwendener on a mechanical model, consisting essentially of a framework which can be opened or closed to correspond with one another of the above series of diagrams*.

The same curious fact, that one Fibonacci series leads to, or involves the rest, is further shewn, in a very simple way, in the following diagrammatic Table (p. 930). It shews, in the first instance, the numerical order of the scales on a fir-cone, in so-called 5/8 phyllotaxis; that is to say, it represents the cone *unwrapped*, with the two principal spirals lying along the axes of a rectangular system. Starting from 0, the abscissae increase by 5, the ordinates by 8; or, in other words, any given number $m = 5x + 8y$; it is

---

* A common form of pail-shaped waste-paper basket, with wide rhomboidal meshes of cane, is well-nigh as good a model as is required.

easy, then, to number the entire system. The generating spiral,
0, 1, 2, 3, ..., and the various secondary Fibonacci spirals, are then
easily recognised.

A Fibonacci series, unwrapped from a cone or cylinder. $m = 5x + 8y$.

1	6	11	**16**	21	26	31	36
$\overline{7}$	$\overline{2}$	3	**8**	13	18	23	28
$\overline{15}$	$\overline{10}$	$\overline{5}$	0	5	10	15	20
$\overline{23}$	$\overline{18}$	$\overline{13}$	$\overline{8}$	$\overline{3}$	2	7	12
$\overline{31}$	$\overline{26}$	$\overline{21}$	$\overline{16}$	$\overline{11}$	$\overline{6}$	$\overline{1}$	4

The place of the first scale in each series is then found to be as
follows:

Series	1	2	3	5	8	13	21	34	55
$\dfrac{x}{y} =$	$\dfrac{-3}{2}$	$\dfrac{2}{-1}$	$\dfrac{-1}{1}$	$\dfrac{1}{0}$	$\dfrac{0}{1}$	$\dfrac{1}{1}$	$\dfrac{1}{2}$	$\dfrac{2}{3}$	$\dfrac{3}{5}$

And this is the Fibonacci series over again. We also see how the
several spirals, of which these are the beginnings, alternate to the
right and left of an asymptotic line, where $x/y = 0{\cdot}618\ldots$
    The Fibonacci numbers, so conspicuous in the fir-cone, make their
appearance also in the flower. The commonest of floral numbers
are 3 and 5; among the Composites we find 8 ray-florets in the
single dahlia, 13 in the ragwort, 21 in the ox-eye daisy or the mari-
gold. In the last two, heads with 34 ray-florets are apt to be
produced at certain times or in certain places*; and in *C. segetum*
these florets are said to vary in a bimodal curve of frequency, with a
high maximum at 13 and a lower at 21†. The simplest explanation
(though perhaps it does not go far) is to suppose that a ligulate
floret terminates, or tends to terminate, each of the principal spiral
series. But among the higher numbers these numerical relations
are only approximate, and the whole matter rests, so far, on somewhat scanty evidence.

The determination of the precise angle of divergence of two consecutive leaves of the generating spiral does not enter into the above
general investigation (though Tait gives, in the same paper, a method

---

* Cf. G. Henslow, On the origin of dimerous and trimerous whorls among the
flowers of Dicotyledons, *Trans. Linn. Soc. (Bot.)* (2), VII, p. 161, 1908.

† Cf. A. Gravis, *Éléments de Physiologie végétale*, 1921, p. 122.

by which it may be easily determined); and the very fact that it does not so enter shews it to be essentially unimportant. The determination of so-called "orthostichies," or precisely vertical successions of leaves, is also unimportant. We have no means, other than observation, of determining that one leaf is vertically above another, and spiral series such as we have been dealing with will appear, whether such orthostichies exist, whether they be near or remote, or whether the angle of divergence be such that no precise vertical superposition ever occurs. And lastly, the fact that the successional numbers, expressed as fractions, 1/2, 2/3, 3/5, represent a convergent series, whose final term is equal to 0·61803..., the *sectio aurea* or "golden mean" of unity, is seen to be a mathematical coincidence, devoid of biological significance; it is but a particular case of Lagrange's theorem that the roots of every numerical equation of the second degree can be expressed by a periodic continued fraction. The same number has a multitude of curious arithmetical properties. It is the final term of all similar series to that with which we have been dealing, such for instance as 1/3, 3/4, 4/7, etc., or 1/4, 4/5, 5/9, etc. It is a number beloved of the circle-squarer, and of all those who seek to find, and then to penetrate, the secrets of the Great Pyramid. It is deep-set in the regular pentagon and dodecahedron, the triumphs of Pythagorean or Euclidean geometry. It enters (as the chord of an angle of 36°) into the thrice-isosceles triangle of which we have spoken on p. 762; it is a number which becomes (by the addition of unity) its own reciprocal—its properties never end. To Kepler (as Naber tells us) it was a symbol of Creation, or Generation. Its recent application to biology and art-criticism by Sir Theodore Cook and others is not new. Naber's book, already quoted, is full of it. Zeising*, in 1854, found in it the key to all

* A. Zeising, *Neue Lehre von der Proportion des menschlichen Körpers aus einem bisher unerkannt gebliebenen die ganze Natur und Kunst durchdringenden morphologischen Grundgesetze entwickelt*, Leipzig, 1854, 457 pp.; *ibid. Deutsche Vierteljahrsschrift*, 1868, p. 261; also, posthumously, *Der Goldene Schnitt*, Leipzig, 1884, 24 pp. Cf. S. Gunther, Adolph Zeising als Mathematiker, *Ztschr. f. Math. u. Physik. (Hist. Lit. Abth.)*, xxi, pp. 157–165, 1876; also F. X. Pfeiffer, Die Proportionen des goldenen Schnittes an den Blättern u. Stengeln der Pflanzen, *Ztschr. f. math. u. naturw. Unterricht*, xv, pp. 325–338, 1885. For other references, see R. C. Archibald, *op. cit.* Among modern books on similar lines, the following are curious, interesting and beautiful (whether we agree with them or not): Jay Hambidge, *Dynamic Symmetry*, Yale, 1920; C. Arthur Coan, *Nature's Harmonic Unity*, New York, 1912.

morphology, and the same writer, later on, declared it to dominate both architecture and music.   But indeed, to use Sir Thomas Browne's words (though it was of another number that he spoke): "To enlarge this contemplation into all the mysteries and secrets accommodable unto this number, were inexcusable Pythagorisme." That this number has any serious claim at all to enter into the biological question of phyllotaxis seems to depend on the assertion, first made by Chauncey Wright*, that, if the successive leaves of the fundamental spiral be placed at the particular azimuth which divides the circle in this "sectio aurea," then no two leaves will ever be superposed†; and thus we are said to have "the most thorough and rapid distribution of the leaves round the stem, each new or higher leaf falling over the angular space between the two older ones which are nearest in direction, so as to divide it in the same ratio (K), in which the first two or any two successive ones divide the circumference.   Now 5/8 and all successive fractions differ inappreciably from K."   To this view there are many simple objections.   In the first place, even 5/8, or 0·625, is but a moderately close approximation to the "golden mean"; and furthermore, the arrangements by which a better approximation is got, such as 8/13, 13/21, and the very close approximations such as 34/55, 55/89, 89/144, etc., are comparatively rare, while the much less close approximations of 3/5 or 2/3, or even 1/2, are extremely common. Again, the general type of argument such as that which asserts that the plant is "aiming at" something which we may call an "ideal angle" is one which cannot commend itself to a plain student of physical science: nor is the hypothesis rendered more acceptable when Sir T. Cook qualifies it by telling us that "all that a plant can do is to vary, to make blind shots at constructions, or to 'mutate' as it is now termed: and the most suitable of these constructions will in the long run be isolated by the action of Natural Selection."   Thirdly, we must not suppose the Fibonacci numbers

---

* On the uses and origin of the arrangement of leaves in plants, *Mem. Amer. Acad.* IX, p. 380, 1871, Cambridge, Mass. Cf. J. Wiesner, *Ueber die Beziehungen der Stellungsverhältnisse der Laubblätter zur Beleuchtung*, Wien, 1902.

† This is what Ruskin spoke of as "the vacant space"; *Mod. Painters*, v, chap. VI, p. 44, 1860. Leonardo had in like manner explained the leaf-arrangement as serving to let air pass between the leaves, keep one from overshadowing another, and let rain-drops fall from the one leaf to the one below.

to have any *exclusive* relation to the Golden Mean; for arithmetic teaches us that, beginning with any two numbers whatsoever, we are led by successive summations toward one out of innumerable series of numbers whose ratios one to another converge to the Golden Mean*. Fourthly, the supposed isolation of the leaves, or their most complete "distribution to the action of the surrounding atmosphere" is manifestly very little affected by any conditions which are confined to the angle of azimuth. For if it be (so to speak) Nature's object to set them farther apart than they actually are, to give them freer exposure to the air or to the sunlight than they actually have, then it is surely manifest that the simple way to do so is to elongate the axis, and to set the leaves farther apart, lengthways on the stem. This has at once a far more potent effect than any nice manipulation of the "angle of divergence."

Lastly, and this seems the simplest, the most cogent and most unanswerable objection of them all, if it be indeed desirable that no leaf should be superimposed above another, the one condition necessary is that the common angle of azimuth should *not* be a rational multiple of a right angle—should not be equivalent to $\dfrac{m}{n}\left(\dfrac{\pi}{2}\right)$. One irrational angle is as good as another: there is no special merit in any one of them, not even in the *ratio divina*. We come then without more ado to the conclusion that while the Fibonacci series stares us in the face in the fir-cone, it does so for mathematical reasons; and its supposed usefulness, and the hypothesis of its introduction into plant-structure through natural selection, are matters which deserve no place in the plain study of botanical phenomena. As Sachs shrewdly recognised years ago, all such speculations as these hark back to a school of mystical idealism.

* Thus, instead of beginning with 1, 1, let us begin 1, 7. The summation-series is then 1, 7, 8, 15, 23, 38, 61, 99, 160, 259, ..., etc.; and $99/160 = 0.618...$ and $259/160 = 1.619...$; and so on. But after all, the old Fibonacci numbers are not far away. For we may write the new series in the form:

$$7 \ (0, 1, 1, 2, 3, 5, 8, \ldots)$$
$$+ \ 1 \ (1, 0, 1, 1, 2, 3, 5, \ldots).$$

CHAPTER XV

ON THE SHAPES OF EGGS, AND OF CERTAIN OTHER HOLLOW STRUCTURES

THE eggs of birds and all other hard-shelled eggs, such as those of the tortoise and the crocodile, are simple solids of revolution; but they differ greatly in form, according to the configuration of the plane curve by the revolution of which the egg is, in a mathematical sense, generated. Some few eggs, such as those of the owl, the penguin, or the tortoise, are spherical or very nearly so; a few more, such as the grebe's, the cormorant's or the pelican's, are approximately ellipsoidal, with symmetrical or nearly symmetrical ends, and somewhat similar are the so-called "cylindrical" eggs of the megapodes and the sand-grouse; the great majority, like the hen's egg, are "ovoid," a little blunter at one end than the other; and some, by an exaggeration of this lack of antero-posterior symmetry, are blunt at one end but characteristically pointed at the other, as is the case with the eggs of the guillemot and puffin, the sandpiper, plover and curlew. It is an obvious but by no means negligible fact that the egg, while often pointed, is never flattened or discoidal; it is a prolate, but never an oblate, spheroid. Its oval outline has one maximal and two minimal radii of curvature, one minimum being less than the other. The evolute to a curve often emphasises, even exaggerates, its features; and the evolutes to a series of eggs (i.e. to their generating curves) are more conspicuously different than the eggs themselves (Fig. 453)*.

The careful study and collection of birds' eggs would seem to have begun with the Count de Marsigli†; the same celebrated naturalist

* Cf. A. Mallock, On the shapes of birds' eggs, *Nature*, CXVI, p. 311, 1925. The evolute may be easily if somewhat roughly drawn by erecting perpendiculars on a sufficient number of tangents to the curve. The evolute then appears as an *envelope*, the perpendiculars all being tangents to it.

† *De avibus circa aquas Danubii vagantibus et de ipsarum nidis* (Vol. v of the *Danubius Panonico-Mysicus*), Hagae Com. 1726. Count Giuseppi Ginanni, or Zinanni, came soon afterwards with his book *Delle uove e dei nidi degli uccelli*, Venezia, 1737.

who first studied the "flowers" of the coral, and who wrote the *Histoire physique de la mer*; and the specific form as well as the colour and other attributes of the egg have been again and again discussed, and not least by the many dilettanti naturalists of the eighteenth century who soon followed in Marsigli's footsteps*.

We need do no more than mention Aristotle's belief, doubtless old in his time, that the more pointed egg produces the male chicken, and the blunter egg the hen; though this theory survived into modern times† and still lingers on (cf. p. 943).   Several naturalists, such as Günther (1772) and Bühle (1818), have taken the trouble to disprove it by experiment.   A more modern and more generally accepted

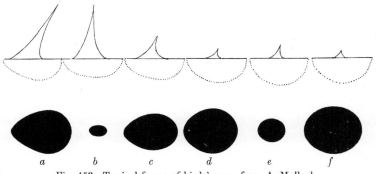

Fig. 453. Typical forms of birds' eggs: from A. Mallock.
The figures below are pinhole photographs of the eggs. The upper figures (drawn to a uniform scale) shew the generating curves and their evolutes.

a Green plover.	*c* Crow.	*e* Kingfisher.
b Humming-bird.	*d* Pheasant.	*f* Owl.

explanation has been that the form of the egg is in direct relation to that of the bird which has to be hatched within—a view that would seem to have been first set forth by Naumann and Bühle, in their great treatise on eggs‡, and adopted by Des Murs§ and many other well-known writers.

In a treatise by de Lafresnaye‖, an elaborate comparison is made

* But Sir Thomas Browne had a collection of eggs at Norwich in 1671, according to Evelyn.

† Cf. Lapierre, in Buffon's *Histoire Naturelle*, ed. Sonnini, 1800.

‡ *Eier der Vögel Deutschlands*, 1818–28 (*cit.* Des Murs, p. 36).

§ *Traité d'Oologie*, 1860.

‖ F. de Lafresnaye, Comparaison des œufs des oiseaux avec leurs squelettes, comme seul moyen de reconnaître la cause de leurs différentes formes, *Rev. Zool.* 1845, pp. 180–187, 239–244.

between the skeleton and the egg of various birds, to shew, for instance, how those birds with a deep-keeled sternum laid rounded eggs, which alone could accommodate the form of the young. According to this view, that "Nature had foreseen*" the form adapted to and necessary for the growing embryo, it was easy to correlate the owl with its spherical egg, the diver with its elliptical one, and in like manner the round egg of the tortoise and the elongated one of the crocodile, with the shape of the creatures which had afterwards to be hatched therein. A few writers, such as Thienemann†, looked at the same facts the other way, and asserted that the form of the egg was determined by that of the bird by which it was laid and in whose body it had been conformed.

In more recent times, other theories, based upon the principles of Natural Selection, have been current and very generally accepted to account for these diversities of form. The pointed, conical egg of the guillemot is generally supposed to be an adaptation, advantageous to the species in the circumstances under which the egg is laid; the pointed egg is less apt than a spherical one to roll off the narrow ledge of rock on which this bird is said to lay its solitary egg, and the more pointed the egg, so much the fitter and likelier is it to survive. The fact that the plover or the sandpiper, breeding in very different situations, lay eggs that are also conical, elicits another explanation, to the effect that here the conical form permits the many large eggs to be packed closely under the mother bird‡. Whatever truth there be in these apparent adaptations to existing circumstances, it is only by a very hasty logic that we can accept them as a *vera causa*, or adequate explanation of the facts; and it is obvious that in the bird's egg we have an admirable case for direct investigation of the mechanical or physical significance of its form§.

---

* Cf. Des Murs, p. 67: "Elle devait encore penser au moment où ce germe aurait besoin de l'espace nécessaire à son accroissement, à ce moment où...il devra remplir exactement l'intervalle circonscrit par sa fragile prison, etc."

† F. A. L. Thienemann, *Syst. Darstellung der Fortpflanzung der Vögel Europas*, Leipzig, 1825–38.

‡ Cf. Newton's *Dictionary of Birds*, 1893, p. 191; Szielasko, Gestalt der Vogeleier, *Journ. f. Ornith.* LIII, pp. 273–297, 1905.

§ Jacob Steiner suggested a Cartesian oval, $r + mr' = c$, as a general formula for all eggs (cf. Fechner, *Ber. sächs. Ges.* 1849, p. 57); but this formula (which fails in such a case as the guillemot) is purely empirical, and has no mechanical foundation.

Of all the many naturalists of the eighteenth and nineteenth centuries who wrote on the subject of eggs, only two (so far as I am aware) ascribed the form of the egg to direct mechanical causes. Günther*, in 1772, declared that the more or less rounded or pointed form of the egg is a mechanical consequence of the pressure of the oviduct at a time when the shell is yet unformed or unsolidified; and that accordingly, to explain the round egg of the owl or the kingfisher, we have only to admit that the oviduct of these birds is somewhat larger than that of most others, or less subject to violent contractions. This statement contains, in essence, the whole story of the mechanical conformation of the egg. A hundred and twenty years after, Dr J. Ryder of Philadelphia gave, as near as may be, the same explanation†.

Let us consider, very briefly, the conditions to which the egg is subject in its passage down the oviduct.

(1) The "egg," as it enters the oviduct, consists of the yolk only, enclosed in its vitelline membrane. As it passes down the first portion of the oviduct the white is gradually superadded, and becomes in turn surrounded by the "shell-membrane." About this latter the shell is secreted, rapidly and at a late period: the egg having meanwhile passed on into a wider portion of the oviducal tube, called (by loose analogy, as Owen says) the "uterus." Here the egg assumes its permanent form, here it ultimately becomes rigid, and it is to this portion of the oviduct that our argument principally refers.

(2) Both the yolk and the entire egg tend to fill completely their respective membranes, and, whether this be due to growth or imbibition on the part of the contents or to contraction on the part of the surrounding membranes, the resulting tendency is for both yolk and egg to be, in the first instance, spherical, unless or until distorted by external pressure.

(3) The egg is subject to pressure within the oviduct, which is an elastic, muscular tube, along the walls of which pass peristaltic

---

* F. C. Günther, *Sammlung von Nestern und Eyern verschiedener Vögel*, Nürnb. 1772. Cf. also Raymond Pearl, Morphogenetic activity of the oviduct, *J. Exp. Zool.* VI, pp. 339–359, 1909.

† J. Ryder, The mechanical genesis of the form of the fowl's egg, *Proc. Amer. Philosoph. Soc.* Philadelphia, XXXI, pp. 203–209, 1893; cf. A. S. Packard, Inheritance of acquired characters, *Proc. Amer. Acad.* 1894, p. 360.

waves of contraction. These muscular contractions may be described as the contraction of successive annuli of muscle, giving annular (or radial) pressure to successive portions of the egg; they drive the egg forward against the frictional resistance of the tube, while tending at the same time to distort its form. While nothing is known, so far as I am aware, of the muscular physiology of the oviduct, it is well known in the case of the intestine that the presence of an obstruction leads to the development of violent contractions in its rear, which waves of contraction die away, and are scarcely if at all propagated in advance of the obstruction; indeed in normal intestinal peristalsis a wave of relaxation travels close ahead of the wave of constriction.

(4) The egg is, to all intents and purposes, a solid of revolution; in other words, its transverse sections are all but perfect circles, so nearly perfect that, chucked in the lathe, an egg "runs true." This may be taken to shew that the direct pressure of the oviduct, whether elastic or muscular, is large compared with the weight of the egg. Even in ostrich eggs, where if anywhere gravitational deformation should be found, the greatest and least equatorial diameters do not differ by 1 per cent., and sometimes by less than one part in a thousand*.

(5) It is known by observation that a hen's egg is always laid blunt end foremost†.

(6) It can be shewn, at least as a very common rule, that those eggs which are most unsymmetrical, or most tapered off posteriorly, are also eggs of a large size relatively to the parent bird. The guillemot is a notable case in point, and so also are the curlews, sandpipers, phaleropes and terns. We may accordingly presume that the more pointed eggs are those that are large relatively to the tube or oviduct through which they have to pass, or, in other words, are those which are subject to the greatest pressure while

---

* Cf. Mallock, op. cit.

† This was known to Albertus Magnus, though his explanation was wrong, "Ova autem habentia duos colores non sunt omnino penitus rotunda, sed ex una parte sunt acuta habentia angulum sphericum acutum, sicut sunt composita ex duobus semispheris, in una parte extensis ad angulum acutum et in alia parte sphericis non extensis in loco ubi est polus ovi. . . . Et in exitu ovi acutus angulus exit ultimo, eo quod ipse porrectus est ad interiora matricis versus parietem ubi ovum cum matrice continuatur in sui generatione" (De animalibus, lib. xvii, tract. 1, c. 3).

being forced along. So general is this relation that we may go still further, and presume with great plausibility in the few exceptional cases (of which the apteryx is the most conspicuous) where the egg is relatively large though not markedly unsymmetrical, that in these cases the oviduct itself is in all probability large (as Günther had suggested) in proportion to the size of the bird. In the case of the common fowl we can trace a direct relation between the size and shape of the egg, for the first eggs laid by a young pullet are usually smaller, and at the same time are much more nearly spherical than the later ones; and, moreover, some breeds of fowls lay proportionately smaller eggs than others, and on the whole the former eggs tend to be rounder than the latter*.

We may now proceed to enquire more particularly how the form of the egg is controlled by the pressures to which it is subjected.

The egg, just prior to the formation of the shell, is, as we have seen, a fluid body, tending to a spherical shape and *enclosed within a membrane*.

Our problem, then, is: Given an incompressible fluid, contained in a deformable capsule, which is either (a) entirely inextensible, or (b) slightly extensible, and which is placed in a long elastic tube the walls of which are radially contractile, to determine the shape under some given distribution of pressure. We may assume, at least to begin with, that the shell-membrane is homogeneous and isotropic— uniform in all parts and in all directions.

If the capsule be spherical, inextensible, and completely filled with the fluid, absolutely no deformation can take place. The few eggs that are actually or approximately spherical, such as those of the tortoise or the owl, may thus be alternatively explained as cases where little or no deforming pressure has been applied prior to the solidification of the shell, or else as cases where the capsule was so

---

* In so far as our explanation involves a shaping or moulding of the egg by the uterus or oviduct (an agency supplemented by the proper tensions of the egg), it is curious to note that this is very much the same as that old view of Telesius regarding the formation of the embryo (*De rerum natura*, vi, cc. 4 and 10), which he had inherited from Galen, and of which Bacon speaks (*Nov. Org.* cap. 50; cf. Ellis's note). Bacon expressly remarks that "Telesius should have been able to shew the like formation in the shells of eggs." This old theory of embryonic modelling survives in our usage of the term "matrix" for a "mould."

little capable of extension and so completely filled as to preclude the possibility of deformation.

If the capsule be not spherical, but be inextensible, then only such deformation can take place as tends to make the shape more nearly spherical; and as the surface area is thereby decreased, the envelope must either shrink or pucker. In other words, an incompressible fluid contained in an inextensible envelope cannot be deformed without puckering of the envelope.

But let us next assume, as the condition by which this result may be avoided, that the envelope is to some extent extensible and that deformation is so far permitted. It is obvious that, on the presumption that the envelope is only moderately extensible, the whole structure can only be distorted to a moderate degree away from the spherical or spheroidal form.

At all points the shape is determined by the law of the distribution of *radial pressure within the given region of the tube*, surface friction helping to maintain the egg in position. If the egg be under pressure from the oviduct, but without any marked component either in a forward or backward direction, the egg will be compressed in the middle, and will tend more or less to the form of a cylinder with spherical ends. The eggs of the grebe, cormorant, or crocodile may be supposed to receive their shape in such circumstances.

When the egg is subject to the peristaltic contraction of the oviduct during its formation, then from the nature and direction of motion of the peristaltic wave the pressure will be greatest somewhere behind the middle of the egg; in other words, the tube is converted for the time being into a more conical form, and the simple result follows that the anterior end of the egg becomes the broader and the posterior end the narrower.

The peristalsis of the oviduct thus plays a double part, in propelling the egg down the oviduct and in impressing on it its ovoid form; but the whole process is a very slow one, for the hen's oviduct is only a few inches long, and the egg is some ten or twelve hours upon its way. We shall consider presently certain shells which may be regarded as so many drops or vesicles deformed by gravity; that is a statical problem. Compared with it the problem of the egg is a dynamical one; and yet it becomes a quasi-statical one, because the action is so very slow. It is an action without lag and without momentum; and the question, common in dynamical problems, of the relation between the period of the application of the force and the free period of response or adjustment to it need not concern us at all.

Again, the case of the egg is somewhat akin to a hydrodynamical problem; for as it lies in the oviduct we may look on it as a stationary body round which waves are flowing, with the same result as when a body moves through a fluid at rest. Thus we may treat it as a hydrodynamical problem, but a very simple one—simplified by the absence of all eddies and every form of turbulence; and we come to look on the egg as a *streamlined* structure, though its streamlines are of a very simple kind.

The mathematical statement of the case begins as follows: In our egg, consisting of an extensible membrane filled with an incompressible fluid and under external pressure, the equation of the envelope is $p_n + T(1/r + 1/r') = P$, where p_n is the normal component of external pressure at a point where r and r' are the radii of curvature, T is the tension of the envelope, and P the internal fluid pressure. This is simply the equation of an elastic surface where T represents the coefficient of elasticity; in other words, a flexible elastic shell has the same mathematical properties as our fluid, membrane-covered egg. And this is the identical equation which we have already had so frequent occasion to employ in our discussion of the forms of cells; save only that in these latter we had chiefly to study the tension T (i.e. the surface-tension of the semi-fluid cell) and had little or nothing to do with the factor of external pressure (p_n), which in the case of the egg becomes of chief importance.

To enquire how an elastic sphere or spheroid will be deformed in passing down a peristaltic tube is an ill-defined and indeterminate problem; but we can study the effect produced in the shape of any particular egg, and so far infer the forces which have been in action. We need only study a single meridian of the egg, inasmuch as we have found it to be a solid of revolution. At successive points along this meridian, let us determine the amount of curvature, that is to say the principal radii of curvature, in latitude and longitude, in the Gaussian formula $P = p_n + T(1/r + 1/r')$: or, as we may write it if we have any reason to doubt the uniformity or isotropy of the membrane, $T/r + T'/r'$. The sum of these curvatures varies from point to point; the internal or hydrodynamical pressure, P, is constant; and therefore the external pressure, p_n, varies from point to point with the curvature, and is a direct function of the shape of the egg.

Some few eggs, such as the owl's and the kingfisher's, are so nearly spherical that we are apt to speak of them as spheres; but they are all prolate more or less, and no egg is so nearly circular in meridional section as all eggs are in their circles of latitude. When the egg is all but spherical that shape may be due (as we have seen) to various causes: to a relatively small size of the egg, allowing it to descend the tube under a minimum of peristaltic pressure; perhaps to an unusually strong shell-membrane, resistant of deformation; in general terms, to a possible diminution of p_n, or a possible increase of T. But all eggs have approximately spherical ends, and the big anterior end of the large conical eggs of plover or curlew or guillemot is conspicuously so. Here the egg projects into the wide cavity of the uncontracted oviduct, external or peristaltic pressure does not exist, the shell-membrane has to resist internal pressure without further external support, and the resultant spherical curvature is an indication of the uniformity, or isotropy, of the membrane. The lesser of the two spherical ends, that is to say the posterior end, has by much the greater curvature, and the tension there is correspondingly great. It would seem that the membrane ought to be thicker or stronger at this pointed end than elsewhere, but it is not known to be so. In any case, it is just here, in this presumably weakest part, that we are most apt to find the irregularities and deformities of misshapen eggs.

Within the egg lies the yolk, and the yolk is invariably spherical or very nearly so, whatever be the form of the entire egg. The reason is simple, and lies in the fact that the fluid yolk is itself enclosed within another membrane, between which and the shell-membrane lies the fluid albumin, which transmits a uniform hydrostatic pressure to the yolk*. The lack of friction between the yolk-membrane and the white of the egg is indicated by the well-known fact that the "germinal spot" on the surface of the yolk is always

---

* In like manner, the cell-nucleus is "usually globular, except in certain specialised tissues, or when it degenerates" (Darlington). Whether it possesses a *membrane* is matter in dispute, but it at all events possesses a *surface*, with a phase-difference between it and the surrounding cytoplasm. Cf. above, p. 295.

found uppermost, however we may place and wherever we may open the egg; that is to say, the yolk easily rotates within the egg, bringing its lighter pole uppermost.

In its passage down the oviduct the egg is not merely thrust but also *screwed* along; and its spiral course leaves traces on wellnigh all its structure save the shell.   When we have broken the shell of a hard-boiled egg the shell-membrane below peels off in spiral strips, and even the white tends to flake off in layers, spirally.   In the fresh unboiled egg two knotted cords—the treadles or *chalazae*—are connected with the yolk, and lie fore-and-aft of it, loose in the albumen.   These represent the free ends of a yolk-membrane, which got caught in the constricted oviduct while the yolk between them was being screwed along: very much as we may wrap an apple in a handkerchief, hold the two ends fast, and twirl the apple round..

These, then, are the general principles involved in, and illustrated by, the configuration of an egg; and they take us as far as we can safely go without actual quantitative determination, in each particular case, of the forces concerned*.

In certain cases among the invertebrates, we again find instances of hard-shelled eggs which have obviously been moulded by the oviduct, or so-called "ootype," in which they have lain: and not merely in such a way as to shew the effects of peristaltic pressure upon a uniform elastic envelope, but so as to impress upon the egg the more or less irregular form of the cavity within which it had been for a time contained and compressed.   After this fashion is explained the curious form of the egg in *Bilharzia* (*Schistosoma*) *haematobium*, a formidable parasitic worm to which is due a disease wide-spread in Africa and Arabia, and an especial scourge of the Mecca pilgrims.   The egg in this worm is provided at one end with a little spine, which is explained as having been moulded within a little funnel-shaped expansion of the uterus, just where it communicates with the common duct leading from the ovary and yolk-gland. Owing to some anatomical difference in the uterus, the little

* It is a common but unfounded belief among poultry-men that shape and size are related to the sex of the egg; the longer eggs producing mostly male chicks. That there is no such correlation between sex on the one hand and weight, length or shape on the other, has been clearly demonstrated. Cf. M. A. Jull and J. P. Quinn, *Journ. Agr. Research*, xxix, pp. 195–201, 1924.

spine may be at the end or towards the side of the egg: and this visible difference has led to the recognition of a new species, *S. mansoni**. In a third species, *S. japonicum*, the egg is described as bulging into a so-called "calotte," or bubble-like convexity at the end opposite to the spine. This, I think, may, with very little doubt, be ascribed to hardening of the egg-shell having taken place just at the time when partial relief from pressure was being experienced by the egg in the neighbourhood of the dilated orifice of the oviduct.

This case of Bilharzia is not, from our present point of view, a very important one, but nevertheless it is interesting. It ascribes to a mechanical cause a curious peculiarity of form; and it shews, by reference to this mechanical principle, how two simple mechanical modifications of the same thing may not only seem very different to the systematic naturalist's eye, but may actually lead to the recognition of a new species, with its own geographical distribution, and its own pathogenic characteristics.

On the form of sea-urchins

As a corollary to the problem of the bird's egg, we may consider for a moment the forms assumed by the shells of the sea-urchins. These latter are commonly divided into two classes—the Regular and the Irregular Echinids. The regular sea-urchins, save in slight details which do not affect our problem, have a complete axial symmetry. The axis of the animal's body is vertical, with mouth below and the intestinal outlet above; and around this axis the shell is built as a symmetrical system. It follows that in horizontal section the shell is everywhere circular, and we need only consider its form as seen in vertical section or projection. The irregular urchins (very inaccurately so-called) have the anal extremity of the body removed from its central, dorsal situation; and it follows that they have now a single plane of symmetry, about which the organism, shell and all, is bilaterally symmetrical. We need not concern ourselves in detail with the shapes of their shells, which may be very simply interpreted, by the help of radial coordinates, as deformations of the circular or "regular" type.

---

* L. W. Sambon, *Proc. Zool. Soc.* 1907 (i), p. 283; also in *Journ. Trop. Med. and Hygiene*, Sept. 15, 1926.

The sea-urchin shell consists of a membrane, stiffened into rigidity by calcareous deposits, which constitute a beautiful skeleton of separate, neatly fitting "ossicles." The rigidity of the shell is more apparent than real, for the entire structure is, in a sluggish way, plastic; inasmuch as each little ossicle is capable of growth, and the entire shell grows by increments to each and all of these multitudinous elements, whose individual growth involves a certain amount of freedom to move relatively to one another; in a few cases the ossicles are so little developed that the whole shell appears soft and flexible. The viscera of the animal occupy but a small part of the space within the shell, the cavity being mainly filled by a large quantity of watery fluid, whose density must be very near to that of the external sea-water.

Apart from the fact that the sea-urchin continues to grow, it is plain that we have here the same general conditions as in the egg-shell, and that the form of the sea-urchin is subject to a similar equilibrium of forces. But there is this important difference, that an external muscular pressure (such as the oviduct administers during the consolidation of the egg-shell) is now lacking. In its place we have the steady continuous influence of gravity, and there is yet another force which in all probability we require to take into consideration.

While the sea-urchin is alive, an immense number of delicate "tub -feet," with suckers at their tips, pass through minute pores in the shell, and, like so many long cables, moor the animal to the ground. They constitute a symmetrical system of forces, with one resultant downwards, in the direction of gravity, and another outwards in a radial direction; and if we look upon the shell as originally spherical, both will tend to depress the sphere into a flattened cake. We need not consider the radial component, but may treat the case as that of a spherical shell symmetrically depressed under the influence of gravity. This is precisely the condition which we have to deal with in a drop of liquid lying on a plate; the form of which is determined by its own uniform surface-tension, plus gravity, acting against the internal hydrostatic pressure. Simple as this system is, the full mathematical investigation of the form of a drop is not easy, and we can scarcely hope that the systematic study of the Echinodermata will ever be conducted by methods based on Laplace's differential equation*; but we have little difficulty in seeing that the various forms represented in a series of sea-urchin shells are no other than those which we may easily and perfectly imitate in drops.

In the case of the drop of water (or of any other particular liquid) the specific surface-tension is always constant, and the pressure varies inversely as the radius of curvature; therefore the smaller the drop the more nearly is it able to conserve the spherical form, and the larger the drop the more does it become flattened under gravity†. We can imitate this phenomenon by using india-rubber balls filled with water, of different sizes; the little ones will remain very nearly spherical, but the larger will fall down "of their own weight," into the form of more and more flattened cakes; and we see the same thing when we let drops of heavy oil (such as the orthotoluidene spoken of on p. 370) fall through a tall column of water, the little ones remaining round, and the big ones getting more and more flattened as they sink. In the case of the sea-urchin, the same series of forms may be assumed to occur, irrespective of size, through variations in T, the specific tension, or "strength" of the enveloping shell. Accordingly we may study, entirely from this point of view, such a series as the following (Fig. 454). In a very few cases, such as the fossil *Palaeechinus*, we have an approximately spherical shell, that is to say a shell so strong that the influence of gravity becomes negligible as a cause of deformation, just as (to compare small things with great) the surface tension of mercury is so high that small drops of it seem perfectly spherical‡. The ordinary species of *Echinus* begin to display a pronounced depression, and this reaches its maximum in such soft-shelled flexible forms as *Phormosoma*. On the general question I took the oppor-

---

* Cf. Bashforth and Adams, *Theoretical Forms of Drops, etc.*, Cambridge, 1883.

† The drops must be spherical, or very nearly so, to produce a rainbow. But the bow is said to be always better defined near the top than down below; which seems to shew that the lower and larger raindrops are the less perfect spheres. (Cf. T. W. Backhouse, *Symons's M. Met. Mag.* 1879, p. 25.) For the small round droplets in the cloud tend to cannon off one another, and remain small and spherical. But when there comes a difference of potential between cloud and cloud, or between earth and sky, then the spherules become distorted, one droplet coalesces with another, and the big drops begin to fall.

‡ Cf. A. Ferguson, On the theoretical shape of large bubbles and drops, *Phil. Mag.* (6), xxv, pp. 507–520, 1913.

tunity of consulting Mr C. R. Darling, who is an acknowledged expert in drops, and he at once agreed with me that such forms as are represented in Fig. 454 are no other than diagrammatic illustrations of various kinds of drops, "most of which can easily be reproduced in outline by the aid of liquids of approximately equal density to water, although some of them are fugitive." He found a difficulty in the case of the outline which represents *Asthenosoma*, but the reason for the anomaly is obvious; the flexible shell has flattened

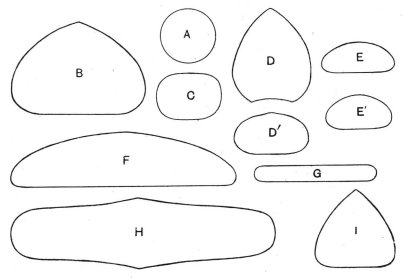

Fig. 454.   Diagrammatic vertical outlines of various sea-urchins: A, *Palaeechinus*;
   B, *Echinus acutus*; C, *Cidaris*; D, D', *Coelopleurus*; E, E', *Genicopatagus*;
   F, *Phormosoma luculenter*; G, *P. tenuis*; H, *Asthenosoma*; I, *Urechinus*.

down until it has come in contact with the hard skeleton of the jaws, or "Aristotle's lantern," within, and the curvature of the outline is accordingly disturbed.   The elevated, conical shells such as those of *Urechinus* and *Coelopleurus* evidently call for some further explanation; for there is here some cause at work to elevate, rather than to depress the shell.   Mr Darling tells me that these forms "are nearly identical in shape with globules I have frequently obtained, in which, on standing, bubbles of gas rose to the summit and pressed the skin upwards, without being able to escape."   The same condition may be at work in the sea-urchin; but a similar tendency would also be manifested by the presence in the upper part of the shell of any accumulation of substance lighter than water, such as is actually present in the masses of fatty, oily eggs.

On the form and branching of blood-vessels

Passing to what may seem a very different subject, we may investigate a number of interesting points in connection with the form and structure of the blood-vessels, and we shall find ourselves helped, at least in the outset, by the same equations as those we have used in studying the egg-shell.

We know that the fluid pressure (P) within the vessel is balanced by (1) the tension (T) of the wall, divided by the radius of curvature, and (2) the external pressure (p_n), normal to the wall: according to our formula

$$P = p_n + T\,(1/r + 1/r').$$

If we neglect the external pressure, that is to say any support which may be given to the vessel by the surrounding tissues, and if we deal only with a cylindrical vein or artery, this formula becomes simplified to the form $P = T/R$. That is to say, under constant pressure, the tension varies as the radius. But the tension, per unit area of the vessel, depends upon the thickness of the wall, that is to say on the amount of membranous and especially of muscular tissue of which it is composed. Therefore, so long as the pressure is constant, the thickness of the wall should vary as the radius, or as the diameter, of the blood-vessel.

But it is not the case that the pressure is constant, for it gradually falls off, by loss through friction, as we pass from the large arteries to the small; and accordingly we find that while, for a time, the cross-sections of the larger and smaller vessels are symmetrical figures, with the wall-thickness proportional to the size of the tube, this proportion is gradually lost, and the walls of the small arteries, and still more of the capillaries, become exceedingly thin, and more so than in strict proportion to the narrowing of the tube.

In the case of the heart we have, within each of its cavities, a pressure which, at any given moment, is constant over the whole wall-area, but the thickness of the wall varies very considerably. For instance, in the left ventricle the apex is by much the thinnest portion, as it is also that with the greatest curvature. We may assume, therefore (or at least suspect), that the formula, $t(1/r + 1/r') = C$, holds good; that is to say, that the thickness (t) of the wall varies inversely as the mean curvature. This may be tested experimentally, by dilating a heart with alcohol under a known pressure, and then measuring the thickness of the walls in various parts after the whole organ has become hardened. By this means it is found that, for each of the cavities, the law holds good with great accuracy[*]. Moreover, if we begin by dilating the right ventricle and then dilate the left in like manner, until the whole heart is equally and symmetrically dilated, we find (1) that we have had to use a pressure in the left ventricle from six to seven times as great as in the right ventricle, and (2) that the thickness of the walls is just in the same proportion[†].

Many problems of a hydrodynamical kind arise in connection with the flow of blood through the blood-vessels; and while these are of primary importance to the physiologist they interest the morphologist in so far as they bear on questions of structure and form. As an example of such mechanical problems we may take the conditions which go to determine the manner of branching of an artery, or the angle at which its branches are given off; for, as John Hunter said[‡], "To keep up a circulation sufficient for the part, and no more, Nature has varied the angle of the origin of the arteries accordingly." This is a vastly important theme, and leads us a deal farther than does the problem, petty in comparison, of the shape of an egg. For the theorem which John Hunter has set forth in these simple words is no other than that "principle of minimal work" which is fundamental in physiology, and which some have deemed the very criterion

[*] R. H. Woods, On a physical theorem applied to tense membranes, *Journ. of Anat. and Phys.* xxvi, pp. 362–371, 1892. A similar investigation of the tensions in the uterine wall, and of the varying thickness of its muscles, was attempted by Haughton in his *Animal Mechanics*, 1873, pp. 151–158.

[†] This corresponds with a determination of the normal pressures (in systole) by Knohl, as being in the ratio of 1 : 6·8.

[‡] *Essays*, edited by Owen, i, p. 134, 1861. The subject greatly interested Keats. See his *Notebook*, edited by M. B. Forman, 1932, p. 7; and cf. *Keats as a Medical Student*, by Sir Wm Hale-White, in *Guy's Hospital Reports*, lxxiii, pp. 249–262, 1925.

of "organisation*." For the principle of Lagrange, the "principle of virtual work," is the key to physiological equilibrium, and physiology itself has been called a problem in maxima and minima †. This principle, overflowing into morphology, helps to bring the morphological and the physiological concepts together. We have dealt with problems of maxima and minima in many simple configurations, where form alone seemed to be in question; and we meet with the same principle again wherever work has to be done and mechanism is at hand to do it. That this mechanism is the best possible under all the circumstances of the case, that its work is done with a maximum of efficiency and at a minimum of cost, may not always lie within our range of quantitative demonstration, but to believe it to be so is part of our common faith in the perfection of Nature's handiwork. All the experience and the very instinct of the physiologist tells him it is true; he comes to use it as a postulate, or *methodus inveniendi*, and it does not lead him astray. The discovery of the circulation of the blood was implicit in, or followed quickly after, the recognition of the fact that the valves of heart and veins are adapted to a one-way circulation; and we may begin likewise by assuming a perfect fitness or adaptation in all the minor details of the circulation.

As part of our concept of organisation we assume that the cost of operating a physiological system is a minimum, what we mean by *cost* being measurable in calories and ergs, units whose dimensions are equivalent to those of *work*. The circulation teems with illustrations of this great and cardinal principle. "To keep up a circulation sufficient for the part and no more" Nature has not only varied the angle of branching of the blood-vessels to suit her purpose, she has regulated the dimensions of every branch and stem and twig and capillary; the normal operation of the heart is perfection itself, even the amount of oxygen which enters and leaves the capillaries is such that the work involved in its exchange and transport is a minimum. In short, oxygen transport is the main object of the circulation, and it seems that through all the trials and errors of

* Cf. Cecil D. Murray, The physiological principle of minimal work, in the vascular system, and the cost of blood-volume, *Proc. Acad. Nat. Sci.* xii, pp. 207–214, 1926; The angle of branching of the arteries, *Journ. Gen. Physiol.* ix, pp. 835–841, 1926; On the branching-angles of trees, *ibid.* x, p. 725, 1927.

† By Dr F. H. Pike, quoted by C. D. Murray.

growth and evolution an efficient mode of transport has been attained. To prove that it is the very best of all possible modes of transport may be beyond our powers and beyond our needs; but to assume that it is *perfectly economical* is a sound working hypothesis*. And by this working hypothesis wt seek to understand the form and dimensions of this structure or that, in terms of the work which it has to do.

The general principle, then, is that the form and arrangement of the blood-vessels is such that the circulation proceeds with a minimum of effort, and with a minimum of wall-surface, the latter condition leading to a minimum of friction and being therefore included in the first. What, then, should be the angle of branching, such that there shall be the least possible loss of energy in the course of the circulation? In order to solve this problem in any particular case we should obviously require to know (1) how the loss of energy depends upon the distance travelled, and (2) how the loss of energy varies with the diameter of the vessel. The loss of energy is evidently greater in a narrow tube than in a wide one, and greater, obviously, in a long journey than a short. If the large artery, *AB*, gives off a comparatively narrow branch leading to *P* (such as *CP*, or *DP*), the route *ACP* is evidently shorter than *ADP*, but on the other hand, by the latter path, the blood has tarried longer in the wide vessel *AB*, and has had a shorter course in the narrow branch.

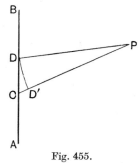

Fig. 455.

The relative advantage of the two paths will depend on the loss of energy in the portion *CD*, as compared with that in the alternative portion *CD'*, the one being short and narrow, the other long and wide. If we ask, then, which factor is the more important, length

* Cf. A. W. Volkmann, *Die Haemodynamik nach Versuchen*, Leipzig, 1850 (a work of great originality); G. Schwalbe, Ueber...die Gestaltung des Arteriensystems, *Jen. Zeitschr.* xii, p. 267, 1878; W. Hess, Eine mechanischbedingte Gesetzmässigkeit im Bau des Blutgefässsystems, *A. f. Entw. Mech.* xvi, p. 632, 1903; Ueber die peripherische Regulierung der Blutzirkulation, *Pfluger's Archiv*, clxviii, pp. 439–490, 1917; R. Thoma, Die mittlere Durchflussmengen der Arterien des Menschen als Funktion des Gefässradius, *ibid.* clxxix, pp. 282–310, cxciii, pp. 385–406, 1921–22; E. Blum, Querschnittsbeziehungen zwischen Stamm u. Ästen im Arteriensystem, *ibid.* clxxv, pp. 1–19, 1919.

or width, we may safely take it that the question is one of degree; and that the factor of width will become the more important of the two wherever artery and branch are markedly unequal in size. In other words, it would seem that for small branches a large angle of bifurcation, and for large branches a small one, is always the better. Roux has laid down certain rules in regard to the branching of arteries, which correspond with the general conclusions which we have just arrived at. The most important of these are as follows: (1) If an artery bifurcates into two equal branches, these branches come off at equal angles to the main stem. (2) If one of the two branches be smaller than the other, then the main branch, or continuation of the original artery, makes with the latter a smaller angle than does the smaller or "lateral" branch. And (3) all branches which are so small that they scarcely seem to weaken or diminish the main stem come off from it at a large angle, from about 70° to 90°.

We may follow Hess in a further investigation of the phenomenon. Let AB be an artery, from which a branch has to be given off so

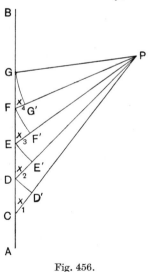

Fig. 456.

as to reach P, and let ACP, ADP, etc., be alternative courses which the branch may follow: CD, DE, etc., in the diagram, being equal distances $(= l)$ along AB. Let us call the angles PCD, PDE, x_1, x_2, etc.: and the distances CD', DE', by which each branch exceeds the next in length, we shall call l_1, l_2, etc. Now it is evident that, of the courses shewn, ACP is the shortest which the blood can take, but it is also that by which its transit through the narrow branch is the longest. We may reduce its transit through the narrow branch more and more, till we come to CGP, or rather to a point where the branch comes off at right angles to the main stem; but in so doing we very considerably increase the whole distance travelled. We may take it that there will be some intermediate point which will strike the balance of advantage.

Now it is easy to shew that if, in Fig. 456, the route ADP and AEP (two contiguous routes) be equally favourable, then any other route on either side of these, such as ACP or AFP, must be less favourable than either. Let ADP and AEP, then, be equally favourable; that is to say, let the loss of energy which the blood suffers in its passage along these two routes be equal. Then, if we make the distance DE very small, the angles x_2 and x_3 are nearly equal, and may be so treated. And again, if DE be very small, then $DE'E$ becomes a right angle, and l_2 (or DE') $= l \cos x_2$. But if L be the loss of energy per unit distance in the wide tube AB, and L' be the corresponding loss of energy in the narrow tube DP, etc., then $lL = l_2 L'$, because, as we have assumed, the loss of energy on the route DP is equal to that on the whole route DEP. Therefore $lL = lL' \cos x_2$, and $\cos x_2 = L/L'$. That is to say, the most favourable angle of branching will be such that the cosine of the angle is equal to the ratio of the loss of energy which the blood undergoes, per unit of length, in the main vessel, as compared with that which it undergoes in the branch. The path of a ray of light from one refractive medium to another is an analogous but much more famous problem; and the analogy becomes a close one when we look upon the branching artery as the special case of "grazing incidence."

After thus dealing with the most suitable angle of branching, we have still to consider the appropriate cross-section of the branches compared with the main trunk, for instance in the special case where a main artery bifurcates into two. That the sectional area of the two branches may together equal the area of the parent trunk, it is (of course) only necessary that the diameters of trunk and branch should be as $\sqrt{2} : 1$, or (say) as $14 : 10$, or (still more roughly) as $10 : 7$; and in the great vessels, this simple ratio comes very nearly true. We have, for instance, the following measurements of the common iliac arteries, into which the abdominal aorta subdivides:

*Internal diameter of abdominal arteries**

Aorta abdom. (mm.)	15·2	12·0	14·1	13·9
Iliaca comm. d. '	10·8	8·8	10·4	8·6
Iliaca comm. s.	10·7	8·6	9·5	10·0
Mean of do.	10·8	8·7	10·0	9·3
Ratio	71	72	69	67 p.c.
			Av. 70 p.c.	

* From R. Thoma, *op. cit.* p. 388.

But the increasing surface of the branches soon means increased friction, and a slower pace of the blood travelling through; and therefore the branches must be more capacious than at first appears. It becomes a question not of capacity but of resistance; and in general terms the answer is that the ratio of resistance to cross-section shall be equal in every part of the system, before and after bifurcation, as a condition of least possible resistance in the whole system; the total cross-section of the branches, therefore, must be greater than that of the trunk in proportion to the increased resistance.

An approximate result, familiar to students of hydrodynamics, is that the resistance is a minimum, and the condition an optimum, when the cross-section of the main stem is to the sum of the cross-sections of the branches as $1 : \sqrt[3]{2}$, or $1 : 1 \cdot 26$. Accordingly, in the case of a blood-vessel bifurcating into two equal branches, the diameter of each should be to that of the main stem (approximately) as

$$\sqrt{\frac{1 \cdot 26}{2}} : 1, \quad \text{or (say) } 8 : 10.$$

While these statements are so far true, and while they undoubtedly cover a great number of observed facts, yet it is plain that, as in all such cases, we must regard them not as a complete explanation, but as *factors* in a complicated phenomenon: not forgetting that (as one of the most learned of all students of the heart and arteries, Dr Thomas Young, said in his Croonian lecture *) all such questions as these, and all matters connected with the muscular and elastic powers of the blood-vessels, "belong to the most refined departments of hydraulics"; and Euler himself had commented on the "in-

---

* On the functions of the heart and arteries, *Phil. Trans.* 1809, pp. 1–31, cf. 1808, pp. 164–186; *Collected Works*, I, pp. 511–534, 1855. The same lesson is conveyed by all such work as that of Volkmann, E. H. Weber and Poiseuille. Cf. Stephen Hales's *Statical Essays*, II, *Introduction*: "Especially considering that they [i.e. animal Bodies] are in a manner framed of one continued Maze of innumerable Canals, in which Fluids are incessantly circulating, some with great Force and Rapidity, others with very different Degrees of rebated Velocity· Hence, etc." Even Leonardo had brought his knowledge of hydrodynamics to bear on the valves of the heart and the vortex-like eddies of the blood. Cf. J. Playfair McMurrich, *L. da Vinci, the Anatomist,* 1930, p. 165; etc. How complicated the physiological aspect of the case becomes may be judged by Thoma's papers quoted above.

superable difficulties" of this sort of problem*. Some other explanation must be sought in order to account for a phenomenon which particularly impressed John Hunter's mind, namely the gradually altering angle at which the successive intercostal arteries are given off from the thoracic aorta: the special interest of this case arising from the regularity and symmetry of the series, for "there is not another set of arteries in the body whose origins are so much the same, whose offices are so much the same, whose distances from their origin to the place of use, and whose uses [? sizes]† are so much the same."

The mechanical and hydrodynamical aspect of the circulation was as plain to John Hunter's mind as it had been to William Harvey or to Stephen Hales, or as it was afterwards to Thomas Young; but it was not always plain to other men. When a turtle's heart has been removed from its body, the blood may still be seen moving in the capillaries for some short while thereafter; and Haller, seeing this, "attributed it to some unknown power which he conceived to be exerted by the solid tissues on the blood and also by the globules of the blood on each other; to which power, until further investigation should elucidate its nature, he gave the name of *attraction*." So said William Sharpey, the father of modern English physiology; and Sharpey went on to say that "many physiologists accordingly maintain the existence of a peculiar propulsive power in the coats of the capillary vessels different from contractility, or that the globules of the blood are possessed of the power of spontaneous motion." Alison, great physician and famous vitalist, "extended this view, in so far as he regards the motion of the blood in the capillaries as one of the effects produced by what he calls vital attraction and repulsion, powers which he conceives to be general attributes of living matter." But Sharpey's own clear insight so far overcame his faith in Alison that he found it "not impossible that a certain degree of agitation might be occasioned in the blood by the elastic resilience of the vessels reacting on it, after the distending force of the heart has been withdrawn"; and, in short, that the evidence in the case did not "warrant the

---

* In a tract entitled *Principia pro motu sanguinis per arterias determinando Op. posth.* xi, pp. 814–823, 1862.

† "Sizes" is Owen's editorial emendation, which seems amply justified.

assumption of a peculiar power acting on the blood, of whose existence in the animal economy we have as yet no other evidence*."

Sir Charles Bell, whose anatomical skill was great but his mathematical insight small, drew the conclusion, of no small historic interest, that "the laws of hydraulics, though illustrative, are not strictly applicable to the explanation of the circulation of the blood, nor to the actions of the living frame." He goes on to say: "Although we perceive admirable mechanism in the heart, and in the adjustment of the tubes on hydraulic principles: and although the arteries and veins have form, calibre and curves suited to the conveyance of fluid, according to our knowledge of hydraulic engines: yet the laws of life, òr of physiology, are essential to the explanation of the circulation of the blood. And this conclusion we draw, not only from the extent and minuteness of the vessels, but also from the peculiar nature of the blood itself. Life is in both, and a mutual influence prevails †." This peculiar form of vitalism savours more of Bichat and the French school than of the teaching of John Hunter or Thomas Young. It is precisely that idea of "organic, control" or "organic coordination," which the physiologists are always reluctant to accept, always unwilling to abandon: which is said to be inherent in every process or operation of the body, and to differentiate biology from all the physical sciences: and of which in our own day Haldane has been the chief and great protagonist. But it is a subject with which this book is not concerned.

To conclude, we may now approach the question of economical size of the blood-vessels in a broader way. They must not be too small, or the work of driving blood through them will be too great;

---

* See Sharpey's article on *Cilia*, in Todd's *Cyclopaedia*, I, p. 637, 1836; also Allen Thomson's admirable article on the *Circulation, ibid.* p. 672. `Alison's views were based not only on Haller, but largely on Dr James Black's *Essay on the Capillary Circulation*, London, 1825.

† *Practical Essays*, 1842, p. 88. When Sir Charles Bell declared that hydraulic principles were not enough, but that "the laws of life" were needed to explain the circulation of the blood, he was right from his point of view. He was slow to see, and unwilling to admit, that hydrodynamical principles suffice to explain a large, essential part of the problem; but as a physiologist he had every reason to know that that part was not the whole. He may have had many things in mind: the arrest of the circulation in inflammation, as we see it in a frog's web; that a cut artery bleeds to death while a torn one does not bleed at all; that blood does not coagulate when stagnant within its own vessels—a fact which, as John Hunter said, "has ever appeared to me the most interesting fact in physiology."

they must not be too large, or they will hold more blood than is needed—and blood is a costly thing. We rely once more on Poiseuille's Law*, which tells us the amount of work done in causing so much fluid to flow through a tube against resistance, the said resistance being measured by the viscosity of the fluid, the coefficient of friction and the dimensions of the tube; but we have also to account for the blood itself, whose maintenance requires a share of the bodily fuel, and whose cost per c.c. may (in theory at least) be expressed in calories, or in ergs per day. The total cost, then, of operating a given section of artery will be measured by (1) the work done in overcoming its resistance, and (2) the work done in providing blood to fill it; we have come again to a differential equation, leading to an equation of maximal efficiency. The general result† is as follows: it can be made a quantitative one by introducing known experimental values. Were blood a cheaper thing than it is we might expect all arteries to be uniformly larger than they are, for thereby the burden on the heart (the flow remaining equal) would be greatly reduced—thus if the blood-vessels were doubled in diameter, and their volume thereby quadrupled, the work of the heart would be reduced to one-sixteenth. On the other hand, were blood a scarcer and still costlier fluid, narrower blood-vessels would hold the available supply; but a larger and stronger heart would be needed to overcome the increased resistance.

---

* Owing to faulty determination of the fall of pressure in the capillaries, Poiseuille's equation used to be deemed inapplicable to them; but Krogh's recent work removes, or tends to remove, the inconsistency (*Anatomy and Physiology of the Capillaries*, Yale University Press, 1922).

† Cf. C. D. Murray, *op. cit.* p. 211.

CHAPTER XVI

ON FORM AND MECHANICAL EFFICIENCY

THERE is a certain large class of morphological· problems of which
we have not yet spoken, and of which we shall be able to say but
little. Nevertheless they are so important, so full of deep theoretical
significance, and so bound up with the general question of form
and its determination as a result of growth, that an essay on
growth and form is bound to take account of them, however imperfectly and briefly. The phenomena which I have in mind are
just those many cases where *adaptation* in the strictest sense is
obviously present, in the clearly demonstrable form of mechanical
fitness for the exercise of some particular function or action which
has become·inseparable from the life and well-being of the organism.

When we discuss certain so-called "adaptations" to outward
circumstance, in the way of form, colour and so forth, we are often
apt to use illustrations convincing enough to certain minds. but
unsatisfying to others—in other words, incapable of demonstration.
With regard to coloration, for instance, it is by colours "cryptic,"
"warning," "signalling," "mimetic," and so on*, that we prosaically
expound, and slavishly profess to justify, the vast Aristotelian
synthesis that Nature makes all things with a purpose and "does
nothing in vain." Only for a moment let us glance at some few
instances by which the modern teleologist accounts for this or that
manifestation of colour, and is led on and on to beliefs and doctrines
to which it becomes more and more difficult to subscribe.

Some dangerous and malignant animals are said (in sober earnest)
to wear a perpetual war-paint, in order to "remind their enemies
that they had better leave them alone†." The wasp and the hornet,

---

* For a more elaborate classification, into colours cryptic, procryptic, anti-
cryptic, apatetic, epigamic, sematic, episematic, aposematic, etc., see Poulton's
Colours of Animals (Int. Scientific Series, LXVIII, 1890; cf. also R. Meldola,
Variable protective colouring in insects, *P.Z.S.* 1873, pp. 153–162; etc. The subject
is well and fully set forth by H. B. Cott, *Adaption coloration in Animals*, 1940.

† Dendy, *Evolutionary Biology*, 1912, p. 336.

in gallant black and gold, are terrible as an army with banners; and the Gila Monster (the poison-lizard of the Arizona desert) is splashed with scarlet—its dread and black complexion stained with heraldry more dismal. But the wasp-like livery of the noisy, idle hover-flies and drone-flies is but stage armour, and in their tinsel suits the little counterfeit cowardly knaves mimic the fighting crew.

The jewelled splendour of the peacock and the humming-bird, and the less effulgent glory of the lyre-bird and the Argus pheasant, are ascribed to the unquestioned prevalence of vanity in the one sex and wantonness in the other*.

The zebra is striped that it may graze unnoticed on the plain, the tiger that it may lurk undiscovered in the jungle; the banded Chaetodont and Pomacentrid fishes are further bedizened to the hues of the coral-reefs in which they dwell†. The tawny lion is yellow as the desert sand; but the leopard wears its dappled hide to blend, as it crouches on the branch, with the sun-flecks peeping through the leaves.

The ptarmigan and the snowy owl, the arctic fox and the polar bear, are white among the snows; but go he north or go he south, the raven (like the jackdaw) is boldly and impudently black.

The rabbit has his white scut, and sundry antelopes their piebald flanks, that one timorous fugitive may hie after another, spying the warning signal. The primeval terrier or collie-dog had brown spots over his eyes that he might seem awake when he was sleeping‡: so that an enemy might let the sleeping dog lie, for the singular reason that he imagined him to be awake. And a flock of flamingos,

---

* Delight in beauty is one of the pleasures of the imagination; there is no limit to its indulgence, and no end to the results which we may ascribe to its exercise. But as for the particular "standard of beauty" which the bird (for instance) admires and selects (as Darwin says in the *Origin*, p. 70, edit. 1884), we are very much in the dark, and we run the risk of arguing in a circle; for wellnigh all we can safely say is what Addison says (in the 412th *Spectator*)—that each different species "is most affected with the beauties of its own kind....Hinc merula in nigro se oblectat nigra marito;...hinc noctua tetram Canitiem alarum et glaucos miratur ocellos."

† Cf. T. W. Bridge, *Cambridge Natural History* (Fishes), VII, p. 173, 1904; also K. v. Frisch, Ueber farbige Anpassung bei Fische, *Zool. Jahrb.* (*Abt. Allg. Zool.*), XXXII, pp. 171–230, 1914. But Reighard, in what Raymond Pearl calls "one of the most beautiful experimental studies of natural selection which has ever been made," found no relation between the colours of coral-reef fishes and their elimination by natural enemies (*Carnegie Inst. Publication* 103, pp. 257–325, 1908).

‡ *Nature*, L, p. 572; LI, pp. 33, 57, 533, 1894–95.

wearing on rosy breast and crimson wings a garment of invisibility, fades away into the sky at dawn or sunset like a cloud incarnadine *.

To buttress the theory of natural selection the same instances of "adaptation" (and many more) are used, as in an earlier but not distant age testified to the wisdom of the Creator and revealed to simple piety the immediate finger of God. In the words of a certain learned theologian †, "The free use of final causes to explain what seems obscure was temptingly easy.... Hence the finalist was often the man who made a liberal use of the *ignava ratio*, or lazy argument: when you failed to explain a thing by the ordinary process of causality, you could 'explain' it by reference to some purpose of nature or of its Creator. This method lent itself with dangerous facility to the well-meant endeavours of the older theologians to expound and emphasise the beneficence of the divine purpose." *Mutatis mutandis*, the passage carries its plain message to the naturalist.

The fate of such arguments or illustrations is always the same. They attract and captivate for awhile; they go to the building of a creed, which contemporary orthodoxy ‡ defends under its severest penalties: but the time comes when they lose their fascination, they somehow cease to satisfy and to convince, their foundations

* They are "wonderfully fitted for 'vanishment' against the flushed, rich-coloured skies of early morning and evening...their chief feeding-times"; and "look like a real sunset or dawn, repeated on the opposite side of the heavens— either east or west as the case may be" (Thayer, *Concealing-coloration in the Animal Kingdom*, New York, 1909, pp. 154–155). This hypothesis, like the rest, is not free from difficulty. Twilight is apt to be short in the homes of the flamingo; moreover, Mr Abel Chapman watched them on the Guadalquivir *feeding by day*, as I also have seen them at Walfisch Bay.

† Principal Galloway, *Philosophy of Religion*, 1914, p. 344.

‡ Professor D. M. S. Watson, addressing the British Association in 1929 on *Adaptation*, parted company with what I had called *contemporary orthodoxy* in 1917. Speaking of such morphological differences as "have commonly been assumed to be of an adaptive nature," he said: "That these structural differences are adaptive is for the most part pure assumption.... There is no branch of zoology in which assumption has played a greater part, or evidence a less part, than in the study of such presumed adaptations." Hume, in his *Dialogue concerning Natural Religion*, shewed similar caution: "Steps of a stair are plainly constructed that human legs may use them in mounting; and this inference is certain and infallible. Human legs are also contrived for walking and mounting; and this inference, I allow, is not altogether so certain."

are discovered to be insecure, and in the end no man troubles to controvert them*.

But of a very different order from all such "adaptations" as these are those very perfect adaptations of form which, for instance, fit a fish for swimming or a bird for flight. Here we are far above the region of mere hypothesis, for we have to deal with questions of mechanical efficiency where statical and dynamical considerations can be applied and established in detail. The naval architect learns a great part of his lesson from the stream-lining of a fish†; the yachtsman learns that his sails are nothing more than a great bird's wing, causing the slender hull to *fly* along‡; and the mathematical study of the stream-lines of a bird, and of the principles underlying the areas and curvatures of its wings and tail, has helped to lay the very foundations of the modern science of aeronautics.

We know, for example, how in strict accord with theory (it was George Cayley who explained it first) the wing, whether of bird or insect, stands stiff along its "leading edge," like the mast before the sail; and how, conversely, it thins out exquisitely fine along its rear or "trailing edge," where sharp discontinuity favours the formation of uplifting eddies. And we see how, alike in the flying wing, in the penguin's swimming wing and in the whale's flipper, the same design of stiff fore-edge and thin fine trailing edge, both curving away evenly to meet at the tip, is continually exemplified.

We learn how lifting power not only depends on area but has a linear factor besides, such that a long narrow wing is more stable and effective both for speedy and for soaring flight than a short and broad one of equal area; and how in this respect the hawkmoth differs from the butterfly, the swallow from the thrush. We are taught how every wing, and every kite or sail, must have a certain

---

* The influence of environment on coloration is one thing, and the hypothesis of protective colouring is quite another. That arctic animals are often white, and desert animals sandy-hued or isabelline, are simple and undisputed facts; but such field-naturalists as Theodore Roosevelt, Selous (in his *African Nature Notes*), Buxton (*Animal Life in Deserts*), and Abel Chapman (*Savage Sudan*, and *Retrospect*) reject with one accord the theory of colour-protection.

† No creature shews more perfect stream-lining than a fur-seal swimming. Every curve is a *continuous* curve, the very ears and eye-slits and whiskers falling into the scheme, and the flippers folding close against the body.

‡ Cf. Manfred Curry, *Yacht Racing, and the Aerodynamics of Sails*, London, 1928.

amount of arch* or "belly," slight in the rapid fliers, deeper in the slow, flattened in the strong wind, bulging in the gentler breeze; and how advantageous is all possible stiffening of sail or wing, and why accordingly the yachtsman inserts "battens" and the Chinaman bamboos in his sail. We are shewn by Lilienthal himself how a powerful eddy, the so-called "ram," forms under the fore-edge, and is sometimes caught in a pocket of the bird's under wing-coverts and made use of as a forward drive.

We have lately learned how the gaps or slots between the primary wing-feathers of a crow, and a slight power of the wing-feathers to twist, like the slats of a Venetian blind, play their necessary

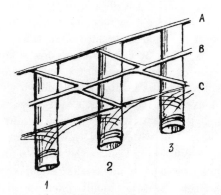

Fig. 457. Ligaments in a swan's wing. 1, 2, 3, remiges; A, B, longitudinal ligaments, with their oblique branches; C, small subcutaneous ligaments. From Marey, after Pettigrew.

part under certain conditions in the perfect working of the machine. Nothing can be simpler than the mechanism by which all this is done. Delicate ligaments run along the base of the wing from feather to feather, and send a branch to every quill (Fig. 457); by these, as the wing extends, the quills are raised into their places, and kept at their due and even distances apart. Not only that, but every separate ligamentous strand curls a little way round its feather where it is inserted into it; and thereby the feather is not only elevated into its place, but is given the little twist which brings it to its proper and precise obliquity.

* On the *curvatura veli*, cf. J. Bernoulli, *Acta Erudit. Lips.* 1692, p. 202. Studied also by Eiffel, *Résistance de l'air et l'aviation*, Paris, 1910.

All this is part of the automatic mechanism of the wing, than which there is nothing prettier in all anatomy.   The triceps muscle, massed on the shoulder, extends the elbow-joint in the usual way.   But another muscle (a long flexor of the wrist) has its origin above the elbow and its insertion below the wrist.   It passes over two joints; it is what German anatomists call a *zweigelenkiger Muskel*.   It transmits to the one joint the movements of the other; and in birds of powerful flight it becomes less and less muscular, more and more tendinous, and so more and more completely automatic.   The wing itself is kept light; its chief muscle is far back on the shoulder; a contraction of that remote muscle throws the whole wing into gear.   The little ligaments we have been speaking of are so linked up with the rest of the mechanism that we have only to hold a bird's wing by the arm-bone and extend its elbow-joint, to see the whole wing spring into action, with every joint extended, and every feather tense and in its place*.

Again on the fore-edge of the wing there lies a tiny mobile "thumb," whose little tuft of stiff, strong feathers forms the so-called "bastard wing."   We used to look on it as a "vestigial organ," a functionless rudiment, a something which from ancient times had "lagged superfluous on the stage"; until a man of genius saw that it was just the very thing required to break the leading vortices, keep the flow stream-lined at a larger angle of incidence than before, and thereby help the plane to land.   So he invented, to his great profit and advantage, the "slotted wing."

We learn many and many another interesting thing.   How a stiff "comb" along the leading edge, a broad soft fringe along the trailing edge (the fringe acting as a damper and preventing "fluttering"), and a soft, downy upper surface of the wing, all help as silencers, and give the owl her noiseless flight.   How the wing-loading of the owls is lower than in any other birds, lower even than in the eagles; and how owl and eagle have power to spare to carry easily their prey of mouse or mountain-hare.   How the deep terminal wing-slots aid the heavy rook or heron in their slow

---

* The mole's forelimb has a somewhat similar action, by which, as the arm and hand are pulled violently backward, the claws are powerfully and automatically flexed for digging.   The "suspensory ligament" of the horse, which is, or was, a short flexor of the digit, is an analogous mechanism.   See a paper of mine On the nature and action of certain ligaments, *Journ. of Anat. and Physiol.* xviii, pp. 406–410, 1884.

flapping flight; how the sea-gull does not need them, for his load is lighter and his wings move slowly. There is never a discovery made in the theory of aerodynamics but we find it adopted already by Nature, and exemplified in the construction of the wing*.

We may illustrate some few of the principles involved in the construction of the bird's wing with a half-sheet of paper, whose laws, as it planes or glides downwards, Clerk Maxwell explained many years ago†.

To improve this first and roughest of models, we see that its leading edge had better be as long as possible, and that sharp

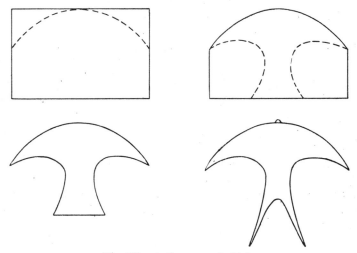

Fig. 458.   A diagrammatic bird.

corners are bound to cause disturbance; let us get rid of the corners and turn the leading edge into a continuous curve (Fig. 458). The leading edge is now doing most of the work, and the area within and behind is doing little good. Vorticoid air-currents are beating down on either side on this inner area; moreover, air is "sliding out" below, and tending to curl round the tip and edges of the

---

* Note that the aeroplane copies the beetle rather than the bird, as Lilienthal himself points out, in *Vom Gleitflug zu Segelflug*, Berlin, 1923.

† On a particular case of the descent of a heavy body in a resisting medium, *Camb. and Dublin Math. Journ.* IX, pp. 145–148, 1854; *Sci. Papers*, I, pp. 115–118. This elegant and celebrated little paper was written by Clerk Maxwell while an undergraduate at Trinity College, Cambridge.

wing, all with so much waste of energy*. In short the broad wing is less efficient than the narrow; and on either side of our sheet of paper we may cut out a portion which is useless and in the way: for the same reasons we may cut out the middle part of the tail, which also is doing more harm than good †.

Hard as the problem is, and harder as it becomes, we may venture on. In aeronautics, as in hydrodynamics, we try to determine the resistances encountered by bodies of various shapes, moving through various fluids at various speeds; and in so doing we learn the enormous, the paramount importance of "stream-lining." There would be no need for stream-lining in a "perfect fluid," but in air or water it makes all the difference in the world. Stream-lining implies a shape round which the medium streams so smoothly that resistance is at last practically *nil*; there only remains the slight "skin-friction," which can be reduced or minimised in various ways. But the least imperfection of the stream-lining leads to whirls and "pockets" of dead water or dead air, which mean large resistance and waste of energy. The converse and more general problem soon emerges, of how in natural objects stream-lining comes to be; and whether or no the more or less stream-lined shape tends to be impressed on a deformable or plastic body by its own steady motion through a fluid ‡. The principle of least action, the "loi de repos," is enough to suggest that the stream will tend to impress its stream-lines on the plastic body, causing it to yield or "give," until it ends by offering a minimum of resistance; and experiment goes some way to support the hypothesis. A bubble of mercury, poised in a tube through which air is blown, assumes a stream-lined shape, in so far as the forces due to the moving current avail against the

---

* This is why "slotting" so improves the broad wing of the crow. See on this and other matters, R. R. Graham's papers on Safety devices in the wings of birds, in *British Birds*, xxiv, 1930.

† Pettigrew shewed long ago (*Tr. R.S.E.* xxvi, p. 361, 1872) that the wing-area (in insects) "is usually greatly in excess of what is absolutely required for flight," and that the posterior or trailing edge could be largely trimmed away without the power of flight being at all diminished. We see how in the swallow this trailing edge is "trimmed away" till a bare minimum is left, and how (at least for a certain kind of flight) the wing is thereby greatly improved.

‡ Cf. Enoch Farrer, The shape assumed by a deformable body immersed in a moving fluid, *Journ. Franklin Inst.* 1921, pp. 737–756; also Vaughan Cornish, on *Waves of Sand and Snow*.

other forces of restraint. We have seen how the egg is automatically stream-lined, after a simple fashion, by the muscular pressure which drives it on its way. The contours of a snowdrift, of a wind-swept sand-dune, even of the flame of a lamp, shew endless illustrations of stream-lines or eddy-curves which the stream itself imposes, and which are oftentimes of great elegance and complexity. Always the stream tends to mould the bodies it streams over, facilitating its own flow; and the same principle must somehow come into play, at least as a contributory factor, in the making of a fish or of a bird. But it is obvious in both of these that even though the stream-lining be perfected in the individual it is also an inheritance of the race; and the twofold problem of accumulated inheritance, and of perfect structural adaptation, confronts us once again and passes all our understanding*.

When, after attempting to comprehend the exquisite adaptation of the swallow or the albatross to the navigation of the air, we try to pass beyond the empirical study and contemplation of such per-fection of mechanical fitness, and to ask how such fitness came to be, then indeed we may be excused if we stand wrapt in wonderment, and if our minds be occupied and even satisfied with the conception of a final cause. And yet all the while, with no loss of wonderment nor lack of reverence, do we find ourselves constrained to believe that somehow or other, in dynamical principles and natural law, there lie hidden the steps and stages of physical causation by which the material structure was so shapen to its ends †.

The problems associated with these phenomena are difficult at every stage, even long before we approach to the unsolved secrets of causation; and for my part I confess I lack the requisite know-ledge for even an elementary discussion of the form of a fish, or of an insect, or of a bird. But in the form of a bone we have a problem

* Mechanical perfection has often little to do with immunity from accident or with capacity to survive. Legs and wings of locust or mayfly are indescribably perfect for their brief spell of life and narrow sphere of toil; but they may be torn asunder in a moment, and whole populations perish in an hour. Careful of the type, but careless of the single life, Nature seems ruthless and indiscriminate in the sacrifice of these little lives.

† Cf. Professor Flint, in his Preface to Affleck's translation of Janet's *Causes finales*: "We are, no doubt, still a long way from a mechanical theory of organic growth, but it may be said to be the *quaesitum* of modern science, and no one can say that it is a chimaera."

of the same kind and order, so far simplified and particularised that we may to some extent deal with it, and may possibly even find, in our partial comprehension of it, a partial clue to the principles of causation underlying this whole class of phenomena.

Before we speak of the form of a bone, let us say a word about the mechanical properties of the material of which it is built*, in relation to the strength it has to manifest or the forces it has to resist: understanding always that we mean thereby the properties of fresh or living bone, with all its organic as well as inorganic constituents, for dead, dry bone is a very different thing. In all the structures raised by the engineer, in beams, pillars and girders of every kind, provision has to be made, somehow or other, for strength of two kinds, strength to resist compression or crushing, and strength to resist tension or pulling asunder. The evenly loaded column is designed with a view to supporting a downward pressure, the wire-rope, like the tendon

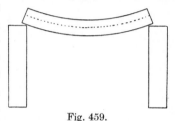

Fig. 459.

of a muscle, is adapted only to resist a tensile stress; but in many or most cases the two functions are very closely inter-related and combined. The case of a loaded beam is a familiar one; though, by the way, we are now told that it is by no means so simple as it looks, and indeed that "the stresses and strains in this log of timber are so complex that the problem has not yet been solved in a manner that reasonably accords with the known strength of the beam as found by actual experiment†." However, be that as it may, we know, roughly, that when the beam is loaded in the middle and supported at both ends, it tends to be bent into an arc, in which condition its lower fibres are being stretched, or are undergoing a tensile

---

* Cf. Sir Donald MacAlister, How a bone is built, *Engl. Ill. Mag.* 1884.

† Professor Claxton Fidler, *On Bridge Construction*, p. 22 (4th ed.), 1909; cf. (*int. al.*) Love's *Elasticity*, p. 20 (*Historical Introduction*), 2nd ed., 1906, where the bending of the beam, and the distortion or warping of its cross-section, are studied after the manner of St Venant, in his Memoir on Torsion (1855). How complex the question has become may be judged from such papers as Price, On the structure of wood in relation to its elastic properties, *Phil. Trans.* (A), ccviii, 1928; or D. B. Smith and R. V. Southwell, On the stresses induced by flexure in a deep rectangular beam, *Proc. R.S.* (A), cxliii, pp. 271–285, 1934.

stress, while its upper fibres are undergoing compression. It follows that in some intermediate layer there is a "neutral zone," where the fibres of the wood are subject to no stress of either kind.

The phenomenon of a compression-member side by side with a tension-member may be illustrated in many simple ways. Ruskin (in *Deucalion*) describes it in a glacier. He then bids us warm a stick of sealing-wax and bend it in a horseshoe: "you will then see, through a lens of moderate power, the most exquisite facsimile of glacier fissures produced by extension on its convex surface, and as faithful an image of glacier surge produced by compression on its concave side." A still more beautiful way of exhibiting the distribution of strain is to use gelatin, into which bubbles of gas have been introduced with the help of sodium bicarbonate. A bar of such gelatin, when bent into a hoop, shews on the one side the bubbles elongated by tension and on the other those shortened by compression*.

In like manner a vertical pillar, if unevenly loaded (as for instance the shaft of our thigh-bone normally is), will tend to bend, and so to endure compression on its concave, and tensile stress upon its convex side. In many cases it is the business of the engineer to separate out, as far as possible, the pressure-lines from the tension-lines, in order to use separate modes of construction, or even different materials for each. In a suspension-bridge, for instance, a great part of the fabric is subject to tensile strain only, and is built throughout of ropes or wires; but the massive piers at either end of the bridge carry the weight of the whole structure and of its load, and endure all the "compression-strains" which are inherent in the system. Very much the same is the case in that wonderful arrangement of struts and ties which constitute, or complete, the skeleton of an animal. The "skeleton," as we see it in a Museum, is a poor and even a misleading picture of mechanical efficiency†. From the engineer's point of view, it is a diagram shewing all the compression-lines, but by no means all the tension-lines of the construction; it shews all the struts, but few of the ties, and perhaps we might even say *none* of the principal

---

* Cf. Emil Hatschek, Gestalt und Orientirung von Gasblasen in Gelen, *Kolloid-Ztschr.* xv, pp. 226–234, 1914.

† In preparing or "macerating" a skeleton, the naturalist nowadays carries on the process till nothing is left but the whitened bones. But the old anatomists, whose object was not the study of "comparative morphology" but the wider theme of comparative physiology, were wont to macerate by easy stages; and in many of their most instructive preparations the ligaments were intentionally left in connection with the bones, and as part of the "skeleton."

ones; it falls all to pieces unless we clamp it together, as best we can, in a more or less clumsy and immobilised way. But in life, that fabric of struts is surrounded and interwoven with a complicated system of ties—"its living mantles jointed strong, With glistering band and silvery thong*": ligament and membrane, muscle and tendon, run between bone and bone; and the beauty and strength of the mechanical construction lie not in one part or in another, but in the harmonious concatenation which all the parts, soft and hard, rigid and flexible, tension-bearing and pressure-bearing, make up together†.

However much we may find a tendency, whether in Nature or art, to separate these two constituent factors of tension and compression, we cannot do so completely; and accordingly the engineer seeks for a material which shall, as nearly as possible, offer equal resistance to both kinds of strain‡.

From the engineer's point of view, bone may seem weak indeed; but it has the great advantage that it is very nearly as good for a tie as for a strut, nearly as strong to withstand rupture, or tearing asunder, as to resist crushing. The strength of timber varies with the kind, but it always stands up better to tension than to compression, and wrought iron, with its greater strength, does much the same; but in cast-iron there is a still greater discrepancy the other way, for it makes a good strut but a very bad tie indeed. Mild steel, which has displaced the old-fashioned wrought iron in all engineering constructions, is not only a much stronger material, but it also possesses, like bone, the two kinds of strength in no very great relative disproportion§.

---

* See Oliver Wendell Holmes' *Anatomist's Hymn*.

† In a few anatomical diagrams, for instance in some of the drawings in Schmaltz's *Atlas der Anatomie des Pferdes*, we may see the system of "ties" diagrammatically inserted in the figure of the skeleton. Cf. W. K. Gregory, On the principles of quadrupedal locomotion, *Ann. N. Y. Acad. of Sciences*, XXII, p. 289, 1912.

‡ The strength of materials is not easy to discuss, and is still harder to tabulate. The wide range of qualities in each material, in timber the wide differences according to the direction in which the block is cut, and in all cases the wide difference between yield-point and fracture-point, are some of the difficulties in the way of a succinct statement.

§ In the modern device of "reinforced concrete," blocks of cement and rods of steel are so combined together as to resist both compression and tension in due or equal measure.

When the engineer constructs an iron or steel girder, to take the place of the primitive wooden beam, we know that he takes advantage of the elementary principle we have spoken of, and saves weight and economises material by leaving out as far as possible all the middle portion, all the parts in the neighbourhood of the "neutral zone"; and in so doing he reduces his girder to an upper and lower "flange," connected together by a "web," the whole resembling, in cross-section, an **I** or an **⊥** .

But it is obvious that, if the strains in the two flanges are to be equal as well as opposite, and if the material be such as cast-iron or wrought-iron, one or other flange must be made much thicker than the other in order that they may be equally strong*; and if at times the two flanges have, as it were, to change places, or play each other's parts, then there must be introduced a margin of safety by making both flanges thick enough to meet that kind of stress in regard to which the material happens to be weakest. There is great economy, then, in any material which is, as nearly as possible, equally strong in both ways; and so we see that, from the engineer's or contractor's point of view, bone is a good and suitable material for purposes of construction.

The I or the H-girder or rail is designed to resist bending in one particular direction, but if, as in a tall pillar, it be necessary to resist bending in all directions alike, it is obvious that the tubular or cylindrical construction best meets the case; for it is plain that this hollow tubular pillar is but the I-girder turned round every way, in a "solid of revolution," so that on any two opposite sides compression and tension are equally met and resisted, and there is now no need for any substance at all in the way of web or "filling" within the hollow core of the tube. And it is not only in the supporting pillar that such a construction is useful; it is appropriate in every case where *stiffness* is required, where bending has to be resisted. A sheet of paper becomes a stiff rod when you roll it up, and hollow tubes of thin bent wood withstand powerful thrusts in aeroplane construction. The long bone of a bird's wing has little or no weight to carry, but it has to withstand powerful bending

* This principle was recognized as soon as iron came into common use as a structural material. The great suspension bridges only became possible, in Telford's hands, when wrought iron became available.

moments; and in the arm-bone of a long-winged bird, such as an albatross, we see the tubular construction manifested in its perfection, the bony substance being reduced to a thin, perfectly cylindrical, and almost empty shell*. The quill of the bird's feather, the hollow shaft of a reed, the thin tube of the wheat-straw bearing its heavy burden in the ear, are all illustrations which Galileo used in his account of this mechanical principle†; and the working of his practical mind is exemplified by this catalogue of varied instances which one demonstration suffices to explain.

The same principle is beautifully shewn in the hollow body and tubular limbs of an insect or a crustacean; and these complicated and elaborately jointed structures have doubtless many constructional lessons to teach us. We know, for instance, that a thin cylindrical tube, under bending stress, tends to flatten before it buckles, and also to become "lobed" on the compression side of the bend; and we often recognise both of these phenomena in the joints of a crab's leg‡.

Two points, both of considerable importance, present themselves here, and we may deal with them before we go further on. In the first place, it is not difficult to see that in our bending beam the stress is greatest at its. middle; if we press our walking-stick hard against the ground, it will tend to snap midway. Hence, if our cylindrical column be exposed to strong bending stresses, it will be prudent and economical to make its walls thickest in the middle and thinning off gradually towards the ends; and if we look at a longitudinal section of a thigh-bone, we shall see that this is just what Nature has done. The presence of a "danger-point" has been avoided, and the thickness of the walls becomes nothing less than

---

* Marsigli (op. cit.) was acquainted with the hollow wing-bones of the pelican; and Buffon deals with the whole subject in his Discours sur la nature des oiseaux.

† Galileo, Dialogues concerning Two New Sciences (1638), Crew and Salvio's translation, New York, 1914, p. 150; Opere, ed. Favaro, VIII, p. 186. (According to R. A. Millikan, "we owe our present day civilisation to Galileo.") Cf. Borelli, De Motu Animalium, I, prop. CLXXX, 1685. Cf. also P. Camper, La structure des os dans les oiseaux, Opp. III, p. 459, ed. 1803; A. Rauber, Galileo über Knochen-formen, Morphol. Jahrb. VII, pp. 327, 328, 1881; Paolo Enriques, Della economia di sostanza nelle osse cave, Arch. f. Entw. Mech. XX, pp. 427–465, 1906. Galileo's views on the mechanism of the human body are also discussed by O. Fischer, in his article on Physiologische Mechanik, in the Encycl. d. mathem. Wissenschaften, 1904.

‡ Cf. L. G. Brazier, On the flexure of thin cylindrical shells, etc., Proc. R.S. (A), CXVI, p. 104, 1927.

a diagram, or "graph," of the bending-moments from one point to another along the length of the bone.

The second point requires a little more explanation. If we imagine our loaded beam to be supported at one end only (for instance, by being built into a wall), so as to form what is called

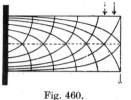

Fig. 460.

a "bracket" or "cantilever," then we can see, without much difficulty, that the lines of stress in the beam run somewhat as in the accompanying diagram. Immediately under the load, the "compression-lines" tend to run vertically downward, but where the bracket is fastened to the wall there is pressure directed horizontally against the wall in the lower part of the surface of attachment; and the vertical beginning and the horizontal end of these pressure-lines must be continued into one another in the form of some even mathematical curve—which, as it happens, is part of a parabola. The tension-lines are identical in form with the compression-lines, of which they constitute the "mirror-image"; and where the two systems intercross they do so at right angles, or "orthogonally" to one another. Such systems of stress-lines as these we shall deal with again; but let us take note here of the important though well-nigh obvious fact, that while in the beam they both unite to carry the load, yet it is often possible to weaken one set of lines at the expense of the other, and in some cases to do altogether away with one set or the other. For example, when we replace our end-supported beam by a curved bracket, bent upwards or downwards as the case may be, we have evidently cut away in the one case the greater part of the tension-lines, and in the other the greater part of the compression-lines. And if instead of bridging a stream with our beam of wood we bridge it with a rope, it is evident that this new construction contains all the tension-lines, but none of the compression-lines of the old. The biological interest connected with this principle lies chiefly in the mechanical construction of the rush or the straw, or any other typically cylindrical stem. The material of which the stalk is constructed is very weak to withstand compression, but parts of it have a very great tensile strength. Schwendener, who was both botanist and engineer, has elaborately investigated the factor of strength in the cylindrical

stem, which Galileo was the first to call attention to. Schwendener*
shewed that its strength was concentrated in the little bundles of
"bast-tissue," but that these bast-fibres had a tensile strength per
square mm. of section not less, up to the limit of elasticity, than
that of steel-wire of such quality as was in use in his day.

For instance, we see in the following table the load which various
fibres, and various wires, were found capable of sustaining, not
up to the breaking-point but up to the "elastic limit," or point
beyond which complete recovery to the original length took place
no longer after release of the load.

	Stress, or load in gms. per sq. mm., at Limit of Elasticity	Do., in tons per sq. inch	Strain, or amount of stretching, per mille
Secale cereale	15–20	9·4–12·5	4·4
Lilium auratum	19	11·8	7·6
Phormium tenax	20	12·5	13·0
Papyrus antiquorum	20	12·5	15·2
Molinia coerulea	22	13·8	11·0
Pincenectia recurvata	25	15·6	14·5
Copper wire	12·1	7·6	1·0
Brass ,,	13·3	8·5	1·35
Iron ,,	21·9	13·7	1·0
Steel ,,	24·6*	15·4	1·2

   * This figure should be considerably higher for the best modern steel.

In other respects, it is true, the plant-fibres were inferior to the
wires; for the former broke asunder very soon after the limit of
elasticity was passed, while the iron-wire could stand, before snapping,
about twice the load which was measured by its limit of elasticity:
in the language of a modern engineer, the bast-fibres had a low
"yield-point," little above the elastic limit. Nature seems content,
as Schwendener puts it, if the strength of the fibre be ensured up
to the elastic limit; for the equilibrium of the structure is lost as
soon as that limit is passed, and it then matters little how far off
the actual breaking-point may be†. But nevertheless, within certain limits, plant-fibre and wire were just as good and strong one

   * S. Schwendener, *Das mechanische Princip im anatomischen Bau der Monocotyleen,*
Leipzig, 1874; Zur Lehre von der Festigkeit der Gewächse, *Sb. Berlin. Akad.* 1884,
pp. 1045–1070.

   † The great extensibility of the plant-fibre is due to the spiral arrangement of
the ultramicroscopic micellae of which the bast-fibre is built up: the spiral untwisting
as the fibre stretches, in a right or left-hand spiral according to the species. Cf.
C. Steinbruck, Die Micellartheorie auf botanischem Gebiete, *Biol. Centralbl.* 1925,
p. 1.

as the other.   And then Schwendener proceeds to shew, in many beautiful diagrams, the various ways in which these strands of strong tensile tissue are arranged in various stems: sometimes, in the simpler cases, forming numerous small bundles arranged in a peripheral ring, not quite at the periphery, for a certain amount of space has to be left for living and active tissue; sometimes in a sparser ring of larger and stronger bundles; sometimes with these bundles further strengthened by radial balks or ridges; sometimes with all the fibres set close together in a continuous hollow cylinder.   In the case figured in Fig. 461, Schwendener calculated that the resistance to bending was at least twenty-five times as great as it would have been had the six main bundles been brought close together in a solid core.   In many cases the centre of the stem is altogether empty; in all other cases it is filled with soft tissue, suitable for various functions, but never such as to confer mechanical rigidity.   In a tall conical stem, such as that of a palm-tree, we can see not only these principles in the construction of the cylindrical trunk, but we can observe, towards the apex, the bundles of fibre curving over and intercrossing orthogonally with one another, exactly after the fashion of our stress-lines in Fig. 460; but of course, in this case, we are still dealing with tensile members, the opposite bundles taking on in turn, as the tree sways, the alternate function of resisting tensile strain*.

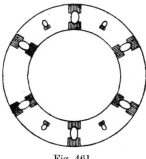

Fig. 461.

* For further botanical illustrations, see (int. al.) R. Hegler, Einfluss der Zug-kräften auf die Festigkeit und die Ausbildung mechanischer Gewebe in Pflanzen, SB. sächs. Ges. d. Wiss. 1891, p. 638; Einfluss des mechanischen Zuges auf das Wachstum der Pflanze, Cohn's Beiträge, VI, pp. 383–432, 1893; O. M. Ball, Einfluss von Zug auf die Ausbildung der Festigkeitsgewebe, Jahrb. d. wiss. Bot. XXXIX, pp. 305–341, 1903; L. Kny, Einfluss von Zug und Druck auf die Richtung der Scheidewände in sich teilenden Pflanzenzellen, Ber. d. bot. Gesellsch. XIV, pp. 378–391, 1896; Sachs, Mechanomorphose und Phylogenie, Flora, LXXVIII, 1894; cf. also Pflüger, Einwirkung der Schwerkraft, etc., über die Richtung der Zelltheilung, Archiv, XXXIV, 1884; G. Haberlandt's Physiological Plant Anatomy, tr. by Montagu Drummond, 1914, pp. 150–213.   On the engineering side of the case, see Angus R. Fulton, Experiments to show how failure under stress occurs in timber, etc., Trans. R.S.E. XLVIII, pp. 417–440, 1912; Fulton shews (int. al.) that "the initial cause of fracture in timbers lies in the medullary rays."

The Forth Bridge, from which the anatomist may learn many
a lesson, is built of tubes, which correspond even in detail to the
structure of a cylindrical branch or stem. The main diagonal struts
are tubes twelve feet in diameter, and within the wall of each of
these lie six T-shaped "stiffeners," corresponding precisely to the
fibro-vascular bundles of Fig. 461; in the same great tubular struts
the tendency to "buckle" is resisted, just as in the jointed stem of
a bamboo, by "stiffening rings," or perforated diaphragms set
twenty feet apart within the tube. We may draw one more curious,
albeit parenthetic, comparison. An engineering construction, no
less than the skeleton of plant or animal, has *to grow*; but the
living thing is in a sense complete during every phase of its existence,
while the engineer is often hard put to it to ensure sufficient strength
in his unfinished and imperfect structure. The young twig stands
more upright than the old, and between winter and summer the
weight of leafage affects all the curving outlines of the tree. A
slight upward curvature, a matter of a few inches, was deliberately
given to the great diagonal tubes of the bridge during their piecemeal
construction; and it was a triumph of engineering foresight to see
how, like the twig, as length and weight increased, they at last came
straight and true.

Let us now come, at last, to the mechanical structure of bone,
of which we find a well-known and classical illustration in the
various bones of the human leg. In the case of the tibia, the bone
is somewhat widened out above, and its hollow shaft is capped by
an almost flattened roof, on which the weight of the body directly
rests. It is obvious that, under these circumstances, the engineer
would find it necessary to devise means for supporting this flat roof,
and for distributing the vertical pressures which impinge upon it
to the cylindrical walls of the shaft.

In the long wing-bones of a bird the hollow of the bone is empty,
save for a thin layer of living tissue lining the cylinder of bone;
but in our own bones, and all weight-carrying bones in general, the
hollow space is filled with marrow, blood-vessels and other tissues;
and amidst these living tissues lies a fine lattice-work of little
interlaced "trabeculae" of bone, forming the so-called "cancellous
tissue." The older anatomists were content to describe this can-

cellous tissue as a sort of spongy network or irregular honeycomb*; but at length its orderly construction began to be perceived, and attempts were made to find a meaning or "purpose" in the arrangement. Sir Charles Bell had a glimpse of the truth when he asserted † that "this minute lattice-work, or the cancelli which constitute the interior structure of bone, have still reference to the forces acting on the bone"; but he did not succeed in shewing what these forces are, nor how the arrangement of the cancelli is related to them.

Jeffries Wyman, of Boston, came much nearer to the truth in a paper long neglected and forgotten ‡. He gives the gist of the whole matter in two short paragraphs: "1. The cancelli of such bones as assist in supporting the weight of the body are arranged either in the direction of that weight, or in such a manner as to support and brace those cancelli which are in that direction. In a mechanical point of view they may be regarded in nearly all these bones as a series of 'studs' and 'braces.' 2. The direction of these fibres in some of the bones of the human skeleton is characteristic and, it is believed, has a definite relation to the erect position which is naturally assumed by man alone." A few years afterwards the story was told again, and this time with convincing accuracy. It was shewn by Hermann Meyer (and afterwards in greater detail by Julius Wolff and others) that the trabeculae, as seen in a longitudinal section of the femur, spread in beautiful curving lines from the head to the hollow shaft of the bone; and that these linear bundles are crossed by others, with so nice a regularity of arrangement that each intercrossing is as nearly as possible an orthogonal one: that is to say, the one set of fibres or cancelli cross the other everywhere at right angles. A great engineer, Professor Culmann of Zürich, to whom by the way we owe the whole modern method of "graphic statics," happened (in the year 1866) to come into his colleague Meyer's dissecting-room, where the anatomist was contemplating

---

* Sir John Herschel described a bone as a "framework of the most curious carpentry: in which occurs not a single straight line nor any known geometrical curve, yet all evidently systematic, and constructed by rules which defy our research" (*On the Study of Natural Philosophy*, 1830, p. 203).

† In *Animal Mechanics, or Proofs of Design in the Animal Frame*, 1827.

‡ Animal mechanics: on the cancellated structure of some of the bones of the human body, *Boston Soc. of Nat. Hist.* 1849. Reprinted, together with Sir C. Bell's work, by Morrill Wyman, Cambridge, Mass., 1902.

The lines of stress are bundled close together along the sides of the shaft, and lost or concealed there in the substance of the solid wall of bone; but in and near the head of the bone, a peripheral shell of bone does not suffice to contain them, and they spread out through the central mass in the actual concrete form of bony trabeculae *.

Mutatis mutandis, the same phenomenon may be traced in any other bone which carries weight and is liable to flexure; and in the *os calcis* and the tibia, and more or less in all the bones of the lower limb, the arrangement is found to be very simple and clear.

Thus, in the *os calcis*, the weight resting on the head of the bone has to be transmitted partly through the backward-projecting heel to the ground, and partly forwards through its articulation with

---

* Among other works on the mechanical construction of bone see: Bourgery, *Traité de l'anatomie (I. Ostéologie)*, 1832 (with admirable illustrations of trabecular structure); L. Fick, *Die Ursachen der Knochenformen*, Göttingen, 1857; H. Meyer, Die Architektur der Spongiosa, *Arch. f. Anat. und Physiol.* XLVII, pp. 615–628, 1867; *Statik u. Mechanik des menschlichen Knochengerüstes*, Leipzig, 1873; H. Wolfermann, Beitrag zur K. der Architektur der Knochen, *Arch. f. Anat. und Physiol.* 1872, p. 312; J. Wolff, Die innere Architektur der Knochen, *Arch. f. Anat. und Phys.* L, 1870; *Das Gesetz der Transformation bei Knochen*, 1892; Y. Dwight, The significance of bone-architecture, *Mem. Boston Soc. N.H.* IV, p. 1, 1886; V. von Ebner, Der feinere Bau der Knochensubstanz, *Wiener Bericht*, LXXII, 1875; Anton Rauber, *Elastizität und Festigkeit der Knochen*, Leipzig, 1876; O. Meserer, *Elast. u. 'Festigk. d. menschlichen Knochen*, Stuttgart, 1880; Sir Donald MacAlister, How a bone is built, *English Illustr. Mag.* 1884, pp. 640–649; Rasumowsky, Architektonik des Fussskelets, *Int. Monatsschr. f. Anat.* 1889, p. 197; Zschokke, *Weitere Unters. über das Verhältnis der Knochenbildung zur Statik und Mechanik des Vertebratenskelets*, Zürich, 1892; W. Roux, *Ges. Abhandlungen über Entwicklungsmechanik der Organismen*, Bd. I, *Funktionelle Anpassung*, Leipzig, 1895; J. Wolff, Die Lehre von der funktionellen Knochengestalt, *Virchow's Archiv*, CLV, 1899; R. Schmidt, Vergl. anat. Studien über den mechanischen Bau der Knochen und seine Vererbung, *Z. f. w. Z.* LXV, p. 65, 1899; B. Solger, Der gegenwärtige Stand der Lehre von der Knochenarchitektur, in Moleschott's *Unters. z. Naturlehre des Menschen.* XVI, p. 187, 1899; H. Triepel, Die Stossfestigkeit der Knochen, *Arch. f. Anat. und Phys.* 1900; Gebhardt, Funktionellwichtige Anordnungsweisen der feineren und gröberen Bauelemente des Wirbelthierknochens, etc., *Arch. f. Entw. Mech.* 1900–10; Revenstorf, Ueber die Transformation der Calcaneus-architektur, *Arch. f. Entw. Mech.* XXIII, p. 379, 1907; H. Bernhardt, *Vererbung der inneren Knochenarchitektur beim Menschen, und die Teleologie bei J. Wolff*, Inaug. Diss., München, 1907; Herm. Triepel, Die trajectoriellen Structuren (in *Einf. in die Physikalische Anatomie*, 1908); A. F. Dixon, Architecture of the cancellous tissue forming the upper end of the femur, *Journ. Anat. and Phys.* (3), XLIV, pp. 223–230, 1910; A. Benninghoff, Ueber Leitsystem der Knochencompacta; Studien zur Architektur der Knochen, *Beitr. z. Anat. funktioneller Systeme*, I, 1930.

the cuboid bone, to the arch of the foot. We thus have, very much as in a triangular roof-tree, two compression-members sloping apart from one another; and these have to be bound together by a "tie" or tension-member, corresponding to the third, horizontal member of the truss.

It is a simple corollary, confirmed by observation, that the trabeculae have a very different distribution in animals whose actions and attitudes are materially different, as in the aquatic mammals, such as the beaver and the seal*. And in much less extreme cases there are lessons to be learned from a study of the

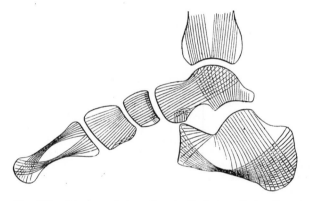

Fig. 464. Diagram of stress-lines in the human foot. From
Sir D. MacAlister, after H. Meyer.

same bone in different animals, as the loads alter in direction and magnitude. The gorilla's heelbone resembles man's, but the load on the heel is much less, for the erect posture is imperfectly achieved: in a common monkey the heel is carried high, and consequently the direction of the trabeculae is still more changed. The bear walks on the sole of his foot, though less perfectly than does man, and the lie of the trabeculae is plainly analogous in the two; but in the bear more powerful strands than in the *os calcis* of man transmit the load forward to the toes, and less of it through the heel to the ground. In the leopard we see the full effect of tip-toe, or digitigrade, progression. The long hind part (or tuberosity) of

* Cf. G. de M. Rudolf, Habit and the architecture of the mammalian femur, *Journ. Anatomy*, LVI, pp. 139–146, 1922.

the heel is now more a mere lever than a pillar of support; it is little more than a stiffened rod, with compression-members and tension-members in opposite bundles, inosculating orthogonally at the two ends*.

In the bird the small bones of the hand, dwarfed as they are in size, have still a deal to do in carrying the long primary flight-feathers, and in forming a rigid axis for the terminal part of the wing. The simple tubular construction, which answers well for the long, slender arm-bones, does not suffice where a still more efficient stiffening is required. In all the mechanical side of anatomy

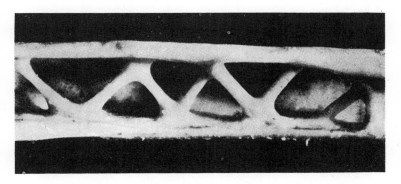

Fig. 465. Metacarpal bone from a vulture's wing; stiffened after the manner of a Warren's truss. From O. Prochnow, *Formenkunst der Natur.*

nothing can be more beautiful than the construction of a vulture's metacarpal bone, as figured here (Fig. 465). The engineer sees in it a perfect Warren's truss, just such a one as is often used for a main rib in an aeroplane. Not only so, but the bone is better than the truss; for the engineer has to be content to set his V-shaped struts all in one plane, while in the bone they are put, with obvious but inimitable advantage, in a three-dimensional configuration.

So far, dealing wholly with the stresses and strains due to tension and compression, we have omitted to speak of a third very important factor in the engineer's calculations, namely what is known as "shearing stress." A shearing force is one which produces

---

* Cf. Fr. Weidenreich, Ueber formbestimmende Ursachen am Skelett, und die Erblichkeit der Knochenform, *Arch. f. Entw. Mech.* LI, pp. 438–481, 1922.

"angular distortion" in a figure, or (what comes to the same thing) which tends to cause its particles to slide over one another. A shearing stress is a somewhat complicated thing, and we must try to illustrate it (however imperfectly) in the simplest possible way. If we build up a pillar, for instance, of flat horizontal slates, or of a pack of cards, a vertical load placed upon it will produce compression, but will have no tendency to cause one card to slide, or shear, upon another; and in like manner, if we make up a cable of parallel wires and, letting it hang vertically, load it evenly with

Fig. 466. Trabecular structure of the os calcis. From MacAlister.

a weight, again the tensile stress produced has no tendency to cause one wire to slip or shear upon another. But the case would have been very different if we had built up our pillar of cards or slates lying obliquely to the lines of pressure, for then at once there would have been a tendency for the elements of the pile to slip and slide asunder, and to produce what the geologists call "a fault" in the structure.

Somewhat more generally, if ·AB be a bar, or pillar, of cross-section a under a direct load P, giving a direct and uniformly distributed stress per unit area $=p$, then the whole pressure $P=pa$. Let CD be an oblique section, inclined at an angle θ to the cross-section; the pressure on CD will evidently be $=pa \cos \theta$. But at any point O in CD, the pressure P may be resolved into the shearing force Q acting along CD, and the direct force N perpendicular to it: where $N=P \cos \theta=pa \cos \theta$, and $Q=P \sin \theta=pa \sin \theta$. The shearing force

Q upon $CD=q$. area of CD, which is $=q.a/\cos\theta$. Therefore $qa/\cos\theta=pa\sin\theta$, therefore $q=p\sin\theta\cos\theta=\frac{1}{2}p\sin 2\theta$. Therefore when $\sin 2\theta=1$, that is, when $\theta=45°$, q is a maximum, and $=p/2$; and when $\sin 2\theta=0$, that is when $\theta=0°$ or $90°$, then q vanishes altogether.

This is as much as to say, that under this form of loading there is no shearing stress along or perpendicular to the lines of principal stress, or along the lines of maximum compression or tension; but shear has a definite value on all other planes, and a maximum value when it is inclined at 45° to the cross-section. This may be further illustrated in various simple ways. When we submit a cubical block of iron to compression in the testing machine, it does not tend to give way by crumbling all to pieces, but always disrupts by shearing, and along some plane approximately at 45° to the axis of compression; this is known as Coulomb's Theory of Fracture, and, while subject to many qualifications, it is still an important first approximation to the truth. Again, in the beam which we have already con-sidered under a bending moment, we

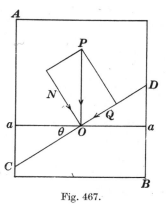

Fig. 467.

know that if we substitute for it a pack of cards, they will be strongly sheared on one another; and the shearing stress is greatest in the "neutral zone," where neither tension nor compression is manifested: that is to say in the line which cuts at equal angles of 45° the orthogonally intersecting lines of pressure and tension.

In short we see that, while shearing *stresses* can by no means be got rid of, the danger of rupture or breaking-down under shearing stress is lessened the more we arrange the materials of our con-struction along the pressure-lines and tension-lines of the system; for *along these lines* there is no shear*.

To apply these principles to the growth and development of our bone, we have only to imagine a little trabecula (or group of trabeculae) being secreted and laid down fortuitously in any direction within the substance of the bone. If it lie in the direction of one of the pressure-lines, for instance, it will be in a position of comparative

---

* It is also obvious that a free surface is always a region of zero-shear.

equilibrium, or minimal disturbance; but if it be inclined obliquely to the pressure-lines, the shearing force will at once tend to act upon it and move it away. This is neither more nor less than what happens when we comb our hair, or card a lock of wool: filaments lying in the direction of the comb's path remain where they were; but the others, under the influence of an oblique component of pressure, are sheared out of their places till they too come into coincidence with the lines of force. So straws show how the wind blows—or rather how it has been blowing. For every straw that lies askew to the wind's path tends to be sheared into it; but as soon as it has come to lie the way of the wind it tends to be disturbed no more, save (of course) by a violence such as to hurl it bodily away.

In the biological aspect of the case, we must always remember that our bone is not only a living, but a highly plastic structure; the little trabeculae are constantly being formed and deformed, demolished and formed anew. Here, for once, it is safe to say that "heredity" need not and cannot be invoked to account for the configuration and arrangement of the trabeculae: for we can see them at any time of life in the making, under the direct action and control of the forces to which the system is exposed. If a bone be broken and so repaired that its parts lie somewhat out of their former place, so that the pressure- and tension-lines have now a new distribution, before many weeks are over the trabecular system will be found to have been entirely remodelled, so as to fall into line with the new system of forces. And as Wolff pointed out, this process of reconstruction extends a long way off from the seat of injury, and so cannot be looked upon as a mere accident of the physiological process of healing and repair; for instance, it may happen that, after a fracture of the *shaft* of a long bone, the trabecular meshwork is wholly altered and reconstructed within the distant *extremities* of the bone. Moreover, in cases of transplantation of bone, for example when a diseased metacarpal is repaired by means of a portion taken from the lower end of the ulna, with astonishing quickness the plastic capabilities of the bony tissue are so manifested that neither in outward form nor inward structure can the old portion be distinguished from the new.

Herein then lies, so far as we can discern it, a great part at least of the physical causation of what at first sight strikes us as a purely functional adaptation: as a phenomenon, in other words, whose physical cause is as obscure as its final cause or end is apparently manifest.

Partly associated with the same phenomenon, and partly to be looked upon (meanwhile at least) as a fact apart, is the very important physiological truth that a condition of *strain*, the result of a *stress*, is a direct stimulus to growth itself. This indeed is no less than one of the cardinal facts of theoretical biology. The soles of our boots wear thin, but the soles of our feet grow thick, the more we walk upon them: for it would seem that the living cells are "stimulated" by pressure, or by what we call "exercise," to increase and multiply. The surgeon knows, when he bandages a broken limb, that his bandage is doing something more than merely keeping the parts together: and that the even, constant pressure which he skilfully applies is a direct encouragement of growth and an active agent in the process of repair. In the classical experiments of Sédillot*, the greater part of the shaft of the tibia was excised in some young puppies, leaving the whole weight of the body to rest upon the fibula. The latter bone is normally about one-fifth or sixth of the diameter of the tibia; but under the new conditions, and under the "stimulus" of the increased load, it grew till it was as thick or even thicker than the normal bulk of the larger bone. Among plant tissues this phenomenon is very apparent, and in a somewhat remarkable way; for a strain caused by a constant or increasing weight (such as that in the stalk of a pear while the pear is growing and ripening) produces a very marked increase of *strength* without any necessary increase of bulk, but rather by some histological, or molecular, alteration of the tissues. Hegler, Pfeffer, and others have investigated this subject, by loading the young shoot of a plant nearly to its breaking point, and then redetermining the breaking-strength after a few days. Some young shoots of the sunflower were found to break with a strain of 160 gm.; but when loaded with 150 gm., and retested after two days, they were able to support 250 gm.; and being again loaded with something short

* Sédillot, De l'influence des fonctions sur la structure et la forme des organes, *C.R.* LIX, p. 539, 1864; cf. LX, p. 97, 1865; LXVIII, p. 1444, 1869.

of this, by next day they sustained 300 gm., and a few days later even 400 gm.*

The kneading of dough is an analogous phenomenon. The viscosity and perhaps other properties of the stuff are affected by the strains to which we have submitted it, and may thus be said to depend not only on the nature of the substance but on its history †. It is a long way from this simple instance, but we stretch across it easily in imagination, to the experimental growth of a nerve-fibre within a mass of clotted lymph: where, when we draw out the clot in one direction or another we lay down traction-lines, or tension-lines, and make of them a path for growth to follow ‡.

Such experiments have been amply confirmed, but so far as I am aware we do not know much more about the matter: we do not know, for instance, how far the change is accompanied by increase in number of the bast-fibres, through transformation of other tissues; or how far it is due to increase in size of these fibres; or whether it be not simply due to strengthening of the original fibres by some molecular change. But I should be much inclined to suspect that this last had a good deal to do with the phenomenon. We know nowadays that a railway axle, or any other piece of steel, is weakened by a constant succession of frequently interrupted strains; it is said to be "fatigued," and its strength is restored by a period of rest. The converse effect of continued strain in a uniform direction may be illustrated by a homely example. The confectioner takes a mass of boiled sugar or treacle (in a particular molecular condition determined by the temperature to which it has been raised), and draws the soft sticky mass out into a rope; and then, folding it up lengthways, he repeats the process again and again. At first the rope is pulled out of the ductile mass without difficulty; but as the work goes on it gets harder to do, until all the man's force is used to stretch the rope. Here we have the phenomenon

---

* *Op. cit.* Hegler's results are criticised by O. M. Ball, Einfluss von Zug auf die Ausbildung der Festigungsgewebe, *Jb. d. wiss. Botanik,* XXXIX, pp. 305–341, 1903, and by H. Keller, Einfluss von Belastung und Lage auf die Ausbildung des Gewebes in Fruchtstielen, *Inaug. Diss.* Kiel, 1904.

† Cf. R. K. Schofield and G. W. S. Blair, On dough, *Proc. R.S.* (A), CXXXVIII, p. 707; CXXXIX, p. 557, 1932–33; also Nadai and Wahl's *Plasticity,* 1931. For analogous properties of hairs and fibres, see Shorter, *Journ. Textile Inst.* XV, 1824; etc.

‡ Cf. Ross Harrison's *Croonian Lecture,* 1933.

of increasing strength, following mechanically on a rearrangement of molecules, as the original isotropic condition is transmuted more and more into molecular asymmetry or anisotropy; and the rope apparently "adapts itself" to the increased strain which it is called on to bear, all after a fashion which at least suggests a parallel to the increasing strength of the stretched and weighted fibre in the plant. For increase of strength by rearrangement of the particles we have already a rough illustration in our lock of wool or hank of tow. The tow will carry but little weight while its fibres are tangled and awry: but as soon as we have carded or "hatchelled" it out, and brought all its long fibres parallel and side by side, we make of it a strong and useful cord*.

But the lessons which we learn from dough and treacle are nowadays plain enough in steel and iron, and become immensely more important in these. For here again plasticity is associated with a certain capacity for structural rearrangement, and increased strength again results therefrom. Elaborate processess of rolling, drawing, bending, hammering, and so on, are regularly employed to toughen and strengthen the material. The "mechanical structure" of solids has become an important subject. And when the engineer talks of repeated loading, of elastic fatigue, of hysteresis, and other phenomena associated with plasticity and strain, the physiological analogues of these physical phenomena are perhaps not far away.

In some such ways as these, then, it would seem that we may coordinate, or hope to coordinate, the phenomenon of growth with certain of the beautiful structural phenomena which present themselves to our eyes as "provisions," or mechanical adaptations†, for the display of strength where strength is most required. That is to say the origin, or causation, of the phenomenon would seem to lie partly in the tendency of growth to be accelerated under strain: and partly in the automatic effect of shearing strain, by which it tends to displace parts which grow obliquely to the direct lines of tension and of pressure, while leaving those in place which happen to lie parallel or perpendicular to those lines: an automatic effect

---

* Cf. Sir Charles Bell's *Animal Mechanics*, chap. v, "Of the tendons compared with cordage."

† So P. Enriques (*op. cit. supra*, p. 5), writing on the economy of material in the construction of a bone, admits that "una certa impronta di teleologismo qua e là è rimasta, mio malgrado, in questo scritto."

which we can probably trace as working on all scales of magnitude, and as accounting therefore for the rearrangement of minute particles in the metal or the fibre, as well as for the bringing into line of the fibres within the plant, or of the trabeculae within the bone.

But we may now attempt to pass from the study of the individual bone to the much wider and not less beautiful problems of mechanical construction which are presented to us by the skeleton as a whole. Certain problems of this class are by no means neglected by writers on anatomy, and many have been handed down from Borelli, and even from older writers. For instance, it is an old tradition of anatomical teaching to point out in the human body examples of the three orders of levers*; again, the principle that the limb-bones tend to be shortened in order to support the weight of a very heavy animal is well understood by comparative anatomists, in accordance with Euler's law, that the weight which a column liable to flexure is capable of supporting varies inversely as the square of its length; and again, the statical equilibrium of the body, in relation for instance to the erect posture of man, has long been a favourite theme of the philosophical anatomist. But the general method, based upon that of graphic statics, to which we have been introduced in our study of a bone, has not, so far as I know, been applied to the general fabric of the skeleton. Yet it is plain that each bone plays a part in relation to the whole body, analogous to that which a little trabecula, or a little group of trabeculae, plays within the bone itself: that is to say, in the normal distribution of forces in the body the bones tend to follow the lines of stress, and especially the pressure-lines. To demonstrate this in a comprehensive way would doubtless be difficult; for we should be dealing with a framework of very great complexity, and should have to take account of

---

* E.g. (1) the head, nodding backwards and forwards on a fulcrum, represented by the atlas vertebra, lying between the weight and the power; (2) the foot, raising on tip-toe the weight of the body against the fulcrum of the ground, where the weight is between the fulcrum and the power, the latter being represented by the *tendo Achillis*; (3) the arm, lifting a weight in the hand, with the power (i.e. the biceps muscle) between the fulcrum and the weight. (The second case, by the way, has been much disputed; cf. Haycraft in Schäfer's *Textbook of Physiology*, 1900, p. 251.) Cf. (*int. al.*) G. H. Meyer, *Statik u. Mechanik der menschlichen Knochengerüste*, 1873, pp. 13–25.

a great variety of conditions*. This framework is complicated as we see it in the skeleton, where (as we have said) it is only, or chiefly, the *struts* of the whole fabric which are represented; but to understand the mechanical structure in detail, we should have to follow out the still more complex arrangement of the *ties*, as represented by the muscles and ligaments, and we should also require much detailed information as to the weights of the various parts and as to the other forces concerned. Without these latter data we can only treat the question in a preliminary and imperfect way. But, to take once again a small and simplified part of a big problem, let us think of a quadruped (for instance, a horse) in a standing posture, and see whether the methods and terminology of the engineer may not help us, as they did in regard to the minute structure of the single bone. And let us note in passing that the "standing posture," whether on two legs or on four, is no very common thing; but is (so to speak), with all its correlated anatomy, a privilege of the few.

Standing four-square upon its fore-legs and hind-legs, with the weight of the body suspended between, the quadruped at once suggests to us the analogy of a bridge, carried by its two piers. And if it occurs to us, as naturalists, that we never look at a standing quadruped without contemplating a bridge, so, conversely, a similar idea has occurred to the engineer; for Professor Fidler, in this *Treatise on Bridge-Construction*, deals with the chief descriptive part of his subject under the heading of "The Comparative Anatomy of Bridges†." The designation is most just, for in studying the various types of bridges we are studying a series of well-planned *skeletons‡*; and (at the cost of a little pedantry)

---

* Our problem is analogous to Thomas Young's problem of the best disposition of the timbers in a wooden ship (*Phil. Trans.* 1814, p. 303). He was not long of finding that the forces which act upon the fabric are very numerous and very variable, and that the best mode of resisting them, or best structural arrangement for ultimate strength, becomes an immensely complicated problem.

† By a bolder metaphor Fontenelle said of Newton that he had "fait l'anatomie de la lumière."

‡ In like manner, Clerk Maxwell could not help employing the term "skeleton" in defining the mathematical conception of a "frame," constituted by points and their interconnecting lines: in studying the equilibrium of which, we consider its different points as mutually acting on each other with forces whose directions are those of the lines joining each pair of points. Hence (says Maxwell), "in order to exhibit the mechanical action of the frame in the most elementary manner, we may

we might go even further, and study (after the fashion of the
anatomist) the "osteology" and "desmology" of the structure, that
is to say the bones which are represented by "struts," and the
ligaments, etc., which are represented by "ties." Furthermore
after the methods of the comparative anatomist, we may classify
the families, genera and species of bridges according to their distinctive mechanical features, which correspond to certain definite
conditions and functions.

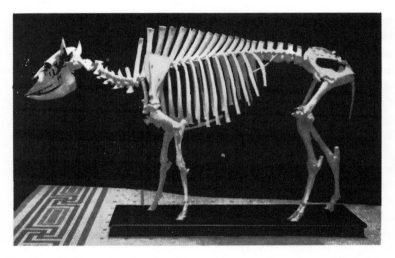

Fig. 468. Skeleton of an American bison. (An unusually well-mounted skeleton,
of American workmanship, now in the Anatomical Museum of Edinburgh
University.)

In more ways than one, the quadrupedal bridge is a remarkable
one; and perhaps its most remarkable peculiarity is that it is a
jointed and flexible bridge, remaining in equilibrium under considerable and sometimes great modifications of its curvature, such
as we see, for instance, when a cat humps or flattens her back.
The fact that *flexibility* is an essential feature in the quadrupedal

draw it as a *skeleton*, in which the different points are joined by straight lines,
and we may indicate by numbers attached to these lines the tensions or compressions in the corresponding pieces of the frame" (*Trans. R.S.E.* XXVI, p. 1,
1870). It follows that the diagram so constructed represents a "diagram of
forces," in this limited sense that it is geometrical as regards the position and
direction of the forces, but arithmetical as regards their magnitude. It is to just
such a diagram that the animal's skeleton tends to approximate.

bridge, while it is the last thing which an engineer desires and the first which he seeks to provide against, will impose certain important limiting conditions upon the design of the skeletal fabric. But let us begin by considering the quadruped at rest, when he stands upright and motionless upon his feet, and when his legs exercise no function save only to carry the weight of the whole body. So far as that function is concerned, we might now perhaps compare the horse's legs with the tall and slender piers of some railway bridge; but it is obvious that these jointed legs are ill-adapted to receive the *horizontal thrust* of any *arch* that may be placed atop of them. Hence it follows that the curved backbone of the horse, which appears to cross like an arch the span between his shoulders and his flanks, cannot be regarded as an *arch*, in the

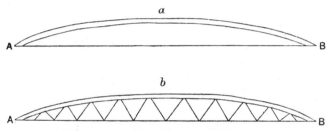

Fig. 469.   *a*, tied arch; *b*, bowstring girder.

engineer's sense of the word. It resembles an arch in *form*, but not in *function*, for it cannot act as an arch unless it be held back at each end (as every arch is held back) by *abutments* capable of resisting the horizontal thrust; and these necessary abutments are not present in the structure. But in various ways the engineer can modify his superstructure so as to supply the place of these *external* reactions, which in the simple arch are obviously indispensable. Thus, for example, we may begin by inserting a straight steel tie, *AB* (Fig. 469), uniting the ends of the curved rib *AaB*; and this tie will supply the place of the external reactions, converting the structure into a "tied arch," such as we may see in the roofs of many railway stations. Or we may go on to fill in the space between arch and tie by a "web-system," converting it into what the engineer describes as a "parabolic bowstring girder" (Fig. 469 *b*). In either case, the structure becomes an independent "detached girder," supported at each end but not otherwise fixed, and consisting essentially of an upper compression-member, *AaB*, and a lower tension-member, *AB*. But again, in the skeleton of the quadruped, *the necessary tie, AB, of the simple bow-girder is not to be found*; and it follows that these comparatively simple types of bridge do not correspond to, nor do they help us to understand, the type of bridge which Nature has designed in the skeleton of the quadruped. Nevertheless if we try to look, as an engineer would look, at the actual design of the animal skeleton and the actual distribution of its load, we find that the one is most admirably adapted to the other, according to the strict principles of engineering construction. The structure is not an arch, nor a tied arch, nor a bowstring girder: but it is strictly and beautifully comparable to the main girder of a double-armed cantilever bridge.

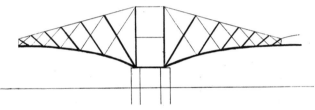

Fig. 470. A two-armed cantilever of the Forth Bridge. Thick lines, compression-members (bones); thin lines, tension-members (ligaments).

Obviously, in our quadrupedal bridge, the superstructure does not terminate (as it did in our former diagram) at the two points of support, but it extends beyond them, carrying the head at one end and sometimes a heavy tail at the other, upon projecting arms or "cantilevers."

In a typical cantilever bridge, such as the Forth Bridge (Fig. 470), a certain simplification is introduced. For each pier carries, in this case, its own double-armed cantilever, linked by a short connecting girder to the next, but so jointed to it that no weight is transmitted from one cantilever to another. The bridge in short is *cut* into separate sections, practically independent of one another; at the joints a certain amount of bending is not precluded, but shearing strain is evaded; and each pier carries only its own load. By this arrangement the engineer finds that design and construction are alike simplified and facilitated. In the horse or the ox, it is obvious that the two piers of the bridge, that is to say the fore-legs and the hind-legs, do not bear (as they do in the Forth Bridge) separate and independent loads, but the whole system forms a continuous structure. In this case, the calculation of the loads will be a little more difficult and the corresponding design of the structure a little more complicated. We shall accordingly simplify our problem very considerably if, to begin with, we look upon the quadrupedal skeleton as constituted of two separate systems, that is to say of two balanced cantilevers, one supported on the fore-legs and the other on the hind; and we may deal afterwards with the fact that these two cantilevers are not independent, but are bound up in one common field of force and plan of construction.

In both horse and ox it is plain that the two cantilever systems into which we may thus analyse the quadrupedal bridge are unequal in magnitude and importance. The fore-part of the animal is much bulkier than its hind-quarters, and the fact that the fore-legs carry, as they so evidently do, a greater weight than the hind-legs has long been known and is easily proved; we have only to walk a horse on to a weigh-bridge, weigh first his fore-legs and then his hind-legs, to discover that what we may call his front half weighs a good deal more than what is carried on his hind feet, say about three-fifths of the whole weight of the animal.

The great (or anterior) cantilever then, in the horse, is constituted by the heavy head and still heavier neck on one side of that pier which is represented by the fore-legs, and by the dorsal vertebrae carrying a large part of the weight of the trunk upon the other side; and this weight is so balanced over the fore-legs that the cantilever, while "anchored" to the other parts of the structure, transmits but little of its weight to the hind-legs, and the amount so transmitted will vary with the attitude of the head and with the position of any artificial load*. Under certain conditions, as when the head is thrust well forward, it is evident that the hind-legs will be actually relieved of a portion of the comparatively small load which is their normal share.

* When the jockey crouches over the neck of his race-horse, and when Tod Sloan introduced the "American seat," the avowed object in both cases is to relieve the hind-legs of weight, and so leave them free for the work of propulsion. On the share taken by the hind-limbs in this latter duty, and other matters, cf. Stillman, *The Horse in Motion*, 1882, p. 69.

But here we pass from the statical problem to the dynamical, from the horse at rest to the horse in motion, from the observed fact that weight lies mainly over the fore-legs to the question of what advantage is gained by such a distribution of the load. Taking the hind-legs as the main propulsive agency, as we may now safely do, the moment of propulsion is about the hind-hooves; then (as we see in Fig. 471) we may take the weight, $W = A \sin \alpha$, and the propulsive force, $f = A \cos \alpha$, and $\dfrac{W}{f} = \dfrac{H}{L}$, $WL = fH$ being the balanced condition. From the statical point of view the load must balance over the fore-legs; from the dynamical point of view it might well lie even farther forward. And when the jockey crouches

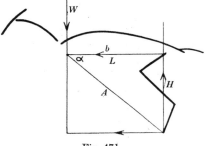

Fig. 471.

over the horse's neck, and when Tod Sloan introduced the "American seat," both shew a remarkable, though perhaps unconscious, insight into the dynamical proposition.

Our next problem is to discover, in a rough and approximate way, some of the structural details which the balanced load upon the double cantilever will impress upon the fabric.

Working by the methods of graphic statics, the engineer's task is, in theory, one of great simplicity. He begins by drawing in outline the structure which he desires to erect; he calculates the stresses and bending-moments necessitated by the dimensions and load on the structure; he draws a new diagram *representing these forces*, and he designs and builds his fabric on the lines of this statical diagram. He does, in short, precisely what we have seen *nature* doing in the case of the bone. For if we had begun, as it were,

by blocking out the femur roughly, and considering its position
and dimensions, its means of support and the load which it has to
bear, we could have proceeded at once to draw the system of
stress-lines which must occupy that field of force: and to precisely

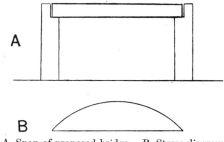

Fig. 472.   A, Span of proposed bridge.   B, Stress diagram, or diagram
of bending-moments*.

those stress-lines has Nature kept in the building of the bone, down
to the minute arrangement of its trabeculae.

The essential function of a bridge is to stretch across a certain
span, and carry a certain definite load; and this being so, the chief

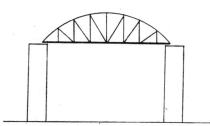

Fig. 473.   The bridge constructed, as a parabolic girder.

problem in the designing of a bridge is to provide due resistance
to the "bending-moments" which result from the load.   These
bending-moments will vary from point to point along the girder,

* This and the following diagrams are borrowed and adapted from Professor
Fidler's *Bridge Construction*.   We may reflect with advantage on Clerk Maxwell's
saying that "the use of diagrams is a particular instance of that method of symbols
which is so powerful an aid in the advancement of science"; and on his explanation
that "a diagram differs from a picture in this respect that in a diagram no attempt
is made to represent those factors of the actual material system which are not the
special objects of our study."

and taking the simplest case of a uniform load, whether supported at one or both ends, they will be represented by points on a parabola. If the girder be of uniform depth and section, that is to say if its two flanges, respectively under tension and compression, be equal and parallel to one another, then the stress upon these flanges will vary as the bending-moments, and will accordingly be very severe in the middle and will dwindle towards the ends. But if we make the *depth* of the girder everywhere proportional to the bending-moments, that is to say if we copy in the girder the outlines of the bending-moment diagram, then our design will automatically meet the circumstances of the case, for the horizontal stress in each flange will now be uniform throughout the length of the girder. In short,

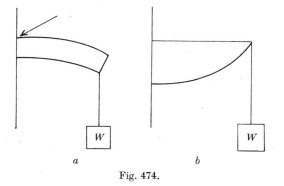

a   b

Fig. 474.

in Professor Fidler's words, "Every diagram of moments represents the outline of a framed structure which will carry the given load with a uniform horizontal stress in the principal members."

In the above diagrams (Fig. 474, a, b) (which are taken from the original ones of Culmann), we see at once that the loaded beam or bracket (a) has a "danger-point" close to its fixed base, that is to say at the point remotest from its load. But in the parabolic bracket (b) there is no danger-point at all, for the dimensions of the structure are made to increase *pari passu* with the bending-moments: stress and resistance vary together. Again in Fig. 475, we have a simple span (A), with its stress diagram (B); and in (C) we have the corresponding parabolic girder, whose stresses are now uniform throughout. In fact we see that, by a process of conversion, the stress diagram in each case becomes the structural diagram in the other *. Now all this is but the modern rendering of one of Galileo's most famous propositions. In the Dialogue which we have already quoted more than once †, Sagredo says "It would be a fine thing if one could discover the proper shape to give a solid in order to make it equally resistant at every point, in which case a load placed at the middle would not produce fracture more easily

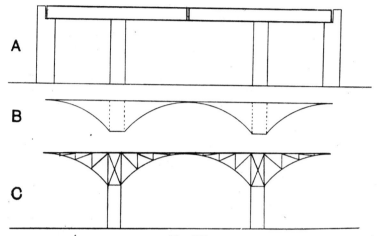

Fig. 475.

than if placed at any other point ‡." And Galileo (in the person of Salviati) first puts the problem into its more general form; and then shews us how, by giving a parabolic outline to our beam, we have its simple and comprehensive solution. It was such teaching as this that led R. A. Millikan to say that "we owe our present-day civilisation to Galileo."

---

* The method of constructing *reciprocal diagrams*, of which one should represent the outlines of a frame and the other the system of forces necessary to keep it in equilibrium, was first indicated in Culmann's *Graphische Statik*; it was greatly developed soon afterwards by Macquorn Rankine (*Phil. Mag.* Feb. 1864, and *Applied Mechanics, passim*), to whom the application of the principle to engineering practice is mainly due. See also Fleeming Jenkin, On the practical application of reciprocal figures to the calculation of strains in framework, *Trans. R.S.E.* xxv, pp. 441–448, 1869; and Clerk Maxwell, *ibid.* xxvi, p. 9, 1870, and *Phil. Mag.* April 1864.

† *Dialogues concerning Two New Sciences* (1638); Crew and Salvio's translation p. 140 *seq.*

‡ As in the great case of the Eiffel Tower, *supra*, p. 29.

In the case of our cantilever bridge, we shew the primitive girder in Fig. 475, A, with its bending-moment diagram (B); and it is evident that, if we turn this diagram upside down, it will still be illustrative, just as before, of the bending-moments from point to point: for as yet it is merely a diagram, or graph, of relative magnitudes.

To either of these two stress diagrams, direct or inverted, we may fit the design of the construction, as in Figs. 475, C and 476.

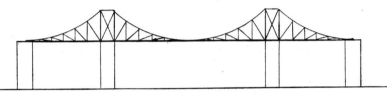

Fig. 476.

Now in different animals the amount and distribution of the load differ so greatly that we can expect no single diagram, drawn from the comparative anatomy of bridges, to apply equally well to all the cases met with in the comparative anatomy of quadrupeds; but nevertheless we have already gained an insight into the general principles of "structural design" in the quadrupedal bridge.

In our last diagram the upper member of the cantilever is under tension; it is represented in the quadruped by the *ligamentum nuchae* on the one side of the cantilever, and by the supraspinous ligaments of the dorsal vertebrae on the other. The compression-member is similarly represented, on both sides of the cantilever, by the vertebral column, or rather by the *bodies* of the vertebrae; while the web, or "filling," of the girders, that is to say the upright or sloping members which extend from one flange to the other, is represented on the one hand by the spines of the vertebrae, and on the other hand by the oblique interspinous ligaments and muscles— that is to say, by compression-members and tension-members inclined in opposite directions to one another. The high spines over the quadruped's withers are no other than the high struts which rise over the supporting piers in the parabolic girder, and correspond to the position of the maximal bending-moments. The fact that these tall vertebrae of the withers usually slope backwards, some-

times steeply, in a quadruped, is easily and obviously explained *.
For each vertebra tends to act as a "hinged lever," and its spine,
acted on by the tensions transmitted by the ligaments on either side,
takes up its position as the diagonal of the parallelogram of forces
to which it is exposed.

It happens that in these comparatively simple types of cantilever
bridge the whole of the parabolic curvature is transferred to one
or other of the principal members, either the tension-member or
the compression-member as the case may be. But it is of course
equally permissible to have both members curved, in opposite
directions. This, though not exactly the case in the Forth Bridge,
is approximately so; for here the main compression-member is
curved or arched, and the main tension-member slopes downwards
on either side from its maximal height above the piers. In short,
the Forth Bridge (Fig. 470) is a nearer approach than either of
the other bridges which we have illustrated to the plan of the
quadrupedal skeleton; for the main compression-member almost
exactly recalls the form of the backbone, while the main tension-
member, though not so closely similar to the supraspinous and
nuchal ligaments, corresponds to the plan of these in a somewhat
simplified form.

We may now pass without difficulty from the two-armed cantilever supported on a single pier, as it is in each separate section of the
Forth Bridge, or as we have imagined it to be in the fore-quarters
of a horse, to the condition which actually exists in a quadruped,
when a two-armed cantilever has its load distributed over two
separate piers. This is not precisely what an engineer calls a
"continuous" girder, for that term is applied to a girder which,
as a continuous structure, has three supports and crosses two or more
spans, while here there is only one. But nevertheless, this girder

* The form and direction of the vertebral spines have been frequently and
elaborately described; cf. (e.g.) H. Gottlieb, Die Anticlinie der Wirbelsäule der
Säugethiere, Morphol. Jahrb. XLIX, pp. 179–220, 1915, and many works quoted
therein. According to Morita, Ueber die Ursachen der Richtung und Gestalt der
thoracalen Dornfortsätze der Säugethierwirbelsäule (ibi cit. p. 201), various changes
take place in the direction or inclination of these processes in rabbits, after section
of the interspinous ligaments and muscles. These changes seem to be very much
what we should expect, on simple mechanical grounds. See also O. Fischer,
Theoretische Grundlagen für eine Mechanik der lebenden Körper, Leipzig, 1906,
pp. x, 372.

is *effectively* continuous from the head to the tip of the tail; and at each point of support (*A* and *B*) it is subjected to the negative bending-moment due to the overhanging load on each of the projecting cantilever arms *AH* and *BT*. The diagram of bending-moments will (according to the ordinary conventions) lie below the base line (because the moments are negative), and must take some such form as that shewn in the diagram: for the girder must suffer its greatest bending stress not at the centre, but at the two points of support *A* and *B*, where the moments are measured by the vertical ordinates. It is plain that this figure only differs from a representation of *two* independent two-armed cantilevers in the fact that there is no point midway in the span where the bending-moment vanishes, but only a region between the two piers in which it tends to diminish.

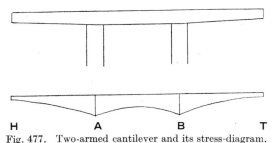

H   A   B   T

Fig. 477.　Two-armed cantilever and its stress-diagram.

The diagram effects a graphic summation of the positive and negative moments, but its form may assume various modifications according to the method of graphic summation which we choose to adopt; and it is obvious also that the form of the diagram may assume many modifications of detail according to the actual distribution of the load. In all cases the essential points to be observed are these: firstly that the girder which is to resist the bending-moments induced by the load must possess its two principal members—an upper tension-member or tie, represented by ligament (whose tension doubtless varies along its length), and a lower compression-member represented by bone: these members being united by a web represented by the vertebral spines with their interspinous ligaments, and being placed one above the other in the order named because the moments are negative; secondly we observe that the depth of the web, or distance apart of the principal members—that is to say the height of the vertebral spines—must be proportional to the bending-moment at each point along the length of the girder.

In the case of an animal carrying most of his weight upon his fore-legs, as the horse or the ox do, the bending-moment diagram will be unsymmetrical, after the fashion of Fig. 478, the precise form depending on the distribution of weights and distances.

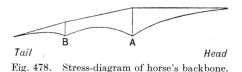

Tail   Head

Fig. 478.   Stress-diagram of horse's backbone.

On the other hand the Dinosaur, with his light head and enormous tail would give us a moment-diagram with the opposite kind of asymmetry, the greatest bending stress being now found over the haunches, at B (Fig. 479). A glance at the skeleton of Diplodocus will shew us the high vertebral spines over the loins, in precise correspondence with the requirements of this diagram: just as in the horse, under the opposite conditions of load, the highest vertebral spines are those of the withers, that is to say those of the posterior cervical and anterior dorsal vertebrae.

We have now not only dealt with the general resemblance, both in structure and in function, of the quadrupedal backbone with its associated ligaments to a double-armed cantilever girder, but we have begun to see how the characters of the vertebral system must differ in different quadrupeds, according to the conditions imposed by the varying distribution of the load: and in particular how the height of the vertebral spines which constitute the web will be in a definite relation, as regards magnitude and position, to the bending-moments induced thereby. We should require much detailed information as to the actual weights of the several parts of the body before we could follow out quantitatively the mechanical efficiency of each type of skeleton; but in an approximate way what we have already learnt will enable us to trace many interesting correspondences between structure and function in this particular part of comparative anatomy. We must, however, be careful to note that the great cantilever system is not of necessity constituted by the vertebral column and its ligaments alone, but that the pelvis, firmly united as it is to the sacral vertebrae, and stretching backwards far beyond the acetabulum, becomes an intrinsic part of the system; and helping (as it does) to carry the load of the abdominal viscera, it constitutes a great portion of the posterior cantilever arm, or even its chief portion in cases where the size and weight of the tail are insignificant, as is the case in the majority of terrestrial mammals.

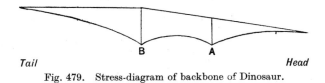

Tail   *Head*
Fig. 479.   Stress-diagram of backbone of Dinosaur.

We may also note here that just as a bridge is often a "combined" or composite structure, exhibiting a combination of principles in its construction, so in the quadruped we have, as it were, another girder supported by the same piers to carry the viscera; and consisting of an inverted parabolic girder, whose compression-member is again constituted by the backbone, its tension-member by the line of the sternum and the abdominal muscles, while the ribs and intercostal muscles play the part of the web or filling.

A very few instances must suffice to illustrate the chief variations in the load, and therefore in the bending-moment diagram, and therefore also in the plan of construction, of various quadrupeds. But let us begin by setting forth, in a few cases, the actual weights which are borne by the fore-limbs and the hind-limbs, in our quadrupedal bridge*.

	Gross weight ton cwt.	On fore-feet cwt.	On hind-feet cwt.	% on fore-feet	% on hind-feet
Camel (Bactrian)	— 14·25	9·25	4·5	67·3	32·7
Llama	— 2·75	1·75	0·875	66·7	33·3
Elephant (Indian)	1 15·75	20·5	14·75	58·2	41·8
Horse	— 8·25	4·75	3·5	57·6	42·4
Horse (large Clydesdale)	— 15·5	8·5	7·0	54·8	45·2

* I owe the first four of these determinations to the kindness of Sir P. Chalmers Mitchell, who had them made for me at the Zoological Society's Gardens; while the great Clydesdale carthorse was weighed for me by a friend in Dundee.

It will be observed that in all these animals the load upon the fore-feet preponderates considerably over that upon the hind, the preponderance being rather greater in the elephant than in the horse, and markedly greater in the camel and the llama than in the other two. But while these weights are helpful and suggestive, it is obvious that they do not go nearly far enough to give us a full insight into the constructional diagram to which the animals are conformed. For such a purpose we should require to weigh the total load, not in two portions but in many; and we should also have to take close account of the general form of the animal, of the relation between that form and the distribution of the load, and of the actual directions of each bone and ligament by which the forces of compression and tension were transmitted. All this lies beyond us for the present; but nevertheless we may consider,

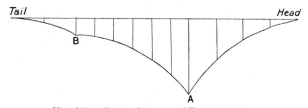

Fig. 480.   Stress-diagram of Titanotherium.

very briefly, the principal cases involved in our enquiry, of which the above animals form a partial and preliminary illustration.

(1) Wherever we have a heavily loaded anterior cantilever arm, that is to say whenever the head and neck represent a considerable fraction of the whole weight of the body, we tend to have large bending-moments over the fore-legs, and correspondingly high spines over the vertebrae of the, withers. This is the case in the great majority of four-footed terrestrial animals, the chief exceptions being found in animals with comparatively small heads but large and heavy tails, such as the anteaters or the Dinosaurian reptiles, and also (very naturally) in animals such as the crocodile, where the "bridge" can scarcely be said to be developed, for the long heavy body sags down to rest upon the ground. The case is sufficiently exemplified by the horse, and still more notably by the stag, the ox, or the pig. It is illustrated in the skeleton of a bison (Fig. 468), or

in the accompanying diagram of the conditions in the great extinct Titanotherium.

(2) In the elephant and the camel we have similar conditions, but slightly modified. In both cases, and especially in the latter, the weight on the fore-quarters is relatively large; and in both cases the bending-moments are all the larger, by reason of the length and forward extension of the camel's neck and the forward position of the heavy tusks of the elephant. In both cases the dorsal spines are large, but they do not strike us as exceptionally so; but in both cases, and especially in the elephant, they slope backwards in a marked degree. Each spine, as already explained, must in all cases assume the position of the diagonal in the parallelogram of forces defined by the tensions acting on it at its extremity; for it constitutes a "hinged lever," by which the bending-moments on either side are automatically balanced; and it is plain that the more the spine slopes backwards the more it indicates a relatively large strain thrown upon the great ligament of the neck, and a relief of strain upon the more directly acting, but weaker, ligaments of the back and loins. In both cases, the bending-moments would seem to be more evenly distributed over the region of the back than, for instance, in the stag, with its light hind-quarters and heavy load of antlers: and in both cases the high "girder" is considerably prolonged, by an extension of the tall spines backwards in the direction of the loins. When we come to such a case as the mammoth, with its immensely heavy and immensely elongated tusks, we perceive at once that the bending-moments over the fore-legs are now very severe; and we see also that the dorsal spines in this region are much more conspicuously elevated than in the ordinary elephant.

(3) In the case of the giraffe we have, without doubt, a very heavy load upon the fore-legs, though no weighings are at hand to define the ratio; but as far as possible this disproportionate load would seem to be relieved by help of a downward as well as backward thrust, through the sloping back to the unusually low hind-quarters. The dorsal spines of the vertebrae are very high and strong, and the whole girder-system very perfectly formed. The elevated rather than protruding position of the head lessens the anterior bending-moment as far as possible, but it leads to a strong

compressional stress transmitted almost directly downwards through the neck: in correlation with which we observe that the bodies of the cervical vertebrae are exceptionally large and strong, and steadily increase in size and strength from the head downwards.

(4) In the kangaroo, the fore-limbs are entirely relieved of their load, and accordingly the tall spines over the withers, which were so conspicuous in all heavy-headed *quadrupeds*, have now completely vanished. The creature has become bipedal, and body and tail form the extremities of *a single* balanced cantilever, whose maximal bending-moments are marked by strong, high lumbar and sacral vertebrae, and by iliac bones of peculiar form, of exceptional strength and nearly upright position.

Precisely the same condition is illustrated in the Iguanodon, and better still by reason of the great bulk of the creature and of the heavy load which falls to be supported by the great cantilever and by the hind-legs which form its piers. The long heavy body and neck require a balance-weight (as in the kangaroo) in the form of a long heavy tail; and the double-armed cantilever, so constituted, shews a beautiful parabolic curvature in the graded heights of the whole series of vertebral spines, which rise to a maximum over the haunches and die away slowly towards the neck and towards the tip of the tail.

(5) In the case of some of the great American fossil reptiles such as Diplodocus, it has always been a more or less disputed question whether or not they assumed, like Iguanodon, an erect, bipedal attitude. In all of them we see an elongated pelvis, and, in still more marked degree, we see elevated spinous processes of the vertebrae over the hind-limbs; in all of them we have a long heavy tail, and in most of them we have a marked reduction in size and weight both of the fore-limb and of the head itself. The great size of these animals is not of itself a proof against the erect attitude; because it might well have been accompanied by an aquatic or partially submerged habitat, and the crushing stress of the creature's huge bulk proportionately relieved. But we must consider each such case in the whole light of its own evidence; and it is easy to see that, just as the quadrupedal mammal may carry the greater part but not all of its weight upon its fore-limbs, so a heavy-tailed reptile may carry the greater part upon its hind-limbs, without

this process going so far as to relieve its fore-limbs of all weight whatsoever. This would seem to be the case in such a form as Diplodocus, and also in Stegosaurus, whose restoration by Marsh is doubtless substantially correct*. The fore-limbs, though comparatively small, are obviously fashioned for support, but the weight which they have to carry is far less than that which the hind-limbs bear. The head is small and the neck short, while on the other hand the hind-quarters and the tail are big and massive. The backbone bends into a great double-armed cantilever, culminating over the pelvis and the hind-limbs, and here furnished with its highest and strongest spines to separate the tension-member from the com-

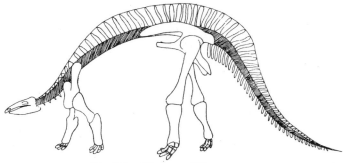

Fig. 481.   Diagram of Stegosaurus.

pression-member of the girder. The fore-legs form a secondary supporting pier to this great continuous cantilever, the greater part of whose weight is poised upon the hind-limbs alone.

(6) In the slender body of a weasel, neither head nor tail is such as to form an efficient cantilever; and though the lithe body is arched in active exercise, our parallel of the bridge no longer works well. What else to compare it with is far from clear; but the mechanism has some resemblance (perhaps) to an elastic spring. Animals of this habit of body are all small; their bodily weight is a light burden, and gravity becomes an ineffectual force.

* This pose of Diplodocus, and of other Sauropodous reptiles, has been much discussed. Cf. (int. al.) O. Abel, Abh. k. k. zool. bot. Ges. Wien, v, 1909–10 (60 pp.); Tornier, SB. Ges. Naturf. Fr. Berlin, 1909, pp. 193–209; O. P. Hay, Amer. Nat. Oct. 1908; Tr. Wash. Acad. Sci. XLII, pp. 1–25, 1910; Holland, Amer. Nat. May 1910, pp. 259–283; Matthew, ibid. pp. 547–560; C. W. Gilmore (Restoration of Stegosaurus), Pr. U.S. Nat. Museum, 1915.

(7) An abnormal and very curious case is that of the sloth, which hangs by hooked hands and feet, head downwards, from high branches in the Brazilian forest. The vertebrae are unusually numerous, they are all much alike one to another, and (as we might well suppose) the whole pensile chain of vertebrae hangs in what closely approximates to a catenary curve*.

(8) We find a highly important corollary in the case of aquatic animals. For here the effect of gravity is neutralised; we have neither piers nor cantilevers; and we find accordingly in all aquatic mammals of whatsoever group—whales, seals or sea-cows—that the high arched vertebral spines over the withers, or corresponding structures over the hind-limbs, have both entirely disappeared.

But in the whale or dolphin (and not less so in the aquatic bird), *stiffness* must be ensured in order to enable the muscles to act against the resistance of the water in the act of swimming; and accordingly Nature must provide against bending-moments irrespective of gravity. In the dolphin, at any rate as regards its tail-end, the conditions will be not very different from those of a column or beam with fixed ends, in which, under deflection, there will be two points of contrary flexure, as at C, D, in Fig. 482.

Here, between C and D we have a varying bending-moment, represented by a continuous curve with its maximal elevation midway between the points of inflection. And correspondingly, in our dolphin, we have a continuous series of high dorsal spines, rising to a maximum about the middle of the animal's

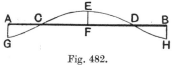

Fig. 482.

body, and falling to nil at some distance from the end of the tail. It is their business (as usual) to keep the tension-member, represented by the strong supraspinous ligaments, wide apart from the compression-member, which is as usual represented by the backbone itself. But in our diagram we see that on the farther side of C and D we have a *negative* curve of bending-moments, or bending-moments in a contrary direction. Without enquiring how these stresses are precisely met

* A *heavy* cord, or a cord carrying equal weights for equal distances along its line, hangs in a catenary: imagine it frozen and inverted, and we have an arch, carrying the same sort of load, and under compression only. On the other hand, a flexible cable (itself of negligible weight), carrying a uniform load along the line of its horizontal projection, hangs in the form of a parabola.

towards the dolphin's head (where the coalesced cervical vertebrae suggest themselves as a partial explanation), we see at once that towards the tail they are met by the strong series of chevron-bones, which in the caudal region, where tall *dorsal* spines are no longer needed, take their place *below* the vertebrae, in precise correspondence with the bending-moment diagram. In many cases other than these aquatic ones, when we have to deal with animals with long and heavy tails (like the Iguanodon and the kangaroo of which we have already spoken), we are apt to meet with similar, though usually shorter chevron-bones; and in all these cases we may see without difficulty that a negative bending-moment in the vertical direction has to be resisted or controlled.

In the dolphin we may find an illustration of the fact that not only is it necessary to provide for rigidity in the vertical direction but often also in the horizontal, where a tendency to bending must be resisted on either side. This function is effected in part by the ribs with their associated muscles, but they extend but a little way and their efficacy for this purpose can be but small. We have, however, behind the region of the ribs and on either side of the backbone a strong series of elongated and flattened transverse processes, forming a web for the support of a tension-member in the usual form of ligament, and so playing a part precisely analogous to that performed by the dorsal spines in the same animal. In an ordinary fish, such as a cod or a haddock, we see precisely the same thing:

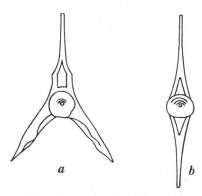

Fig. 483. *a*, dorsal and *b*, caudal vertebrae of haddock.

the backbone is stiffened by the indispensable help of its *three series* of ligament-connected processes, the dorsal and the two transverse series; but there are no such stiffeners in the eel. When we come to the region of the tail, where rigidity gives place to lateral flexibility, the three stiffeners give place to two—the dorsal and haemal spines of the caudal vertebrae. And here we see that the three series of processes, or struts, tend (when all three are present) to be arranged well-nigh at equal angles, of 120°, with one another, giving the greatest and most uniform strength of which such a system is capable. On the other hand, in a flat fish, such as a plaice, where from the natural mode of progression it is necessary that the backbone should be flexible in one direction while stiffened in another, we find the whole outline of the fish comparable to that of a double bowstring girder, the compression-member being (as usual) the backbone itself, the tension-member on either side being constituted by the interspinous ligaments and muscles, while the web or filling is very beautifully represented by the long and evenly graded neural and haemal spines, which spring symmetrically up and down from each individual vertebra.

In the skeleton of the flat fishes, the web of the otherwise perfect parabolic girder has to be cut away and encroached on to make room for the viscera. When the body is long and the vertebrae many, as in the sole, the space required is small compared with the length of the girder, and the strength of the latter is not much impaired. In the shorter, rounder kinds with fewer vertebrae, like the turbot, the visceral cavity is large compared with the length of the fish, and its presence would seem to weaken the girder very seriously. But Nature repairs the breach by framing in the hinder part of the space with a strong curved bracket or angle-iron, which takes the place very efficiently of the bony struts which have been cut away.

The main result at which we have now arrived, in regard to the construction of the vertebral column and its associated parts, is that we may look upon it as a certain type of *girder*, whose depth is everywhere very nearly proportional to the height of the corresponding ordinate in the diagram of moments: just as it is in a girder designed by a modern engineer. In short, after the nineteenth or twentieth century engineer has done his best in framing the design of a big cantilever, he may find that some of his best ideas had, so to speak, been anticipated ages ago in the fabric of the great saurians and the larger mammals.

But it is possible that the modern engineer might be disposed to criticise the skeleton girder at two or three points; and in particular he might think the girder, as we see it for instance in Diplodocus or Stegosaurus, not deep enough for carrying the animal's enormous weight of some twenty tons. If we adopt a much greater depth (or ratio of depth to length) as in the modern cantilever, we shall greatly increase the *strength* of the structure; but at the same time we should greatly increase its *rigidity*, and this is precisely what, in the circumstances of the case, it would seem that Nature is bound to avoid. We need not suppose that the great saurian was by any means active and limber; but a certain amount of activity and flexibility he was bound to have, and in a thousand ways he would find the need of a backbone that should be *flexible* as well as *strong*. Now this opens up a new aspect of the matter and is the beginning of a long, long story, for in every direction this double requirement of strength and flexibility imposes new conditions upon the design. To represent all the correlated quantities we should have to construct not only a diagram of moments but also a diagram of elastic deflection and its so-called "curvature"; and the engineer would want to know something more about the *material* of the ligamentous tension-member—its flexibility, its modulus of elasticity in direct tension, its elastic limit, and its safe working stress.

In various ways our structural problem is beset by "limiting conditions." Not only must rigidity be associated with flexibility, but also stability must be ensured in various positions and attitudes; and the primary function of support or weight-carrying must be combined with the provision of *points d'appui* for the muscles concerned in locomotion. We cannot hope to arrive at a numerical or quantitative solution of this complicate problem, but we have found it possible to trace it out in part towards a qualitative solution. And speaking broadly we may certainly say that in each case the problem has been solved by Nature herself, very much as she solves the difficult problems of minimal areas in a system of soap-bubbles; so that each animal is fitted with a backbone adapted to his own individual needs, or (in other words) corresponding to the mean resultant of the many stresses to which as a mechanical system it is exposed.

The mechanical construction of a bird is a more elaborate affair than a quadruped's, inasmuch as it has a double part to play, the bird's whole weight being borne now by its legs and now by its wings. As it stands on the ground our bird is a balanced cantilever, carried

on two legs as on a pier, the cantilever being constituted by the
pelvic bones, drawn out fore and aft and firmly welded to a long
stretch of vertebral column.   The centre of gravity is kept in a line
passing through the acetabulum, and the long toes help to preserve
an unstable but well-adjusted equilibrium.   One arm of the cantilever
carries head, neck and wings, the other, the shorter arm, carries the
abdomen; but the whole weight of the viscera hangs in the abdomen
as in a bag, and on the other hand head and neck are kept small and
light, and their purchase on the fulcrum is under constant modification and control.   A stork or a heron is continually balancing itself;

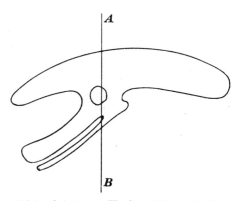

Fig. 484.   Pelvis of *Apteryx*.   The line *AB* is vertical, or nearly so,
in the standing posture of the bird.

as the beak is thrust forward a leg stretches back, as the bird walks
along its whole body sways in keeping.   No less elegant is the
perfect balance of the same birds at rest—the heron standing on
one leg, even on a tree top, the flamingo also on one long leg,
with its neck close coiled and its head tucked amongst the
feathers.

The approximately parabolic form of the great pelvic cantilever
is best seen in the ostrich and other running birds, but more
commonly the strength of the cantilever is got in other ways.
Usually, as in the fowl, it consists of a thin shell of bone curved
over like the bonnet of a motor car and stiffened, or "cambered," by
ridges converging on the acetabulum.   A doubled sheet of paper,
cut roughly to the shape of the pelvis and then pinched up into folds

on either side, as in Fig. 485, will serve as a model of the skeletal cantilever and shew how its limp surface is stiffened by the folds.

Save in the ostrich, and a few other flightless birds, the breast-bone or sternum is a broad, flat bone, produced into a deep, descending ridge or "keel." Very firmly fixed to the sternum on either side is a short strong bone, the coracoid; attached to it again, and bending backwards over the ribs, is the scapula; and at the junction of scapula and coracoid is the socket, or glenoid cavity, for the wing. The clavicles, fused into a "merry-thought," run from near the glenoid cavity to the front end of the keel; in strong-flying birds they are stout and curved, and a continuous curve sweeps

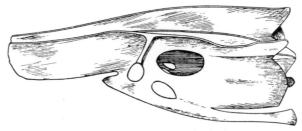

Fig. 485. Rough paper model of a fowl's pelvis.

round from scapula to sternal keel. The keel is commonly explained as necessary to give space enough for the attachment of the muscles of flight, but this explanation is inadequate, even untrue; for one thing, the great pectoral muscle springs from the edge, not from the broad surface of the keel. The keel is essentially *a flange*, and, as in a piece of T iron, adds immensely to the strength and stiffness of the construction*; that it tends to give the fibres of the muscle more stretch and play, and a straighter pull on the arm-bone to which they run, is a secondary advantage. Strong as they are, these bones are exquisitely light and thin. A great frigate-bird, with a 7-foot span of wings, weighs a little over a couple of pounds, and all its bones weigh about four ounces. The bones weigh less than the feathers †.

---

* T irons, if I am not mistaken, were among the many inventions of Robert Stephenson, in his construction of the Menai tubular bridge a hundred years ago.

† Cf. R. C. Murphy, *Natural History*, Oct. 1939.

While the bird stands on the ground its backbone is, as in ourselves, its skeletal axis, and it, including the great cantilever associated with

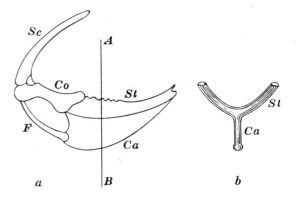

Fig. 485. *a*, Sternum and shoulder-girdle of a skua gull. *b*, section of do., through the line *AB*. *St*, sternum; *Ca*, its carina or keel; *Co*, coracoid bone; *Sc*, scapula; *F*, merry-thought or furcula.

it, carries, and transmits to the legs, the whole weight of the body. But as soon as a bird spreads its wings and rests upon the air, legs,

Fig. 487. A flying bird. Sternum, shoulder-girdle and wings combine to support the body; and all the rest lies as a dead weight thereon.

backbone, cantilever and all become merely so much weight to be carried; and the whole rests, as on a floor, on the strong, stiff platform made of sternum and shoulder-girdle, which the wings (so to speak) take hold of and support, and which is now, mechanically speaking, the axis of the body. The bird has two points of suspension, which it uses alternately: the one through the two acetabula, the other through the glenoid cavities and the outstretched wings. Glenoid cavity and acetabulum are but a little way apart, and the bird swings its weight over from one to the other easily and smoothly. At first sight it seems a curious feature of the bird's skeleton that breast-bone, shoulder-girdle, wings and all are but very slightly attached to the rest of the body, and to what we look on, usually, as its main axis of support; the only skeletal attachment is by the framework of the ribs, and these are slight and slender. The fact is that the two skeletal axes, the backbone and the breast-bone, have their separate and independent roles, and each is but loosely connected with the other.

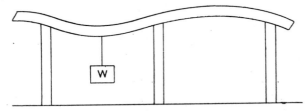

Fig. 488.   Diagram of a continuous girder.

The curvature of the bird's neck is very beautiful: one curve leads on to another; and indeed the bird's whole axial skeleton, from head to tail, is one even and continuous curve. Where a bridge crosses the gap between two piers, it sags as the load passes over; where successive girders cross successive gaps, each sags in its turn under the travelling load. But suppose one *continuous* girder to cross two gaps; it bends in a more complicated way, and one half tends to bend up while the other is sagging down. We cannot analyse the whole field of force to which the bird is subject, but we realise that it is a *continuous* field, in which what the engineer calls a "continuous girder" has its great part to play. The continuous girder is apt to sag and bend and sway in an erratic fashion unless its ends be firm and secure, and the bird's head must, of necessity, be under some analogous control; the semicircular canals are the potent factors in equilibrium, and the bird "keeps a level head" by their help and guidance. Then, between the level of the skull and the level of the great pelvic cantilever a continuous field of force governs and defines the S-shaped curvature. Man's vertebral column shews, *mutatis mutandis*, the same phenomenon of continuous but alternating curvature. The dorsal region is, of necessity, concave towards the cavity of the chest, and as a simple consequence the cervical and lumbar regions curve the other way.

The typically aquatic birds, such as swim under water as penguins and divers do, have characteristic features and adaptations of their own. Just as the cantilever girder becomes obsolete in the aquatic mammal so does it tend to weaken and disappear in the aquatic bird. There is a marked contrast between the high-arched strongly built pelvis in the ostrich or the hen, and the long, thin, comparatively straight and apparently weakly bone which represents it in a diver, a grebe or a penguin. Wings large enough for air would be an obstruction under water, and small wings are enough; for they have to produce thrust only, not lift, and the former is but a small fraction of the latter load. The feet also are now mainly concerned with the same forward thrust, and we begin to see how the long narrow pelvis gives just the *point d'appui* which that thrust requires.

The woodcock, as ornithologists are aware, shews us an osteological paradox, which is commonly described by saying that this bird's ear is in front of its eye! If we hold a woodcock's skull level, *beak and all*, this indeed seems to be the case, but no woodcock does so. Standing or flying, the woodcock holds its beak pointing downwards, and its skull is then level, like that of other birds; in other words, its beak is not in a line with the basi-cranial axis, as a guillemot's is, but bends sharply downwards. When the axis of the skull is horizontal, the beak points downwards at an angle of nearly 60°, and the auditory aperture is then as much behind the eye as in other birds.

There is a certain other principle much to the fore in the construction of the skeleton, well known to the designer of a hydroplane or "flying boat," and not wholly neglected by the bridge-building engineer; it is the principle of non-rigid, flexible or *elastic stability**. A homely comparison between a basket and a tin-can tells us in a moment what it means, and shews us some at least of its peculiar advantages. This method of construction helps to *distribute* the load,

---

* India-rubber has great elastic stability. It is *not* compressible, but is almost as incompressible as water itself, as J. D. Forbes discovered a century ago.

bridges over points or areas where pressure might be unduly concentrated or confined, adapts itself to a sudden impact or concentrated
stress, helps to lessen or to guard against *shock*, and imparts to the
whole structure a quality which we may call, for short, *resiliency*.

The engineer finds it easiest of attainment when his principal
members are in tension; hence elastic movement and resilience are
apt to be conspicuous in a suspension-bridge. One way and another,
resilience shews to perfection in a bird. The **S**-shaped curve of the
neck carrying the light weight of the head, the zig-zag flexures of
the legs bearing the balanced burden of the body, the supple basket

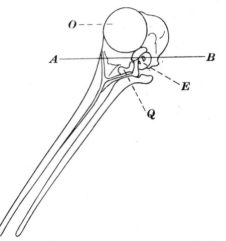

Fig. 489. A woodcock's skull, in (or nearly in) the natural attitude. *A*, *B*, the
basis cranii; *E*, auditory meatus; *O*, orbit; *Q*, quadrate bone.

of the ribs, each rib in two halves one flexed on the other, all these
are such as to make the whole framework act like an elastic spring,
absorbing every shock as the bird lights on or rises from the ground.
Bird, beast and man exhibit this resilience, each in its degree;
a springy step is part of the joy of youth, and its loss is one of the
first infirmities of age.

Nature's engineering is marvellous in our eyes, and our finest
work is narrow in scope and clumsy in execution compared to her
construction and design. But following her example; wittingly or
unwittingly, our own problems evolve and our ambitions enlarge
towards the conception of an "organised structure." In such

triumphs of modern mechanism as a torpedo, a racing aeroplane, a high-speed railway-train, the whole construction is knit together in a new way. It finds its streamlined outline in what seems to be a simple and natural way; it is solid and robust, it is graceful as well as strong; it is no longer a bundle of parts, it has become an organic whole: its likeness, even its outward likeness, to a living organism has become patent and clear.

Throughout this short discussion of the principles of construction, we see the same general principles at work in the skeleton as a whole as we recognised in the plan and construction of an individual bone. That is to say, we see a tendency for material to be laid down just in the lines of *stress*, and so to evade thereby the distortions and disruptions due to *shear*. In these phenomena there lies a definite law of growth, whatever its ultimate expression or explanation may come to be. Let us not press either argument or hypothesis too far: but be content to see that skeletal form, as brought about by growth, is to a very large extent determined by mechanical considerations, and tends to manifest itself as a diagram, or reflected image, of mechanical stress. If we fail, owing to the immense complexity of the case, to unravel all the mathematical principles involved in the construction of the skeleton, we yet gain something, and not a little, by applying this method to the familiar objects of anatomical study: *obvia conspicimus, nubem pellente mathesi**.

Before we leave this subject of mechanical adaptation, let us dwell once more for a moment upon the considerations which arise from our conception of a field of force, or field of stress, in which tension and compression (for instance) are inevitably combined, and are met by the materials naturally fitted to resist them. It has been remarked over and over again how harmoniously the whole organism hangs together, and how throughout its fabric one part is related and fitted to another in strictly functional correlation. But this conception, though never denied, is sometimes apt to be forgotten in the course of that process of more and more minute

---

* The motto was Macquorn Rankine's, in 1857; cf. *Trans. R.S.E.* xxvi, p. 715, 1872.

analysis by which, for simplicity's sake, we seek to unravel the intricacies of a complex organism.

As we analyse a thing into its parts or into its properties, we tend to magnify these, to exaggerate their apparent independence, and to hide from ourselves (at least for a time) the essential integrity and individuality of the composite whole. We divide the body into its organs, the skeleton into its bones, as in very much the same fashion we make a subjective analysis of the mind, according to the teachings of psychology, into component factors: but we know very well that judgment and knowledge, courage or gentleness, love or fear, have no separate existence, but are, somehow mere manifestations, or imaginary coefficients, of a most complex integral. And likewise, as biologists, we may go so far as to say that even the bones themselves are only in a limited and even a deceptive sense, separate and individual things. The skeleton begins as a *continuum*, and a *continuum* it remains all life long. The things that link bone with bone, cartilage, ligaments, membranes, are fashioned out of the same primordial tissue, and come into being *pari passu* with the bones themselves. The entire fabric has its soft parts and its hard, its rigid and its flexible parts; but until we disrupt and dismember its bony, gristly and fibrous parts one from another, it exists simply as a "skeleton," as one integral and individual whole.

A bridge was once upon a time a loose heap of pillars and rods and rivets of steel. But the identity of these is lost, just as if they were fused into a solid mass, when once the bridge is built; their separate functions are only to be recognised and analysed in so far as we can analyse the stresses, the tensions and the pressures, which affect this part of the structure or that; and these forces are not themselves separate entities, but are the resultants of an analysis of the whole field of force. Moreover when the bridge is broken it is no longer a bridge, and all its strength is gone. So is it precisely with the skeleton. In it is reflected a field of force: and keeping pace, as it were, in action and interaction with this field of force, the whole skeleton and every part thereof, down to the minute intrinsic structure of the bones themselves, is related in form and in position to the lines of force, to the resistances it has to encounter; for by one of the mysteries of biology, resistance begets resistance, and where pressure falls there growth springs up in strength to meet it. And, pursuing the same train of thought, we see that all this is true not of the skeleton alone but of the whole fabric of the body. Muscle and bone, for instance, are inseparably associated and connected; they are moulded one with another; they come into being together, and act and react together*. We may study them apart, but it is as a concession to our weakness and to the narrow outlook of our minds. We see, dimly perhaps but yet with all the assurance of conviction, that between muscle and bone there can be no change in the one but it is correlated with changes in the other; that through and through they are linked in indissoluble association; that they are only separate entities in this limited and subordinate sense, that they are *parts* of a whole which, when it loses its composite integrity, ceases to exist.

The biologist, as well as the philosopher, learns to recognise that the whole is not merely the sum of its parts. It is this, and much more than this. For it is not a bundle of parts but an organisation of parts, of parts in their mutual arrangement, fitting one with another, in what Aristotle calls "a single and indivisible principle of unity"; and this is no merely metaphysical conception, but is in biology the fundamental truth which lies at the basis of Geoffroy's (or Goethe's) law of "compensation," or "balancement of growth."

Nevertheless Darwin found no difficulty in believing that "natural selection will tend in the long run to reduce *any part* of the organisation, as soon as, through changed habits, it becomes superfluous: without by any means causing some other part to be largely developed in a corresponding degree. And conversely, that natural selection may perfectly well succeed in largely developing an organ without requiring as a necessary compensation the reduction of some adjoining part†." This view has been developed into a doctrine of the "independence of single characters" (not to be confused with the germinal "unit characters" of Mendelism), especially by the palaeontologists. Thus Osborn asserts a "principle of hereditary correlation," combined with a "principle of *hereditary separability*,

---

* John Hunter was seldom wrong; but I cannot believe that he was right when he said (*Scientific Works*, ed. Owen, I, p. 371), "The bones, in a mechanical view, appear to be the first that are to be considered. We can study their shape, connections, number, uses, etc., *without considering any other part of the body*."

† *Origin of Species*, 6th ed. p. 118.

whereby the body is a colony, a mosaic, of single individual and separable characters*." I cannot think that there is more than a very small element of truth in this doctrine. As Kant said, "die Ursache der Art der Existenz bei jedem Theile eines lebenden Körpers *ist im Ganzen enthalten.*" And, according to the trend or aspect of our thought, we may look upon the coordinated parts, now as related and fitted *to the end or function of* the whole, and now as related to or resulting *from the physical causes* inherent in the entire system of forces to which the whole has been exposed, and under whose influence it has come into being†.

In John Hunter's day the anatomist studied every bone of the skeleton in its own place, in order to discover its useful purpose and understand its mechanical perfection. The morphologist of a hundred years later preferred to study an isolated bone from many animals, collar-bones or shoulder-blades by themselves, apart from the field of force in which their work was done, in the search for signs of blood-relationship and common ancestry. Truth lies both ways; immediate use and old inheritance are blended in Nature's handiwork as in our own. In the marble columns and architraves of a Greek temple we still trace the timbers of its wooden prototype, and see beyond these the tree-trunks of a primeval sacred grove; roof and eaves of a pagoda recall the sagging mats which roofed an

---

* *Amer. Naturalist,* April, 1915, p. 198, etc. Cf. *infra,* p. 1036.

† Driesch saw in "Entelechy" that something which differentiates the whole from the sum of its parts in the case of the organism: "The organism, we know, is a system the single constituents of which are inorganic in themselves; only the whole constituted by them in their typical order or arrangement owes its specificity to 'Entelechy'" (*Gifford Lectures,* 1908, p. 229): and I think it could be shewn that many other philosophers have said precisely the same thing. So far as the argument goes, I fail to see how *this* Entelechy is shewn to be peculiarly or specifically related to the *living* organism. The conception (at the bottom of General Smuts's '*Holism*') that the whole is *always* something very different from its parts is a very ancient doctrine. The reader will perhaps remember how, in another vein, the theme is treated by Martinus Scriblerus (Huxley quoted it once, for his own ends): "In every Jack there is a *meat-roasting* Quality, which neither resides in the fly, nor in the weight, nor in any particular wheel of the Jack, but is the result of the whole composition; etc., etc." Indeed it was at that very time, in the early eighteenth century, that the terms *organism* and *organisation* were coming into use, to connote that harmonious combination of parts "qui conspirent toutes ensembles à produire cet effet général que nous nommons la vie" (Buffon). Cf. Ch. Robin, Recherches sur l'origine et le sens des termes organisme et organisation, *Jl. de l'Anat.* LX, pp. 1–55, 1880.

earlier edifice; Anglo-Saxon land-tenure influences the planning of
our streets, and the cliff-dwelling and the cave-dwelling linger on
in the construction of our homes! So we see enduring traces of
the past in the living organism—landmarks which have lasted on
through altered functions and altered needs; and yet at every stage
new needs are met and new functions effectively performed.

When we consider (for instance) the several bones in a fish's
shoulder-girdle—clavicle, supra-clavicle, post-clavicle, post-temporal
and so on—and recognise these in this fish or that under countless
minor transformations, we have something which is not only wide-

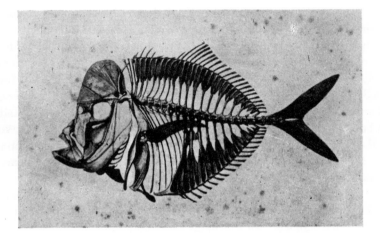

Fig. 490. Skeleton of moonfish, *Vomer* sp. From L. Agassiz.

spread but is rooted in antiquity, and whose full significance seems
beyond our reach. But take the skeleton of some particular fish, a
moonfish or a John Dory will do very well, and look at its shoulder-
girdle from the mechanical point of view. It is a deal more than is
needed for the support of the small, weak pectoral fin; but another
function, and its perfect adaptation for that function, are not hard to
see. The flattened body of the fish is built (as we have seen also in the
plaice) on the plan of a parabolic girder; but out of this girder a
great gap has had to be cut, to hold the viscera. The great shoulder-
girdle serves to strengthen and complete the girder, to bind its
upper and lower members together, and to compensate for the part

which has been taken away. It fulfils this function by various means; by the way in which the two sides of the girdle are conjoined into a single arch; by its strong attachment to the head, and again to the pelvis, and through the latter to the chain of ossicles which bound or constitute the abdominal border of the fish; and a large part of the stress upon the shoulder-girdle proper is taken up, or relieved, by the strong post-clavicular bones, which form a supplementary arch running downwards from the clavicle (just where it begins to incline forward), straight to the ventral border, to be firmly attached there to the ventral ossicles. Similarly we notice at the hinder border of the abdominal cavity, a strong curved bone running from the anterior part of the ventral fin to a solid attachment with the vertebral column, stiffening the ventral part, and helping the shoulder-girdle to restore full strength to the girder after it had been reduced, so to speak, to the brink of inevitable collapse. The skull itself is not only streamlined with the rest of the body, but is an intrinsic part of the whole engineering construction. The lines of stress run simply and clearly through the skeleton, and a bone can no longer teach us its full and proper lesson after we have taken it apart. To look on the hereditary or evolutionary factor as *the guiding principle* in morphology is to give to that science a one-sided and fallacious simplicity*

It would seem to me that the mechanical principles and phenomena which we have dealt with in this chapter are of no small importance to the morphologist, all the more when he is inclined to direct his study of the skeleton exclusively to the problem of phylogeny; and especially when, according to the methods of modern comparative morphology, he is apt to take the skeleton to pieces, and to draw from the comparison of a series of scapulae, humeri, or individual vertebrae, conclusions as to the descent and relationship of the animals to which they belong.

It would, I dare say, be an exaggeration to see in every bone nothing more than a resultant of immediate and direct physical or mechanical conditions; for to do so would be to deny the existence,

* The extreme evolutionary, or phylogenetic, aspect of morphology was being questioned even forty years ago. "Where we once thought we detected relationships we now know we were often being misled, and the old-time supposition that mere community of structure is necessarily an index of community of origin has gone to the wall" (G. B. Howes, in *Nature*, Jan. 10, 1901).

in this connection, of a principle of heredity. And though I have tried throughout this book to lay emphasis on the direct action of causes other than heredity, in short to circumscribe the employment of the latter as a working hypothesis in morphology, there can still be no question whatsoever but that heredity is a vastly important as well as a mysterious thing; it is *one* of the great factors in biology, however we may attempt to figure to ourselves, or howsoever we may fail even to imagine, its underlying physical explanation. But I maintain that it is no less an exaggeration if we tend to neglect these direct physical and mechanical modes of causation altogether, and to see in the characters of a bone merely the results of variation and of heredity, and to trust, in consequence, to those characters as a sure and certain and unquestioned guide to affinity and phylogeny. Comparative anatomy has its physiological side, which filled men's minds in John Hunter's day, and in Owen's day; it has its classificatory and phylogenetic aspect, which all but filled men's minds in the early days of Darwinism; and we can lose sight of neither aspect without risk of error and misconception.

It is certain that the question of phylogeny, always difficult, becomes especially so in cases where a great change of physical or mechanical conditions has come about, and where accordingly the former physical and physiological constraints are altered or removed. The great depths of the sea differ from other habitations of the living, not least in their eternal quietude. The fishes which dwell therein are quaint and strange; their huge heads, prodigious jaws, and long tails and tentacles are, as it were, gross exaggerations of the common and conventional forms. We look in vain for any purposeful cause or physiological explanation of these enormities; and are left under a vague impression that life has been going on in the security of all but perfect equilibrium, and that the resulting forms, liberated from many ordinary constraints, have grown with unusual freedom*.

To discuss these questions at length would be to enter on a discussion of Lamarck's philosophy of biology, and of many other things besides. But let us take one single illustration. The affinities of the whales constitute, as will be readily admitted, a very hard problem in phylogenetic classification. We know now that the

* Cf. *supra*, p. 423.

extinct Zeuglodons are related to the old Creodont carnivores, and thereby (though distantly) to the seals*; and it is supposed, but it is by no means so certain, that in turn they are to be considered as representing, or as allied to, the ancestors of the modern toothed whales†. The proof of any such a contention becomes, to my mind, extraordinarily difficult and complicated; and the arguments commonly used in such cases may be said (in Bacon's phrase) to allure, rather than to extort assent. Though the Zeuglodons were aquatic animals, we do not know, and we have no right to suppose or to assume, that they swam after the fashion of a whale (any more than the seal does), that they dived like a whale, or leaped like a whale. But the fact that the whale does these things, and the way in which he does them, is reflected in many parts of his skeleton—perhaps more or less in all: so much so that the lines of stress which these actions impose are the very plan and working-diagram of great part of his structure. That the Zeuglodon has a scapula like that of a whale is to my mind no necessary argument that he is akin by blood-relationship to a whale: that his dorsal vertebrae are very different from a whale's is no conclusive argument that such blood-relationship is lacking. The former fact goes a long way to prove that he used his flippers very much as a whale does; the latter goes still farther to prove that his general movements and equilibrium in the water were totally different. The whale may be descended from the Carnivora, or might for that matter, as an older school of naturalists believed, be descended from the Ungulates; but whether or no, we need not expect to find in him the scapula, the pelvis or the vertebral column of the lion or of the cow, for it would be physically impossible that he could live the life he does with any one of them. In short, when we hope to find the missing links between a whale and his terrestrial ancestors, it must be not by means of conclusions drawn from a scapula, an axis, or

---

* See (*int. al.*) my paper On the affinities of *Zeuglodon* in *Studies from the Museum of University College, Dundee,* 1889.

† "There can be no doubt that Fraas is correct in regarding this type (*Procetus*) as an annectant form between the Zeuglodonts and the Creodonta, but. although the origin of the Zeuglodonts is thus made clear, it still seems to be by no means so certain as that author believes, that they may not themselves be the ancestral forms of the Odontoceti" (Andrews, *Tertiary Vertebrata of the Fayum,* 1906, p. 235).

even from a tooth, but by the discovery of forms so intermediate in their general structure as to indicate an organisation and, *ipso facto*, a mode of life, intermediate between the terrestrial and the Cetacean form. There is no valid syllogism to the effect that A has a flat curved scapula like a seal's, and B has a flat curved scapula like a seal's: and therefore A and B are related to the seals and to each other; it is merely a flagrant case of an "undistributed middle." But there is validity in an argument that B shews in its general structure, extending over this bone and that bone, resemblances both to A and to the seals: and that therefore he may be presumed to be related to both, in his hereditary habits of life and in actual kinship by blood. It is cognate to this argument that (as every palaeontologist knows) we find clues to affinity more easily, that is to say with less confusion and perplexity, in certain structures than in others. The deep-seated rhythms of growth which, as I venture to think, are the chief basis of morphological heredity, bring about similarities of form which endure in the absence of conflicting forces; but a new system of forces, introduced by altered environment and habits, impinging on those particular parts of the fabric which lie within this particular field of force, will assuredly not be long of manifesting itself in notable and inevitable modifications of form. And if this be really so, it will further imply that modifications of form will tend to manifest themselves, not so much in small and *isolated* phenomena, in this part of the fabric or in that, in a scapula for instance or a humerus: but rather in some slow, *general*, and more or less uniform or graded modification, spread over a number of correlated parts, and at times extending over the whole, or over great portions, of the body. Whether any such general tendency to widespread and correlated transformation exists, we shall attempt to discuss in the following chapter.

CHAPTER XVII

ON THE THEORY OF TRANSFORMATIONS, OR THE COMPARISON OF RELATED FORMS

In the foregoing chapters of this book we have attempted to study the inter-relations of growth and form, and the part which the physical forces play in this complex interaction; and, as part of the same enquiry, we have tried in comparatively simple cases to use mathematical methods and mathematical terminology to describe and define the forms of organisms. We have learned in so doing that our own study of organic form, which we call by Goethe's name of Morphology, is but a portion of that wider Science of Form which deals with the forms assumed by matter under all aspects and conditions, and, in a still wider sense, with forms which are theoretically imaginable.

The study of form may be descriptive merely, or it may become analytical. We begin by describing the shape of an object in the simple words of common speech: we end by defining it in the precise language of mathematics; and the one method tends to follow the other in strict scientific order and historical continuity. Thus, for instance, the form of the earth, of a raindrop or a rainbow, the shape of the hanging chain, or the path of a stone thrown up into the air, may all be described, however inadequately, in common words; but when we have learned to comprehend and to define the sphere, the catenary, or the parabola, we have made a wonderful and perhaps a manifold advance. The mathematical definition of a "form" has a quality of precision which was quite lacking in our earlier stage of mere description; it is expressed in few words or in still briefer symbols, and these words or symbols are so pregnant with meaning that thought itself is economised; we are brought by means of it in touch with Galileo's aphorism (as old as Plato, as old as Pythagoras, as old perhaps as the wisdom of the Egyptians), that "the Book of Nature is written in characters of Geometry*."

---

* Cf. Plutarch, *Symp*. viii, 2, on the meaning of Plato's aphorism ("if it actually was Plato's"): πῶς Πλάτων ἔλεγε τὸν θεὸν ἀεὶ γεωμετρεῖν.

We are apt to think of mathematical definitions as too strict and rigid for common use, but their rigour is combined with all but endless freedom. The precise definition of an ellipse introduces us to all the ellipses in the world; the definition of a "conic section" enlarges our concept, and a "curve of higher order" all the more extends our range of freedom*. By means of these large limitations, by this controlled and regulated freedom, we reach through mathematical analysis to mathematical synthesis. We discover homologies or identities which were not obvious before, and which our descriptions obscured rather than revealed: as for instance, when we learn that, however we hold our chain, or however we fire our bullet, the contour of the one or the path of the other is always mathematically homologous.

Once more, and this is the greatest gain of all, we pass quickly and easily from the mathematical concept of form in its statical aspect to form in its dynamical relations: we rise from the conception of form to an understanding of the forces which gave rise to it; and in the representation of form and in the comparison of kindred forms, we see in the one case a diagram of forces in equilibrium, and in the other case we discern the magnitude and the direction of the forces which have sufficed to convert the one form into the other. Here, since a change of material form is only effected by the movement of matter†, we have once again the support of the schoolman's and the philosopher's axiom, *Ignorato motu, ignoratur Natura.*"

---

* So said Gustav Theodor Fechner, the author of Fechner's Law, a hundred years ago. (Ueber die mathematische Behandlung organischer Gestalten und Processe, *Berichte d. k. sächs. Gesellsch., Math.-phys. Cl.,* Leipzig, 1849, pp. 50–64.) Fechner's treatment is more purely mathematical and less physical in its scope and bearing than ours, and his paper is but a short one, but the conclusions to which he is led differ little from our own. Let me quote a single sentence which, together with its context, runs precisely on the lines which we have followed in this book: "So ist also die mathematische Bestimmbarkeit im Gebiete des Organischen eben so gut vorhanden als in dem des Unorganischen, und in letzterem eben solchen oder äquivalenten Beschränkungen unterworfen als in ersterem; und nur sofern die unorganischen Formen und das unorganische Geschehen sich einer einfacheren Gesetzlichkeit mehr nähern als die organischen, kann die Approximation im unorganischen Gebiet leichter und weiter getrieben werden als im organischen. Dies wäre der ganze, sonach rein relative, Unterschied." Here, in a nutshell, is the gist of the whole matter.

† "We can *move* matter, that is all we can do to it" (Oliver Lodge).

There is yet another way—we learn it of Henri Poincaré—to regard the function of mathematics, and to realise why its laws and its methods *are bound* to underlie all parts of physical science. Every natural phenomenon, however simple, is really composite, and every visible action and effect is a summation of countless subordinate actions. Here mathematics shews her peculiar power, to combine and to generalise. The concept of an average, the equation to a curve, the description of a froth or cellular tissue, all come within the scope of mathematics for no other reason than that they are summations of more elementary principles or phenomena. Growth and Form are throughout of this composite nature; therefore the laws of mathematics are bound to underlie them, and her methods to be peculiarly fitted to interpret them.

In the morphology of living things the use of mathematical methods and symbols has made slow progress; and there are various reasons for this failure to employ a method whose advantages are so obvious in the investigation of other physical forms. To begin with, there would seem to be a psychological reason, lying in the fact that the student of living things is by nature and training an observer of concrete objects and phenomena and the habit of mind which he possesses and cultivates is alien to that of the theoretical mathematician. But this is by no means the only reason; for in the kindred subject of mineralogy, for instance, crystals were still treated in the days of Linnaeus as wholly within the province of the naturalist, and were described by him after the simple methods in use for animals and plants: but as soon as Haüy shewed the application of mathematics to the description and classification of crystals, his methods were immediately adopted and a new science came into being.

A large part of the neglect and suspicion of mathematical methods in organic morphology is due (as we have partly seen in our opening chapter) to an ingrained and deep-seated belief that even when we seem to discern a regular mathematical figure in an organism, the sphere, the hexagon, or the spiral which we so recognise merely resembles, but is never entirely explained by, its mathematical analogue; in short, that the details in which the figure differs from its mathematical prototype are more important and more interesting than the features in which it agrees; and even that the peculiar aesthetic pleasure with which we regard a living thing is somehow bound up with the departure from mathematical regularity which it manifests as a peculiar attribute of life. This view seems to me to involve a misapprehension. There is no such essential difference between these phenomena of organic form and those which are manifested in portions of inanimate matter*. The mathematician knows better than we do the value of an approximate result†. The child's skipping-rope is but an approximation to Huygens's catenary curve—but in the catenary curve lies the whole gist of the matter. We may be dismayed too easily by contingencies which are nothing short of irrelevant compared to the main issue; there is a *principle of negligibility*. Someone has said that if Tycho Brahé's instruments had been ten times as exact there would have been no Kepler, no Newton, and no astronomy.

If no chain hangs in a perfect catenary and no raindrop is a perfect sphere, this is for the reason that forces and resistances other than the main one are inevitably at work. The same is true of organic form, but it is for the mathematician to unravel the conflicting forces which are at work together. And this process of investigation may lead us on step by step to new phenomena, as it has done in physics, where sometimes a knowledge of form leads us to the interpretation of forces, and at other times a knowledge of the forces at work guides us towards a better insight into form. After the fundamental advance had been made which taught us that the world

---

* M. Bergson repudiates, with peculiar confidence, the application of mathematics to biology; cf. *Creative Evolution*, p. 21, "Calculation touches, at most, certain phenomena of organic destruction. Organic creation, on the contrary, the evolutionary phenomena which properly constitute life, we cannot in any way subject to a mathematical treatment." Bergson thus follows Bichat: "C'est peu connaître les fonctions animales que de vouloir les soumettre au moindre calcul, parceque leur instabilité est extrême. Les phénomènes restent toujours les mêmes, et c'est ce qui nous importe; mais leurs variations, en plus ou en moins, sont sans nombre" (*La Vie et la Mort*, p. 257).

† When we make a 'first approximation' to the solution of a physical problem, we usually mean that we are solving one part while neglecting others. Geometry deals with *pure forms* (such as a straight line), defined by a single law; but these are few compared with the *mixed forms*, like the surface of a polyhedron, or a segment of a sphere, or any ordinary mechanical construction or any ordinary physical phenomenon. It is only in a purely mathematical treatment of physics that the "single law" can be dealt with alone, and the approximate solution dispensed with accordingly.

was round, Newton shewed that the forces at work upon it must lead to its being imperfectly spherical, and in the course of time its oblate spheroidal shape was actually verified. But now, in turn, it has been shewn that its form is still more complicated, and the next step is to seek for the forces that have deformed the oblate spheroid. As Newton somewhere says, "Nature delights in transformations."

The organic forms which we can define more or less precisely in mathematical terms, and afterwards proceed to explain and to account for in terms of force, are of many kinds, as we have seen; but nevertheless they are few in number compared with Nature's all but infinite variety. The reason for this is not far to seek. The living organism represents, or occupies, a field of force which is never simple, and which as a rule is of immense complexity. And just as in the very simplest of actual cases we meet with a departure from such symmetry as could only exist under conditions of *ideal* simplicity, so do we pass quickly to cases where the interference of numerous, though still perhaps very simple, causes leads to a resultant complexity far beyond our powers of analysis. Nor must we forget that the biologist is much more exacting in his requirements, as regards form, than the physicist; for the latter is usually content with either an ideal or a general description of form, while the student of living things must needs be specific. Material things, be they living or dead, shew us but a shadow of mathematical perfection*. The physicist or mathematician can give us perfectly satisfying expressions for the form of a wave, or even of a heap of sand; but we never ask him to define the form of any particular wave of the sea, nor the actual form of any mountain-peak or hill.

In this there lies a certain justification for a saying of Minot's, of the greater part of which, nevertheless, I am heartily inclined to disapprove. "We biologists," he says, "cannot deplore too frequently or too emphatically the great mathematical delusion by which men often of great if limited ability have been misled into becoming advocates of an erroneous conception of accuracy. The delusion is that no science is accurate until its results can be expressed mathematically. The error comes from the assumption that mathematics can express complex relations. Unfortunately mathematics have a very limited scope, and are based upon a few extremely rudimentary

---

* Cf. Haton de la Goupillière, *op. cit.*: "On a souvent l'occasion de saisir dans la nature *un reflet* des formes rigoureuses qu'étudie la géometrie."

experiences, which we make as very little children and of which no adult has
any recollection. The fact that from this basis men of genius have evolved
wonderful methods of dealing with numerical relations should not blind us
to another fact, namely, that the observational basis of mathematics is,
psychologically speaking, very minute compared with the observational basis
of even a single minor branch of biology.... While therefore here and there
the mathematical methods may aid us, *we need a kind and degree of accuracy
of which mathematics is absolutely incapable*.... With human minds constituted
as they actually are, we cannot anticipate that there will ever be a mathematical expression for any organ or even a single cell, although formulae will
continue to be useful for dealing now and then with isolated details..."
(*op. cit.* p. 19, 1911). It were easy to discuss and criticise these sweeping
assertions, which perhaps had their origin and parentage in an *obiter dictum*
of Huxley's, to the effect that "Mathematics is that study which knows nothing
of observation, nothing of experiment, nothing of induction, nothing of
causation" (*cit.* Cajori, *Hist. of Elem. Mathematics*, p. 283). But Gauss,
"rex mathematicorum," called mathematics "a science of the eye"; and
Sylvester assures us that "most, if not all, of the great ideas of modern
mathematics have had their origin in observation" (*Brit. Ass. Address*, 1869,
and *Laws of Verse*, p. 120, 1870.

Réaumur said the same thing two hundred years ago (*Mém.* I, p. 49, 1734).
Maupertuis, he said, was both naturalist and mathematician; and all his
mathematics "n'ont en rien affaibli son gout pour les insectes, personne
peut-être n'a plus d'amour pour eux." He goes on to say: "L'esprit
d'observation qu'on regarde comme le caractere d'esprit essentiel aux
naturalistes, est également necessaire pour faire des progrès en quelque science
que ce soit. C'est l'esprit d'observation qui fait appercevoir ce qui a
échappé aux autres, qui fait saisir des rapports qui sont entre des choses
qui semblent differentes, ou qui fait trouver les differences qui sont entre
celles qui paroissent semblables. On ne résoud les problemes les plus épineux
de Geometrie qu'après avoir sçû observer des rapports qui ne se découvrent
qu'à un esprit penetrant, et extrêmement attentif. Ce sont des observations
qui mettent en état de résoudre les problemes de physique comme ceux
d'histoire naturelle, car l'histoire naturelle a ses problemes à résoudre, et
elle n'en a même que trop qui ne sont pas résolus." It is in a deeper sense
than this, however, that the modern physicist looks on mathematics as an
"empirical" science, and no longer a matter of pure intuition, or "reine
Anschauung." Cf. Max Born, on Some philosophical aspects of modern
physics, *Proc. R.S.E.* LVII, pp. 1–18, 1936.

For one reason or another there are very many organic forms which
we cannot describe, still less define, in mathematical terms: just as
there are problems even in physical science beyond the mathematics
of our age. We never even seek for a formula to define this fish or
that, or this or that vertebrate skull. But we may already use
mathematical language to describe, even to define in general terms,

the shape of a snail-shell, the twist of a horn, the outline of a leaf, the texture of a bone, the fabric of a skeleton, the stream-lines of fish or bird, the fairy lace-work of an insect's wing. Even to do this we must learn from the mathematician to eliminate and to discard; to keep the type in mind and leave the single case, with all its accidents, alone; and to find in this sacrifice of what matters little and conservation of what matters much one of the peculiar excellences of the method of mathematics*.

In a very large part of morphology, our essential task lies in the comparison of related forms rather than in the precise definition of each; and the *deformation* of a complicated figure may be a phenomenon easy of comprehension, though the figure itself have to be left unanalysed and undefined. This process of comparison, of recognising in one form a definite permutation or *deformation* of another, apart altogether from a precise and adequate understanding of the original "type" or standard of comparison, lies within the immediate province of mathematics, and finds its solution in the elementary use of a certain method of the mathematician. This method is the Method of Coordinates, on which is based the Theory of Transformations†.

I imagine that when Descartes conceived the method of co-ordinates, as a generalisation from the proportional diagrams of the artist and the architect, and long before the immense possibilities of this analysis could be foreseen, he had in mind a very simple purpose; it was perhaps no more than to find a way of translating the *form* of a curve (as well as the position of a point) into *numbers* and into *words*. This is precisely what we do, by the method of coordinates, every time we study a statistical curve; and conversely, we translate numbers into form whenever we "plot a curve," to illustrate a table of mortality, a rate of growth, or the daily variation of temperature or barometric pressure. In precisely the same way

---

* Cf. W. H. Young, The mathematical method and its limitations, *Congresso dei Matematici*, Bologna, 1928.

† The mathematical Theory of Transformations is part of the Theory of Groups, of great importance in modern mathematics. A distinction is drawn between Substitution-groups and Transformation-groups, the former being discontinuous, the latter continuous—in such a way that within one and the same group each transformation is infinitely little different from another. The distinction among biologists between a mutation and a variation is curiously analogous.

it is possible to inscribe in a net of rectangular coordinates the outline, for instance, of a fish, and so to translate it into a table of numbers, from which again we may at pleasure reconstruct the curve.

But it is the next step in the employment of coordinates which is of special interest and use to the morphologist; and this step consists in the alteration, or deformation, of our system of coordinates, and in the study of the corresponding transformation of the curve or figure inscribed in the coordinate network.

Let us inscribe in a system of Cartesian coordinates the outline of an organism, however complicated, or a part thereof: such as a fish, a crab, or a mammalian skull. We may now treat this complicated figure, in general terms, as a function of x, y. If we submit our rectangular system to deformation on simple and recognised lines, altering, for instance, the direction of the axes, the ratio of x/y, or substituting for x and y some more complicated expressions, then we obtain a new system of coordinates, whose deformation from the original type the inscribed figure will precisely follow. In other words, we obtain a new figure which represents the old figure under a more or less homogeneous *strain*, and is a function of the new coordinates in precisely the same way as the old figure was of the original coordinates x and y.

The problem is closely akin to that of the cartographer who transfers identical data to one projection or another*; and whose object is to secure (if it be possible) a complete correspondence, *in each small unit of area*, between the one representation and the other. The morphologist will not seek to draw his organic forms in a new and artificial projection; but, in the converse aspect of the problem, he will enquire whether two different but more or less obviously related forms can be so analysed and interpreted that each may be shewn to be a transformed representation of the other. This once demonstrated, it will be a comparatively easy task (in all probability) to postulate the direction and magnitude of the force capable of effecting the required transformation. Again, if such a simple alteration of the system of forces can be proved adequate to·meet the case, we may find ourselves able to dispense with many

---

* Cf. (e.g.) Tissot, *Mémoire sur la représentation des surfaces, et les projections des cartes géographiques*, Paris, 1881.

T G F

widely current and more complicated hypotheses of biological causation. For it is a maxim in physics that an effect ought not to be ascribed to the joint operation of many causes if few are adequate to the production of it: *Frustra fit per plura, quod fieri potest per pauciora.*

We might suppose that by the combined action of appropriate forces any material form could be transformed into any other: just as out of a "shapeless" mass of clay the potter or the sculptor models his artistic product; or just as we attribute to Nature herself the power to effect the gradual and successive transformation of the simple germ into the complex organism. But we need not let these considerations deter us from our method of comparison of *related* forms. We shall strictly limit ourselves to cases where the transformation necessary to effect a comparison shall be of a simple kind, and where the transformed, as well as the original, coordinates shall constitute an harmonious and more or less symmetrical system. We should fall into deserved and inevitable confusion if, whether by the mathematical or any other method, we attempted to compare organisms separated far apart in Nature and in zoological classification. We are limited, both by our method and by the whole nature of the case, to the comparison of organisms such as are manifestly related to one another and belong to the same zoological class. For it is a grave sophism, in natural history as in logic, to make a transition into another kind*.

Our enquiry lies, in short, just within the limits which Aristotle himself laid down when, in defining a "genus," he shewed that (apart from those superficial characters, such as colour, which he called "accidents") the essential differences between one "species" and another are merely differences of proportion, of relative magnitude, or (as he phrased it) of "excess and defect." "Save only for a difference in the way of excess or defect, the parts are identical in the case of such animals as are of one and the same genus; and by 'genus' I mean, for instance, Bird or Fish." And again: "Within the limits of the same genus, as a general rule, most of the parts exhibit differences...in the way of multitude or fewness, magnitude

---

* The saying *heterogenea comparari non possunt* is discussed by Coleridge in his *Aids to Reflexion.*

or parvitude, in short, in the way of excess or defect. For 'the more' and 'the less' may be represented as 'excess' and 'defect'*." It is precisely this difference of relative magnitudes, this Aristotelian "excess and defect" in the case of form, which our coordinate method is especially adapted to analyse, and to reveal and demonstrate as the main cause of what (again in the Aristotelian sense) we term "specific" differences.

The applicability of our method to particular cases will depend upon, or be further limited by, certain practical considerations or qualifications. Of these the chief, and indeed the essential, condition is, that the form of the entire structure under investigation should be found to vary in a more or less uniform manner, after the fashion of an approximately homogeneous and isotropic body. But an imperfect isotropy, provided always that some "principle of continuity" run through its variations, will not seriously interfere with our method; it will only cause our transformed coordinates to be somewhat less regular and harmonious than are those, for instance, by which the physicist depicts the motions of a perfect fluid, or a theoretic field of force in a uniform medium.

Again, it is essential that our structure vary in its entirety, or at least that "independent variants" should be relatively few. That independent variations occur, that localised centres of diminished or exaggerated growth will now and then be found, is not only probable but manifest; and they may even be so pronounced as to appear to constitute new formations altogether. Such independent variants as these Aristotle himself clearly recognised: "It happens further that some have parts which others have not; for instance, some [birds] have spurs and others not, some have crests, or combs, and others not; but, as a general rule, most parts and those that go to make up the bulk of the body are either identical with one another, or differ from one another in the way of contrast and of excess and defect. For 'the more' and 'the less' may be represented as 'excess' or 'defect'†."

If, in the evolution of a fish, for instance, it be the case that its

---

* *Historia Animalium* I, 1.

† Aristotle's argument is even more subtle and far-reaching; for the differences of which he speaks are not merely those between one bird and another, but between them all and the very type itself, or Platonic "idea" of a bird.

several and constituent parts—head, body and tail, or this fin and that fin—represent so many independent variants, then our coordinate system will at once become too complex to be intelligible; we shall be making not one comparison but several separate comparisons, and our general method will be found inapplicable. Now precisely this independent variability of parts and organs—here, there, and everywhere within the organism—would appear to be implicit in our ordinary accepted notions regarding variation; and, unless I am greatly mistaken, it is precisely on such a conception of the easy, frequent, and normally independent variability of parts that our conception of the process of natural selection is fundamentally based. For the morphologist, when comparing one organism with another, describes the differences between them point by point, and "character" by "character*." If he is from time to time constrained to admit the existence of "correlation" between characters (as a hundred years ago Cuvier first shewed the way), yet all the while he recognises this fact of correlation somewhat vaguely, as a phenomenon due to causes which, except in rare instances, he can hardly hope to trace; and he falls readily into the habit of thinking and talking of evolution as though it had proceeded on the lines of his own descriptions, point by point, and character by character†.

With the "characters" of Mendelian genetics there is no fault to be found; tall and short, rough and smooth, plain or coloured are opposite tendencies or contrasting qualities, in plain logical contradistinction. But when the morphologist compares one animal with another, point by point or character by character, these are too often the mere outcome of artificial dissection and analysis. Rather is the living body one integral and indivisible whole, in

---

* Cf. *supra*, p. 1020.

† Cf. H. F. Osborn, On the origin of single characters, as observed in fossil and living animals and plants, *Amer. Nat.* XLIX, pp. 193–239, 1915 (and other papers); *ibid.* p. 194, "Each individual is composed of a vast number of somewhat similar new or old characters, each character has its independent and separate history, each character is in a certain stage of evolution, each character is correlated with the other characters of the individual....The real problem has always been that of the origin and development of characters. Since the *Origin of Species* appeared, the terms variation and variability have always referred to single characters; if a species is said to be variable, we mean that a considerable number of the single characters or groups of characters of which it is composed are variable," etc.

which we cannot find, when we come to look for it, any strict dividing line even between the head and the body, the muscle and the tendon, the sinew and the bone. Characters which we have differentiated insist on integrating themselves again; and aspects of the organism are seen to be conjoined which only our mental analysis had put asunder. The coordinate diagram throws into relief the integral solidarity of the organism, and enables us to see how simple a certain kind of *correlation* is which had been apt to seem a subtle and a complex thing.

But if, on the other hand, diverse and dissimilar fishes can be referred as a whole to identical functions of very different coordinate systems, this fact will of itself constitute a proof that variation has proceeded on definite and orderly lines, that a comprehensive "law of growth" has pervaded the whole structure in its integrity, and that some more or less simple and recognisable system of forces has been in control. It will not only shew how real and deep-seated is the phenomenon of "correlation," in regard to form, but it will also demonstrate the fact that a correlation which had seemed too complex for analysis or comprehension is, in many cases, capable of very simple graphic expression. This, after many trials, I believe to be in general the case, bearing always in mind that the occurrence of independent or localised variations must sometimes be considered.

We are dealing in this chapter with the forms of related organisms, in order to shew that the differences between them are as a general rule simple and symmetrical, and just such as might have been brought about by a slight and simple change in the system of forces to which the living and growing organism was exposed. Mathematically speaking, the phenomenon is identical with one met with by the geologist, when he finds a bed of fossils squeezed flat or otherwise symmetrically deformed by the pressures to' which they, and the strata which contain them, have been subjected. In the first step towards fossilisation, when the body of a fish or shellfish is silted over and buried, we may take it that the wet sand or mud exercises, approximately, a hydrostatic pressure— that is to say a pressure which is uniform in all directions, and by which the form of the buried object will not be appreciably changed. As the strata consolidate and accumulate, the fossil organisms which they contain will tend to be flattened by the vast superincumbent load, just as the stratum which contains them will also be compressed and will have its molecular arrangement more or less modified*. But the deformation due to direct vertical pressure in a horizontal stratum is not nearly so striking as are the deformations produced by the oblique or shearing stresses to which inclined

* Cf. Sorby, *Quart. Journ. Geol. Soc. (Proc.)*, 1879, p. 88.

and folded strata have been exposed, and by which their various "dislocations" have been brought about.   And especially in mountain regions, where these dislocations are especially numerous and complicated, the contained fossils are apt to be so curiously and yet so symmetrically deformed (usually by a simple shear) that they may easily be interpreted as so many distinct and separate "species *."   A great number of described species, and here and there a new genus (as the genus *Ellipsolithes* for an obliquely deformed *Goniatite* or *Nautilus*), are said to rest on no other foundation †.

If we begin by drawing a net of rectangular equidistant coordinates (about the axes x and y), we may alter or *deform* this network in various ways, several of which are very simple indeed.   Thus (1) we may alter the dimensions of our system, extending it along one or

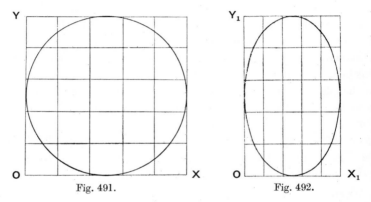

Fig. 491.   Fig. 492.

other axis, and so converting each little square into a corresponding and proportionate oblong (Figs. 491, 492).   It follows that any figure which we may have inscribed in the original net, and which we transfer to the new, will thereby be *deformed* in strict proportion to the deformation of the entire configuration, being still defined by corresponding points in the network and being throughout in conformity with the original figure.   For instance, a circle inscribed

---

* Cf. Alc. D'Orbigny, *Cours élém. de Paléontologie*, etc., i, pp. 144–148, 1849; see also Daniel Sharpe, On slaty cleavage, *Q.J.G.S.* iii, p. 74, 1847.

† Thus *Ammonites erugatus*, when compressed, has been described as *A. planorbis*: cf. J. F. Blake, *Phil. Mag.* (5), vi, p. 260, 1878.   Wettstein has shewn that several species of the fish-genus *Lepidopus* have been based on specimens artificially deformed in various ways: Ueber die Fischfauna des Tertiären Glarnerschiefers, *Abh. Schw. Palaeont. Gesellsch.* xiii, 1886 (see especially pp. 23–38, pl. i).   The whole subject, interesting as it is, has been little studied; both Blake and Wettstein deal with it mathematically.

in the original "Cartesian" net will now, after extension in the y-direction, be found elongated into an ellipse. In elementary mathematical language, for the original x and y we have substituted x_1 and cy_1, and the equation to our original circle, $x^2 + y^2 = a^2$, becomes that of the ellipse, $x_1^2 + c^2y_1^2 = a^2$.

If I draw the cannon-bone of an ox (Fig. 493, A), for instance, within a system of rectangular coordinates, and then transfer the same drawing, point for point, to a system in which for the x of the original diagram we substitute $x' = 2x/3$, we obtain a drawing (B) which is a very close approximation to the cannon-bone of the sheep. In other words, the main (and perhaps the only) difference

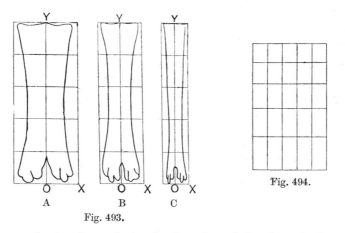

Fig. 494.

A   B   C

Fig. 493.

between the two bones is simply that that of the sheep is elongated along the vertical axis as compared with that of the ox, in the proportion of 3/2. And similarly, the long slender cannon-bone of the giraffe (C) is referable to the same identical type, subject to a reduction of breadth, or increase of length, corresponding to $x'' = x/3$.

(2) The second type is that where extension is not equal or uniform at all distances from the origin: but grows greater or less, as, for instance, when we stretch a *tapering* elastic band. In such cases, as I have represented it in Fig. 494, the ordinate increases logarithmically, and for y we substitute ϵ^y. It is obvious that this logarithmic extension may involve both abscissae and ordinates, x becoming ϵ^x while y becomes ϵ^y. The circle in our original figure is now deformed into some such shape as that of Fig. 495. This

method of deformation is a common one, and will often be of use to us in our comparison of organic forms.

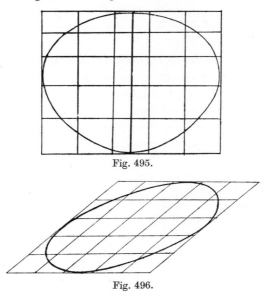

Fig. 495.

Fig. 496.

(3) Our third type is the "simple shear," where the rectangular coordinates become "oblique," their axes being inclined to one another at a certain angle ω. Our original rectangle now becomes such a figure as that of Fig. 496. The system may now be described in terms of the oblique axes X, Y; or may be directly referred to new rectangular coordinates ξ, η by the simple transposition $x = \xi - \eta \cot \omega$, $y = \eta \csc \omega$.

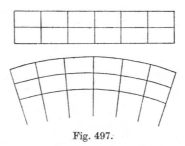

Fig. 497.

(4) Yet another important class of deformations may be represented by the use of radial coordinates, in which one set of lines are

represented as radiating from a point or "focus," while the other set are transformed into circular arcs cutting the radii orthogonally. These radial coordinates are especially applicable to cases where there exists (either within or without the figure) some part which is supposed to suffer no deformation; a simple illustration is afforded by the diagrams which illustrate the flexure of a beam (Fig. 497). In biology these coordinates will be especially applicable in cases where the growing structure includes a "node," or point where growth is absent or at a minimum; and about which node the rate of growth may be assumed to increase symmetrically. Precisely

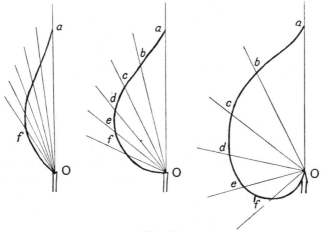

Fig. 498.

such a case is furnished us in a leaf of an ordinary dicotyledon. The leaf of a typical monocotyledon—such as a grass or a hyacinth, for instance—grows continuously from its base, and exhibits no node or "point of arrest." Its sides taper off gradually from its broad base to its slender tip, according to some law of decrement specific to the plant; and any alteration in the relative velocities of longitudinal and transverse growth will merely make the leaf a little broader or narrower, and will effect no other conspicuous alteration in its contour. But if there once come into existence a node, or "locus of no growth," about which we may assume growth—which in the hyacinth leaf was longitudinal and trans-verse—to take place radially and transversely to the radii, then we shall soon see the sloping sides of the hyacinth leaf give place to a more typical and "leaf-like" shape.   If we alter the ratio between the radial and tangential velocities of growth—in other words, if we increase the angles between corresponding radii—we pass successively through the various configurations which the botanist describes as the lanceolate, the ovate, and the cordiform leaf.   These successive changes may to some extent, and in appropriate cases, be traced as the individual leaf grows to maturity; but as a much more general rule, the balance of forces, the ratio between radial and tangential velocities of growth, remains so nicely and constantly balanced that the leaf increases in size without conspicuous modification of form. It is rather what we may call a long-period variation, a tendency for the relative velocities to alter from one generation to another, whose result is brought into view by this method of illustration.

There are various corollaries to this method of describing the form of a leaf which may be here alluded to.   For instance, the so-called

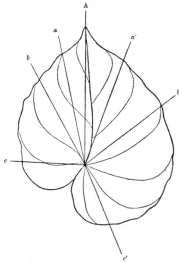

Fig. 499   *Begonia daedalea.*

unsymmetrical leaf* of a begonia, in which one side of the leaf may be merely ovate while the other has a cordate outline, is seen to be really a case of *unequal*, and not truly asymmetrical, growth on either side of the midrib.   There is nothing more mysterious in its conformation than, for instance, in that of a forked twig in which one limb of the fork has grown longer than the other. The case of the begonia leaf is of sufficient interest to deserve illustration, and in Fig. 499 I have outlined a leaf of the large *Begonia daedalea*.   On the smaller left-hand side of the leaf I have taken at random three points *a*, *b*, *c*, and have measured the angles, *AOa*, etc.,

* Cf. Sir Thomas Browne, in *The Garden of Cyrus*: "But why ofttimes one side of the leaf is unequall unto the other, as in Hazell and Oaks, why on either side the master vein the lesser and derivative channels stand not directly opposite, not at equall angles, respectively unto the adverse side, but those of one side do often exceed the other, as the Wallnut and many more, deserves another enquiry."

which the radii from the hilus of the leaf to these points make with the median axis. On the other side of the leaf I have marked the points a', b', c', such that the radii drawn to this margin of the leaf are equal to the former, Oa' to Oa, etc. Now if the two sides of the leaf are mathematically similar to one another, it is obvious that the respective angles should be in continued proportion, i.e. as AOa is to AOa', so should AOb be to AOb'. This proves to be very nearly the case. For I have measured the three angles on one side, and one on the other, and have then compared, as follows, the calculated with the observed values of the other two:

	AOa	AOb	AOc	AOa'	AOb'	AOc'
Observed values	12°	28·5°	88°	—	—	157°
Calculated ,,	—	—	—	21·5°	51·1°	—
Observed ,,	—	—	—	20	52	—

The agreement is very close, and what discrepancy there is may be amply accounted for, firstly, by the slight irregularity of the sinuous margin of the leaf; and secondly, by the fact that the true axis or midrib of the leaf is not straight but slightly curved, and therefore that it is curvilinear and not rectilinear triangles which we ought to have measured. When we understand these few points regarding the peripheral curvature of the leaf, it is easy to see that its principal veins approximate closely to a beautiful system of isogonal coordinates. It is also obvious that we can easily pass, by a process of shearing, from those cases where the principal veins start from the base of the leaf to those where they arise successively from the midrib, as they do in most dicotyledons.

It may sometimes happen that the node*, or "point of arrest," is at the upper instead of the lower end of the leaf-blade; and occasionally there is a node at both ends. In the former case, as we have it in the daisy, the form of the leaf will be, as it were, inverted, the broad, more or less heart-shaped, outline appearing at the upper end, while below the leaf tapers gradually downwards to an ill-defined base. In the latter case, as in *Dionaea*, we obtain a leaf equally expanded, and similarly ovate or cordate, at both ends. We may notice, lastly, that the shape of a solid fruit, such as an apple or a cherry, is a solid of revolution, developed from similar curves and to be explained on the same principle. In the

---

* "Node," in the botanical, not the mathematical, sense.

cherry we have a "point of arrest" at the base of the berry, where it joins its peduncle, and about this point the fruit (in imaginary section) swells out into a cordate outline; while in the apple we have two such well-marked points of arrest, above and below, and about both of them the same conformation tends to arise. The bean and the human kidney owe their "reniform" shape to precisely the same phenomenon, namely, to the existence of a node or "hilus," about which the forces of growth are radially and symmetrically arranged. When the seed is small and the pod roomy, the seed may grow round, or nearly so, like a pea; but it is flattened and bean-shaped, or elliptical like a kidney-bean, when compressed within a narrow and elongated pod. If the original seed have any simple pattern, of the nature for instance of meridians or parallels of latitude, it is easy to see how these will suffer a conformal transformation, corresponding to the deformation of the sphere*.

We might go farther, and farther than we have room for here, to illustrate the shapes of leaves by means of radial coordinates, and even to attempt to define them by polar equations. In a former chapter we learned to look upon the curve of sines as an easy, gradual and natural transition—perhaps the simplest and most natural of all—from minimum to corresponding maximum, and so on alternately and continuously; and we found the same curve going round like the hands of a clock, when plotted on radial coordinates and (so to speak) prevented from leaving its place. Either way it represents a "simple harmonic motion." Now we have just seen an ordinary dicotyledonous leaf to have a "point of arrest," or zero-growth in a certain direction, while in the opposite direction towards the tip it has grown with a maximum velocity. This progress from zero to maximum suggests one-half of the sine-curve; in other words, if we look on the outline of the leaf as a vector-diagram of its own growth, at rates varying from zero to zero in a complete circuit of $360°$, this suggests, as a possible and very simple case, the plotting of $r = \sin \theta/2$. Doing so, we obtain a curve (Fig. 500) closely resembling what the botanists call a *reniform* (or kidney-shaped) leaf, that is to say, with a cordate outline at the base formed of two "auricles," one on either side, and then rounded

* *Vide supra*, p. 524.

off with no projecting apex*. The ground-ivy and the dog-violet (Fig. 501) illustrate such a leaf; and sometimes, as in the violet, the veins of the leaf show similar curves congruent with the outer edge. Moreover the violet is a good example of how the reniform leaf may be drawn out more and more into an acute and ovate form.

From $\sin \theta/2$ we may proceed to any other given fraction of θ, and plot, for instance, $r = \sin 5\theta/3$, as in Fig. 502; which now no longer represents a single leaf but has become a diagram of the

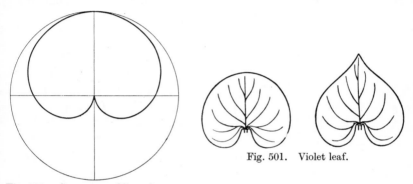

Fig. 501.   Violet leaf.

Fig. 500.   Curve resembling the outline of a reniform leaf: $r = \sin \theta/2$.

five petals of a pentamerous flower. Abbot Guido Grandi, a Pisan mathematician of the early eighteenth century, drew such a curve and pointed out its botanical analogies; and we still call the curves of this family "Grandi's curves†."

The gamopetalous corolla is easily transferred to polar coordinates, in which the radius vector now consists of two parts, the one a constant, the other expressing the amplitude (or half-amplitude) of the sine-curve; we may write the formula $r = a + b \cos n\theta$. In Fig. 503 $n = 5$; in this figure, if the radius of the outermost circle be taken as unity, the outer of the two sinuous curves has $a:b$ as

* Fig. 500 illustrates the whole leaf, but only shows one-half of the sine-curve. The rest is got by reflecting the moiety already drawn in the horizontal axis $(\theta = \pi/2)$.

† Dom. Guido Grandus, *Flores geometrici ex rhodonearum et cloeliarum curvarum descriptione resultantes...*, Florentiae, 1728. Cf. Alfred Lartigue, *Biodynamique générale*, Paris, 1930—a curious but eccentric book.

9:1, and the inner curve as 3:1; while the five petals become separate when $a = b$, and the formula reduces to $r = \cos^2 \dfrac{5\theta'}{2}$.

In Fig. 504 we have what looks like a first approximation to a horse-chestnut leaf. It consists of so many separate leaflets, akin to the five petals in Fig. 503; but these are now inscribed in (or have a *locus* in) the cordate or reniform outline of Fig. 500. The new curve is, in short, a composite one; and its general formula is $r = \sin \theta/2 . \sin n\theta$. The small size of the two leaflets adjacent to the petiole is characteristic of the curve, and helps to explain the development of "stipules."

Fig. 502.   Grandi's curves based on $r = \sin \frac{5}{3}\theta$, and illustrating the five petals of a simple flower.

Fig. 503.   Diagram illustrating a corolla of five petals, or of five lobes, are based on the equation $r = a + b \cos \theta$.

In this last case we have combined one curve with another, and the doing so opens out a new range of possibilities. On the outline of the simple leaf, whether ovate, lanceolate or cordate, we may superpose secondary sine-curves of lesser period and varying amplitude, after the fashion of a Fourier series; and the results will vary from a mere crenate outline to the digitate lobes of an ivy-leaf, or to separate leaflets such as we have just studied in the horse-chestnut. Or again, we may inscribe the separate petals of Fig. 505 within a spiral curve, equablé or equiangular as the case may be; and then, continuing the series on and on, we shall obtain a figure resembling the clustered leaves of a stonecrop, or the petals of a water-lily or other polypetalous flower.

Most of the transformations which we have hitherto considered (other than that of the simple shear) are particular cases of a general transformation, obtainable by the method of conjugate functions and equivalent to the projection of the original figure on a new plane. Appropriate transformations, on these general lines, provide for the cases of a coaxial system where the Cartesian coordinates are replaced by coaxial circles, or a confocal system in which they are replaced by confocal ellipses and hyperbolas.

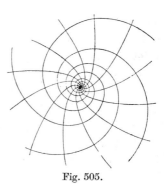

Fig. 504. Outline of a compound leaf, like a horse-chestnut, based on a composite sine-curve, of the form $r = \sin \theta/2 . \sin n\theta$.

Fig. 505.

Yet another curious and important transformation, belonging to the same class, is that by which a system of straight lines becomes transformed into a conformal system of logarithmic spirals: the straight line $Y - AX = c$ corresponding to the logarithmic spiral $\theta - A \log r = c$ (Fig. 505). This beautiful and simple transformation lets us at once convert, for instance, the straight conical shell of the Pteropod or the *Orthoceras* into the logarithmic spiral of the Nautiloid; it involves a mathematical symbolism which is but a slight extension of that which we have employed in our elementary treatment of the logarithmic spiral.

These various systems of coordinates, which we have now briefly considered, are sometimes called "isothermal coordinates," from the fact that, when employed in this particular branch of physics, they perfectly represent the phenomena of the conduction of heat, the contour lines of equal temperature appearing, under appropriate conditions, as the orthogonal lines of the coordinate system. And it follows that the "law of growth" which our biological analysis by means of orthogonal coordinate systems presupposes, or at least foreshadows, is one according to which the organism grows or develops along stream-lines, which may be defined by a suitable mathematical transformation.

When the system becomes no longer orthogonal, as in many of the following illustrations—for instance, that of *Orthagoriscus* (Fig. 526)—then the transformation is no longer within the reach of comparatively simple mathematical analysis. Such departure from the typical symmetry of a "stream-line" system is, in the first instance, sufficiently accounted for by the simple fact that the developing organism is very far from being homogeneous and isotropic, or, in other words, does not behave like a perfect fluid. But though under such circumstances our coordinate systems may be no longer capable of strict mathematical analysis, they will still indicate *graphically* the relation of the new coordinate system to the old, and conversely will furnish us with some guidance as to the "law of growth," or play of forces, by which the transformation has been effected.

Before we pass from this brief discussion of transformations in general, let us glance at one or two cases in which the forces applied are more or less intelligible, but the resulting transformations are, from the mathematical point of view, exceedingly complicated.

The "marbled papers" of the bookbinder are a beautiful illustration of visible "stream-lines." On a dishful of a sort of semi-liquid gum the workman dusts a few simple lines or patches of colouring matter; and then, by passing a comb through the liquid, he draws the colour-bands into the streaks, waves, and spirals which constitute the marbled pattern, and which he then transfers to sheets of paper laid down upon the gum. By some such system of shears, by the effect of unequal traction or unequal growth in various directions and superposed on an originally simple pattern, we may account for the not dissimilar marbled patterns which we recognise, for instance, on a large serpent's skin. But it must be remarked, in the case of the marbled paper, that though the method of application of the forces is simple, yet in the aggregate the system of forces set up by the many teeth of the comb is exceedingly complex, and its complexity is revealed in the complicated "diagram of forces" which constitutes the pattern.

To take another and still more instructive illustration. To turn one circle (or sphere) into two circles (or spheres) would be, from the point of view of the mathematician, an extraordinarily difficult transformation; but, physically speaking, its achievement may be extremely simple. The little round gourd grows naturally, by its symmetrical forces of expansive growth, into a big, round, or some-what oval pumpkin or melon*. But the Moorish husbandman ties a rag round its middle, and the same forces of growth, unaltered save for the presence of this trammel, now expand the globular structure into two superposed and connected globes. And again, by varying the position of the encircling band, or by applying several such ligatures instead of one, a great variety of artificial forms of "gourd" may be, and actually are, produced. It is clear, I think, that we may account for many ordinary biological processes of development or transformation of form by the existence of trammels or lines of constraint, which limit and determine the action of the expansive forces of growth that would otherwise be uniform and symmetrical. This case has a close parallel in the operations of the glass-blower, to which we have already, more than once, referred in passing†. The glass-blower starts his operations with a *tube*, which he first closes at one end so as to form a hollow vesicle, within which his blast of air exercises a uniform pressure on all sides; but the spherical conformation which this uniform expansive force would naturally tend to produce is modified into all kinds of forms by the trammels or resistances set up as the workman lets one part or another of his bubble be unequally heated or cooled. It was Oliver Wendell

---

* Analogous structural differences, especially in the fibrovascular bundles, help to explain the differences between (e.g.) a smooth melon and a cantelupe, or between various elongate, flattened and globular varieties. These breed true to type, and obey, when crossed, the laws of Mendelian inheritance. Cf. E. W. Sinnott, Inherit-ance of fruit-shape in Cucurbita, *Botan. Gazette*, LXXIV, pp. 95–103, 1922, and other papers.

† Where gourds are common, the glass-blower is still apt to take them for a prototype, as the prehistoric potter also did. For instance, a tall, annulated Florence oil-flask is an exact but no longer a conscious imitation of a gourd which has been converted into a bottle in the manner described.

Holmes who first shewed this curious parallel between the operations of the glass-blower and those of Nature, when she starts, as she so often does, with a simple tube*. The alimentary canal, the arterial system including the heart, the central nervous system of the vertebrate, including the brain itself, all begin as simple tubular structures. And with them Nature does just what the glass-blower does, and, we might even say, no more than he. For she can expand the tube here and narrow it there; thicken its walls or thin them; blow off a lateral offshoot or caecal diverticulum; bend the tube, or twist and coil it; and infold or crimp its walls as, so to speak, she pleases. Such a form as that of the human stomach is easily explained when it is regarded from this point of view; it is simply an ill-blown bubble, a bubble that has been rendered lopsided by a trammel or restraint along one side, such as to prevent its symmetrical expansion—such a trammel as is produced if the glass-blower lets one side of his bubble get cold, and such as is actually present in the stomach itself in the form of a muscular band.

The Florence flask, or any other handiwork of the glass-blower, is always beautiful, because its graded contours are, as in its living analogues, a picture of the graded forces by which it was conformed. It is an example of mathematical beauty, of which the machine-made, moulded bottle has no trace at all. An alabaster bottle is different again. It is no longer an unduloid figure of equilibrium. Turned on a lathe, it is a solid of revolution, and not without beauty; but it is not near so beautiful as the blown flask or bubble.

The gravitational field is part of the complex field of force by which the form of the organism is influenced and determined. Its share is seldom easy to define, but there is a resultant due to gravity in hanging breasts and tired eyelids and all the sagging wrinkles of the old. Now and then we see gravity at work in the normal construction of the body, and can describe its effect on form in a general, or qualitative, way. Each pair of ribs in man forms a hoop which droops of its own weight in front, so flattening the chest, and at the same time twisting the rib on either hand near its point of suspension†. But in the dog each costal hoop is dragged

---

* Cf. *Elsie Venner*, chap. II.

† See T. P. Anderson Stuart, How the form of the thorax is partly determined by gravitation, *Proc. R.S.* XLIX, p.143, 1891.

straight downwards, into a vertical instead of a transverse ellipse, and is even narrowed to a point at the sternal border.

We may now proceed to consider and illustrate a few permutations or transformations of organic form, out of the vast multitude which are equally open to this method of enquiry.

We have already compared in a preliminary fashion the metacarpal or cannon-bone of the ox, the sheep, and the giraffe (Fig. 493); and we have seen that the essential difference in form between these three bones is a matter of relative length and breadth, such that, if we reduce the figures to an identical standard of length (or identical values of y), the breadth (or value of x) will be approximately two-thirds that of the ox in the case of the sheep and one-third that of the ox in the case of the giraffe. We may easily, for the sake of closer comparison, determine these ratios more accurately, for instance, if it be our purpose to compare the different racial varieties within the limits of a single species. And in such cases, by the way, as when we compare with one another various breeds or races of cattle or of horses, the ratios of length and breadth in this particular bone are extremely significant*.

If, instead of limiting ourselves to the cannon-bone, we inscribe the entire foot of our several Ungulates in a coordinate system, the same ratios of x that served us for the cannon-bones still give us a first approximation to the required comparison; but even in the case of such closely allied forms as the ox and the sheep there is evidently something wanting in the comparison. The reason is that the relative elongation of the several parts, or individual bones, has not proceeded equally or proportionately in all cases; in other words, that the equations for x will not suffice without some simultaneous modification of the values of y (Fig. 506). In such a case it may be found possible to satisfy the varying values of y by some logarithmic

---

* This significance is particularly remarkable in connection with the development of speed, for the metacarpal region is the seat of very important leverage in the propulsion of the body. In a certain Scottish Museum there stand side by side the skeleton of an immense carthorse (celebrated for having drawn all the stones of the Bell Rock Lighthouse to the shore), and a beautiful skeleton of a racehorse, long supposed to be the actual skeleton of Eclipse. When I was a boy my grandfather used to point out to me that the cannon-bone of the little racer is not only relatively, but actually, longer than that of the great Clydesdale.

or other formula; but, even if that be possible, it will probably be somewhat difficult of discovery or verification in such a case as the present, owing to the fact that we have too few well-marked points of correspondence between the one object and the other, and that especially along the shaft of such long bones as the cannon-bone of the ox, the deer, the llama, or the giraffe there is a complete lack of easily recognisable corresponding points. In such a case a brief tabular statement of apparently corresponding values of y, or of those obviously corresponding values which coincide with the boundaries of the several bones of the foot, will, as in the following example, enable us to dispense with a fresh equation.

			a	b	c	d
y (Ox)	...	0	18	27	42	100
y' (Sheep)	...	0	10	19	36	100
y'' (Giraffe)	...	0	5	10	24	100

This summary of values of y', coupled with the equations for the value of x, will enable us, from any drawing of the ox's foot, to construct a figure of that of the sheep or of the giraffe with remarkable accuracy.

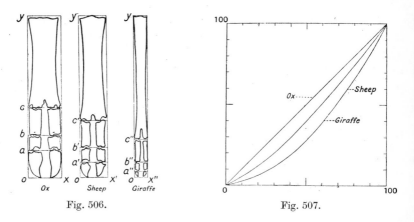

Fig. 506.   Fig. 507.

That underlying the varying amounts of extension to which the parts or segments of the limb have been subject there is a law, or principle of continuity, may be discerned from such a diagram as the above (Fig. 507), where the values of y in the case of the ox are plotted as a straight line, and the corresponding values for the sheep (extracted from the above table) are seen to form a more or less regular and even curve. This simple graphic result implies the existence of a comparatively simple equation between y and y'.

An elementary application of the principle of coordinates to the study of proportion, as we have here used it to illustrate the varying proportions of a bone, was in common use in the sixteenth and seventeenth centuries by artists in their study of the human form. The method is probably much more ancient, and may even be classical *; it is fully described and put in practice by Albert Dürer in his *Geometry*, and especially in his *Treatise on Proportion* †. In this latter work, the manner in which the human figure, features, and facial expression are all transformed and modified by slight variations in the relative magnitude of the parts is admirably and copiously illustrated (Fig. 508).

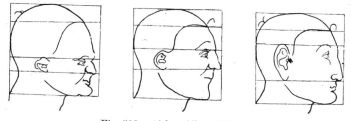

Fig. 508. (After Albert Dürer.)

In a tapir's foot there is a striking difference, and yet at the same time there is an obvious underlying resemblance, between the middle toe and either of its unsymmetrical lateral neighbours. Let us take the median terminal phalanx and inscribe its outline in a net of rectangular equidistant coordinates (Fig. 509, a). Let us then make a similar network about axes which are no longer at right angles, but inclined to one another at an angle of about 50° (b).

---

* Cf. Vitruvius, iii, 1.

† *Les quatres livres d'Albert Dürer de la proportion des parties et pourtraicts des corps humains*, Arnheim, 1613, folio (and earlier editions). Cf. also Lavater, *Essays on Physiognomy*, iii, p. 271, 1799; also H. Meige, La géométrie des visages d'après Albert Dürer, *La Nature*, Dec. 1927. On Dürer as mathematician, cf. Cantor, ii, p. 459; S. Günther, *Die geometrische Näherungsconstructione Albrecht Dürers*, Ansbach, 1866; H. Staigmuller, *Dürer als Mathematiker*, Stüttgart, 1891.

If into this new network we fill in, point for point, an outline precisely corresponding to our original drawing of the middle toe, we shall find that we have already represented the main features of the adjacent lateral one. We shall, however, perceive that our new diagram looks a little too bulky on one side, the inner side, of the lateral toe. If now we substitute for our equidistant ordinates,

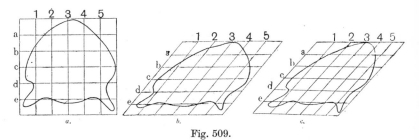

Fig. 509.

ordinates which get gradually closer and closer together as we pass towards the median side of the toe, then we shall obtain a diagram which differs in no essential respect from an actual outline copy of the lateral toe (c). In short, the difference between the outline of the middle toe of the tapir and the next lateral toe may be almost completely expressed by saying that if the one be represented by rectangular equidistant coordinates, the other will be represented by oblique coordinates, whose axes make an angle of 50°, and in

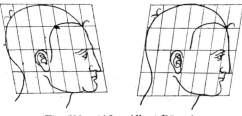

Fig. 510.   (After Albert Dürer.)

which the abscissal interspaces decrease in a certain logarithmic ratio. We treated our original complex curve or projection of the tapir's toe as a function of the form $F(x, y) = 0$. The figure of the tapir's lateral toe is a precisely identical function of the form $F(e^x, y_1) = 0$, where x_1, y_1 are oblique coordinate axes inclined to one another at an angle of 50°.

Dürer was acquainted with these oblique coordinates also, and I have copied two illustrative figures from his book*.

In Fig. 511 I have sketched the common Copepod *Oithona nana*, and have inscribed it in a rectangular net, with abscissae three-fifths the length of the ordinates. Side by side (Fig. 512) is drawn a very different Copepod, of the genus *Sapphirina*; and about it is drawn a network such that each coordinate passes (as nearly as possible) through points corresponding to those of the former figure. It will be seen that two differences are apparent. (1) The values of y in Fig. 512 are large in the upper part of the figure, and diminish

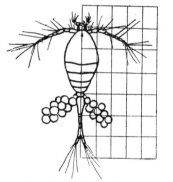

Fig. 511. *Oithona nana*.   Fig. 512. *Sapphirina*.

rapidly towards its base. (2) The values of x are very large in the neighbourhood of the origin, but diminish rapidly as we pass towards either side, away from the median vertical axis; and it is probable that they do so according to a definite, but somewhat complicated,

* It was these very drawings of Dürer's that gave to Peter Camper his notion of the "facial angle." Camper's method of comparison was the very same as ours, save that he only drew the axes, without filling in the network, of his coordinate system; he saw clearly the essential fact, that the skull *varies as a whole*, and that the "facial angle" is the index to a general deformation. "The great object was to shew that natural differences might be reduced to rules, of which the direction of the facial line forms the *norma* or canon; and that these directions and inclinations are always accompanied by correspondent form, size and position of the other parts of the cranium," etc.; from Dr T. Cogan's preface to Camper's work *On the Connexion between the Science of Anatomy and the Arts of Drawing, Painting and Sculpture* (1768?), quoted in Dr R. Hamilton's Memoir of Camper, in *Lives of Eminent Naturalists (Nat. Libr.)*, Edinburgh, 1840. See also P. Camper, *Dissertation sur les différences réelles que presentent les Traits du Visage chez les hommes de différents pays et de différents âges*, Paris, 1791 (*op. posth.*); cf. P. Topinard, Études sur Pierre Camper, et sur l'angle facial dit de Camper, *Rev. d'Anthropol.* II. 1874.

ratio. If, instead of seeking for an actual equation, we simply tabulate our values of x and y in the second figure as compared with the first (just as we did in comparing the feet of the Ungulates), we get the dimensions of a net in which, by simply projecting the figure of *Oithona*, we obtain that of *Sapphirina* without further trouble, e.g.:

x (*Oithona*)	0	3	6	9	12	15	—
x' (*Sapphirina*)	0	8	10	12	13	14	—
y (*Oithona*)	0	5	10	15	20	25	30
y' (*Sapphirina*)	0	2	7	3	23	32	40

In this manner, with a single model or type to copy from, we may record in very brief space the data requisite for the production of approximate outlines of a great number of forms. For instance, the difference, at first sight immense, between the attenuated body of a *Caprella* and the thick-set body of a *Cyamus* is obviously little, and is probably nothing more than a difference of relative magnitudes, capable of tabulation by numbers and of complete expression by means of rectilinear coordinates.

The Crustacea afford innumerable instances of more complex deformations. Thus we may compare various higher Crustacea with one another, even in the case of such dissimilar forms as a lobster and a crab. It is obvious that the whole body of the former is elongated as compared with the latter, and that the crab is relatively broad in the region of the carapace, while it tapers off rapidly towards its attenuated and abbreviated tail. In a general way, the elongated rectangular system of coordinates in which we may inscribe the outline of the lobster becomes a shortened triangle in the case of the crab. In a little more detail we may compare the outline of the carapace in various crabs one with another: and the comparison will be found easy and significant, even, in many cases, down to minute details, such as the number and situation of the marginal spines, though these are in other cases subject to independent variability.

If we choose, to begin with, such a crab as *Geryon* (Fig. 513, 1) and inscribe it in our equidistant rectangular coordinates, we shall see that we pass easily to forms more elongated in a transverse direction, such as *Matuta* or *Lupa* (5), and conversely, by transverse compression, to such a form as *Corystes* (2). In certain other cases the carapace conforms to a triangular diagram, more or less curvilinear, as in Fig. 513, 4, which represents the genus *Paralomis*. Here we can easily see that the posterior border is transversely elongated as compared with that of *Geryon*, while at the same time the anterior

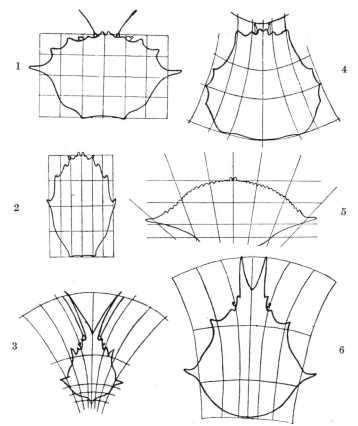

Fig. 513. Carapaces of various crabs. 1, *Geryon*; 2, *Corystes*; 3, *Scyramathia*; 4, *Paralomis*; 5, *Lupa*; 6, *Chorinus*.

part is longitudinally extended as compared with the posterior. A system of slightly curved and converging ordinates, with orthogonal and logarithmically interspaced abscissal lines, as shewn in the figure, appears to satisfy the conditions.

In an interesting series of cases, such as the genus *Chorinus*, or *Scyramathia*, and in the spider-crabs generally, we appear to have just the converse of this.   While the carapace of these crabs presents a somewhat triangular form, which seems at first sight more or less similar to those just described, we soon see that the actual posterior border is now narrow instead of broad, the broadest part of the carapace corresponding precisely, not to that which is broadest in *Paralomis*, but to that which was broadest in *Geryon*; while the most striking difference from the latter lies in an antero-posterior lengthening of the forepart of the carapace, culminating in a great elongation of the frontal region, with its two spines or "horns." The curved ordinates here converge posteriorly and diverge widely in front (Fig. 513, 3 and 6), while the decremental interspacing of the abscissae is very marked indeed.

We put our method to a severer test when we attempt to sketch an entire and complicated animal than when we simply compare corresponding parts such as the carapaces of various Malacostraca, or related bones as in the case of the tapir's toes.   Nevertheless, up to a certain point, the method stands the test very well.   In other words, one particular mode and direction of variation is often (or even usually) so prominent and so paramount throughout the entire organism, that one comprehensive system of coordinates suffices to give a fair picture of the actual phenomenon.   To take another illustration from the Crustacea, I have drawn roughly in Fig. 514, 1 a little amphipod of the family Phoxocephalidae (*Harpinia* sp.). Deforming the coordinates of the figure into the curved orthogonal* system in Fig. 514, 2, we at once obtain a very fair representation of an allied genus, belonging to a different family of amphipods, namely *Stegocephalus*.   As we proceed further from our type our coordinates will require greater deformation, and the resultant figure will usually be somewhat less accurate.   In Fig. 514, 3 I shew a network, to which, if we transfer our diagram of *Harpinia* or of *Stegocephalus*, we shall obtain a tolerable representation of the aberrant genus *Hyperia*†, with its narrow abdomen, its reduced pleural lappets, its great eyes, and its inflated head.

---

* Similar coordinates are treated of by Lamé, *Leçons sur les coordonnées curvilignes*, Paris, 1859.

† For an analogous, but more detailed comparison, see H. Mogk, Versuch einer Formanalyse bei Hyperiden, *Int. Rev. d. ges. Hydrobiol.*, etc., XIV, pp. 276–311, 1923; XVII, pp. 1–98, 1926.

The hydroid zoophytes constitute a "polymorphic" group, within which a vast number of species have already been distinguished; and the labours of the systematic naturalist are constantly adding to the number. The specific distinctions are for the most part based, not upon characters directly presented by the living animal, but upon the form, size and arrangement of the little cups, or "calycles," secreted and inhabited by the little individual polyps

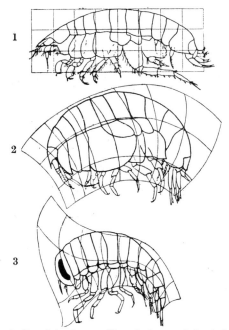

Fig. 514. 1, *Harpinia plumosa* Kr.; 2, *Stegocephalus inflatus* Kr.;
3, *Hyperia galba.*

which compose the compound organism. The variations, which are apparently infinite, of these conformations are easily seen to be a question of relative magnitudes, and are capable of complete expression, sometimes by very simple, sometimes by somewhat more complex, coordinate networks.

For instance, the varying shapes of the simple wineglass-shaped cups of the Campanularidae are at once sufficiently represented and compared by means of simple Cartesian coordinates (Fig. 515). In the two allied families of Plumulariidae and Aglaopheniidae the calycles are set unilaterally upon a jointed stem, and small cup-like structures (holding rudimentary polyps) are associated with the large calycles in definite number and position.   These small calyculi

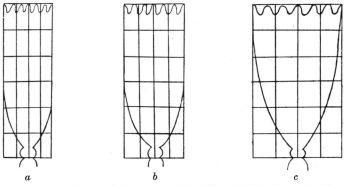

Fig. 515.   a, *Campanularia macroscyphus* Allm.; b, *Gonothyraea hyalina* Hincks; c, *Clytia Johnstoni* Alder.

are variable in number, but in the great majority of cases they accompany the large calycle in groups of three—two standing by its upper border, and one, which is especially variable in form and magnitude, lying at its base.   The stem is liable to flexure and,

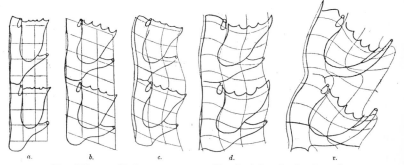

Fig. 516.   a, *Cladocarpus crenatus* F.; b, *Aglaophenia pluma* L.; c, *A. rhynchocarpa* A.; d, *A. cornuta* K.; e, *A. ramulosa* K.

in a high degree, to extension or compression; and these variations extend, often on an exaggerated scale, to the related calycles.   As a result we find that we can draw various systems of curved or sinuous coordinates, which express, all but completely, the configuration of the various hydroids which we inscribe therein (Fig. 516).

The comparative smoothness of denticulation of the margin of the calycle, and the number of its denticles, constitutes an independent variation, and requires separate description; we have already seen (p. 391) that this denticulation is in all probability due to a particular physical cause.

Among countless other invertebrate animals which we might illustrate, did space and time permit, we should find the bivalve molluscs shewing certain things extremely well. If we start with a more or less oblong shell, such as *Anodon* or *Mya* or *Psammobia*, we can see how easily it may be transformed into a more circular or orbicular, but still closely related form; while on the other hand a simple shear is well-nigh all that is needed to transform the oblong *Anodon* into the triangular, pointed *Mytilus*, *Avicula* or *Pinna*. Now suppose we draw the shell of *Anodon* in the usual rectangular coordinates, and deform this network into the corresponding oblique coordinates of *Mytilus*, we may then proceed to draw within the same two nets the anatomy of the same two molluscs. Then of the two adductor muscles, coequal in *Anodon*, one becomes small, the other large, when transferred to the oblique network of *Mytilus*; at the same time the foot becomes stunted and the siphonal aperture enlarged. In short, having "transformed" one shell into the other we may perform an identical transformation on their contained anatomy: and so (provided the two are not too distantly related) deduce the bodily structure of the one from our knowledge of the other, to a first but by no means negligible approximation.

Among the fishes we discover a great variety of deformations, some of them of a very simple kind, while others are more striking and more unexpected. A comparatively simple case, involving a simple shear, is illustrated by Figs. 517 and 518. The one represents, within Cartesian coordinates, a certain little oceanic fish known as *Argyropelecus Olfersi*. The other represents precisely the same outline, transferred to a system of oblique coordinates whose axes are inclined at an angle of 70°; but this is now (as far as can be seen on the scale of the drawing) a very good figure of an allied fish, assigned to a different genus, under the name of *Sternoptyx diaphana*. The deformation illustrated by this case of *Argyropelecus* is precisely analogous to the simplest and commonest kind of deformation to which fossils are subject (as we have seen on p. 811) as the result of shearing-stresses in the solid rock.

Fig. 519 is an outline diagram of a typical Scaroid fish. Let us deform its rectilinear coordinates into a system of (approximately) coaxial circles, as in Fig. 520, and then filling into the new system,

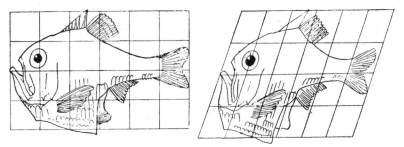

Fig. 517. *Argyropelecus Olfersi.*   Fig. 518. *Sternoptyx diaphana.*

space by space and point by point, our former diagram of *Scarus*, we obtain a very good outline of an allied fish, belonging to a neighbouring family, of the genus *Pomacanthus*. This case is all the more interesting, because upon the body of our *Pomacanthus* there are striking colour bands, which correspond in direction very closely

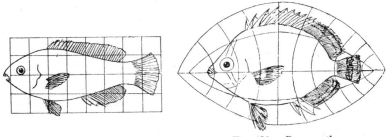

Fig. 519. *Scarus* sp.   Fig. 520. *Pomacanthus.*

to the lines of our new curved ordinates. In like manner, the still more bizarre outlines of other fishes of the same family of Chaetodonts will be found to correspond to very slight modifications of similar coordinates; in other words, to small variations in the values of the constants of the coaxial curves.

In Figs. 521–524 I have represented another series of Acanthopterygian fishes, not very distantly related to the foregoing. If we

start this series with the figure of *Polyprion*, in Fig. 521, we see that
the outlines of *Pseudopriacanthus* (Fig. 522) and of *Sebastes* or
Scorpaena (Fig. 523) are easily derived by substituting a system

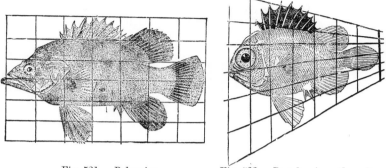

Fig. 521.   *Polyprion.*   Fig. 522.   *Pseudopriacanthus altus.*

of triangular, or radial, coordinates for the rectangular ones in which
we had inscribed *Polyprion*.   The very curious fish *Antigonia capros*,
an oceanic relative of our own boar-fish, conforms closely to the
peculiar deformation represented in Fig. 524.

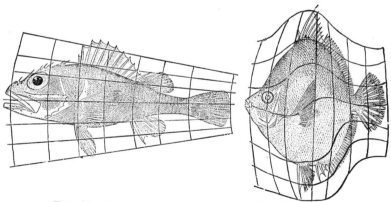

Fig. 523.   *Scorpaena* sp.   Fig. 524.   *Antigonia capros.*

Fig. 525 is a common, typical *Diodon* or porcupine-fish, and in
Fig. 526 I have deformed its vertical coordinates into a system of
concentric circles, and its horizontal coordinates into a system of
curves which, approximately and provisionally, are made to resemble

a system of hyperbolas*.   The old outline, transferred in its integrity
to the new network, appears as a. manifest representation of the
closely allied, but very different looking, sunfish, *Orthagoriscus mola*.
This is a particularly instructive case of deformation or transformation.   It is true that, in a mathematical sense, it is not a perfectly
satisfactory or perfectly regular deformation, for the system is no

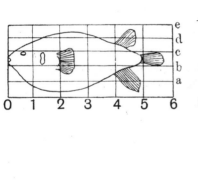

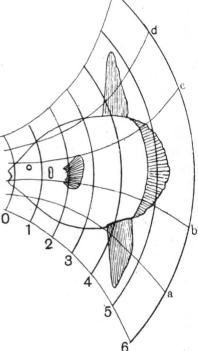

Fig. 525.   *Diodon*.   Fig. 526.   *Orthagoriscus*.

longer isogonal; but nevertheless, it is symmetrical to the eye, and
obviously approaches to an isogonal system under certain conditions
of friction or constraint.   And as such it accounts, by one single
integral transformation, for all the apparently separate and distinct
external differences between the two fishes.   It leaves the parts

* The coordinate system of Fig. 526 is somewhat different from that which
I first drew and published.   It is not unlikely that further investigation will
further simplify the comparison, and shew it to involve a still more symmetrical
system.

near to the origin of the system, the whole region of the head, the opercular orifice and the pectoral fin, practically unchanged in form, size and position; and it shews a greater and greater apparent modification of size and form as we pass from the origin towards the periphery of the system.

In a word, it is sufficient to account for the new and striking contour in all its essential details, of rounded body, exaggerated dorsal and ventral fins, and truncated tail. In like manner, and using precisely the same coordinate networks, it appears to me possible to shew the relations, almost bone for bone, of the skeletons of the two fishes; in other words, to reconstruct the skeleton of the one from our knowledge of the skeleton of the other, under the guidance of the same correspondence as is indicated in their external configuration.

The family of the crocodiles has had a special interest for the evolutionist ever since Huxley pointed out that, in a degree only second to the horse and its ancestors, it furnishes us with a close and almost unbroken series of transitional forms, running down in continuous succession from one geological formation to another. I should be inclined to transpose this general statement into other terms, and to say that the Crocodilia constitute a case in which, with unusually little complication from the presence of independent variants, the trend of one particular mode of transformation is visibly manifested. If we exclude meanwhile from our comparison a few of the oldest of the crocodiles, such as *Belodon*, which differ more fundamentally from the rest, we shall find a long series of genera in which we can refer not only the changing contours of the skull, but even the shape and size of the many constituent bones and their intervening spaces or "vacuities," to one and the same simple system of transformed coordinates. The manner in which the skulls of various Crocodilians differ from one another may be sufficiently illustrated by three or four examples.

Let us take one of the typical modern crocodiles as our standard of form, e.g. *C. porosus*, and inscribe it, as in Fig. 527, *a*, in the usual Cartesian coordinates. By deforming the rectangular network into a triangular system, with the apex of the triangle a little way in front of the snout, as in *b*, we pass to such a form as *C. americanus*.

By an exaggeration of the same process we at once get an approximation to the form of one of the sharp-snouted, or longirostrine,
crocodiles, such as the genus *Tomistoma*; and, in the species figured,
the oblique position of the orbits, the arched contour of the occipital
border, and certain other characters suggest a certain amount of
curvature, such as I have represented in the diagram (Fig. 527, *b*),
on the part of the horizontal coordinates. In the still more elongated
skull of such a form as the Indian Gavial, the whole skull has undergone a great longitudinal extension, or, in other words, the ratio
of x/y is greatly diminished; and this extension is not uniform, but
is at a maximum in the region of the nasal and maxillary bones.

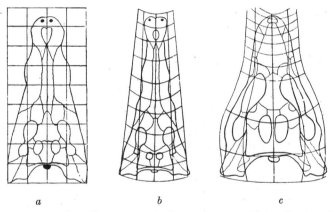

a   b   c

Fig. 527.   *a, Crocodilus porosus*; *b, C. americanus*; *c, Notosuchus terrestris*.

This especially elongated region is at the same time narrowed in an
exceptional degree, and its excessive narrowing is represented by
a curvature, convex towards the median axis, on the part of the
vertical ordinates. Let us take as a last illustration one of the
Mesozoic crocodiles, the little *Notosuchus*, from the Cretaceous formation. This little crocodile is very different from our type in the
proportions of its skull. The region of the snout, in front of and
including the frontal bones, is greatly shortened; from constituting
fully two-thirds of the whole length of the skull in *Crocodilus*, it
now constitutes less than half, or, say, three-sevenths of the whole;
and the whole skull, and especially its posterior part, is curiously
compact, broad, and squat. The orbit is unusually large. If in

the diagram of this skull we select a number of points obviously corresponding to points where our rectangular coordinates intersect particular bones or other recognisable features in our typical crocodile, we shall easily discover that the lines joining these points in *Notosuchus* fall into such a coordinate network as that which is represented in Fig. 527, *c*. To all intents and purposes, then, this not very complex system, representing one harmonious "deformation," accounts for *all* the differences between the two figures, and is sufficient to enable one at any time to reconstruct a detailed drawing, bone for bone, of the skull of *Notosuchus* from the model furnished by the common crocodile.

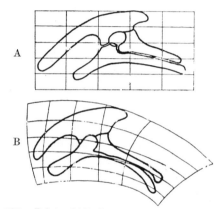

Fig. 528.   Pelvis of (A) *Stegosaurus*; (B) *Camptosaurus*.

The many diverse forms of Dinosaurian reptiles, all of which manifest a strong family likeness underlying much superficial diversity, furnish us with plentiful material for comparison by the method of transformations. As an instance, I have figured the pelvic bones of *Stegosaurus* and of *Camptosaurus* (Fig. 528, *a*, *b*) to shew that, when the former is taken as our Cartesian type, a slight curvature and an approximately logarithmic extension of the x-axis brings us easily to the configuration of the other. In the original specimen of *Camptosaurus* described by Marsh*, the anterior portion of the iliac bone is missing; and in Marsh's restoration this part of the bone is drawn as though it came somewhat abruptly to a sharp point. In my figure I have completed this missing part

* *Dinosaurs of North America*, pl. LXXXI, etc., 1896.

of the bone in harmony with the general coordinate network which
is suggested by our comparison of the two entire pelves; and I
venture to think that the result is more natural in appearance, and
more likely to be correct than was Marsh's conjectural restoration.
It would seem, in fact, that there is an obvious field for the employment of the method of coordinates in this task of reproducing missing

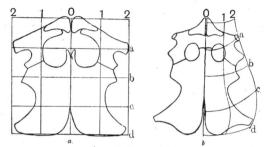

Fig. 529. Shoulder-girdle of *Cryptocleidus*. *a*, young; *b*, adult.

portions of a structure to the proper scale and in harmony with
related types.   To this subject we shall presently return.

In Fig. 529, *a, b*, I have drawn the shoulder-girdle of *Cryptocleidus*,
a Plesiosaurian reptile, half-grown in the one case and full-grown
in the other.   The change of form during growth in this region of
the body is very considerable, and its nature is well brought out

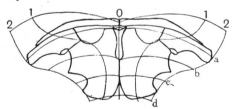

Fig. 530. Shoulder-girdle of *Ichthyosaurus*.

by the two coordinate systems.   In Fig. 530 I have drawn the
shoulder-girdle of an Ichthyosaur, referring it to *Cryptocleidus* as
a standard of comparison.   The interclavicle, which is present in
Ichthyosaurus, is minute and hidden in *Cryptocleidus*; but the
numerous other differences between the two forms, chief among
which is the great elongation in *Ichthyosaurus* of the two clavicles,
are all seen by our diagrams to be part and parcel of one general
and systematic deformation.

Before we leave the group of reptiles we may glance at the very strangely modified skull of *Pteranodon*, one of the extinct flying reptiles, or Pterosauria. In this very curious skull the region of the jaws, or beak, is greatly elongated and pointed; the occipital bone is drawn out into an enormous backwardly directed crest; the posterior part of the lower jaw is similarly produced backwards; the orbit is small; and the quadrate bone is strongly inclined downwards and forwards. The whole skull has a configuration which stands, apparently, in the strongest possible contrast to that of a more normal Ornithosaurian such as *Dimorphodon*. But if we inscribe the latter in Cartesian coordinates (Fig. 531, *a*), and refer

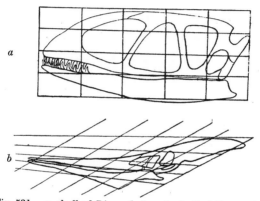

Fig. 531. *a*, skull of *Dimorphodon*; *b*, skull of *Pteranodon*.

our *Pteranodon* to a system of oblique coordinates (*b*), in which the two coordinate systems of parallel lines become each a pencil of diverging rays, we make manifest a correspondence which extends uniformly throughout all parts of these very different-looking skulls.

We have dealt so far, and for the most part we shall continue to deal, with our coordinate method as a means of comparing one known structure with another. But it is obvious, as I have said, that it may also be employed for drawing hypothetical structures, on the assumption that they have varied from a known form in some definite way. And this process may be especially useful, and will be most obviously legitimate, when we apply it to the particular case of representing intermediate stages between two forms which are actually known to exist, in other words, of reconstructing the transitional stages through which the course of evolution must have successively travelled if it has brought about the change from some ancestral type to its presumed descendant. Some years ago I sent my friend, Mr Gerhard Heilmann of Copenhagen, a few of

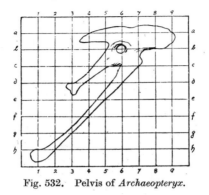

Fig. 532.   Pelvis of *Archaeopteryx*.

my own rough coordinate diagrams, including some in which the pelves of certain ancient and primitive birds were compared one with another. Mr Heilmann, who is both a skilled draughtsman and an able morphologist, returned me a set of diagrams which are

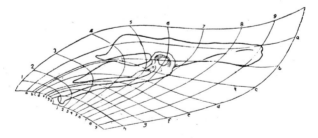

Fig. 533.   Pelvis of *Apatornis*.

a vast improvement on my own, and which are reproduced in Figs. 532–537. Here we have, as extreme cases, the pelvis of *Archaeopteryx*, the most ancient of known birds, and that of *Apatornis*, one of the fossil "toothed" birds from the North American Cretaceous formations—a bird shewing some resemblance to the modern terns. The pelvis of *Archaeopteryx* is taken as our type, and referred accordingly to Cartesian coordinates (Fig. 532); while the corresponding coordinates of the very different pelvis of *Apatornis* are represented in Fig. 533. In Fig. 534 the outlines of these two

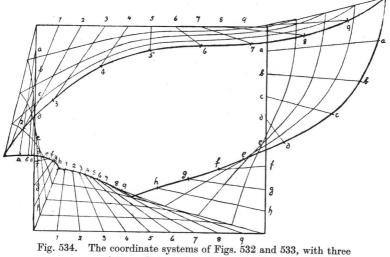

Fig. 534. The coordinate systems of Figs. 532 and 533, with three intermediate systems interpolated.

coordinate systems are superposed upon one another, and those of three intermediate and equidistant coordinate systems are interpolated between them. From each of these latter systems,

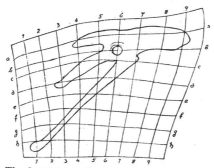

Fig. 535. The first intermediate coordinate network, with its corresponding inscribed pelvis.

so determined by direct interpolation, a complete coordinate diagram is drawn, and the corresponding outline of a pelvis is found from each of these systems of coordinates, as in Figs. 535, 536. Finally, in Fig. 537 the complete series is represented, beginning with the known pelvis of *Archaeopteryx*, and leading up by our three intermediate hypothetical types to the known pelvis of *Apatornis*.

Among mammalian skulls I will take two illustrations only, one drawn from a comparison of the human skull with that of the higher apes, and another from the group of Perissodactyle Ungulates, the group which includes the rhinoceros, the tapir, and the horse.

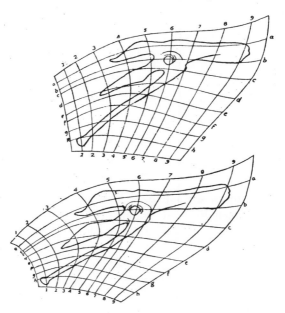

Fig. 536.   The second and third intermediate coordinate networks, with their corresponding inscribed pelves.

Let us begin by choosing as our type the skull of *Hyrachyus agrarius* Cope, from the Middle Eocene of North America, as figured by Osborn in his Monograph of the Extinct Rhinoceroses* (Fig. 538). The many other forms of primitive rhinoceros described in the monograph differ from *Hyrachyus* in various details—in the characters of the teeth, sometimes in the number of the toes, and so forth; and they also differ very considerably in the general appearance of the skull. But these differences in the conformation

* *Mem. Amer. Mus. of Nat. Hist.* I, III, 1898.

of the skull, conspicuous as they are at first sight, will be found easy to bring under the conception of a simple and homogeneous transformation, such as would result from the application of some not very complicated stress. For instance, the corresponding coordinates of *Aceratherium tridactylum*, as shewn in Fig. 539, indicate that the

Fig. 537. The pelvis of *Archaeopteryx* and of *Apatornis*, with three transitional types interpolated between them.

essential difference between this skull and the former one may be summed up by saying that the long axis of the skull of *Aceratherium* has undergone a slight double curvature, while the upper parts of the skull have at the same time been subject to a vertical expansion, or to growth in somewhat greater proportion than the lower parts. Precisely the same changes, on a somewhat greater scale, give us the skull of an existing rhinoceros.

Among the species of *Aceratherium,* the posterior, or occipital, view of the skull presents specific differences which are perhaps more conspicuous than those furnished by the side view; and these

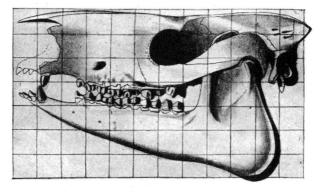

Fig. 538.   Skull of *Hyrachyus agrarius.*   After Osborn.

differences are very strikingly brought out by the series of conformal transformations which I have represented in Fig. 540.   In this case it will perhaps be noticed that the correspondence is not always quite accurate in small details.   It could easily have been made

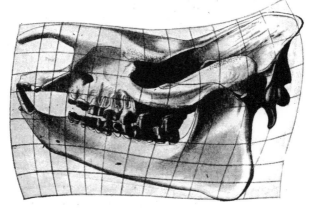

Fig. 539.   Skull of *Aceratherium tridactylum.* · After Osborn.

much more accurate by giving a slightly sinuous curvature to certain of the coordinates.   But as they stand, the correspondence indicated is very close, and the simplicity of the figures illustrates all the better the general character of the transformation.

By similar and not more violent changes we pass easily to such allied forms as the Titanotheres (Fig. 541); and the well-known series of species of *Titanotherium*, by which Professor Osborn has

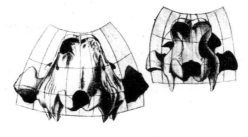

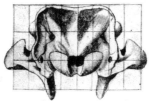

Fig. 540. Occipital view of the skulls of various extinct rhinoceroses
(*Aceratherium* spp.). After Osborn.

illustrated the evolution of this genus, constitutes a simple and suitable case for the application of our method.

But our method enables us to pass over greater gaps than these, and to discern the general, and to a very large extent even the

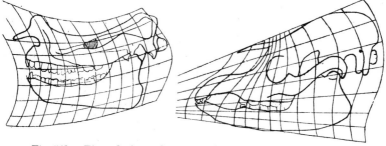

Fig. 541. *Titanotherium robustum.*   Fig. 542. Tapir's skull.

detailed, resemblances between the skull of the rhinoceros and those of the tapir or the horse. From the Cartesian coordinates in which we have begun by inscribing the skull of a primitive rhinoceros, we pass to the tapir's skull (Fig. 542), firstly, by converting the rectangular into a triangular network, by which we represent the depression of the anterior and the progressively increasing elevation of the posterior part of the skull; and secondly, by giving to the vertical ordinates a curvature such as to bring about a certain longitudinal compression, or condensation, in the forepart of the skull, especially in the nasal and orbital regions.

The conformation of the horse's skull departs from that of our primitive Perissodactyle (that is to say our early type of rhinoceros, *Hyrachyus*) in a direction that is nearly the opposite of that taken by *Titanotherium* and by the recent species of rhinoceros. For we perceive, by Fig. 543, that the horizontal coordinates, which in these latter cases become transformed into curves with the concavity

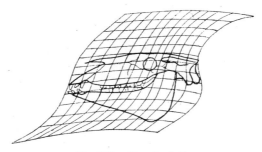

Fig. 543.   Horse's skull.

upwards, are curved, in the case of the horse, in the opposite direction. And the vertical ordinates, which are also curved, somewhat in the same fashion as in the tapir, are very nearly equidistant, instead of being, as in that animal, crowded together anteriorly. Ordinates and abscissae form an oblique system, as is shewn in the figure. In this case I have attempted to produce the network beyond the region which is actually required to include the diagram of the horse's skull, in order to shew better the form of the general transformation, with a part only of which we have actually to deal.

It is at first sight not a little surprising to find that we can pass, by a cognate and even simpler transformation, from our Perissodactyle skulls to that of the rabbit; but the fact that we can easily do so is a simple illustration of the undoubted affinity which exists between the Rodentia, especially the family of the Leporidae, and the more primitive Ungulates. For my part, I would go further; for I think there is strong reason to believe that the Perissodactyles are more closely related to the Leporidae than the former are to the other Ungulates, or than the Leporidae are to the rest of the Rodentia. Be that as it may, it is obvious from Fig. 544 that the rabbit's skull

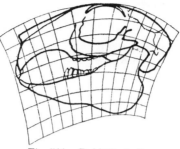

Fig. 544. Rabbit's skull.

conforms to a system of coordinates corresponding to the Cartesian coordinates in which we have inscribed the skull of *Hyrachyus*, with the difference, firstly, that the horizontal ordinates of the latter are

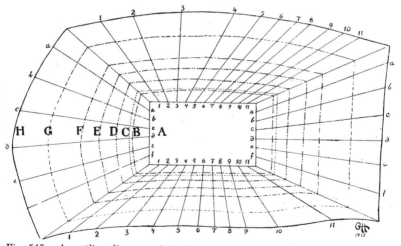

Fig. 545. A, outline diagram of the Cartesian coordinates of the skull of *Hyracotherium* or *Eohippus*, as shewn in Fig. 546, A. H, outline of the corresponding projection of the horse's skull. B–G, intermediate, or interpolated, outlines.

transformed into equidistant curved lines, approximately arcs of circles, with their concavity directed downwards; and secondly, that the vertical ordinates are transformed into a pencil of rays approxi-

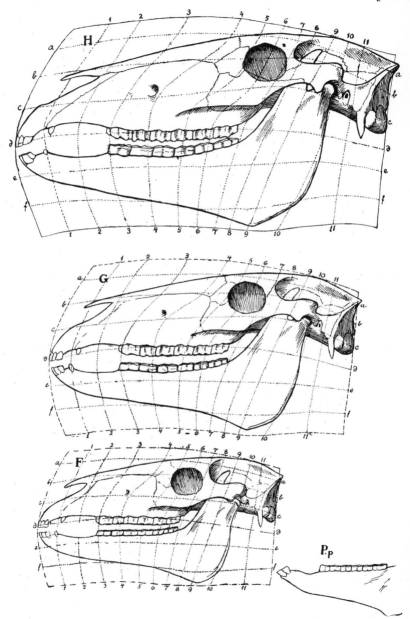

Fig. 546. A, skull of *Hyracotherium*, from the Eocene, after W. B. Scott; H, skull of horse, represented as a coordinate transformation of that of *Hyracotherium*,

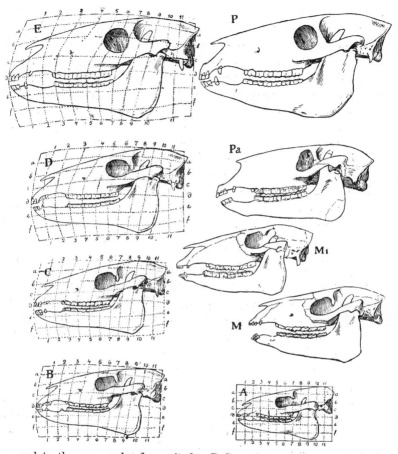

and to the same scale of magnitude; B–G, various artificial or imaginary types, reconstructed as intermediate stages between A and H; M, skull of *Mesohippus*, from the Oligocene, after Scott, for comparison with C; P, skull of *Protohippus*, from the Miocene, after Cope, for comparison with E; Pp, lower jaw of *Protohippus placidus* (after Matthew and Gidley), for comparison with F; Mi, *Miohippus* (after Osborn), Pa, *Parahippus* (after Peterson), shewing resemblance, but less perfect agreement, with C and D.

mately orthogonal to the circular arcs.   In short, the configuration
of the rabbit's skull is derived from that of our primitive rhinoceros
by the unexpectedly simple process of submitting the latter to a
strong and uniform flexure in the downward direction (cf. Fig. 538,
p. 1074).   In the case of the rabbit the configuration of the individual
bones does not conform quite so well to the general transformation
as it does when we are comparing the several Perissodactyles one
with another; and the chief departures from conformity will be
found in the size of the orbit and in the outline of the immediately
surrounding bones.   The simple fact is that the relatively enormous
eye of the rabbit constitutes an independent variation, which cannot
be brought into the general and fundamental transformation, but
must be dealt with separately.   The enlargement of the eye, like
the modification in form and number of the teeth, is a separate
phenomenon, which supplements but in no way contradicts our
general comparison of the skulls taken in their entirety.

Before we leave the Perissodactyla and their allies, let us look
a little more closely into the case of the horse and its immediate
relations or ancestors, doing so with the help of a set of diagrams
which I again owe to Mr Gerard Heilmann *.   Here we start afresh,
with the skull (Fig. 546, A) of *Hyracotherium* (or *Eohippus*), inscribed
in a simple Cartesian network.   At the other end of the series (H)
is a skull of *Equus*, in its own corresponding network; and the
intermediate stages (B–G) are all drawn by direct and simple interpolation, as in Mr Heilmann's former series of drawings of *Archæop-
teryx* and *Apatornis*.   In this present case, the relative magnitudes
are shewn, as well as the forms, of the several skulls.   Alongside
of these reconstructed diagrams are set figures of certain extinct
"horses" (Equidae or Palaeotheriidae), and in two cases, viz. *Mesohippus* and *Protohippus* (M, P), it will be seen that the actual
fossil skull coincides in the most perfect fashion with one of the
hypothetical forms or stages which our method shews to be implicitly
involved in the transition from *Hyracotherium* to *Equus* †.   T·· a third
case, that of *Parahippus* (Pa), the correspondence (as Mr Heilmann

* These and also other coordinate diagrams will be found in Mr G. Heilmann's
beautiful and original book *Fuglenes Afstamning*, 398 pp., Copenhagen, 1916; see
especially pp. 368–380.

† Cf. Zittel, *Grundzüge d. Palaeontologie*, 1911, p. 463.

points out) is by no means exact. The outline of this skull comes nearest to that of the hypothetical transition stage D, but the "fit" is now a bad one; for the skull of *Parahippus* is evidently a longer, straighter and narrower skull, and differs in other minor characters besides. In short, though some writers have placed *Parahippus* in the direct line of descent between *Equus* and *Eohippus*, we see at once that there is no place for it there, and that it must, accordingly, represent a somewhat divergent branch or offshoot of the Equidae *. It may be noticed, especially in the case of *Protohippus* (P), that the configuration of the angle of the jaw does not tally quite so accurately with that of our hypothetical diagrams as do other parts of the skull. As a matter of fact, this region is somewhat variable, in different species of a genus, and even in different individuals of the same species; in the small figure (Pp) of *Protohippus placidus* the correspondence is more exact.

In considering this series of figures we cannot but be struck, not only with the regularity of the succession of "transformations," but also with the slight and inconsiderable differences which separate each recorded stage from the next, and even the two extremes of the whole series from one another. These differences are no greater (save in regard to actual magnitude) than those between one human skull and another, at least if we take into account the older or remoter races; and they are again no greater, but if anything less, than the range of variation, racial and individual, in certain other human bones, for instance the scapula †.

The variability of this latter bone is great, but it is neither surprising nor peculiar; for it is linked with all the considerations of

* Cf. W. B. Scott (*Amer. Journ. of Science*, xlviii, pp. 335–374, 1894), "We find that any mammalian series at all complete, such as that of the horses, is remarkably continuous, and that the progress of discovery is steadily filling up what few gaps remain. So closely do successive stages follow upon one another that it is sometimes extremely difficult to arrange them all in order, and to distinguish clearly those members which belong in the main line of descent, and those which represent incipient branches. Some phylogenies actually suffer from an embarrassment of riches."

† Cf. T. Dwight, The range of variation of the human scapula, *Amer. Nat.* xxi, pp. 627–638, 1887. Cf. also Turner, *Challenger Rep.* xlvii, on Human Skeletons, p. 86, 1886: "I gather both from my own measurements, and those of other observers, that the range of variation in the relative length and breadth of the scapula is very considerable in the same race, so that it needs a large number of bones to enable one to obtain an accurate idea of the mean of the race."

T G F

mechanical efficiency and functional modification which we dealt with in our last chapter. The scapula occupies, as it were, a focus in a very important field of force; and the lines of force converging on it will be very greatly modified by the varying development of

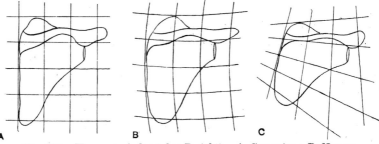

Fig. 547.   Human scapulae (after Dwight).   A, Caucasian; B, Negro;
C, North American Indian (from Kentucky Mountains).

the muscles over a large area of the body and of the uses to which they are habitually put.

Let us now inscribe in our Cartesian coordinates the outline of a human skull (Fig. 548), for the purpose of comparing it with the skulls of some of the higher apes. We know beforehand that the main differences between the human and the simian types depend

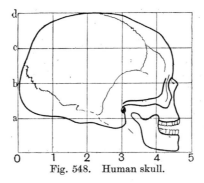

Fig. 548.   Human skull.

upon the enlargement or expansion of the brain and braincase in man, and the relative diminution or enfeeblement of his jaws. Together with these changes, the "facial angle" increases from an oblique angle to nearly a right angle in man, and the configuration of every constituent bone of the face and skull undergoes an altera-

tion. We do not know to begin with, and we are not shewn by the
ordinary methods of comparison, how far these various changes
form part of one harmonious and congruent transformation, or
whether we are to look, for instance, upon the changes undergone
by the frontal, the occipital, the maxillary, and the mandibular

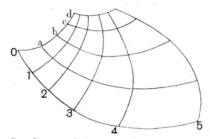

Fig. 549.   Coordinates of chimpanzee's skull, as a projection of
the Cartesian coordinates of Fig. 548.

regions as a congeries of separate modifications or independent
variants.   But as soon as we have marked out a number of points
in the gorilla's or chimpanzee's skull, corresponding with those which
our coordinate network intersected in the human skull, we find that
these corresponding points may be at once linked up by smoothly
curved lines of intersection, which form a new system of coordinates

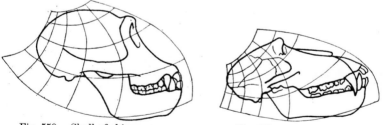

Fig. 550.   Skull of chimpanzee.   Fig. 551.   Skull of baboon.

and constitute a simple "projection" of our human skull. The
network represented in Fig. 549 constitutes such a projection of
the human skull on what we may call, figuratively speaking, the
"plane" of the chimpanzee; and the full diagram in Fig. 550
demonstrates the correspondence. In Fig. 551 I have shewn the
similar deformation in the case of a baboon, and it is obvious that
the transformation is of precisely the same order, and differs only

in an increased intensity or degree of deformation*. These anthropoid skulls, then, which we can transform one into another by a "continuous transformation," are admirable examples of what Listing called "topological similitude."

In both dimensions, as we pass from above downwards and from behind forwards, the corresponding areas of the network are seen to increase in a gradual and approximately logarithmic order in the lower as compared with the higher type of skull; and, in short, it becomes at once manifest that the modifications of jaws, brain-case, and the regions between are all portions of one continuous and integral process. It is of course easy to draw the inverse diagrams, by which the Cartesian coordinates of the ape are transformed into curvilinear and non-equidistant coordinates in man†.

From this comparison of the gorilla's or chimpanzee's with the human skull we realise that an inherent weakness underlies the anthropologist's method of comparing skulls by reference to a small number of axes. The most important of these are the "facial" and "basicranial" axes, which include between them the "facial angle." But it is, in the first place, evident that these axes are merely the principal axes of a system of coordinates, and that their restricted and isolated use neglects all that can be learned from the filling in of the rest of the coordinate network. And, in the second place, the "facial axis," for instance, as ordinarily used in the anthropological comparison of one human skull with another, or of the human skull with the gorilla's, is in all cases treated as a straight line; but our investigation has shewn that rectilinear axes only meet the case in the simplest and most closely related transformations; and that, for instance, in the anthropoid skull no rectilinear axis is homologous

---

* The empirical coordinates which I have sketched in for the chimpanzee as a conformal transformation of the Cartesian coordinates of the human skull look as if they might find their place in an equipotential elliptic field. They are indeed closely analogous to some already figured by MM. Y. Ikada and M. Kuwaori, Some conformal representations by means of the elliptic integrals, *Sci. Papers Inst. Phys. Research, Tokyo*, XXVI, pp. 208–215, 1936: e.g. pl. XXXI*b*.

† Speaking of "diagrams in pairs," and doubtless thinking of his own "reciprocal diagrams," Clerk Maxwell says (in his article *Diagrams* in the *Encyclopaedia Britannica*): "The method in which we simultaneously contemplate two figures, and recognise a correspondence between certain points in the one figure and certain points in the other, is one of the most powerful and fertile methods hitherto known in science....It is sometimes spoken of as the method or principle of duality."

with a rectilinear axis in a man's skull, but what is a straight line in the one has become a certain definite curve in the other.

Mr Heilmann tells me that he has tried, but without success, to obtain a transitional series between the human skull and some prehuman, anthropoid type, which series (as in the case of the Equidae) should be found to contain other known types in direct linear sequence. It appears impossible, however, to obtain such a series, or to pass by successive and continuous gradations through such forms as Mesopithecus, Pithecanthropus, *Homo neanderthalensis*, and the lower or higher races of modern man. The failure is not the fault of our method. It merely indicates that no one straight line of descent, or of consecutive transformation, exists; but on the contrary, that among human and anthropoid types, recent and extinct, we have to do with a complex problem of divergent, rather than of continuous, variation. And in like manner, easy as it is to correlate the baboon's and chimpanzee's skulls severally with that of man, and easy as it is to see that the chimpanzee's skull is much nearer to the human type than is the baboon's, it is also not difficult to perceive that the series is not, strictly speaking, continuous, and that neither of our two apes lies *precisely* on the same direct line or sequence of deformation by which we may hypothetically connect the other with man.

After easily transforming our coordinate diagram of the human skull into a corresponding diagram of ape or of baboon, we may effect a further transformation of man or monkey into dog no less easily; and we are thereby encouraged to believe that any two mammalian skulls may be compared with, or transformed into, one another by this method. There is something, an essential and indispensable something, which is common to them all, something which is the subject of all our transformations, and remains *invariant* (as the mathematicians say) under them all. In these transformations of ours every point may change its place, every line its curvature, every area its magnitude; but on the other hand every point and every line continues to exist, and keeps its relative order and position throughout all distortions and transformations. A series of points, a, b, c, along a certain line persist as corresponding points a', b', c', however the line connecting them may lengthen or bend; and as with points, so with lines, and so also with areas. Ear, eye and nostril, and all the other great landmarks of cranial anatomy, not only continue to exist but retain their relative order and position throughout all our transformations.

We can discover a certain *invariance*, somewhat more restricted than before, between the mammalian skull and that of fowl, frog or even herring. We have still something common to them all; and using another mathematical term (somewhat loosely perhaps) we may speak of the *discriminant characters* which persist unchanged, and continue to form the subject of our transformation. But the method, far as it goes, has its limitations. We cannot fit both beetle and cuttlefish into the same framework, however we distort it; nor by any coordinate transformation can we turn either of them into one another or into the vertebrate type. They are

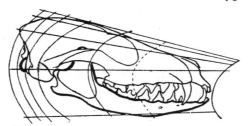

Fig. 552.   Skull of dog, compared with the human skull of Fig. 548.

essentially different; there is nothing about them which can be legitimately compared. Eyes they all have, and mouth and jaws; but what we call by these names are no longer in the same order or relative position; they are no longer the same thing, there is no *invariant* basis for transformation. The cuttlefish eye seems as perfect, optically, as our own; but the lack of an invariant relation of position between them, or lack of true homology between them (as we naturalists say), is enough to shew that they are unrelated things, and have come into existence independently of one another.

As a final illustration I have drawn the outline of a dog's skull (Fig. 552), and inscribed it in a network comparable with the Cartesian network of the human skull in Fig. 548. Here we attempt to bridge over a wider gulf than we have crossed in any of our former comparisons. But, nevertheless, it is obvious that our method still holds good, in spite of the fact that there are various specific differences, such as the open or closed orbit, etc., which have to be separately described and accounted for. We see that the chief essential differences in plan between the dog's skull and the man's lie in the fact that, relatively speaking, the former tapers away in front, a triangular taking the place of a rectangular conformation; secondly, that, coincident with the tapering off, there is a progressive elongation, or pulling out, of the whole forepart of the skull; and lastly, as a minor difference, that the straight vertical ordinates of the human skull become curved, with their convexity directed forwards, in the dog. While the net result is that in the dog, just as in the chimpanzee, the brain-pan is smaller and the jaws are larger than in man, it is now conspicuously evident that the coordinate network of the ape is by no means intermediate between those which fit the other two. The mode of deformation is on different lines; and, while it may be correct to say that the chimpanzee and the baboon are more brute-like, it would be by no means accurate to assert that they are more dog-like, than man.

In this brief account of coordinate transformations and of their morphological utility I have dealt with plane coordinates only, and have made no mention of the less elementary subject of coordinates in three-dimensional space. In theory there is no difficulty whatsoever in such an extension of our method; it is just as easy to refer the form of our fish or of our skull to the rectangular coordinates x, y, z, or to the polar coordinates ξ, η, ζ, as it is to refer their plane projections to the two axes to which our investigation has been confined. And that it would be advantageous to do so goes without saying, for it is the shape of the solid object, not that of the mere drawing of the object, that we want to understand; and already we have found some of our easy problems in solid geometry leading us (as in the case of the form of the bivalve and even of the univalve shell) quickly in the direction of coordinate analysis and the theory of conformal transformations. But this extended theme I have not attempted to pursue, and it must be left to other times, and to other hands. Nevertheless, let us glance for a moment at the sort of simple cases, the simplest possible cases, with which such an investigation might begin; and we have found our plane coordinate systems so easily and effectively applicable to certain fishes that we may seek among them for our first and tentative introduction to the three-dimensional field.

It is obvious enough that the same method of description and analysis which we have applied to one plane, we may apply to another: drawing by observation, and by a process of trial and error, our various cross-sections and the coordinate systems which seem best to correspond. But the new and important problem which now emerges is to *correlate* the deformation or transformation which we discover in one plane with that which we have observed in another: and at length, perhaps, after grasping the general principles of such correlation, to forecast approximately what is likely to take place in the third dimension when we are acquainted with two, that is to say, to determine the values along one axis in terms of the other two.

Let us imagine a common "round" fish, and a common "flat" fish, such as a haddock and a plaice. These two fishes are not as nicely adapted for comparison by means of plane coordinates as some which we have studied, owing to the presence of essentially unimportant, but yet conspicuous differences in the position of the eyes, or in the number of the fins—that is to say in the manner in which the continuous dorsal fin of the plaice appears in the haddock to be cut or scolloped into a number of separate fins. But speaking broadly, and apart from such minor differences as these, it is manifest that the chief factor in the case (so far as we at present see) is simply the broadening out of the plaice's body, as compared with the haddock's, in the dorso-ventral direction, that is to say, along the y axis; in other words, the ratio x/y is much less (and indeed little more than half as great) in the haddock than in the plaice. But we also recognise at once that while the plaice (as compared with the haddock) is expanded in one direction, it is also flattened, or thinned out, in the other: y increases, but z diminishes, relatively to x. And furthermore, we soon see that this is a common or even a general phenomenon. The high, expanded body in our Antigonia or in our sun-fish or in a John Dory is at the same time flattened or *compressed* from side to side, in comparison with the related fishes which we have chosen as standards of reference or comparison; and conversely, such a fish as the skate, while it is expanded from side to side in comparison with a shark or dogfish, is at the same time flattened or *depressed* in its vertical section. We hasten to enquire whether there be any simple relation of *magnitude* discernible between these twin factors of expansion and compression, and the very fact that the two dimensions of breadth and depth tend to vary inversely assures us that, in the general process of deformation, the volume and the area of cross-section are less affected than are those two linear dimensions. Some years ago, when I was studying the weight-length coefficient in fishes (of which we have already spoken in chapter III), that is to say the coefficient k in the formula $W = kL^3$, I was not a little surprised to find that k (let us call it in this case k_l) was all but identical in two such different looking fishes as the haddock and the plaice: thus indicating that these two fishes have approximately the same *volume* when they are equal in *length*; or, in other words, that the extent to which the plaice has broadened is *just about compensated for* by the extent to which it has also got flattened or thinned. In short, if we might conceive of a haddock being transformed directly into a plaice, a very large part of the change would be accounted for by supposing the round fish to be "rolled out" into the flat one, as a baker rolls a piece of dough. This is, as it were, an extreme case of the *balancement des organes*, or "compensation of parts."

We must not forget, while we consider the "deformation" of a fish, that the fish, like the bird, is subject to certain strict limitations of form. What we happen to have found in a particular case was observed fifty years ago, and brought under a general rule, by a naval engineer studying fishes from the shipbuilder's point of view. Mr Parsons* compared the contours and the sectional areas of a number of fishes and of several whales; and he found the sectional areas to be always very much the same at the same proportional distances from the front end of the body†. Increase in depth was balanced (as we also have found) by diminution of breadth; and the magnitude of the "entering angle" presented to the water by the advancing fish was fairly constant. Moreover, according to Parsons, the position of the greatest cross-section is fixed for all species, being situated at 36 per cent. of the length behind the

* H. de B. Parsons, Displacements and area-curves of fish, *Trans. Amer. Soc. of Mechan. Engineers*, IX, pp. 679–695, 1888.

† That is to say, if the areas of cross-section be plotted against their distances from the front end of the body, the results are very much alike for all the species examined. See also Selig Hecht, Form and growth in fishes, *Journ. Morph.* XXVII, pp. 379–400, 1916.

snout.   We need not stop to consider such extreme cases as the eel or the globefish (*Diodon*), whose ways of propulsion and locomotion are materially modified.   But it is certainly curious that no sooner do we try to correlate deformation in one direction with deformation in another, than we are led towards a broad generalisation, touching on hydrodynamical conditions and the limitations of form and structure which are imposed thereby.

Our simple, or simplified, illustrations carry us but a little way, and only half prepare us for much harder things.   But interesting as the whole subject is we must meanwhile leave it alone; recognising, however, that if the difficulties of description and representation could be overcome, it is by means of such coordinates in space that we should at last obtain an adequate and satisfying picture of the processes of deformation and the directions of growth.

A Note on Pattern

We have had so much to do with the study of Form that *pattern* has been wellnigh left out of the account, although it is part of the same story.   Like any other aspect of form, pattern is correlated with growth, and even determined by it.   A feather, for example, which is equally and equidistantly striped to begin with, may have this simple striping transformed into a more complex pattern by the unequal but *graded* elongation of the feather.   We need not go farther than the zebra for a characteristic pattern of stripes, nor need we seek a better illustration of how a common pattern may vary in related species.

A zebra's stripes may be broad or narrow, uniform or alternately dark and pale—these are minor or secondary diversities; but the pattern of the stripes shews more conspicuous differences than these, though the differences remain of a simple kind.   A zebra's stripes fall into several series.   One set covers the neck, including the mane, and extends backwards over the body and forwards on to the face; and these "body-stripes" are all that the extinct Quagga possessed. On the head they are interrupted by the ears and eyes, and end at a definite vertex on the forehead: from which, however, they run down the face in pairs, of which the first pair of all may or may not coalesce into a single median stripe (Fig. 553).   A second series runs up the foreleg. and where it meets the body we have the problem of how best to fit the horizontal leg-stripes and the vertical body-stripes together. There is only one way. A pair of body-stripes diverge apart and the upper leg-stripes fit in between, becoming at the same time chevron-shaped so as to adapt themselves to the space they have come to occupy. The stripes of the forelegs, and their manner of fitting on to the body-stripes, vary very little in the several species or varieties.

A third series of stripes ascends the hindlegs, in a fashion identical to begin with for all, but open to modification where these leg-stripes spread over the haunches; for here there may be great

Fig. 553.   Zebra's head, to shew how the body-stripes
extend to the face.   From A. Rzasnicki.

differences in the extent to which the leg-stripes compete with and interfere with, or (so to speak) encroach upon, the stripes of the body. The typical *Equus zebra* is easily recognised by the so-called "gridiron" on its rump; this is a dorsal continuation of the body-stripes, extending to the tail, but sharply cut off on either side by the stripes ascending from the leg (Fig. 554, C). In Burchell's zebra the hindleg-stripes encroach still farther on the body, and even reach up to the rump, so that the gridiron is entirely cut away*.

* Ward's zebra and Grant's zebra are varieties of *Equus zebra*, the former with a very strong "gridiron," the latter with a mere vestige of the same: which is as much as to say that the leg-stripes encroach little in the one, and much in the other, on the hindmost body-stripes. Chapman's zebra is a form of *E. Burchelli*, with well-striped legs and faint intermediate striping. Cf. W. Ridgeway, on The differentiation of the three species of Zebra, *P.Z.S.* 1909, pp. 547–563; also (*int. al.*) Adolf Rzasnicki, *Zebry*, Warsaw, 1931.

In the Abyssinian *Equus Grevyi*, all the stripes are very numerous, narrow and close-set. The body-stripes refuse, as it were, to be encroached on or obliterated by those of the hindlegs; which latter are merely intercalated between them, chevron fashion, wedging

Fig. 554. Zebra patterns. A, B, *Equus Burchelli*; C, *E. zebra*; D, *E. Grevyi*.

in between the body-stripes as the foreleg-stripes are wont to do. It follows that in the middle of the haunch, over the region of the hip-joint, there is in this species a characteristic "focus," where the leg-stripes fit in between the lumbar and the caudal sections of the body-stripes. We may now add, as a fourth and last series, common to all kinds, the few stripes which surround the lips on either side, and wedge in between the stripes upon the face.

Conclusion

There is one last lesson which coordinate geometry helps us to learn; it is simple and easy, but very important indeed. In the study of evolution, and in all attempts to trace the descent of the animal kingdom, fourscore years' study of the *Origin of Species* has had an unlooked-for and disappointing result. It was hoped to begin with, and within my own recollection it was confidently believed, that the broad lines of descent, the relation of the main branches to one another and to the trunk of the tree, would soon be settled, and the lesser ramifications would be unravelled bit by bit and later on. But things have turned out otherwise. We have long known, in more or less satisfactory detail, the pedigree of horses, elephants, turtles, crocodiles and some few more; and our conclusions tally as to these, again more or less to our satisfaction, with the direct evidence of palaeontological succession. But the larger and at first sight simpler questions remain unanswered; for eighty years' study of Darwinian evolution has not taught us how birds descend from reptiles, mammals from earlier quadrupeds, quadrupeds from fishes, nor vertebrates from the invertebrate stock. The invertebrates themselves involve the selfsame difficulties, so that we do not know the origin of the echinoderms, of the molluscs, of the coelenterates, nor of one group of protozoa from another. The difficulty is not always quite the same. We may fail to find the actual links between the vertebrate groups, but yet their resemblance and their relationship, real though indefinable, are plain to see; there are gaps between the groups, but we can see, so to speak, across the gap. On the other hand, the breach between vertebrate and invertebrate, worm and coelenterate, coelenterate and protozoon, is in each case of another order, and is so wide that we cannot see across the intervening gap at all.

This failure to solve the cardinal problem of evolutionary biology is a very curious thing; and we may well wonder why the long pedigree is subject to such breaches of continuity. We used to be told, and were content to believe, that the old record was of necessity imperfect—we could not expect it to be otherwise; the story was hard to read because every here and there a page had been lost or torn away, like some *hiatus valde deflendus* in an ancient manuscript. But there is a deeper reason. When we begin to draw comparisons between our algebraic curves and attempt to transform one into another, we find ourselves limited by the very nature of the case to curves having some tangible degree of relation to one another; and these "degrees of relationship" imply a *classification* of mathematical forms, analogous to the classification of plants or animals in another part of the *Systema Naturae*.

An algebraic curve has its fundamental formula, which defines the family to which it belongs; and its parameters, whose quantitative variation admits of infinite variety within the limits which the formula prescribes.   With some extension of the meaning of parameters, we may say the same of the families, or genera, or other classificatory groups of plants and animals.   We cross a boundary every time we pass from family to family, or group to group.   The passage is easy at first, and we are led, *along definite lines*, to more and more subtle and elegant comparisons.   But we come in time to forms which, though both may still be simple, yet stand so far apart that direct comparison is no longer legitimate.   We never think of "transforming" a helicoid into an ellipsoid, or a circle into a frequency-curve.   So it is with the forms of animals.   We *cannot* transform an invertebrate into a vertebrate, nor a coelenterate into a worm, by any simple and legitimate deformation, nor by anything short of reduction to elementary principles.

A "principle of discontinuity," then, is inherent in all our classifications, whether mathematical, physical or biological; and the infinitude of possible forms, always limited, may be further reduced and discontinuity further revealed by imposing conditions—as, for example, that our parameters must be whole numbers, or proceed by *quanta*, as the physicists say.   The lines of the spectrum, the six families of crystals, Dalton's atomic law, the chemical elements themselves, all illustrate this principle of discontinuity.   In short, nature proceeds *from one type to another* among organic as well as inorganic forms; and these types vary according to their own parameters, and are defined by physico-mathematical conditions of possibility.   In natural history Cuvier's "types" may not be perfectly chosen nor numerous enough, but *types* they are; and to seek for stepping-stones across the gaps between is to seek in vain, for ever.

This is no argument against the theory of evolutionary descent. It merely states that formal resemblance, which we depend on as our trusty guide to the affinities of animals within certain bounds or grades of kinship and propinquity, ceases in certain other cases to serve us, because under certain circumstances it ceases to exist.   Our geometrical analogies weigh heavily against Darwin's conception of endless small continuous variations; they help to show that discontinuous variations are a natural thing, that "mutations"—or sudden changes, greater or less—are bound to have taken place, and new "types" to have arisen, now and then. Our argument indicates, if it does not prove, that such mutations, occurring on a comparatively few definite lines, or plain alternatives, of physico-mathematical possibility, are likely to repeat themselves: that the "higher" protozoa, for instance, may have sprung not from or through one another, but severally from the simpler forms; or that the worm-type, to take another example, may have come into being again and again.

EPILOGUE

In the beginning of this book I said that its scope and treatment were of so prefatory a kind that of other preface it had no need; and now, for the same reason, with no formal and elaborate conclusion do I bring it to a close. The fact that I set little store by certain postulates (often deemed to be fundamental) of our present-day biology the reader will have discovered and I have not endeavoured to conceal. But it is not for the sake of polemical argument that I have written, and the doctrines which I do not subscribe to I have only spoken of by the way. My task is finished if I have been able to shew that a certain mathematical aspect of morphology, to which as yet the morphologist gives little heed, is interwoven with his problems, complementary to his descriptive task, and helpful, nay essential, to his proper study and comprehension of Growth and Form. *Hic artem remumque repono.*

And while I have sought to shew the naturalist how a few mathematical concepts and dynamical principles may help and guide him, I have tried to shew the mathematician a field for his labour—a field which few have entered and no man has explored. Here may be found homely problems, such as often tax the highest skill of the mathematician, and reward his ingenuity all the more for their trivial associations and outward semblance of simplicity. *Haec utinam excolant, utinam exhauriant, utinam aperiant nobis Viri mathematice docti* *.

That I am no skilled mathematician I have had little need to confess. I am "advanced in these enquiries no farther than the threshold"; but something of the use and beauty of mathematics I think I am able to understand. I know that in the study of material things, number, order and position are the threefold clue to exact knowledge; that these three, in the mathematician's hands, furnish the "first outlines for a sketch of the Universe"; that by square and circle we are helped, like Emile Verhaeren's carpenter, to conceive "Les lois indubitables et fécondes Qui sont la règle et la clarté du monde."

For the harmony of the world is made manifest in Form and

* So Boerhaave, in his *Oratio de Usu Ratiocinii Mechanici in Medicina* (1703).

Number, and the heart and soul and all the poetry of Natural Philosophy are embodied in the concept of mathematical beauty. A greater than Verhaeren had this in mind when he told of "the golden compasses prepared In God's eternal store." A greater than Milton had magnified the theme and glorified Him "that sitteth upon the circle of the earth," saying: He hath measured the waters in the hollow of his hand, and meted out heaven with the span, and comprehended the dust of the earth in a measure.

Moreover, the perfection of mathematical beauty is such (as Colin Maclaurin learned of the bee), that whatsoever is most beautiful and regular is also found to be most useful and excellent.

Not only the movements of the heavenly host must be determined by observation and elucidated by mathematics, but whatsoever else can be expressed by number and defined by natural law. This is the teaching of Plato and Pythagoras, and the message of Greek wisdom to mankind. So the living and the dead, things animate and inanimate, we dwellers in the world and this world wherein we dwell—πάντα γα μὰν τὰ γιγνωσκόμενα— are bound alike by physical and mathematical law. "Conterminous with space and coeval with time is the kingdom of Mathematics; within this range her dominion is supreme; otherwise than according to her order nothing can exist, and nothing takes place in contradiction to her laws." So said, some sixty years ago, a certain mathematician*; and Philolaus the Pythagorean had said much the same.

But with no less love and insight has the science of Form and Number been appraised in our own day and generation by a very great Naturalist indeed†—by that old man eloquent, that wise student and pupil of the ant and the bee, who died while this book was being written; who in his all but saecular life had tasted of the firstfruits of immortality; who curiously conjoined the wisdom of antiquity with the learning of today; whose Provençal verse seems set to Dorian music; in whose plainest words is a sound as of bees' industrious murmur; and who, being of the same blood and marrow with Plato and Pythagoras, saw in Number *le comment et le pourquoi des choses*, and found in it *la clef de voûte de l'Univers*.

---

* William Spottiswoode, in his presidential address to the British Association at Dublin in 1878. † Henri Fabre.

INDEX